Student Solutions Manual

to accompany

Physical Chemistry

Sixth Edition

Ira N. Levine
Department of Chemistry
Brooklyn College

McGraw Hill Higher Education

Boston Burr Ridge, IL Dubuque, IA New York San Francisco St. Louis
Bangkok Bogotá Caracas Kuala Lumpur Lisbon London Madrid Mexico City
Milan Montreal New Delhi Santiago Seoul Singapore Sydney Taipei Toronto

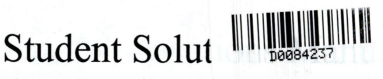

The McGraw·Hill Companies

Higher Education

Student Solutions Manual to accompany
PHYSICAL CHEMISTRY, SIXTH EDITION
IRA N. LEVINE

Published by McGraw-Hill Higher Education, an imprint of The McGraw-Hill Companies, Inc., 1221 Avenue of the Americas, New York, NY 10020.

This book is printed on acid-free paper.

3 4 5 6 7 8 9 0 QDB/QDB 0 14 13 12 11

ISBN: 978-0-07-253863-2
MHID: 0-07-253863-5

www.mhhe.com

To the Student

The purpose of this solutions manual is to help you learn physical chemistry. This purpose will be defeated if you use this manual to avoid working homework problems. You cannot learn how to play the guitar solely by reading a book titled "How to Play the Guitar" or by watching others play the guitar. Rather, most of your time is best spent actually practicing the guitar. Likewise, you won't learn how to solve physical chemistry problems solely by reading the solutions in this manual. Rather, most of your time is best spent working problems.

Do not look up the solution to a problem until you have made a substantial effort to work the problem on your own. When you work a problem, you learn a lot more then when you only read the solution. You can learn a lot by working on a problem even if you don't succeed in solving it. True learning requires active participation on your part. After you have looked up the solution to a problem you could not solve, close the manual and work the problem on your own.

Use the solutions manual as an incentive to work problems and not as a way to avoid working problems.

Ira N. Levine

INLevine@brooklyn.cuny.edu

Before using this manual, please read the preceding note To the Student.

Chapter 1

1.1 **(a)** F. **(b)** T. **(c)** T. **(d)** F. **(e)** F. (A mixture of ice and liquid water has one substance.)

1.2 **(a)** Closed, nonisolated; **(b)** open, nonisolated; **(c)** open; nonisolated.

1.3 **(a)** Three. **(b)** Three; solid AgBr; solid AgCl, and the solution.

1.4 So that pressure or composition differences between systems A and B won't cause changes in the properties of A and B. Such changes can then result only from a temperature difference between A and B.

1.5 **(a)** $$19.3\frac{\text{g}}{\text{cm}^3}\frac{1\text{ kg}}{1000\text{ g}}\frac{(100\text{ cm})^3}{(1\text{ m})^3} = 19300\text{ kg/m}^3$$

(b) $$\frac{\$800}{\text{troy oz}}\frac{1\text{ troy oz}}{480\text{ grains}}\frac{7000\text{ grains}}{1\text{ pound}}\frac{1\text{ pound}}{453.59\text{ g}} = \$25.72/\text{g}$$

$m = \rho V = (19.3\text{ g/cm}^3)(10^6\text{ cm}^3) = 1.93 \times 10^7\text{ g}$

$(1.93 \times 10^7)(\$25.72/\text{g}) = \$4.96 \times 10^8$

1.6 **(a)** T. **(b)** F. **(c)** T. **(d)** T.

1.7 **(a)** 32.0. **(b)** 32.0 amu. **(c)** 32.0. **(d)** 32.0 g/mol.

1.8 100 g of solution contains 12.0 g of HCl and 88.0 g of water. $n_{HCl} = (12.0\text{ g})(1\text{ mol}/36.46\text{ g}) = 0.329\text{ mol}$; $n_{H_2O} = (88.0\text{ g})(1\text{ mol}/18.015\text{ g}) = 4.88_5\text{ mol}$. $x_{HCl} = 0.329/(0.329 + 4.88_5) = 0.0631$; $x_{H_2O} = 1 - x_{HCl} = 0.9369$.

1.9 **(a)** $$\frac{12.0\text{ g/mol}}{6.022 \times 10^{23}\text{ atoms/mol}} = 1.99 \times 10^{-23}\text{ g/atom}$$

(b) $\dfrac{18.0\text{ g/mol}}{6.022\times10^{23}\text{ molecules/mol}} = 2.99\times10^{-23}$ g/molecule

1.10 **(a)** $V_m = M/\rho = (209\text{ g/mol})/(9.20\text{ g/cm}^3) = 22.7\text{ cm}^3\text{/mol}$.

(b) $(22.7\text{ cm}^3\text{/mol})/(6.02\times10^{23}\text{ atoms/mol}) = 3.77\times10^{-23}\text{ cm}^3\text{/atom}$.

(c) The diameter d of the Po atom equals the edge length of the surrounding cube, so $d^3 = 3.77\times10^{-23}\text{ cm}^3$ and $d = 3.35\times10^{-8}$ cm.

(d) $V = \frac{4}{3}\pi r^3 = \frac{4}{3}\pi(1\times10^{-9}\text{ m})^3 = 4.2\times10^{-27}\text{ m}^3 = 4.2\times10^{-21}\text{ cm}^3$. We have $(4.2\times10^{-21}\text{ cm}^3)/(3.77\times10^{-23}\text{ cm}^3\text{/atom}) = 111$ atoms.

(e) $V = \frac{4}{3}\pi(50\times10^{-9}\text{ m})^3 = 5.2\times10^{-22}\text{ m}^3 = 5.2\times10^{-16}\text{ cm}^3$.
$(5.2\times10^{-16}\text{ cm}^3)/(3.77\times10^{-23}\text{ cm}^3\text{/atom}) = 1.4\times10^7$ atoms.

(f) The particle's volume is $(2\times10^{-9}\text{ nm})^3 = 8\times10^{-27}\text{ m}^3 = 8\times10^{-21}\text{ cm}^3$. The particle has $(8\times10^{-21}\text{ cm}^3)/(3.77\times10^{-23}\text{ cm}^3\text{/atom}) = 212$ atoms. The edge length of the particle is 2 nm = 20×10^{-8} cm and use of the atomic diameter from (c) shows that the number of atoms along an edge of the particle is $(20\times10^{-8}\text{ cm})/(3.35\times10^{-8}\text{ cm}) = 6$ atoms. With 6 atoms along each edge of the cube, the top face of the cube has 36 atoms and the cubic particle contains 6 layers of atoms with 36 atoms in each layer. (This comes to $36\times6 = 216$ atoms, and the slight discrepancy with the number 212 will be ignored.) The 36 atoms in the top layer and the 36 atoms in the bottom layer lie on the surface of the cube. If you draw a square array of 36 circles with 6 circles along each of the 4 sides of the square, you will see that 20 circles lie on the sides of the square and 16 circles lie in the interior of the square. [We have $4(6) - 4 = 20$, where 4 is subtracted to allow for the fact that circles at the corners of the square lie on two sides and should not be counted twice.] Thus the four layers of atoms between the top and bottom layers each contain 16 interior atoms and the total number of atoms not on the surface is $16\times4 = 64$. The percentage of interior atoms is $(64/216)\times100\% = 30\%$ and 70% of the atoms lie on the surface.

(g) Repetition of the reasoning of part (f) shows the particle has a volume of $1\times10^{-15}\text{ cm}^3$, contains 2.65×10^7 atoms, and has 299 atoms along each edge. With 299 atoms on an edge, the top layer of the cube has 299^2 = 89401 atoms. For one of the 297 layers between the top and bottom layers, there are $4(299) - 4 = 1192$ atoms on the surface and the total number of atoms on the surface is 297(1192) = 354024. The percentage of atoms lying on the surface is $[(3.54 \times 10^5)/(2.65 \times 10^7)] \times 100\% =$ 1.3%. (See also Fig. 7.13.)

1.11 **(a)** T. **(b)** T. **(c)** F. **(d)** T. **(e)** F. **(f)** T.

1.12 **(a)** $(5.5\ \text{m}^3)(100\ \text{cm})^3/(1\ \text{m})^3 = 5.5 \times 10^6\ \text{cm}^3$.

(b) $(1.0 \times 10^9\ \text{Pa})(1\ \text{bar})/(10^5\ \text{Pa}) = 1.0 \times 10^4$ bar.

(c) $(1\ \text{hPa})(100\ \text{Pa}/1\ \text{hPa})(1\ \text{bar}/10^5\ \text{Pa})(750.062\ \text{torr}/1\ \text{bar}) = 0.750062$ torr.

(d)

$$\frac{1.5\ \text{g}}{\text{cm}^3}\frac{1\ \text{kg}}{10^3\ \text{g}}\frac{10^6\ \text{cm}^3}{1\ \text{m}^3} = 1.5\times 10^3\ \text{kg/m}^3.$$

1.13 The system pressure is less than the barometric pressure by 304.3 torr – 202.1 torr = 102.2 torr. So P_{system} = 754.6 torr – 102.2 torr = 652.4 torr.

1.14 **(a)** $P = \rho_{\text{Hg}} g h_{\text{Hg}} = \rho_{\text{H}_2\text{O}} g h_{\text{H}_2\text{O}}$, so $\rho_{\text{Hg}} h_{\text{Hg}} = \rho_{\text{H}_2\text{O}} h_{\text{H}_2\text{O}}$.

$$h_{\text{H}_2\text{O}} = \frac{(13.53\ \text{g/cm}^3)(30.0\ \text{in.})}{0.997\ \text{g/cm}^3} = 407\ \text{in.} = 33.9\ \text{ft}$$

where the vapor pressure of water was neglected.

(b) Use of $P = \rho gh$ and Eq. (2.8) gives P as

$$(13.53\ \text{g/cm}^3)(978\ \text{cm/s}^2)(30.0 \times 2.54\ \text{cm})\frac{1\ \text{atm}}{1013250\ \text{dyn/cm}^2} =$$

0.995 atm

1.15 For m constant, n is constant, so (1.18) becomes $PV/T = nR =$ const, which is (1.17).

1.16 **(a)** $n = (24.0\ \text{g})(1\ \text{mol}/44.0\ \text{g}) = 0.545$ mol. $P = nRT/V =$ $(0.545\ \text{mol})(82.06\ \text{cm}^3\text{-atm/mol-K})(273.1\ \text{K})/(5000\ \text{cm}^3) = 2.44$ atm.

(b) $V = nRT/P = (1\ \text{mol})(82.06\ \text{cm}^3\text{-atm/mol-K})(298\ \text{K})/(1\ \text{atm}) =$ $24500\ \text{cm}^3$. One $\text{ft}^3 = (12\ \text{in.})^3 = (12 \times 2.54\ \text{cm})^3 = 28300\ \text{cm}^3$. Percent error = $[(28300 - 24500)/24500] \times 100\% = 16\%$

1.17 Use of $P_1V_1/T_1 = P_2V_2/T_2$ gives $P_2 = (V_1/V_2)(T_2/T_1)P_1 = (V_1/2V_1)(3T_1/T_1)P_1 = 1.5P_1 = 1.5(0.800 \text{ bar}) = 1.200 \text{ bar}$.

1.18 $P = nRT/V = mRT/MV$, so $M = mRT/PV$ and

$$M = \frac{(0.0200\text{ g})(82.06\text{ cm}^3\text{-atm/mol-K})(298.1\text{ K})}{(24.7/760)\text{ atm }(500\text{ cm}^3)} = 30.1\text{ g/mol}$$

The only hydrocarbon with molecular weight 30 is C_2H_6.

1.19 At this T and P, N_2 is a gas that behaves nearly ideally. From $PV = nRT = (m/M)RT$, we get $m/V = PM/RT$, so

$$\rho = \frac{PM}{RT} = \frac{[(500/760)\text{ atm}](28.01\text{ g/mol})}{(82.06\text{ cm}^3\text{-atm/mol-K})(293\text{ K})} = 0.000766\text{ g/cm}^3$$

since 0.667 bar = 0.667(750 torr) = 500 torr.

1.20

PV/nT	82.025	81.948	81.880	cm^3-atm/mol-K
P	1.0000	3.0000	5.0000	atm

Plotting these data and extrapolating to $P = 0$, we find $\lim_{P\to 0}(PV/nT) = 82.06\text{ cm}^3\text{-atm/mol-K}$.

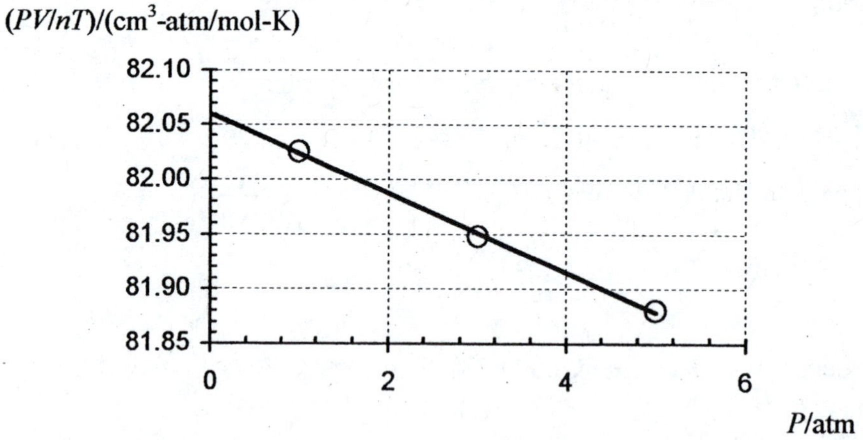

1.21 The P/ρ values are 715.3, 706.2, and 697.1 cm^3 atm/g. A plot of P/ρ vs. P is a straight line with intercept $721._4$ cm^3 atm/g. We have $PV = mRT/M$, so $M = RT/(P/\rho)$, and

$$M = \frac{(82.06\ \text{cm}^3\text{-atm/mol-K})(273.15\ \text{K})}{721._4\ \text{cm}^3\ \text{atm/g}} = 31.0_7\ \text{g/mol}$$

The only amine with molecular weight 31 is CH_3NH_2.

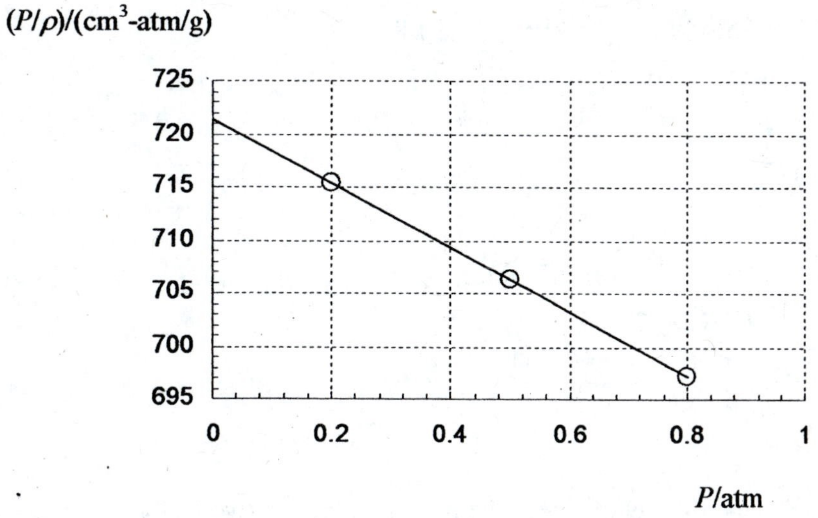

1.22 Use of $n_{tot} = PV/RT$ gives $n_{tot} = \frac{(4.85 \times 10^6\ \text{Pa})[1600(10^{-2}\ \text{m})^3]}{(8.314\ \text{m}^3\text{-Pa/mol-K})(500\ \text{K})} = 1.867$ mol

The reaction is $2NH_3 = N_2 + 3H_2$. Let x moles of N_2 be formed. The numbers of moles of NH_3, N_2, and H_2 present at equilibrium are $1.60 - 2x$, x, and $3x$, respectively. Thus $n_{tot}/\text{mol} = 1.60 - 2x + x + 3x = 1.867$, and $x = 0.13_3$. Then $n(N_2) = 0.13_3$ mol, $n(H_2) = 0.40_0$ mol, $n(NH_3) = 1.33$ mol

1.23 Boyle's law and Charles' law apply under different conditions (constant T, m vs. constant P, m); such equations cannot be combined.

1.24 Consider the processes

$$(P_1, V_1, T_1, m_1) \xrightarrow{a} (P_1, V_a, T_1, m_2) \xrightarrow{b} (P_2, V_2, T_2, m_2)$$

For step (a), P and T are constant, so $V_1/m_1 = V_a/m_2$. For step (b), m is constant, so $P_1V_a/T_1 = P_2V_2/T_2$. Substitution for V_a in this last equation gives $P_2V_2/T_2 = P_1V_1m_2/T_1m_1$ or $P_2V_2/m_2T_2 = P_1V_1/m_1T_1$.

1.25 $P_i = x_iP$. $n_{CO_2} = (30.0\ \text{g})(1\ \text{mol}/44.0\ \text{g}) = 0.682$ mol. $n_{O_2} = 0.625$ mol. $x_{CO_2} = 0.682/(0.682 + 0.625) = 0.522$. $P_{CO_2} = 0.522(3450\ \text{kPa}) = 1800$ kPa.

1.26 **(a)** At constant temperature, $P_2 = P_1V_1/V_2$ for each gas. Therefore

$$P_{H_2} = \frac{(20.0\text{ kPa})(3.00\text{ L})}{4.00\text{ L}} \qquad P_{CH_4} = \frac{(10.0\text{ kPa})(1.00\text{ L})}{4.00\text{ L}}$$

$$P_{H_2} = 15.0\text{ kPa} \qquad P_{CH_4} = 2.5\text{ kPa}$$

$$P_{tot} = 15.0\text{ kPa} + 2.5\text{ kPa} = 17.5\text{ kPa}$$

(b) $P_i = n_iRT/V$ and $P_{tot} = n_{tot}RT/V$. Hence $P_i/P_{tot} = n_i/n_{tot} = x_i$.
We get $x_{H_2} = 15.0\text{ kPa}/17.5\text{ kPa} = 0.857$ and $x_{CH_4} = 2.5\text{ kPa}/17.5\text{ kPa} = 0.143$.

1.27 $P(O_2) = 751\text{ torr} - 21\text{ torr} = 730\text{ torr}$. The equation $P_1V_1/T_1 = P_2V_2/T_2$ gives $V_2 = V_1P_1T_2/P_2T_1$ and

$$V_2 = \frac{(36.5\text{ cm}^3)(730\text{ torr})(273\text{ K})}{(760\text{ torr})(296\text{ K})} = 32.3\text{ cm}^3$$

1.28 When a steady state is reached, the pressures in the two bulbs are equal. From $P_1 = P_2$, we get $n_1RT_1/V_1 = n_2RT_2/V_2$. Since $V_1 = V_2$, we have $n_1T_1 = n_2T_2$. Thus $n_1(200\text{ K}) = (1.00\text{ mol} - n_1)(300\text{ K})$; solving, we get $n_1 = 0.60$ mole in the 200-K bulb and $n_2 = 0.40$ mole in the 300-K bulb.

1.29 We have $PV = nRT = NRT/N_A$, so $N/V = PN_A/RT$ and

$$\frac{N}{V} = \frac{(6.02\times10^{23}\text{ mol}^{-1})P}{(82.06\text{ cm}^3\text{-atm/mol-K})(298\text{ K})} = 2.46\times10^{19}\text{ cm}^{-3}\ \frac{P}{\text{atm}}$$

(a) For $P = 1$ atm, we get $N/V = 2.5\times10^{19}\text{ cm}^{-3}$;

(b) for $P = (1/760)10^{-6}$ atm, we get $N/V = 3.2\times10^{10}\text{ cm}^{-3}$;

(c) for $P = (1/760)10^{-11}$ atm, $N/V = 3.2\times10^{5}\text{ cm}^{-3}$.

1.30 Substitution in $PV = n_{tot}RT$ gives $n_{tot} = 0.01456$ mol. Also

$m_{tot} = m_1 + m_2 = n_1M_1 + n_2M_2$

$0.1480\text{ g} = n_{He}(4.003\text{ g/mol}) + (0.01456\text{ mol} - n_{He})(20.18\text{ g/mol})$.

$n_{He} = 0.00902\text{ mol}, \quad n_{Ne} = 0.00554\text{ mol}$

$x_{He} = 0.00902/0.01456 = 0.619, \quad m_{He} = 0.0361\text{ g}$

1.31 The downward force of the atmosphere on the earth's surface equals the weight W of the atmosphere, so $P = W/A = mg/A$ and $m = AP/g = 4\pi r^2P/g$, where r is the earth's radius and $P = 1$ atm = 101325 N/m^2. Thus

$$m = \frac{4\pi(6.37\times10^6 \text{ m})^2(1.013\times10^5 \text{ N/m}^2)}{9.807 \text{ m/s}^2} = 5.3\times10^{18} \text{ kg}$$

1.32 **(a)** Multiplication of both sides of the equation by 10^{-5} bar gives $P = 9.4 \times 10^{-5}$ bar.

(b) 460 K. **(c)** 1.2×10^3 bar. **(d)** 312 K.

1.33 Take one liter of gas. This volume has $m = 1.185 \text{ g} = m_{N_2} + m_{O_2}$.
We have $n_{tot} = PV/RT =$
(1.000 atm)(1000 cm^3)/(82.06 cm^3-atm/mol-K)(298.1 K) = 0.04087 mol.
$n_{tot} = n_{N_2} + n_{O_2} = m_{N_2}/M_{N_2} + m_{O_2}/M_{O_2} =$
m_{N_2}/(28.01 g/mol) + (1.185 g − m_{N_2})/(32.00 g/mol) = 0.0487 mol.
Solving, we get $m_{N_2} = 0.862$ g; hence $m_{O_2} = 0.323$ g. Then $n_{N_2} =$ 0.0308 mol and $n_{O_2} = 0.0101$ mol; $x_{O_2} = 0.0101/0.0409 = 0.247$.

1.34 **(a)** Use of $P_i = x_i P$ gives $P_{N_2} = 0.78(1.00 \text{ atm}) = 0.78$ atm,
$P_{O_2} = 0.21$ atm, $P_{Ar} = 0.0093$ atm, $P_{CO_2} = 0.0004$ atm.

(b) $V = 3000$ ft^3. 1 ft = 12 in. = 12 × 2.54 cm = 30.48 cm.
V = (3000 ft^3)(30.48 cm)^3/ft^3 = 8.5 × 10^7 cm^3. $n_{tot} = PV/RT =$
[(740/760) atm](8.5 × 10^7 cm^3)/[(82.06 cm^3-atm/mol-K)(293 K)] =
$3.4_4 \times 10^3$ mol. $n_{N_2} = x_{N_2} n_{tot} = 0.78(3.4_4 \times 10^3 \text{ mol}) = 2.6_8 \times 10^3$ mol.
m_{N_2} = (2680 mol)(28.0 g/mol) = 75 kg. Similarly, $m_{O_2} = 23$ kg,
$m_{Ar} = 1.3$ kg, $m_{CO_2} = 60$ g. We have $\rho = m_{tot}/V =$
($99._4$ kg)/(8.5 × 10^7 cm^3) = 0.00117 g/cm^3.

1.35 $f'(x)$ is zero at the two points where f is a local minimum and where f is a local maximum. $f'(x)$ is negative for the portion of the curve between the maximum and the minimum.

1.36 $dy/dx = 2x + 1$. At $x = 1$, the slope is 2(1) + 1 = 3.

1.37 **(a)** $6x^2e^{-3x} - 6x^3e^{-3x}$; **(b)** $-24xe^{-3x^2}$; **(c)** $1/x$ (not $1/2x$); **(d)** $1/(1-x)^2$;
(e) $1/(x+1) - x/(x+1)^2 = 1/(x+1)^2$; **(f)** $2e^{-2x}/(1-e^{-2x})$; **(g)** $6 \sin 3x \cos 3x$.

1.38 **(a)** $y = 2/(1-x)$ and $dy/dx = 2/(1-x)^2$.

(b) $d(x^2e^{3x})/dx = 2xe^{3x} + 3x^2e^{3x}$;

$d^2(x^2e^{3x})/dx^2 = 2e^{3x} + 6xe^{3x} + 6xe^{3x} + 9x^2e^{3x} = 2e^{3x} + 12xe^{3x} + 9x^2e^{3x}$.

(c) $dy = (10x - 3 - 2/x^2)\,dx$.

Reminder: Work the problems before looking up their solutions.

1.39 **(a)**

x	0.1	0.01	0.001	0.0001	0.00001
x^x	0.794	0.955	0.9931	0.9991	0.99988

This indicates (but does not prove) that the limit is 1.

(b)

x	10^{-3}	-10^{-3}	10^{-4}	-10^{-4}	10^{-5}	-10^{-5}
$(1+x)^{1/x}$	2.717	2.720	2.7181	2.7184	2.71827	2.71828

This suggests that the limit is $e = 2.7182818\ldots$

1.40 **(a)** Results on a calculator with 8-digit display and 11 internal digits are: $\Delta y/\Delta x = 277$, 223.4, 218.88, 218.44, 218.398, 218.393, 218.4 for $\Delta x = 10^{-1}, 10^{-2}, 10^{-3}, 10^{-4}, 10^{-5}, 10^{-6}, 10^{-7}$, respectively. The best estimate is 218.393.

(b) $dy/dx = 2xe^{x^2}$, and at $x = 2$, $dy/dx = 218.3926$.
A BASIC program for part (a) is

```
5 CX = 0.1
10 FOR N = 1 TO 7
20 X = 2
30 CY = EXP((X + CX)^2) – EXP(X^2)
40 R = CY/CX
50 PRINT "DELTAX=";CX;
        " RATIO=";R
60 CX = CX/10
70 NEXT N
80 END
```

1.41 **(a)** $1 + ax\cos(axy)$; **(b)** $-2byz\sin(by^2z)$; **(c)** $-(x^2/y^2)e^{x/y}$;
(d) 0; **(e)** $-ae^{-a/y}/y^2(e^{-a/y}+1)^2$.

1.42 **(a)** nR/P; **(b)** $-2P/nRT^3$.

1.43 Equation (1.30) gives $dz = 2axy^3\,dx + 3ax^2y^2\,dy$.

1.44 **(a)** $P\,dV + V\,dP$; **(b)** $-T^{-2}\,dT$; **(c)** $2cT\,dT$; **(d)** $dU + P\,dV + V\,dP$.

1.45 Partial differentiation of $z = x^5/y^3$ gives

$$\frac{\partial z}{\partial x} = \frac{5x^4}{y^3}, \quad \frac{\partial^2 z}{\partial x^2} = \frac{20x^3}{y^3}, \quad \frac{\partial z}{\partial y} = -\frac{3x^5}{y^4}, \quad \frac{\partial^2 z}{\partial y^2} = \frac{12x^5}{y^5}$$

$$\frac{\partial^2 z}{\partial x\,\partial y} = \frac{\partial}{\partial x}\left(-\frac{3x^5}{y^4}\right) = -\frac{15x^4}{y^4}, \quad \frac{\partial^2 z}{\partial y\,\partial x} = \frac{\partial}{\partial y}\frac{5x^4}{y^3} = -\frac{15x^4}{y^4} = \frac{\partial^2 z}{\partial x\,\partial y}$$

1.46 **(a)** P is a function of n, T, and V, so $dP = (\partial P/\partial n)_{T,V}\,dn + (\partial P/\partial T)_{V,n}\,dT + (\partial P/\partial V)_{T,n}\,dV$. Partial differentiation of $P = nRT/V$ gives $(\partial P/\partial n)_{T,V} = RT/V = P/n$ (where $PV = nRT$ was used), $(\partial P/\partial T)_{V,n} = nR/V = P/T$, and $(\partial P/\partial V)_{T,n} = -nRT/V^2 = -P/V$. Substitution into the above equation for dP gives the desired result. (Note that from $P = nRT/V$, we have $\ln P = \ln n + \ln R + \ln T - \ln V$, from which $d\ln P = d\ln n + d\ln T - d\ln V$ follows at once.)

(b) Approximating small changes by infinitesimal changes, we have $dn \approx \Delta n = 0$, $dt \approx \Delta T = 1.00$ K, $dV \approx \Delta V = 50\text{ cm}^3$. The original pressure is $P = nRT/V = 0.8206$ atm. Then $\Delta P \approx dP \approx (0.8206\text{ atm})[0 + (1.00\text{ K})/(300\text{ K}) - (50\text{ cm}^3)/(30000\text{ cm}^3)] = 0.0013_7$ atm.

(c) The accurate final pressure is $(1.0000\text{ mol})(82.06\text{ cm}^3\text{-atm/mol-K}) \times (301.00\text{ K})/(30050\text{ cm}^3) = 0.8219_7$ atm. The accurate ΔP is $0.8219_7\text{ atm} - 0.8206\text{ atm} = 0.0013_7$ atm.

1.47 1.000 bar = 750 torr = (750 torr)(1 atm/760 torr) = 0.987 atm.
$V_m = V/n = (nRT/P)/n = RT/P =$
$(82.06\text{ cm}^3\text{-atm/mol-K})(293.1\text{ K})/(0.987\text{ atm}) = 2.44 \times 10^4\text{ cm}^3\text{/mol}$.

1.48 **(a)** Division by n gives $(P + a/V_m^2)(V_m - b) = RT$.

(b) The units of b are the same as those of V_m, namely, cm^3/mol.
P and a/V_m^2 have the same units, so the units of a are bar · cm^6/mol^2.

1.49 $\alpha = (1/V_m)(\partial V_m/\partial T)_P = (1/V_m)(c_2 + 2c_3T - c_5P)$, where V_m is given by (1.40).
$\kappa = -(1/V_m)(\partial V_m/\partial P)_T = -(1/V_m)(-c_4 - c_5T) =$
$(c_4 + c_5T)/(c_1 + c_2T + c_3T^2 - c_4P - c_5PT)$.

1.50 **(a)** $\rho = m/V = (m/n)/(V/n) = M/V_m$, so $V_m = M/\rho =$
(18.0153 g/mol)/(0.98804 g/cm^3) = 18.233 cm^3/mol.

(b) $\kappa = -(1/V_m)(\partial V_m/\partial P)_T$ and $dV_m/V_m = -\kappa\, dP$ at constant T. Integration gives $\ln (V_{m2}/V_{m1}) = -\kappa(P_2 - P_1)$ at constant T.
$\kappa = (4.4 \times 10^{-10}\ \text{Pa}^{-1})(101325\ \text{Pa}/1\ \text{atm}) = 4.4_6 \times 10^{-5}\ \text{atm}^{-1}$ and
$\ln[V_{m2}/(18.233\ \text{cm}^3/\text{mol})] = -(4.4_6 \times 10^{-5}\ \text{atm}^{-1})(100\ \text{atm} - 1\ \text{atm}) =$
-0.0044, so $V_{m2}/(18.233\ \text{cm}^3/\text{mol}) = e^{-0.0044} = 0.9956$ and $V_{m2} =$
18.15 cm^3/mol.

1.51 **(a)** At constant P, the equation $PV_m = RT$ gives $V_m = aT$, where $a = R/P$ is a positive constant. The isobars on a V_m vs. T diagram are straight lines that start at the origin and have positive slopes. (As P increases, the slope decreases.)

(b) For V_m constant, $PV_m = RT$ gives $P = bT$, where $b = R/V_m$ is constant. The isochores on a P vs. T diagram are straight lines that start at the origin and have positive slopes.

1.52 **(a)** Partial differentiation of $V = nRT(1 + aP)/P$ gives $(\partial V/\partial T)_{P,n} =$
$nR(1 + aP)/P$. The equation of state gives $nR(1 + aP) = PV/T$, so
$(\partial V/\partial T)_{P,n} = V/T$. Then $\alpha = (1/V)(\partial V/\partial T)_{P,n} = 1/T$.
Partial differentiation of $V = nRT(1/P + a)$ gives $(\partial V/\partial P)_{T,n} =$
$-nRT/P^2 = -[PV/(1 + aP)]/P^2 = -V/P(1 + aP)$, where the equation of state was used. Then $\kappa = -(1/V)(\partial V/\partial P)_T = 1/P(1 + aP)$.

(b) Solving the equation of state for P, we get $P = nRT/(V - anRT)$; partial differentiation gives $(\partial P/\partial T)_V = nR/(V - anRT) + an^2R^2T/(V - anRT)^2 =$
$P/T + aP^2/T$, where $P = nRT/(V - anRT)$ was used. From (a), we have α/κ
$= P(1 + aP)/T = P/T + aP^2/T$, which agrees with Eq. (1.45).

1.53 For small ΔT, we have

$$\alpha = \frac{1}{V}\left(\frac{\partial V}{\partial T}\right)_P \approx \frac{1}{V}\left(\frac{\Delta V}{\Delta T}\right)_P$$

Since α is an intensive property, we can take any quantity of water. For 1 g, the equation $V = m/\rho$ gives $V = 1.002965\ \text{cm}^3$ at 25°C, 1 atm and $V = 1.003227\ \text{cm}^3$ at 26°C, 1 atm.

$$\text{Hence } \alpha \approx \frac{1}{1.003\ \text{cm}^3}\,\frac{1.003227\ \text{cm}^3 - 1.002965\ \text{cm}^3}{26°\text{C} - 25°\text{C}} = 0.00026\ \text{K}^{-1}$$

Similarly, $\kappa = -(1/V)(\partial V/\partial P)_T \approx -(1/V)(\Delta V/\Delta P)_T$. At 25°C and 2 atm, we calculate $V = 1.002916\ \text{cm}^3$ for 1 g of water.

$$\text{Thus } \kappa \approx -\frac{1}{1.003\ \text{cm}^3}\,\frac{1.002916\ \text{cm}^3 - 1.002965\ \text{cm}^3}{2\ \text{atm} - 1\ \text{atm}} = 4.9\times 10^{-5}\ \text{atm}^{-1}$$

$$\text{Eq. (1.45) gives } \left(\frac{\partial P}{\partial T}\right)_{V_m} = \frac{\alpha}{\kappa} \approx \frac{2.6\times 10^{-4}\ \text{K}^{-1}}{4.9\times 10^{-5}\ \text{atm}^{-1}} = 5.3\ \text{atm/K}$$

1.54 **(a)** Drawing the tangent line to the 500-bar V_m-vs.-T curve at 100°C, one finds its slope to be (20.8 cm^3/mol – 17 cm^3/mol)/(300°C – 0°C) = 0.013 cm^3/mol-K = $(\partial V_m/\partial T)_P$ at this T and P. The figure gives V_m = 18.2 cm^3/mol at 500 bar and 100°C, so $\alpha = (1/V_m)\,(\partial V_m/\partial T)_P$ = (0.013 cm^3/mol-K)/(18.2 cm^3/mol) = 0.0007 K^{-1}.

(b) Drawing the tangent line to the 300°C V_m-vs.-P curve at 2000 bar, one finds its slope to be –0.0011 cm^3/mol-bar. The figure gives V_m = 20.5 cm^3/mol at this T and P, so $\kappa = -(1/V_m)(\partial V_m/\partial P)_T$ = (0.0011 cm^3/mol-bar)/(20.5 cm^3/mol) = 5 × 10^{-5} bar^{-1}.

1.55 Equation (1.45) gives $\alpha/\kappa = (\partial P/\partial T)_V \approx (\Delta P/\Delta T)_V$, so

$$\Delta P \approx \frac{\alpha}{\kappa}\Delta T = \frac{1.7\times 10^{-4}\ \text{K}^{-1}}{4.7\times 10^{-5}\ \text{atm}^{-1}}(6\ \text{K}) = 22\ \text{atm};\qquad P \approx 23\ \text{atm}$$

1.56 **(a)** As P increases, the molecules are forced closer together; the decrease in empty space between the molecules makes it harder to compress the substance, and κ is smaller.

(b) Most substances expand as T increases. The increased space between molecules makes it easier to compress the substance, and κ increases.

1.57 **(a)** $\kappa = -(1/V)(\partial V/\partial P)_T \approx -(1/V)(\Delta V/\Delta P)_T$ and $\Delta P \approx -\Delta V/V\kappa$ at constant T. For a 1% volume decrease, $\Delta V = -0.01V$ and we have $\Delta P \approx 0.01V/V\kappa = 0.01/\kappa = 0.01/(5 \times 10^{-6} \text{ atm}^{-1}) = 2000$ atm.

(b) $\Delta P \approx 0.01/\kappa = 0.01/(1 \times 10^{-4} \text{ atm}^{-1}) = 100$ atm.

1.58 **(a)** $\sum_{J=0}^{4} 2J + 1 = (0 + 1) + (2 + 1) + (4 + 1) + (6 + 1) + (8 + 1) = 25$.

(b) $\sum_{i=1}^{s} x_i V_i$.

(c) $\sum_{i=1}^{2} (b_{i4} + b_{i5} + b_{i6}) = b_{14} + b_{15} + b_{16} + b_{24} + b_{25} + b_{26}$.

1.59 $\sum_{i=1}^{n} ca_i = ca_1 + ca_2 + \cdots + ca_n = c(a_1 + a_2 + \cdots + a_n) = c\sum_{i=1}^{n} a_i$. Q.E.D.

$\sum_{i=1}^{n} (a_i + b_i) = (a_1 + b_1) + (a_2 + b_2) + \cdots + (a_n + b_n) =$

$a_1 + a_2 + \cdots + a_n + b_1 + b_2 + \cdots + b_n = \sum_{i=1}^{n} a_i + \sum_{i=1}^{n} b_i$. Q.E.D.

The left side of (1.51) is $\sum_{i=1}^{n} \sum_{j=1}^{m} a_i b_j = \sum_{i=1}^{n} (a_i b_1 + a_i b_2 + \cdots + a_i b_m) =$

$\sum_{i=1}^{n} a_i (b_1 + b_2 + \cdots + b_m) = \left(\sum_{i=1}^{n} a_i\right)(b_1 + b_2 + \cdots + b_m) =$

$\sum_{i=1}^{n} a_i \sum_{j=1}^{m} b_j$, which is the right side of (1.51).

1.60 **(a)** $\int_3^{-2} (2V + 5V^2)\, dV = (V^2 + 5V^3/3)|_3^{-2} = (4 - 40/3) - (9 + 45) = -190/3$.

(b) $\int_2^4 V^{-1}\, dV = \ln V\big|_2^4 = \ln 4 - \ln 2 = \ln 2 = 0.693$.

(c) $\int_1^\infty V^{-3}\, dV = -\tfrac{1}{2}V^{-2}\,|_1^\infty = 0 - (-\tfrac{1}{2}) = \tfrac{1}{2}$.

(d) Let $z = x^3$. Then $dz = 3x^2\, dx$ and $\int_0^{\pi/2} x^2 \cos x^3\, dx = (1/3)\int_0^{\pi^3/8} \cos z\, dz =$

$(1/3) \sin z \Big|_0^{\pi^3/8} = (1/3)[\sin(\pi^3/8) - 0] = -0.2233$.

1.61 **(a)** $-a^{-1} \cos ax + C$.

(b) $-a^{-1} \cos ax\, |_0^\pi = (1 - \cos a\pi)/a$.

(c) Differentiation of the (b) answer gives $-a^{-2} + a^{-2} \cos a\pi + a^{-1}\pi \sin a\pi$.

(d) $-a/T + C$.

1.62 **(a)** $\alpha = (1/V_m)(\partial V_m/\partial T)_P$; $dV_m/V_m = \alpha\, dT$; $\int_1^2 V_m^{-1}\, dV_m = \int_1^2 \alpha\, dT$; $\ln(V_{m2}/V_{m1}) = \alpha(T_2 - T_1)$ at constant P, where the T dependence of α was neglected over the short range of T. $\ln(V_{m2}/18.2334 \text{ cm}^3 \text{ mol}^{-1}) = (4.576 \times 10^{-4}/\text{K})(2.00 \text{ K})$. $\ln(V_{m2}/\text{cm}^3 \text{ mol}^{-1}) = \ln 18.2334 + 0.0009152$ and $V_{m2} = 18.2501 \text{ cm}^3/\text{mol}$.

(b) $\kappa = -(1/V_m)(\partial V_m/\partial P)_T$; $dV_m/V_m = -\kappa\, dP$; $\ln(V_{m2}/V_{m1}) = -\kappa(P_2 - P_1)$ at constant T, where the P dependence of κ was neglected. $\ln(V_{m2}/18.2334 \text{ cm}^3 \text{ mol}^{-1}) = -(44.17 \times 10^{-6} \text{ bar}^{-1})(199 \text{ bar})$ and $V_{m2} = 18.074 \text{ cm}^3/\text{mol}$.

1.63 **(a)** Function; **(b)** number; **(c)** number.

1.64 In (b) and (c).

1.65 **(a)** $x^4/2 + 3e^{5x}/5 + C$; **(b)** $24x^7$.

1.66 **(a)** $\int_2^3 x^2 dx \approx \sum_{i=1}^{n} x_i^2\, \Delta x$. For $\Delta x = 0.1$ and x_i at the left end of each subinterval, $\sum_{i=1}^{n} x_i^2\, \Delta x = 0.1[2^2 + (2.1)^2 + (2.2)^2 + \cdots + (2.9)^2] = 6.085$. For $\Delta x = 0.01$, we get 6.30835. For $\Delta x = 0.001$, we get 6.33083. The exact value is $(x^3/3)\big|_2^3 = 27/3 - 8/3 = 6.33333\ldots$.

(b) $\int_0^1 e^{-x^2} dx \approx 0.01[e^{-0^2} + e^{-(0.01)^2} + e^{-(0.02)^2} + \cdots + e^{-(0.99)^2}] = 0.74998$.

A BASIC program for part (a) is

```
10 N = 10
15 FOR J = 1 TO 3
20 X = 2
25 DX = 1/N
30 S = 0
35 FOR I = 1 TO N
40 S = S + X*X*DX
45 X = X + DX
50 NEXT I
55 PRINT "DELTAX="; DX; "SUM=";S
60 N = 10*N
65 NEXT J
70 END
```

1.67 **(a)** $\log(4.2 \times 10^{1750}) = \log 4.2 + \log 10^{1750} = 0.62 + 1750 = 1750.62$.

(b) $\ln(6.0 \times 10^{-200}) = 2.3026 \log(6.0 \times 10^{-200}) =$
$2.3026 \log 6.0 + 2.3026 \log 10^{-200} = 1.79 - 460.52 = -458.73.$

(c) $\log y = -138.265;\ \ y = 10^{-138.265} = 10^{-0.265}10^{-138} = 0.543 \times 10^{-138}.$

(d) $\ln z = 260.433 = 2.3026 \log z;\ \ \log z = 113.10;$
$z = 10^{0.10}10^{113} = 1.2_6 \times 10^{113}.$

1.68 **(a)** 5, since $2^5 = 32$.

(b) 0.

1.69 **(a)** intensive; **(b)** extensive; **(c)** intensive; **(d)** intensive; **(e)** intensive; **(f)** intensive;

1.70 One finds that a plot of PV_m vs. P is approximately linear with an intercept of $PV_m = 58.90$ L atm/mol at $P = 0$. The ideal-gas law $PV_m = RT$ applies to O_2 in the limit of zero pressure, so

$$T = \lim_{P\to 0} \frac{PV_m}{R} = \frac{58900 \text{ cm}^3\text{-atm/mol}}{82.06 \text{ cm}^3\text{-atm/mol-K}} = 717.8 \text{ K}$$

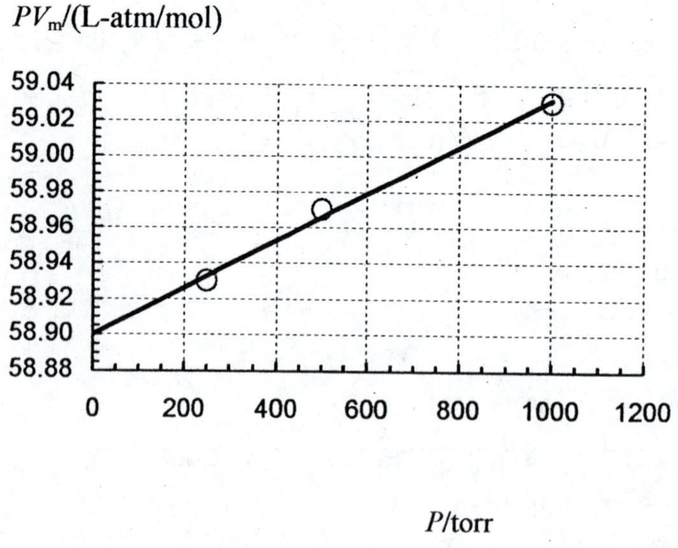

1.71 **(a)** T. **(b)** F. **(c)** F. **(d)** T. **(e)** F. **(f)** T. **(g)** T. **(h)** T. **(i)** T. **(j)** F. **(k)** F.

Chapter 2

2.1 **(a)** T. **(b)** F.

2.2 **(a)** J; **(b)** J; **(c)** m^3; **(d)** N; **(e)** m/s; **(f)** kg.

2.3 **(a)** $1\ J = 1\ N\ m = 1\ kg\ m\ s^{-2}\ m = 1\ kg\ m^2\ s^{-2}$

(b) $1\ Pa = 1\ N/m^2 = 1\ kg\ m\ s^{-2}\ m^{-2} = 1\ kg\ m^{-1}\ s^{-2}$

(c) $1\ L = 10^3\ cm^3 = 10^3\ (10^{-2}\ m)^3 = 10^{-3}\ m^3$

(d) $1\ N = 1\ kg\ m\ s^{-2}$ **(e)** $1\ W = 1\ J/s = 1\ kg\ m^2\ s^{-3}$

2.4 **(a)** $w = \int_1^2 F\,dx = \int_1^2 mg\,dx = mg\,\Delta x = (0.155\text{ kg})(9.81\text{ m/s}^2)(10.0\text{ m}) = 15.2\text{ J}$

(b) $w = \Delta K = K_2 - K_1 = K_2 = 15.2$ J.

(c) $\frac{1}{2}m\upsilon^2 = K$ and $\upsilon = (2K/m)^{1/2} = [2(15.2\text{ J})/(0.155\text{ kg})]^{1/2} = 14.0$ m/s, since $1\ J = 1\ kg\ m^2/s^2$.

2.5 $P = F/A = mg/A = (0.102\text{ kg})(9.81\text{ m/s}^2)/(1.00\text{ m}^2) = 1.00\text{ N/m}^2 = 1.00$ Pa.

2.6 From $F = ma$, 1 dyne = $1\ g\ cm/s^2 = 10^{-3}\ kg\ 10^{-2}\ m/s^2 = 10^{-5}$ N, where (2.7) was used. From $dw = F_x\,dx$, 1 erg = 1 dyne cm = $10^{-5}\ N\ 10^{-2}\ m = 10^{-7}$ J, where (2.12) was used.

2.7 **(a)** F; **(b)** T; **(c)** T; **(d)** F; **(e)** F; **(f)** F; **(g)** F.

2.8 **(a)** area = length × height = $(V_2 - V_1)P_1 = (5000 - 2000)\text{cm}^3(0.230\text{ atm}) = 690\text{ cm}^3\text{ atm}$. $w_{rev} = -\text{area} = -(690\text{ cm}^3\text{ atm})(8.314\text{ J}/82.06\text{ cm}^3\text{ atm}) = -69.9$ J.

(b) $w_{rev} = -\int_1^2 P_1\,dV = -P_1(V_2 - V_1) =$ etc.

2.9 $w_{rev} = -\int_1^2 P\,dV = -P(V_2 - V_1) = -(275/760)\text{ atm} \times (875 - 385)\text{ cm}^3 =$

$-177\text{ cm}^3\text{ atm }(8.314\text{ J}/82.06\text{ cm}^3\text{-atm}) = -18.0$ J.

2.10 **(a)** The area under the curve is the sum of the areas of a rectangle and a right triangle. The rectangle's area is $(V_2 - V_1)P_2 = (2000 - 500)\text{cm}^3(1.00\text{ atm}) = 1500\text{ cm}^3\text{ atm}$.
The triangle's area is ½(base)(altitude) = $\frac{1}{2}(V_2 - V_1)(P_1 - P_2) = \frac{1}{2}(2000 - 500)\text{cm}^3(3.00 - 1.00)\text{atm} = 1500\text{ cm}^3\text{ atm}$.
Thus $w_{\text{rev}} = -3000\text{ cm}^3\text{ atm }(8.314\text{ J}/82.06\text{ cm}^3\text{-atm}) = -304\text{ J}$.

(b) Replacement of y and x with P and V in the straight-line equation gives $(P - P_1)/(V - V_1) = (P_2 - P_1)/(V_2 - V_1)$.

$$w = -\int_1^2 P\,dV = -\int_{V_1}^{V_2}\{P_1 + [(P_2 - P_1)/(V_2 - V_1)](V - V_1)\}\,dV =$$

$$-P_1(V_2 - V_1) - [(P_2 - P_1)/(V_2 - V_1)][(\tfrac{1}{2}V_2^2 - V_1V_2) - (\tfrac{1}{2}V_1^2 - V_1^2)] =$$

$$-P_1(V_2 - V_1) + \tfrac{1}{2}(P_1 - P_2)(V_2 - V_1) = \text{as in } (a).$$

2.11 **(a)**

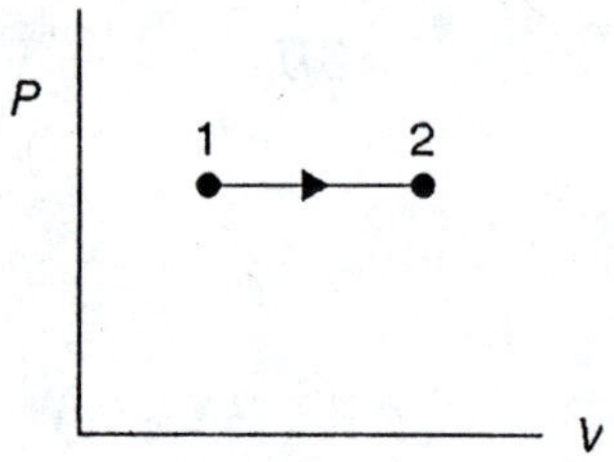

(b)

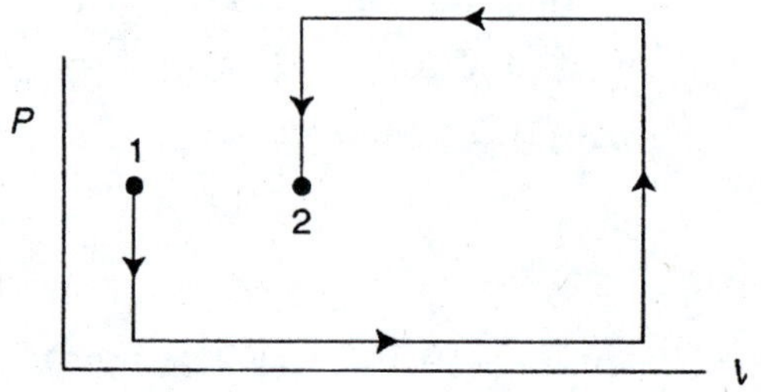

2.12 Neglecting the dependence of specific heat on T, we equate the heat gained by the water to that lost by the metal. The heat gained by the H_2O is
(24.0 g)(1.00 cal/g-°C)(10.0°C) = 240 cal.
Thus 240 cal = (45.0 g)c_{metal}(70.0 – 20.0)°C and c_{metal} = 0.107 cal/g-°C.

2.13 **(a)** T; **(b)** T; **(c)** F; **(d)** T; **(e)** F.

2.14 Only (c).

2.15 **(a)** $2200 \times 10^3 \dfrac{\text{cal}}{\text{day}} \dfrac{4.184\text{ J}}{1\text{ cal}} \dfrac{1\text{ day}}{24\text{ hr}} \dfrac{1\text{ hr}}{3600\text{ s}} = 107\text{ J/s} = 107\text{ W}$

(b) $(7 \times 10^9)(107\text{ J/s})(3600\text{ s/hr})(24\text{ hr/day})(365\text{ days/yr}) = 2._4 \times 10^{19}\text{ J}$

2.16 Since the process is cyclic, $\Delta U = 0$. Hence $q = -w = -338$ J.

2.17 **(a)** The bulk kinetic energy acquired by the 167-ft fall is converted into internal energy, thereby warming the water by ΔT. The bulk kinetic energy equals the potential-energy decrease $mg\,\Delta h$. The ΔU for a temperature increase of ΔT can be set equal to the heat $q = mc_P\,\Delta T$ that would be needed to increase the temperature by ΔT, since the expansion work is negligible. Therefore $mg\,\Delta h = mc_P\,\Delta T$ and

$$\Delta T = \frac{g\,\Delta h}{c_P} = \frac{(9.80\text{ m/s}^2)(167\times12\times2.54\times10^{-2}\text{ m})}{(1.00\text{ cal/g-°C})(10^3\text{ g/kg})}\frac{1\text{ cal}}{4.184\text{ J}} = 0.12\text{ °C}$$

(b) $mg\,\Delta h = (2.55\times10^9\text{ cm}^3)(1.00\text{ g/cm}^3)(1\text{ kg}/10^3\text{ g})(9.80\text{ m/s}^2)(50.9\text{ m}) = 1.27\times10^9\text{ J}$

2.18 We have $0 = \Delta U_1 + \Delta U_2 = q_1 + w_1 + q_2 + w_2 = q_1 + q_2$ (since the wall is rigid); therefore $q_2 = -q_1$.

2.19 This notation might seem to imply that q and w are state functions, which is not so. There is no such thing as the change in heat for a system. There is only an amount of heat transfer for a process.

2.20 Cool the water to some temperature below 25°C and then do enough stirring work to raise its T to 30°C.

2.21 $V = \tfrac{1}{2}kx^2 = \tfrac{1}{2}(125\text{ N/m})(0.100\text{ m})^2 = 0.625\text{ J} = 0.149\text{ cal}$. $\Delta U = (m_1c_1 + m_2c_2)\Delta T$ and $\Delta T = \dfrac{0.149\text{ cal}}{(1.00\text{ cal/g-°C})(112\text{ g}) + (20\text{ g})(0.30\text{ cal/g-°C})}$

$\Delta T = 0.00126$ °C and the final temperature is 18.001°C

2.22 **(a)** $0 = dE_{syst} + dE_{surr} = dE_{syst} + dK_{surr} + dV_{surr} + dU_{surr} = dE_{syst} + dK_{pist} + mg\,dh + 0$, so $dE_{syst} = -mg\,dh - dK_{pist}$.

(b) $dE_{syst} = dq + dw_{irrev} = 0 + dw_{irrev}$, so $dw_{irrev} = dE_{syst} = -mg\,dh - dK_{pist}$. But $mg\,dh = (mg/A)A\,dh$, where A is the piston's area. Since $mg/A = P_{ext}$ and $A\,dh = dV$, we have $mg\,dh = P_{ext}\,dV$ and $dw_{irrev} = -P_{ext}\,dV - dK_{pist}$.

2.23 Eq. (2.33) gives $w_{irrev} = -\int_1^2 P_{ext}\, dV - \Delta K_{pist} = -P_{ext} \int_1^2 dV - 0 =$

$-P_{ext}(V_2 - V_1) = -(0.500 \text{ bar})(4.00 \text{ dm}^3) = -2.00 \text{ dm}^3 \text{ bar}.$

1 dm^3 = 1000 cm^3 and 1 bar = 750 torr = (750/760) atm = 0.987 atm,

so w_{irrev} = –1974 cm^3 atm × (8.314 J/82.06 cm^3-atm) = –200 J.

2.24 **(a)** T; **(b)** F; **(c)** F.

2.25 All except force, mass, and pressure.

2.26 **(a)** From the equation $\Delta H = q_P$.

(b) It can mislead one into thinking that heat is a state function.

2.27 No. For example, in a cyclic process, ΔH is zero but q need not be zero, since q is not a state function.

2.28 $\Delta H = q_P = 0$ for the entire system. Since H is extensive, $H = H_1 + H_2$ and $\Delta H = \Delta H_1 + \Delta H_2 = q_1 + q_2$. Since $\Delta H = 0$, $q_1 + q_2 = 0$.

2.29 **(a)** T; **(b)** T.

2.30 **(a)** $C_{P,m} = C_P/n$ and $C_P = nC_{P,m}$ = (586 g/16.04 g mol^{-1})(94.4 J/mol-K) = 3.45 kJ/K.

(b) (10.0 carat)(0.2 g/carat) = 2.00 g and $C_P = nC_{P,m}$ =
(2.00 g/12.01 g mol^{-1})(6.115 J/mol-K) = 1.018 J/K.
$c_P = C_P/m$ = (1.018 J/K)/(2.00 g) = 0.509 J/g-K.

2.31 $\upsilon = V/m = (m/V)^{-1} = \rho^{-1}$ = (0.958 g/cm^3)$^{-1}$ = 1.044 cm^3/g.

2.32 **(a)** U; **(b)** H.

2.33 $\mu_{JT} = (\partial T/\partial P)_H$, $\Delta T = \int_1^2 \mu_{JT}\, dP$, and $\Delta T \approx \mu_{JT}\, \Delta P$ for H constant. Thus $\Delta T \approx$ (0.2 °C/bar)(–49 bar) = –10 °C. The final temperature is approximately 15°C.

2.34 Eq. (1.35) gives $(\partial U_m/\partial V_m)_T = (\partial U_m/\partial P)_T(\partial P/\partial V_m)_T$. Partial differentiation of $P = RT/V_m$ gives $(\partial P/\partial V_m)_T = -RT/V_m^2 = -P^2/RT$. Hence, $(\partial U_m/\partial V_m)_T = -(\partial U_m/\partial P)_T\, P^2/RT$.

(a) $(\partial U_m/\partial V_m)_T$ = (6.08 J/mol-atm)(1 atm)^2/(82.06 cm^3-atm/mol-K)(301 K) = 2.46 × 10^{-4} J/cm^3.

(b) Doubling P multiplies $(\partial U_m/\partial V_m)_T$ by 4 to give 9.84 × 10^{-4} J/cm^3.

2.35 **(a)** Use of (1.34), (1.32), (2.64), and (2.53) gives

$$-1 = \left(\frac{\partial T}{\partial P}\right)_H \left(\frac{\partial P}{\partial H}\right)_T \left(\frac{\partial H}{\partial T}\right)_P = \frac{\mu_{JT} C_P}{(\partial H/\partial P)_T}$$ and (2.65) follows.

(b) Partial differentiation of $H = U + PV$ and use of (2.63), (1.35), (1.44), and (2.65) give

$$\left(\frac{\partial H}{\partial P}\right)_T = \left(\frac{\partial U}{\partial P}\right)_T + P\left(\frac{\partial V}{\partial P}\right)_T + V = \left(\frac{\partial U}{\partial V}\right)_T \left(\frac{\partial V}{\partial P}\right)_T - PV\kappa + V$$

$-C_P\,\mu_{JT} = C_V\,\mu_J V\kappa - PV\kappa + V$ and the desired result follows.

2.36 **(a)** $\mu_J = (\partial T/\partial V)_U = -(\partial U/\partial V)_T/C_V$. $(\partial U/\partial V)_T$ is intensive, since it is the ratio of the changes in two extensive quantities. C_V is extensive. Therefore, doubling the size of the system at constant T, P, and composition will double C_V, will not affect $(\partial U/\partial V)_T$, and hence will cut μ_J in half. Therefore, μ_J is neither intensive nor extensive, since it is not independent of the size of the system and is not equal to the sum of the μ_J's of the parts of the system.

2.37 **(a)** T; **(b)** F; **(c)** F; **(d)** T; **(e)** T.

2.38 **(a)** Since T is constant, $\Delta U = 0$ and $\Delta H = 0$. (U and H of a perfect gas depend on T only.) $w = -\int_1^2 P\,dV = -nRT\int_1^2 V^{-1}\,dV = -nRT\ln(V_2/V_1) =$

–(2.00 mol) (8.314 J/mol-K)(300 K) ln (1500/500) = –5.48 kJ.
$\Delta U = q + w = 0$, so $q = -w = 5.48$ kJ.

(b) Since U and H are state functions, ΔU and ΔH are still zero. The work w is zero.

2.39 **(a)** $T_2 = T_1 = 300$ K. $P_2 = nRT_2/V_2 =$
(1.00 mol)(82.06 cm^3-atm/mol-K)(300 K)/(49200 cm^3) =
0.500 atm. (An alternative solution uses $P_2V_2 = P_1V_1$.)

(b) $\gamma = C_{P,m}/C_{V,m} = (C_{V,m} + R)/C_{V,m} = 2.5R/1.5R = 1.667$.
$P_1 = nRT_1/V_1 = 1.00$ atm. For a reversible adiabatic process with C_V constant, Eq. (2.77) gives $P_1V_1^\gamma = P_2V_2^\gamma$ and $P_2 = (V_1/V_2)^\gamma P_1 =$
(24.6 L/49.2 L)$^{1.667}$(1.00 atm) = 0.315 atm. $T_2 = P_2V_2/nR = 189$ K.

(c)

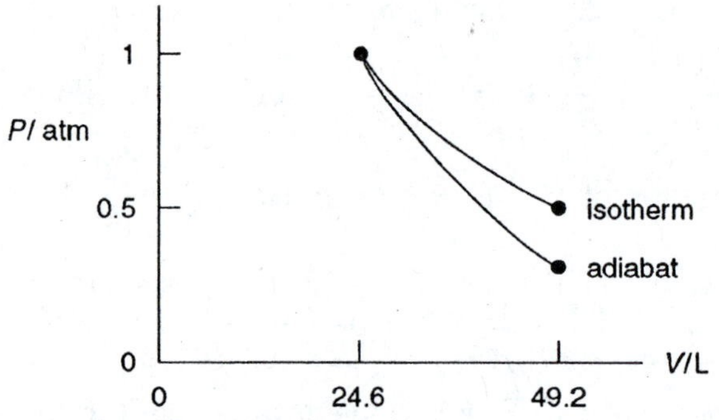

2.40 **(a)** $q = 0$ since the process is adiabatic.
$C_{V,m} = C_{P,m} - R = 2.5R$. $T_1 = P_1V_1/nR = 160$ K.
$T_2 = T_1(V_1/V_2)^{R/C_{V,m}} = (160\text{ K})(4.00)^{0.4} = 279\text{ K}.$
$dU = C_V\,dT$ and $\Delta U = C_V\,\Delta T$, since C_V is constant. Thus
$\Delta U = (0.0400\text{ mol})2.5(8.314\text{ J/mol-K})(119\text{ K}) = 98.9$ J.
$\Delta U = q + w = w = 98.9$ J; $\Delta H = C_P\,\Delta T = 138½$ J.

(b) From $P_1V_1^\gamma = P_2V_2^\gamma$, we get $P_1(nRT_1/P_1)^\gamma = P_2(nRT_2/P_2)^\gamma$, which becomes $P_1^{1-\gamma}T_1^\gamma = P_2^{1-\gamma}T_2^\gamma$. Hence $(T_1/T_2)^\gamma = (P_2/P_1)^{1-\gamma}$ and $P_2/P_1 = (T_1/T_2)^{\gamma/(1-\gamma)}$.
$\gamma = C_{P,m}/C_{V,m} = 3.5R/2.5R = 1.40$. $P_2/(1\text{ atm}) = (298\text{ K}/100\text{ K})^{1.40/(-0.40)} =$
$(2.98)^{-3.5} = 0.0219$, so $P_2 = 0.0219$ atm $= 2.22$ kPa.

2.41 **(a)** $n = 0.500$ mol. 0.800 bar = 0.789 atm. $T_1 = P_1V_1/nR = 384._6$ K,
$T_2 = P_2V_2/nR = 769._2$ K. $dU = C_V\,dT$ and $\Delta U = C_V\,\Delta T = nC_{V,m}\,\Delta T =$
(0.500 mol)(1.5R)(384.$_6$ K) = 2.40 kJ. $\Delta H = C_P\,\Delta T =$
(0.500 mol)(2.5R)(384.$_6$ K) = 4.00 kJ. $w = -\int_1^2 P\,dV =$

$-P(V_2 - V_1) = -(0.789\text{ atm})(2000\text{ cm}^3)(8.314\text{ J}/82.06\text{ cm}^3\text{-atm}) =$
-1.60 kJ. $q = q_P = \Delta H = 4.00$ kJ.

(b) $w = 0$ at constant V. $T_1 = P_1V_1/nR = 216._5$ K; $T_2 = 324._8$ K. $\Delta U = C_V\,\Delta T = 0.675$ kJ. $\Delta H = C_P\,\Delta T = 1.13$ kJ. $\Delta U = q + w = q$, so $q = 0.675$ kJ.

2.42 **(a)** F; **(b)** T; **(c)** F; **(d)** F; **(e)** F; **(f)** F.

2.43 **(a)** Process; **(b)** system property; **(c)** process; **(d)** process; **(e)** system property; **(f)** system property; **(g)** system property.

2.44 $C_{pr} = dq_{pr}/dT_{pr}$. **(a)** $dq_{pr} > 0$ and $dT_{pr} = 0$, so $C_{pr} = \infty$. **(b)** $-\infty$, since $dq_{pr} < 0$. **(c)** $dU = 0 = dq_{pr} + dw_{pr}$; $dq_{pr} = -dw_{pr}$; $dw_{pr} < 0$ and $dq_{pr} > 0$. $dT_{pr} = 0$. Hence $C_{pr} = \infty$. **(d)** 0, since $dq_{pr} = 0$.

2.45 **(a)** Heat is required to melt the benzene, so $q > 0$. $\Delta H = q_p > 0$. The constant-P work is $w = -P\,\Delta V$; since the benzene expands on melting, $w < 0$. The volume change is slight, so $|w| << |q|$ and $\Delta U = q + w \approx q$; hence $\Delta U > 0$.

(b) The same as (a) except that $w > 0$ since the system contracts on melting.

(c) $q = 0$ for this adiabatic process. w is negative for an expansion. We have $\Delta U = q + w = w$, so $\Delta U < 0$. $\Delta H = \Delta U + \Delta(PV) = \Delta U + nR\,\Delta T$. We have $dU = C_V\,dT$, where $dU < 0$ and $C_V > 0$; hence $dT < 0$ and $\Delta T < 0$. Therefore $\Delta H < 0$.

(d) With T constant, we have $\Delta H = 0 = \Delta U$ for the perfect gas. w is negative for the expansion. $\Delta U = 0 = q + w$, so $q = -w$ and q is positive.

(e) $q = 0$, $w = 0$, $\Delta U = q + w = 0$, $\Delta H = \Delta U + \Delta(PV) = \Delta U + nR\,\Delta T = 0$. ΔT is zero because $\mu_J = 0$ for the perfect gas.

(f) For Joule–Thomson throttling, $\Delta H = 0$. Since $\mu_{JT} = (\partial T/\partial P)_H = 0$ for a perfect gas, T is constant. Hence $\Delta U = 0$, since $dU = C_V\,dT = 0$. The process is adiabatic, so $q = 0$. Hence $w = \Delta U - q = 0$; this also follows from the equation $w = P_1V_1 - P_2V_2$ in the text, since $T_2 = T_1$.

(g) $q > 0$ for heating. From $dq_P = C_P\,dT$, it follows that $\Delta T > 0$, since $C_P > 0$. From $dU = C_V\,dT$, it follows that $\Delta U > 0$, since $C_V > 0$. From $PV = nRT$, it follows that $\Delta V > 0$. Hence $w = -P\,\Delta V < 0$. $\Delta H = \Delta U + \Delta(PV) = \Delta U + nR\,\Delta T > 0$.

(h) $q < 0$ and $\Delta T < 0$. Hence $\Delta U < 0$. $w = 0$ since $dw = -P\,dV = 0$.
$\Delta H = \Delta U + nR\,\Delta T < 0$.

2.46 (a) $q = 0$ (since adiabatic), $w = 0$ (since constant V), $\Delta U = q + w = 0$.

(b) $w = 0$ (const. V). The combustion is an exothermic process that releases heat to the surrounding bath; hence $q < 0$. $\Delta U = q + w < 0$.

(c) $q = 0$ (since adiabatic), $w = 0$ (expansion into vacuum), $\Delta U = q + w = 0$.

2.47 The process is adiabatic, so $q = 0$. The volume change is being neglected, so $w \approx 0$. Hence $\Delta U = q + w \approx 0$. $\Delta H = \Delta U + \Delta(PV) \approx 0 + V\,\Delta P$, since V is constant. Thus

$$\Delta H \approx (18\text{ cm}^3)(9.0\text{ atm})\,\frac{1.987\text{ cal/mol-K}}{82.06\text{ cm}^3\text{-atm/mol-K}} = 3.9\text{ cal} = 16\text{ J}$$

2.48 (a) $dq_P = C_P\,dT$; $q_P = \int_1^2 C_P\,dT = \int_1^2 n(a + bT)\,dT =$

$n[a(T_2 - T_1) + \tfrac{1}{2}b(T_2^2 - T_1^2)]$. Hence

$q = (2.00\text{ mol})[6.15\text{ cal/mol-K})(100\text{ K}) +$
$\tfrac{1}{2}(0.00310\text{ cal/mol-K}^2)(400^2 - 300^2)\text{K}^2]$. $q = 1447$ cal

$w = -\int_1^2 P\,dV = -P\,\Delta V = -nR\,\Delta T =$

$-(2.00\text{ mol})(1.987\text{ cal/mol-K})(100\text{ K}) = -397$ cal

$\Delta U = q + w = 1447\text{ cal} - 397\text{ cal} = 1050$ cal. $\Delta H = q_P = 1447$ cal

(b) $dw = -P\,dV = 0$ and $w = 0$. ΔU and ΔH are the same as in (a), since the final and initial temperatures are the same as in (a) and U and H are functions of T only for a perfect gas. Hence $\Delta H = 1447$ cal and $\Delta U = 1050$ cal. $\Delta U = q + w = q = 1050$ cal.

2.49 (a) $q = (333.6\text{ J/g})(18.015\text{ g}) = 6010$ J

$w = -\int_1^2 P\,dV = -P\,\Delta V = -Pm(1/\rho_2 - 1/\rho_1) =$

$$-(1\text{ atm})(18.0\text{ g})(1.000\text{ cm}^3\text{/g} - 1.0905\text{ cm}^3\text{/g})\,\frac{8.314\text{ J}}{82.06\text{ cm}^3\text{ atm}}$$

$w = 0.165$ J; $\Delta U = q + w = 6010$ J; $\Delta H = q_P = 6010$ J.

(b) $q_P = \int_1^2 C_P\, dT = C_P\,\Delta T = (4.19 \text{ J/g-°C})(18.01 \text{ g})(100 \text{ °C}) = 7.55 \text{ kJ}$

$w = -P\,\Delta V =$

$$(1 \text{ atm})(18.0 \text{ g})(1.044 \text{ cm}^3/\text{g} - 1.000 \text{ cm}^3/\text{g})\ \frac{8.314 \text{ J}}{82.06 \text{ cm}^3 \text{ atm}}$$

$w = -0.080$ J

$\Delta U = q + w = 7.55$ kJ. $\Delta H = q_p = 7.55$ kJ

(c) $q = (18.015 \text{ g})(2256.7 \text{ J/g}) = 40654 \text{ J}$

$$V_2 = \frac{(1 \text{ mol})(82.06 \text{ cm}^3\text{-atm/mol-K})(373.15 \text{ K})}{1 \text{ atm}} = 30620 \text{ cm}^3$$

$V_1 = (18.01 \text{ g})/(0.958 \text{ g/cm}^3) = 19 \text{ cm}^3,\ \ \Delta V = 30600 \text{ cm}^3$

$$w = -P\,\Delta V = -(1 \text{ atm})(30600 \text{ cm}^3)\ \frac{8.314 \text{ J}}{82.06 \text{ cm}^3 \text{ atm}} = -3100 \text{ J}$$

$\Delta U = q + w = 37554 \text{ J},\ \ \Delta H = q_p = 40654 \text{ J}$

2.50 For $C_{V,\text{m}}$ independent of T, we have $dU = C_V\,dT$ and $\Delta U = C_V\,\Delta T = 1.5nR\,\Delta T$; also $dH = C_P\,dT$ and $\Delta H = (C_V + nR)\Delta T = 2.5nR\,\Delta T$.

(a) $\Delta T = 200$ K and substitution of numerical values gives $\Delta U = 6240$ J and $\Delta H = 10400$ J.

(b) Use of $PV = nRT$ gives $T_2 = 292._5$ K and $T_1 = 243._7$ K, so $\Delta T = 48._8$ K. We find $\Delta U = 1520$ J and $\Delta H = 2450$ J.

(c) Since $\Delta T = 0$, we have $\Delta U = 0$ and $\Delta H = 0$

2.51 No, since q and w are not state functions. Problem 2.50 tells us the change in state for each process but does not specify the path of each process. q and w depend on the path.

2.52 $1 \text{ dm}^3 = (10^{-1} \text{ m})^3 = 10^{-3} \text{ m}^3 = 10^{-3}(10^2 \text{ cm})^3 = 1000 \text{ cm}^3$

(a) $$w = -P\,\Delta V = -(1 \text{ atm})(20000 \text{ cm}^3)\frac{8.314 \text{ J/mol -K}}{82.06 \text{ cm}^3\text{-atm/mol-K}} = -2.02_6 \text{ kJ}$$

$T_1 = P_1V_1/nR = 121._9 \text{ K},\ \ T_2 = P_2V_2/nR = 243._7 \text{ K}$

$q_p = C_p\,\Delta T = (2.00 \text{ mol})3.5(8.314 \text{ J/mol-K})(121._8 \text{ K})$

$q = 7.09 \text{ kJ};\ \ \Delta U = q + w = 5.05_4 \text{ kJ};\ \ \Delta H = q_p = 7.09 \text{ kJ}$

(b) $w = 0$ since V is constant. $q_V = C_V\,\Delta T$; T goes from $243._7$ K to $121._9$ K, so $\Delta T = -121._8$ K and $q = (2.00 \text{ mol})2.5(8.314 \text{ J/mol-K})(-121._8 \text{ K}) = -5.06_5$ kJ; $\Delta U = q + w = -5.06_5$ kJ; $\Delta H = C_p\,\Delta T = -7.09$ kJ

(c) Since T is constant, $\Delta U = 0 = \Delta H$. $w = -\int_1^2 P\,dV = -\int_1^2 nRT\,dV/V =$

$-nRT \ln (V_2/V_1) = 1.40_4$ kJ. $\Delta U = 0 = q + w$ and $q = -w = -1.40_4$ kJ. For the cycle, $\Delta U = 0 = \Delta H$, $q = 7.90 \text{ kJ} - 5.06_5 \text{ kJ} - 1.40_4 \text{ kJ} = 0.62$ kJ, $w = -2.02_6 \text{ kJ} + 0 + 1.40_4 \text{ kJ} = -0.62$ kJ. On the P-V diagram, step (a) is a horizontal line, step (b) a vertical line, and step (c) a hyperbolic line.

2.53 **(a)** Kinetic; **(b)** kinetic; **(c)** kinetic and potential; **(d)** kinetic and potential.

2.54 At the low temperature of 10 K, the 1-atm gas density is rather high, the average intermolecular distance is rather small, and intermolecular interactions are of significant magnitude. These interactions make $C_{P,\text{m}}$ deviate from the monatomic ideal-gas $C_{P,\text{m}}$.

2.55 **(a)** $V_{\text{gas}} = nRT/P = 24500 \text{ cm}^3$. Each hypothetical cube has a volume $V_{\text{cube}} = (24500 \text{ cm}^3)/(6.02 \times 10^{23}) = 4.1 \times 10^{-20} \text{ cm}^3$ and an edge length $(4.1 \times 10^{-20} \text{ cm}^3)^{1/3} = 3.4 \times 10^{-7} \text{ cm} = 34$ Å.

(b) The distance between the uniformly distributed molecules equals the distance between the cube centers, which is 3.4×10^{-7} cm.

(c) At 40 atm and 25°C, $V_{\text{gas}} = 610 \text{ cm}^3$, $V_{\text{cube}} = 1.0 \times 10^{-21} \text{ cm}^3$ and the distance between centers is 10 Å.

2.56 At 300 K and 1 atm, the contribution of intermolecular interactions to C_P and C_V is small and can be neglected. At room T, $C_{V,\text{vib}}$ is negligible for diatomic molecules that are not heavy. We thus consider only $C_{V,\text{tr}}$ and $C_{V,\text{rot}}$.

(a) $C_{V,\text{m}} = C_{V,\text{tr,m}} = 3R/2$; $C_{P,\text{m}} = C_{V,\text{m}} + R = 5R/2$

(b) $C_{V,\text{m}} = C_{V,\text{tr,m}} + C_{V,\text{rot,m}} = 3R/2 + R = 5R/2$; $C_{P,\text{m}} = 7R/2$.

2.57 The contribution of intermolecular interactions to C_P of the liquid can be estimated by taking $C_{P,\text{liq}} - C_{P,\text{gas}}$, since intermolecular interactions are quite small in the gas. Figure 2.15 shows that $C_{P,\text{liq}} - C_{P,\text{gas}}$ is positive. Therefore $C_{P,\text{intermol}} = (\partial U_{\text{intermol}}/\partial T)_P$ is positive and U_{intermol} must increase as T increases at constant P. (Recall from Sec. 2.11 that U_{intermol} is negative. An increase in U_{intermol} means a less negative U_{intermol} and corresponds to a decrease in intermolecular attractions as T increases.)

2.58 **(a)** Using $q = mc\,\Delta T$ and w = power × time, we have

$q = (27 \text{ lb})(454 \text{ g/lb})(1 \text{ cal/g-°C})(100°\text{C}) = 12.3 \times 10^5 \text{ cal}$

$w = (746 \text{ J/s})(3600 \text{ s/hr})(2.5 \text{ hr}) = 67.1 \times 10^5 \text{ J}$

$12.3 \times 10^5 \text{ cal} = 67.1 \times 10^5 \text{ J}$ and $1 \text{ cal} = 5.5 \text{ J}$

(b) Using $V = mgh$ to find the work needed to raise a one-pound weight by one foot, we have

$1 \text{ ft-lb} = (454 \text{ g})(980 \text{ cm/s}^2)(12 \times 2.54 \text{ cm})(1 \text{ J}/10^7 \text{ ergs})$

$1 \text{ ft-lb} = 1.356 \text{ J}$ and $772 \text{ ft-lb} = 1047 \text{ J}$

$q = (454 \text{ g})(1 \text{ cal/g-°C})(1°\text{F})(\frac{5}{9}\,°\text{C/°F}) = 252 \text{ cal}$

$252 \text{ cal} = 1047 \text{ J}$ and $1 \text{ cal} = 4.15 \text{ J}$

2.59 **(a)** $T = 273.15° + 1.8° = 274.9_5$ K, which has 4 significant figures. Calculations of $1/T$ to 4 significant figures gives $1/T = 0.003637 \text{ K}^{-1}$.

(b) Ignoring significant figures, we have log 4.83 = 0.68395, log 4.84 = 0.68485, log (4.83×10^{20}) = log 4.83 + log 10^{20} = 20.68395, log (4.84×10^{20}) = 20.68485. The numbers 4.83 and 4.84 differ by 1 in their third significant digit and their logs differ by 1 in the third significant digit after the decimal point. The logs of 4.83×10^{20} and 4.84×10^{20} differ by 1 in the third significant digit after the decimal point. The portion of the log that precedes the decimal point should not be considered (since this portion comes from the power of 10) and the log should have as many significant digits after the decimal point as there are significant digits in the number. Thus the logs should be expressed as 0.684, 0.685, 20.684, and 20.685.

(c) $(210.6 \text{ K})^{-1} - (211.5 \text{ K})^{-1} = 0.004748 \text{ K}^{-1} - 0.004728 \text{ K}^{-1} = 0.000020 \text{ K}^{-1}$.

2.60 **(a)** Solving (1.39) for P, we get $P = nRT/(V - nb) - an^2/V^2$. For a reversible isothermal process, $w = -\int_1^2 P\,dV = -\int_1^2 [nRT/(V - nb) - an^2/V^2]\,dV =$

$-nRT \ln (V - nb) \Big|_1^2 - an^2/V \Big|_1^2 = nRT \ln [(V_1 - nb)/(V_2 - nb)] +$
$an^2(1/V_1 - 1/V_2)$. For $a = 0 = b$, we get $w = nRT \ln (V_1/V_2)$, which is (2.74).

(b) $w = (0.500 \text{ mol})(8.314 \text{ J/mol-K})(300 \text{ K}) \ln N +$
$(1.35 \times 10^6 \text{ cm}^6 \text{ atm/mol}^2)(0.500 \text{ mol})^2(8.314 \text{ J/82.06 cm}^3\text{-atm}) \times$
$[1/(400 \text{ cm}^3) - 1/(800 \text{ cm}^3)]$, where $N =$
$[400 \text{ cm}^3 - (½ \text{ mol})(38.6 \text{ cm}^3/\text{mol})]/[800 \text{ cm}^3 - (½ \text{ mol})(38.6 \text{ cm}^3/\text{mol})]$.
$w = -895._7 \text{ J} + 42._7 \text{ J} = -853 \text{ J}$. $w_{\text{ideal}} = nRT \ln (V_1/V_2) = 864._5 \text{ J}$.

2.61 **(a)** –8.0 K. **(b)** 5.00 J/K

2.62 PV = constant holds only when T is constant, and T is not constant in a reversible adiabatic perfect-gas expansion.

2.63 $dq = C_P\, dT$ holds only in a constant-P process, and P is not constant in a reversible isothermal perfect-gas expansion.

2.64 Genevieve erroneously applied an equation for a reversible adiabatic process to an irreversible adiabatic process.

2.65 $dU = C_V\, dT$. $T = PV/nR$ and $dT = (P/nR)\, dV$. Since $dV > 0$, dT is positive. Hence dU is positive and U increases.

2.66 **(a)** Intensive; kg/m^3. **(b)** Extensive; J. **(c)** Intensive; J/mol.
(d) Extensive; J/K. **(e)** Intensive; J/kg-K. **(f)** Intensive; J/mol-K.
(g) Intensive; $\text{Pa} = \text{N/m}^2$. **(h)** Intensive; kg/mol. **(i)** Intensive; K.

2.67 Insertion of the dimensions of each physical quantity in the equation gives

$$\frac{\text{energy}}{\text{temperature}} - \frac{\text{energy}}{\text{temperature}} = (\text{temp.})\,(\text{vol.})\frac{(\text{press.})^n}{(\text{temp.})^m}$$

Equating the powers of temperature on each side of the equation, we get $-1 = 1 - m$ and $m = 2$. Also, $n = 1$, since the product pressure × volume has the dimensions $(\text{force/length}^2)\text{length}^3 = \text{force} \times \text{length} = \text{energy}$.

2.68 $\gamma = C_{P,m}/C_{V,m} = 1.13$. Also, the gas is nearly ideal under these conditions, so $C_{P,m} = C_{V,m} + R$. Hence $\gamma = (C_{V,m} + R)/C_{V,m} = 1 + R/C_{V,m} = 1.13$; $R/C_{V,m} = 0.13$ and $C_{V,m} = R/0.13 = 64$ J/mol-K; $C_{P,m} = C_{V,m} + R = 72$ J/mol-K.

2.69 **(a)** m^3/K; extensive. **(b)** K^{-1}; intensive. **(c)** m^3/mol-Pa; intensive **(d)** J/m^3; intensive.

2.70 **(a)** No. **(b)** Yes. **(c)** No. **(d)** Yes.

2.71 **(a)** False; ΔH is the *change* in a state function.

(b) False; for a perfect gas, C_V is a function of T only, but need not be independent of T.

(c) False; the system must be closed.

(d) False; for an isothermal process, T must remain constant *throughout* the process.

(e) False; U must be replaced by ΔU.

(f) False; For a perfect gas, U depends on T only, but this is not true for other kinds of systems. An isothermal pressure change changes the average intermolecular distance and hence changes the contribution of intermolecular interactions to the internal energy.

(g) False; see the first example in Sec. 2.8.

(h) True.

(i) False; for example, T drops in a reversible adiabatic perfect-gas expansion.

(j) False; the entire path of the process must be specified.

(k) False; for example, a real gas expanding adiabatically into vacuum can undergo a change in T.

(l) True.

(m) False; work can be done.

(n) False; when heat flows into a system of ice and liquid water in equilibrium, T remains constant as long as some ice remains.

(o) False. **(p)** True. **(q)** True. **(r)** False [see (i)].

(s) False. **(t)** False. **(u)** False. **(v)** False.

Chapter 3

3.1 **(a)** T; **(b)** T; **(c)** T; **(d)** F.

3.2 **(a)** $e_{rev} = 1 - T_C/T_H = 1 - (273 \text{ K})/(1073 \text{ K}) = 0.746$

(b) $e_{rev} = -w_{max}/q_H$, $-w_{max} = 0.746(1000 \text{ J}) = 746 \text{ J}$,

$-q_{C,\text{min}} = 1000 \text{ J} - 746 \text{ J} = 254 \text{ J}$

3.3 $e_{rev} = 1 - T_C/T_H = 0.90 = 1 - (283 \text{ K})/T_H$ and $T_H = 2830 \text{ K}$

3.4 $w = -2.50$ kJ. $e = -w/q_H = 0.45 = 2.50 \text{ kJ}/q_H$ and $q_H = 5.56$ kJ.
$\Delta U = 0 = q + w = q_C + q_H + w = q_C + 5.56 \text{ kJ} - 2.50 \text{ kJ}$ and $q_C = -3.06$ kJ.

3.5 **(a)** Since a Carnot cycle is a cyclic process, the first law gives $\Delta U = q + w = 0$, so $0 = q_C + q_H + w$ [Eq. (1)]. Since a Carnot cycle is reversible, we have $\Delta S = \oint dq_{rev}/T = 0 = q_C/T_C + q_H/T_H$, which can be rearranged to $q_C/q_H = -T_C/T_H$ [Eq. (2)]. Use of Eqs. (1) and (2) gives $K_{rev} = q_C/w = -q_C/(q_C + q_H)$. Division of numerator and denominator by q_C gives $K_{rev} = -1/(1 + q_H/q_C) = -1/(1 - T_H/T_C) = T_C/(T_H - T_C)$. Also, $\varepsilon_{rev} = -q_H/w = q_H/(q_C + q_H) = 1/(q_C/q_H + 1) = 1/(-T_C/T_H + 1) = T_H/(T_H - T_C)$.

(b) From (a), $\varepsilon_{rev} = 1/(1 - T_C/T_H)$; the denominator is less than 1 and greater than 0, so $\varepsilon_{rev} > 1$.

(c) $\varepsilon_{rev} = T_H/(T_H - T_C) = (293 \text{ K})/(20 \text{ K}) = 15 = |q_H|/w$, so $|q_H| = 15w = 15$ J. (This indicates why heat pumps are attractive devices for heating homes in winter.)

(d) $K_{rev} = T_C/(T_H - T_C)$, which goes to 0 as T_C goes to 0 K.

3.6

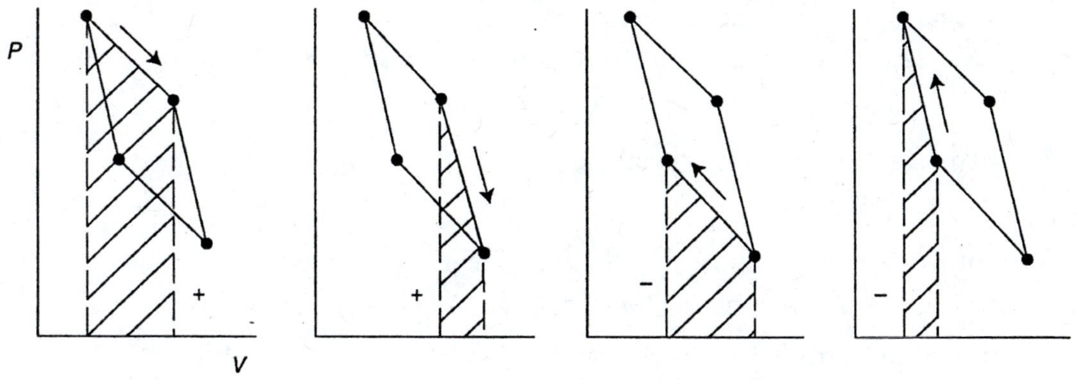

Combining the shaded areas using the marked signs, we get the area enclosed by the cycle.

3.7 **(a)** F; **(b)** T; **(c)** T; **(d)** T; **(e)** T; **(f)** F; **(g)** F; **(h)** F; **(i)** F; **(j)** T; **(k)** T.

3.8 **(a)** $\Delta S_{vap} = q/T = (1560 \text{ cal})/(87.3 \text{ K}) = 17.9 \text{ cal/K}$.

(b) $(5.00 \text{ g})(1 \text{ mol}/39.95 \text{ g}) = 0.125 \text{ mol}$.
$q = -(0.125 \text{ mol})(1560 \text{ cal/mol}) = -195 \text{ cal}$.
$\Delta S = q/T = (-195 \text{ cal})/(87.3 \text{ K}) = -2.24 \text{ cal/K}$.

3.9 $\Delta S = \int_1^2 dq_{rev}/T = \int_1^2 (C_P/T)\, dT = \int_1^2 n(a/T + b)\, dT =$
$n[a \ln (T_2/T_1) - b(T_2 - T_1)]$. Thus

$$\Delta S = 2.00 \text{ mol}\left(6.15\frac{\text{cal}}{\text{mol K}}\ln\frac{400 \text{ K}}{300 \text{ K}} + 0.00310\frac{\text{cal}}{\text{mol K}^2}100 \text{ K}\right) = 4.16 \text{ cal/K}$$

3.10 For melting the ice to liquid water at 0°C, $\Delta S_a = q_{rev}/T =$ (18.015 g)(333.6 J/g)/(273.15 K) = 22.00 J/K.
For heating water from 0 to 100°C at 1 atm, $\Delta S_b = \int_1^2 C_P\, dT/T = C_P \ln (T_2/T_1)$ = (18.01 g)(4.19 J/g-K) ln(373.15/273.15) = 23.54 J/K.
For vaporization of the liquid to vapor at 100°C and 1 atm, $\Delta S_c = q_{rev}/T =$ (18.015 g)(2256.7 J/g)/373.15 = 108.95 J/K. For the isothermal expansion of

the vapor (assumed ideal), Eq. (3.30) gives $\Delta S_d = nR \ln (V_2/V_1) =$ (1.00 mol)(8.3145 J/mol-K) ln 2 = 5.76 J/K.
We have for the overall process:
ΔS = 22.00 J/K + 23.54 J/K + 108.95 J/K + 5.76 J/K = 160.25 J/K.

3.11 Zero, since the process is cyclic.

3.12 For an ideal gas with $C_{V,m} = 1.5R$ at all T's, Eq. (3.30) gives
$\Delta S = C_V \ln (T_2/T_1) + nR \ln (V_2/V_1) = nR[1.5 \ln (T_2/T_1) + \ln (V_2/V_1)]$.

(a) (2.50 mol)(8.314 J/mol-K)(1.5 ln 1.5 + ln 0.75) = 6.66 J/K

(b) (2.50 mol)(8.314 J/mol-K)(1.5 ln 1.2 + ln 1.5) = 14.1 J/K

(c) (2.50 mol)(8.314 J/mol-K)(0 + ln 1.474) = 8.06 J/K

3.13 0, since $dq_{rev} = 0$.

3.14 Use the reversible path in Fig. 3.7. For the first step, $\Delta S_1 = C_P \ln (T_2/T_1) =$ (10 g)(1.01 cal/g-K) ln(273.1/263.1) = 0.37_7 cal/K = 1.58 J/K.
For the second step $\Delta S_2 = q_{rev}/T$ = –(333.6 J/g)(10 g)/(273.1 K) = –12.2 J/K.
For the third step, $\Delta S_3 = C_P \ln (T_2/T_1)$ = (10 g)(0.50 cal/g-K) ln(263.1/273.1) = -0.18_7 cal/K = –0.78 J/K. For the complete process,
ΔS = 1.58 J/K – 12.2 J/K – 0.78 J/K = –11.4 J/K.

3.15 For the step 1 → 2 in Fig. 3.4b: $q > 0$, $w < 0$, $\Delta U = 0$ (since U depends only on T for a perfect gas), $\Delta S = q_H/T_H > 0$. For the step 2 → 3: $q = 0$, $w < 0$, $\Delta U < 0$, $\Delta S = 0$ (since reversible and adiabatic). For the step 3 → 4: $q < 0$, $w > 0$, $\Delta U = 0$, $\Delta S = q_C/T_C < 0$. For the step 4 → 1: $q = 0$, $w > 0$, $\Delta U > 0$, $\Delta S = 0$.

3.16 **(a)** $m_1c_1|\Delta T_1| = m_2c_2|\Delta T|$
(200 g)(0.0313 cal/g-K)(120°C – x) = (25.0 g)(1.00 cal/g-K)(x – 10°C)
x = 32.0°C

(b) $\Delta S = C_P \ln (T_2/T_1)$. For Au, ΔS is
(0.0313 cal/g-K)(200 g) ln (305.1/393.1) = –1.59 cal/K

(c) For the water, ΔS is
(1.00 cal/g-K)(25.0 g) ln (305.1/283.1) = 1.87 cal/K

(d) −1.59 cal/K + 1.87 cal/K = 0.28 cal/K

3.17 $\Delta S = n_a R \ln [(n_a + n_b)/n_a] + n_b R \ln [(n_a + n_b)/n_b]$
$n_a = 2.50$ mol, $n_b = 0.312_5$ mol
$\Delta S = (8.314 \text{ J/mol-K})[2.50 \ln (2.81/2.50) + 0.312_5 \ln (2.81/0.312_5)] =$
8.14 J/K

3.18 ΔS for the unmixing is the negative of ΔS in Eq. (3.33),
so $\Delta S = n_a R \ln x_a + n_b R \ln x_b = (n_a + n_b)R(x_a \ln x_a + x_b \ln x_b)$. We have

$$n_a + n_b = \frac{0.00100 \text{ g}}{(18.998 + 35.453) \text{ g/mol}} = 1.83_7 \times 10^{-5} \text{ mol}$$

$\Delta S = (1.83_7 \times 10^{-5} \text{ mol})(1.987 \text{ cal/mol-K})(0.758 \ln 0.758 + 0.242 \ln 0.242) =$
-2.02×10^{-5} cal/K $= -8.45 \times 10^{-5}$ J/K

Reminder: Do not look up the solution to a problem until you have made a serious effort to solve it.

3.19 To carry out the change of state reversibly, we put the part at T_2 in contact with a heat reservoir whose temperature is infinitesimally less than T_2 and wait until heat dq flows into the reservoir. Then we remove the reservoir from this part of the system and put the part at T_1 in contact with a heat reservoir whose temperature is infinitesimally greater than T_1 and wait until heat dq flows into the system part at T_1. Since these two heat flows are reversible, we can use $dS = dq_{\text{rev}}/T$ to write the system's entropy change as $dS = dq/T_1 - dq/T_2$. (The entropy changes of the reservoirs are irrelevant to dS of the system.)

3.20 **(a)** F; **(b)** F; **(c)** T; **(d)** T; **(e)** T; **(f)** F; **(g)** T; **(h)** F.

3.21 Using $dS = dq_{\text{rev}}/T$, we have for the signs of ΔS:
(a) $dq_{\text{rev}} > 0$ so $\Delta S > 0$. **(b)** $dq_{\text{rev}} > 0$ so $\Delta S > 0$.
(c) $dq_{\text{rev}} = 0$, so $\Delta S = 0$. **(d)** $dq_{\text{rev}} > 0$, so $\Delta S > 0$.
(e) This is an irreversible process in an isolated system, so $\Delta S > 0$.

(f) This is an irreversible adiabatic process, so $\Delta S > 0$. This also follows from Eq. (3.30) with $dT = 0$ and $V_2 > V_1$.

(g) $dq_{rev} > 0$, so $\Delta S > 0$. (h) $dq_{rev} < 0$, so $\Delta S < 0$.

(i) This is an irreversible process in an isolated system, so $\Delta S > 0$.

(j) This is an irreversible process in an isolated system, so $\Delta S > 0$.

ΔS_{univ} is 0 for a reversible process and positive for an irreversible process. Hence the ΔS_{univ} values are: **(a)** 0; **(b)** 0; **(c)** 0; **(d)** 0; **(e)** positive; **(f)** positive; **(g)** 0; **(h)** 0; **(i)** positive; **(j)** positive.

3.22 **(a)** q_H/T_H, 0, q_C/T_C, 0. **(b)** 0, 0, 0, 0, since each step is reversible.

3.23 Assume that $dS = dq_{rev}/T$ is the differential of a state function such that $\Delta S_{univ} \geq 0$ for any process. Let an anti-Clausius device exist. Such a device extracts heat $q > 0$ from a cold reservoir and deposits an equal amount of heat in the hot reservoir, with no other effects. For one cycle of such a device, $\Delta S_{\text{hot res.}} = q/T_H$, $\Delta S_{\text{cold res.}} = -q/T_C$, and $\Delta S_{\text{device}} = 0$ (since the process is cyclic). We have $\Delta S_{univ} = q/T_H - q/T_C = q(T_C - T_H)/T_C T_H < 0$, which violates the assumption that $\Delta S_{univ} \geq 0$. Hence an anti-Clausius device cannot exist. This completes the proof.

3.24 $-q_C/q_H = \tau_C^{1/2}/\tau_H^{1/2}$ and $\tau_C/\tau_H = q_C^2/q_H^2$. Thus $\tau/200.00°\text{M} = q^2/q_{tr}^2$. But $|q|/|q_{tr}| = T/(273.16\text{ K})$. Hence the Melvin temperature τ is given by $\tau/200.00°\text{M} = T^2/(273.16\text{ K})^2$.

(a) For the steam point, $T = 373.13$ K and we find $\tau = 373.2°$M.

(b) For the ice point, $T = 273.15$ K and $\tau = 199.99°$M.

3.25 $e_{rev} = 1 + q_C/q_H$ and $g = 1 - e_{rev} = -q_C/q_H$. We have $g(\tau_2, \tau_3) = -q_{2A}/q_3$ and $g(\tau_1, \tau_2) = -q_1/q_{2B} = q_1/q_{2A}$, so $g(\tau_2, \tau_3)g(\tau_1, \tau_2) = -q_1/q_3$.

3.26 **(a)** Substitution of $x = 3$ gives 0.002736 as the probability that $x \geq 3$. Hence 1 – 0.002736 = 0.9973 is the probability that an observation is within 3 standard deviations from the mean.

(b) Substitution of $x = 10^6$ gives the desired probability as $p = 8 \times 10^{-7} \times e^{-0.5\times10^{12}}$. If $e^z = 10^y$, then taking logs gives $z \log e = y$; hence

$y = -0.5 \times 10^{12} \times 0.434 = -2.2 \times 10^{11}$ and we have
$p = 8 \times 10^{-7} \times 10^{-2.2\times10^{11}} \approx 10^{-2.2\times10^{11}}$.

3.27 **(a)** The probability p of observing a deviation $\geq 10^6$ standard deviations in one observation is $10^{-2\times10^{11}}$. The probability of not observing such a deviation in n trials is $(1 - 10^{-2\times10^{11}})^n$. For $(1 - 10^{-2\times10^{11}})^n = 0.5$, the probability of not observing such a deviation is 0.5 and the probability of observing such a deviation is 0.5. We have $\ln 0.5 = n \ln (1 - 10^{-2\times10^{11}}) \approx -n \times 10^{-2\times10^{11}}$, where Eq. (8.36) was used. We get $n = 0.7 \times 10^{2\times10^{11}}$. There are 3×10^7 seconds in a year, so it takes $(0.7 \times 10^{200,000,000,000})/(3 \times 10^7)$ years of measurements at the rate of one per second to reach the 50% probability.

(b) We want $1 - (1 - p)^n \geq 0.99$. Since $p = ½$, we want $0.01 \geq 1/2^n$ or $2^n \geq 100$. The minimum value of n is 7.

3.28 **(a)** q is not a state function.;

(b) T, since all the others refer to processes.

(c) V, since all the others have dimensions of energy.

(d) V, since the others are intensive properties.

(e) dV is the only infinitesimal quantity.

(f) ΔH is the only one that refers to a process.

3.29 In one minute of operation, $|w| = (1000 \times 10^6 \text{ J/s})(60 \text{ s}) = 6 \times 10^{10}$ J. Then $e = 0.40 = |w|/q_H$ and $q_H = (6 \times 10^{10} \text{ J})/0.40 = 15 \times 10^{10}$ J in one minute. Hence $q_C = 15 \times 10^{10} \text{ J} - 6 \times 10^{10} \text{ J} = 9 \times 10^{10}$ J per minute. Use of $q = mc_P \Delta T$ gives $m = (9 \times 10^{10} \text{ J})/(4.184 \text{ J/g-K})(10 \text{ K}) = 2 \times 10^9$ g. The density of water is 1 g/cm^3, so 2×10^9 cm^3 = 2 million liters are used per minute.

3.30 **(a)** q cannot be calculated, since q depends on the path and the path is not specified.

(b) The path-dependent quantity w cannot be calculated.

(c) $dU = C_V\,dT = nC_{V,\text{m}}\,dT$; $\Delta U = \int_1^2 nC_{V,\text{m}}\,dT = \int_1^2 n(a + bT)\,dT =$
$na(T_2 - T_1) + \tfrac{1}{2}nb(T_2^2 - T_1^2) = (4.00\text{ mol})(25.0\text{ J/mol-K})(500 - 300)\text{K} +$
$\tfrac{1}{2}(4.00\text{ mol})(0.0300\text{ J/mol-K}^2)(500^2 - 300^2)\text{K}^2 = 29.6\text{ kJ}$.

(d) $\Delta H = \Delta U + \Delta(PV) = \Delta U + nR\,\Delta T =$
$29600\text{ J} + (4.00\text{ mol})(8.314\text{ J/mol-K})(200\text{ K}) = 36.3\text{ kJ}$.

(e) From (3.30), $\Delta S = \int_1^2 (C_V/T)\,dT + nR\ln(V_2/V_1)$. But $C_V/T =$
$nC_{V,\text{m}}/T = n(a/T + b)$, so $\Delta S = n\int_1^2 (a/T + b)\,dT + nR\ln(V_2/V_1) =$
$na\ln(T_2/T_1) + nb(T_2 - T_1) + nR\ln[(nRT_2/P_2)/(nRT_1/P_1)] =$
$(4.00\text{ mol})(25.0\text{ J/mol-K})\ln(500/300) +$
$(4.00\text{ mol})(0.0300\text{ J/mol-K}^2)(500 - 300)\text{K} +$
$(4.00\text{ mol})(8.314\text{ J/mol-K})\ln[(500\text{ K})(2.00\text{ atm})/(300\text{ K})(3.00\text{ atm})] =$
78.6 J/K.

3.31 **(a)** Rev; **(b)** irrev; **(c)** irrev; **(d)** irrev; **(e)** irrev; **(f)** irrev; **(g)** rev.

3.32 **(a)** Since U and S are extensive, the 10 g has the higher U and the higher S.

(b) The vapor; the vapor.

(c) The 40°C benzene; the 40°C benzene.

(d) If the system at 300 and 310 K is adiabatically enclosed, it will spontaneously go to the state at 305 K. Since S increases in a spontaneous adiabatic process, the 305-K system has the higher S. Since $q = 0$ and w is negligible for this process, the two systems have the same U..

(e) Neither; the 1-atm gas.

3.33 $P\,dV + V\,dP = d(PV)$, $V\,dV = d(\tfrac{1}{2}V^2)$, $dq_{\text{rev}}/T = dS$, $dq_P = dH$, and $dw_{\text{rev}}/P = -dV$, so b, c, d, f, h, and $j = 0$.

3.34 **(a)** C_P, k, U/T; **(b)** $C_{P,\text{m}}$, R

3.35 The second law of thermodynamics is "hardly ever" violated.

3.36 No. (A pilot plant using this method has been built. For further information, enter OTEC into a search engine.)

3.37 Suppose we could prepare a reservoir at $T_C = 0$ and could reduce the engine's temperature to absolute zero. Then e_{rev} would become equal to 1. The Carnot cycle would convert the heat q_H completely into work. But this would violate the Kelvin–Planck statement of the second law. Hence we can't achieve absolute zero. (See also Sec. 5.11 in the text.)

3.38 **(a)** $\Delta H = q_P = 0$.

(b) Suppose the final state consisted of ice at 0°C, with no liquid present. A hypothetical path to attain this state is to warm the supercooled liquid from –10°C to 0°C and then to freeze all the liquid at 0°C. ΔH for warming the liquid is (1.00 cal/g-K)(10.0 g)(10.0 K) = 100 cal = 418 J, and ΔH for freezing all the liquid is –(333.6 J/g)(10.0 g) = –3336 J. The overall ΔH is –2918 J, which is not 0. Hence the equilibrium state is not ice at 0°C. If the equilibrium state were all ice below 0°C, ΔH would be even more negative than –2918 J, and hence this is not the equilibrium state. Therefore the equilibrium state must consist of ice and liquid water in equilibrium at 0°C. To satisfy the condition that $\Delta H = 0$, the mass m_{ice} of ice produced must satisfy –(333.6 J/g)m_{ice} = –418 J and m_{ice} = 1.25 g. The mass of liquid remaining is 10.0 g – 1.25 g = 8.7$_5$ g.

(c) A reversible path for the process is

$$\text{liq. at } -10°\text{C} \xrightarrow{a} \text{liq. at } 0°\text{C} \xrightarrow{b} m_{\text{ice}} + m_{\text{liq}} \text{ at } 0°\text{C}$$

$\Delta S = \Delta S_a + \Delta S_b$ = (10.0 g)(4.184 J/g-K) ln(273/263) – (333.6 J/g)(1.25 g)/(273 K) = 1.56 J/K – 1.53 J/K = 0.03 J/K, where (3.28) and (3.25) were used.

3.39 **(a)** J/K; **(b)** J/mol-K; **(c)** J; **(d)** Pa = N/m^2; **(e)** no units; **(f)** kg/mol.

3.40 **(a)** False. **(b)** False. **(c)** True. **(d)** False. **(e)** True.

3.41 **(a)** False. (For example, a nonideal gas expanding into vacuum can undergo a change in T—the Joule experiment.)

(b) True.

(c) False; S increases in an irreversible process in an isolated system.

(d) False. For example, a Carnot cycle is reversible, has $\Delta V = 0$ and $w \neq 0$.

(e) True, since S is a state function.

(f) False, e.g., $\Delta S \neq 0$ for (the irreversible) adiabatic expansion of an ideal gas into vacuum.

(g) False (unless T and P are constant).

(h) False. (A counterexample is the melting of ice at 0°C.)

(i) False. (Reversibility and constant pressure are required.)

(j) False. (A counterexample is the melting of ice at 0°C.)

(k) False. (Only the system has to return to its initial state.)

R3.1 **(a)** Melting of ice at 0° and 1 atm. Reversible isothermal expansion of an ideal gas.

(b) Reversible adiabatic expansion of an ideal gas. Combustion of a hydrocarbon in a sealed, adiabatically enclosed container.

(c) Melting of ice at 0° and 1 atm. An exothermic chemical reaction in a solution carried out isothermally.

(d) Impossible.

(e) Adiabatic expansion of an ideal gas into vacuum. Adiabatic mixing of ideal gases at constant T and P.

(f) A Carnot cycle. The process of Fig. 2.10.

R3.2 **(a)** Heat of fusion of ice. Density of ice and of water at 0°C and 1 atm. (However the contribution of w is negligible.)

(b) The heat of fusion of Na at the normal melting point nmp, and T_{nmp}.

(c) Either C_P or C_V as a function of T.

(d) Either C_P or C_V as a function of T.

(e) C_P as a function of T, and the densities at 20°C and 50°C. (However, the contribution of w is negligible.)

R3.3 $\rho = PM/RT$ and $M = \rho RT/P$

$$M = \frac{6.39 \text{ g}}{3450 \text{ cm}^3} \frac{(82.06 \text{ cm}^3 \text{ atm mol}^{-1} \text{ K}^{-1})(283 \text{ K})}{(0.888 \text{ bar})(750 \text{ torr/1 bar})(1 \text{ atm/760 torr})} = 49.1 \text{ g/mol}$$

R3.4 **(a)** False. (It is ΔS_{univ} that cannot be negative.)

(b) False. (P must be constant for this to hold

(c) False. (ΔU is zero. The work associated with the volume change when added to q gives a zero ΔU.)

(d) True.

(e) True.

(f) False. For example, when an ideal gas expands adiabatically and reversibly, the work done by the gas decreases its U and so decreases T.

(g) False. For example, reversible melting of ice at 0° and 1 atm is isothermal but not adiabatic.

(h) True.

(i) False, since q is not a state function. For example, q is not zero for a Carnot cycle.

(j) False. For example, S increases for the adiabatic irreversible expansion of an ideal gas into vacuum.

R3.5 **(a)** kg; **(b)** kg/m^3; **(c)** J/mol-K; **(d)** K^{-1}; **(e)** J/m^3; **(f)** kg/mol; **(g)** Pa = N/m^2; **(h)** J/K.

R3.6 The problem specifies the change of state that occurred but does not specify the path of the process. Therefore, q and w cannot be found since they are not state functions. Since C_V is constant,

$\Delta U = C_V \,\Delta T$ = (2.50 mol)1.5(8.314 J/mol-K)(35 K) = 1.09 kJ.

$\Delta H = \Delta U + \Delta(PV) = \Delta U + nR\,\Delta T$ = 1.09 kJ + 0.73 kJ = 1.82 kJ

(Or ΔH can be found from $C_{P,\text{m}}$, which equals $C_{V,\text{m}} + R$.) Use of (3.30) gives

$\Delta S = C_V \ln(T_2/T_1) + nR\ln(V_2/V_1) = C_V \ln(T_2/T_1) + nR\ln(P_1T_2/P_2T_1) =$

(2.50 mol)(8.314 J/mol-K){1.5ln(333.1/298.1) + ln[(1.00·333.1)/(2.00·298.1)]}
= –8.64 J/K.

R3.7 We add the quantities for melting the ice at 0°C (process 1) and warming the water to 50°C (process 2). We have

$q_1 = (2.00 \text{ mol})(18.01 \text{ g/mol})(333.6 \text{ J/g}) = 12.0 \text{ kJ}$. $\Delta H_1 = q_1 = 12.0 \text{ kJ}$.

$\Delta S_1 = (12.0 \text{ kJ})/(273.15 \text{ K}) = 44.0 \text{ J/K}$.

$w_1 = -P\,\Delta V_1 = -(1 \text{ atm})[(36.0 \text{ g})/(1.000 \text{ g/cm}^3) - (36.0 \text{ g})/(0.917 \text{ g/cm}^3) =$
$3.3 \text{ cm}^3 \text{ atm} \times (8.314 \text{ J}/82.06 \text{ cm}^3 \text{ atm}) = 0.33 \text{ J}$. $\Delta U_1 = q_1 + w_1 = 12.0 \text{ kJ}$.

$C_P = (2.00 \text{ mol})(18.01 \text{g/mol})(4.19 \text{ J g}^{-1} \text{ K}^{-1}) = 151 \text{ J/K}$.

$q_2 = \Delta H_2 = (151 \text{ J/K})(50°\text{C}) = 7.5_5 \text{ kJ}$.

$\Delta S_2 = \int_1^2 (C_P/T)\,dT = (151 \text{ J/K}) \ln(323.1/273.1) = 25.4 \text{ J/K}$.

A handbook gives the density of water at 50°C and 1 atm as 0.988 g/cm^3 and
$w_2 = -P\,\Delta V_2 = -(1 \text{ atm})(36.0 \text{ g})[(0.988 \text{ g/cm}^3)^{-1} - (1.000 \text{ g/cm}^3)^{-1}] =$
$-0.44 \text{ cm}^3 \text{ atm} \times (8.314 \text{ J}/82.06 \text{ cm}^3 \text{ atm}) = -0.044 \text{ J}$. $\Delta U_2 = q_2 + w_2 = 7.5_5 \text{ kJ}$.
$q = q_1 + q_2 = 19.5 \text{ kJ}$. $\Delta H = \Delta H_1 + \Delta H_2 = 19.5 \text{ kJ}$. $\Delta U = \Delta U_1 + \Delta U_2 = 19.5 \text{ kJ}$.
$\Delta S = \Delta S_1 + \Delta S_2 = 69.4 \text{ J/K}$.

R3.8 See the derivation of Eq. (2.74) in the textbook.

R3.9 See paragraphs 6 and 7 on page 89 of the textbook.

R3.10 $P_{O_2} = x_{O_2} P = 0.21(1.01 \text{ bar}) = 0.21 \text{ bar}$.

$$\frac{x_{O_2}}{x_{N_2}} = \frac{n_{O_2}/n_{\text{tot}}}{n_{N_2}/n_{\text{tot}}} = \frac{n_{O_2}}{n_{N_2}} \quad \text{so} \quad n_{O_2} = \frac{0.21}{0.78}(52500 \text{ g})(1 \text{ mol}/28.0 \text{ g}) = 505 \text{ mol}$$

and $m_{O_2} = 16.2 \text{ kg}$. It was not necessary to assume gas ideality.

R3.11 (a) Adiabatic, so $q = 0$. Expansion into vacuum, so $w = 0$. $\Delta U = q + w = 0$. Irreversible process in an isolated system, so $\Delta S > 0$.

(b) $q > 0$. V decreases, so $w > 0$ (although very small). $\Delta U = q + w > 0$. $\Delta S > 0$.

(c) $q < 0$. V decreases, so $w > 0$ (although very small). $\Delta U \approx q$ and $\Delta U < 0$. $\Delta H = q_P < 0$.

(d) $q = 0$, $w = 0$, $\Delta U = 0$, $\Delta S > 0$, since this is an irreversible process in an isolated system.

(e) Since the process is adiabatic, $q = 0$. since the walls are rigid, $w = 0$. So $\Delta U = q + w = 0$. This is an irreversible process in an isolated system, so $\Delta S > 0$.

(f) Since U and H of a perfect gas depend only on T and T does not change, $\Delta U = 0$ and $\Delta H = 0$. For a reversible expansion, $w < 0$. Then $q = \Delta U - w = -w$ and $q > 0$. For an isothermal reversible process, $\Delta S = q_{rev}/T > 0$.

R3.12 $\Delta H = \int_1^2 C_P\,dT = n\int_1^2 C_{P,\mathrm{m}}\,dT = n\int_1^2 (a + bT + c/T^2)\,dT =$

$n[a(T_2 - T_1) + \tfrac{1}{2}b(T_2^2 - T_1^2) + c(T_1^{-1} - T_2^{-1})]$, since $dH/dT = C_P$ for an ideal gas.

From (3.30) and $C_P - C_V = nR$, we have

$$\Delta S = n\int_1^2 \frac{C_{P,\mathrm{m}} - R}{T}\,dT + nR\ln\frac{V_2}{V_1} = n\int_1^2 \frac{C_{P,\mathrm{m}}}{T}\,dT - nR\ln\frac{T_2}{T_1} + nR\ln\frac{V_2}{V_1}$$

$$= n\int_1^2 \frac{C_{P,\mathrm{m}}}{T}\,dT + nR\ln\frac{T_1V_2}{T_2V_1} = n\int_1^2 \frac{C_{P,\mathrm{m}}}{T}\,dT + nR\ln\frac{P_1}{P_2}$$

$$= n\int_1^2 (a/T + b + c/T^3)\,dT + nR\ln(P_1/P_2)$$

$$= n[a\ln(T_2/T_1) + b(T_2 - T_1) + \tfrac{1}{2}(T_1^{-2} - T_2^{-2})] + nR\ln(P_1/P_2)$$

$$\Delta U = \Delta H - \Delta(PV) = \Delta H - nR\,\Delta T$$

$$= n[a(T_2 - T_1) + \tfrac{1}{2}b(T_2^2 - T_1^2) + c(T_1^{-1} - T_2^{-1})] - nR(T_2 - T_1)$$

R3.13 $(\partial V_\mathrm{m}/\partial T)_P = (\partial[RTg(P)]/\partial T)_P = Rg(P)$.

$\alpha \equiv (1/V_\mathrm{m})(\partial V_\mathrm{m}/\partial T)_P = [RTg(P)]^{-1}Rg(P) = T^{-1}$.

R3.14 The temperature of the water increases and that of the metal decreases, so $\Delta S_{\text{water}} > 0$ and $\Delta S_{\text{metal}} < 0$. The process is an irreversible process in an isolated system, so $\Delta S_{\text{metal+water}} > 0$.

R3.15 **(a)** $P\,dV + V\,dP$; **(b)** $dU + P\,dV + V\,dP$; **(c)** $T^{-1}\,dP - PT^{-2}\,dT$.

Chapter 4

4.1 **(a)** T; **(b)** F; **(c)** T; **(d)** F; **(e)** F; **(f)** T.

4.2 Processes a and b are reversible, isothermal, and isobaric and have $\Delta S = q_P/T$.

(a) $\Delta G = \Delta H - T\,\Delta S = q_P - T(q_P/T) = 0$. $\Delta A = \Delta U - T\,\Delta S = q_P + w - T(q_P/T)$ $= w = -\int_1^2 P\,dV = -P\,\Delta V =$
$-(1\text{ atm})[(36.0\text{ g})/(1.000\text{ g/cm}^3) - (36.0\text{ g})/(0.917\text{ g/cm}^3)] =$
$(3.26\text{ cm}^3\text{ atm})(8.314\text{ J}/82.06\text{ cm}^3\text{-atm}) = 0.330\text{ J}$.

(b) As in (a), $\Delta G = 0$ and $\Delta A = w = -P\,\Delta V = -P(V_{\text{gas}} - V_{\text{liq}}) \approx -PV_{\text{gas}} =$
$-nRT = -(0.50\text{ mol})(8.314\text{ J/mol-K})(353.2\text{ K}) = -1.47\text{ kJ}$.

(c) Since the gas is perfect, the final T is 300 K. $\Delta H = 0$ and $\Delta U = 0$, since U and H depend only on T for a perfect gas. From (3.30), $\Delta S = nR\ln(V_2/V_1) = (0.100\text{ mol})(8.314\text{ J/mol-K})\ln(6.00/2.00) =$ 0.91_3 J/K. $\Delta G = \Delta H - T\,\Delta S = 0 - (300\text{ K})(0.91_3\text{ J/K}) = -274$ J. $\Delta A = \Delta U - T\,\Delta S = -274$ J. Processes a and b are reversible and so have $\Delta S_{\text{univ}} = 0$. For process c, there is no change in the surroundings, so $\Delta S_{\text{univ}} = \Delta S = 0.91_3$ J/K.

4.3 **(a)** C_V; **(b)** C_P; **(c)** C_P/T.

4.4 Setting $dU = 0$ in $dU = T\,dS - P\,dV$, we get $0 = T\,dS_U - P\,dV_U$ so $dS_U/dV_U \equiv (\partial S/\partial V)_U = P/T$.

4.5 For $dH = T\,dS + V\,dP = M\,dx + N\,dy$, we have $M = T$, $N = V$, $x = S$, $y = P$, and $(\partial M/\partial y)_x = (\partial N/\partial x)_y$ gives $(\partial T/\partial P)_S = (\partial V/\partial S)_P$. The equations in (4.45) are derived similarly from Eqs. (4.35) and (4.36).

4.6 **(a)** Use of $(\partial U/\partial V)_T = \alpha T/\kappa - P$ gives

$$\left(\frac{\partial U}{\partial V}\right)_T = \frac{(3.04\times10^{-4}\text{ K}^{-1})(303.15\text{ K})}{4.52\times10^{-5}\text{ atm}^{-1}} - 1\text{ atm} = 2040\text{ atm}$$

(b) Use of $\mu_{JT} = V(\alpha T - 1)/C_P$ gives μ_{JT} as

$$\frac{18.1\ \text{cm}^3/\text{mol}}{75.3\ \text{J/mol-K}}\ [(3.04 \times 10^{-4}\ \text{K}^{-1})(303.1\ \text{K}) - 1] \times$$

$$(8.314\ \text{J})/(82.06\ \text{cm}^3\text{-atm}) = -0.0221\ \text{K/atm}$$

4.7 From (4.53), $C_{P,\text{m}} - C_{V,\text{m}} = TV_\text{m}\alpha^2/\kappa$.
$V_\text{m} = M/\rho = (119.4\ \text{g/mol})/(1.49\ \text{g/cm}^3) = 80.1\ \text{cm}^3/\text{mol}$.
$TV_\text{m}\alpha^2/\kappa = (298\ \text{K})(80.1\ \text{cm}^3/\text{mol})(0.00133\ \text{K}^{-1})^2/(9.8 \times 10^{-5}\ \text{atm}^{-1}) =$
$(431\ \text{cm}^3\text{-atm/mol-K})(8.314\ \text{J}/82.06\ \text{cm}^3\text{-atm}) = 44\ \text{J/mol-K}$.
Then $C_{V,\text{m}} = (116 - 44)\ \text{J/mol-K} = 72\ \text{J/mol-K}$.

4.8 Division of (4.30) by n gives $(\partial H_\text{m}/\partial T)_P = C_{P,\text{m}}$. For (b), we use (4.48) divided by n; for (c), (4.47); for (d), (4.49)/n; for (e), (4.50)/n; for (f), (4.53)/n; for (g), we use $(\partial A/\partial V)_T = -P$, which follows from (4.35). Substitution of numerical values gives:

(a) $(\partial H_\text{m}/\partial T)_P = C_{P,\text{m}} = 150\ \text{J/mol-K}$;

(b) $(\partial H_\text{m}/\partial P)_T = V_\text{m} - TV_\text{m}\alpha = 50\ \text{cm}^3/\text{mol} - (298\ \text{K})(50\ \text{cm}^3/\text{mol})(10^{-3}\ \text{K}^{-1})$
$= 35\ \text{cm}^3/\text{mol} = (35\ \text{cm}^3/\text{mol})(8.31\ \text{J}/82.1\ \text{cm}^3\text{-atm}) = 3.5\ \text{J/atm-mol}$;

(c) $(\partial U/\partial V)_T = \alpha T/\kappa - P = (10^{-3}\ \text{K}^{-1})(298\ \text{K})/(10^{-4}\ \text{atm}^{-1}) - 1\ \text{atm} =$
$3000\ \text{atm} = (3000\ \text{atm})(8.31\ \text{J})/(82\ \text{cm}^3\text{-atm}) = 304\ \text{J/cm}^3$;

(d) $(\partial S_\text{m}/\partial T)_P = C_{P,\text{m}}/T = (150\ \text{J/mol-K})/(298\ \text{K}) = 0.50\ \text{J/mol-K}^2$;

(e) $(\partial S_\text{m}/\partial P)_T = -\alpha V_\text{m} = -(10^{-3}\ \text{K}^{-1})(50\ \text{cm}^3/\text{mol})(8.31\ \text{J})/(82\ \text{cm}^3\text{-atm}) =$
$-0.005\ \text{J/mol-K-atm}$;

(f) $C_{V,\text{m}} = C_{P,\text{m}} - TV_\text{m}\alpha^2/\kappa =$
$(150\ \text{J/mol-K}) - (298\ \text{K})(50\ \text{cm}^3/\text{mol})(10^{-6}\ \text{K}^{-2})/(10^{-4}\ \text{atm}^{-1}) =$
$150\ \text{J/mol-K} - (150\ \text{cm}^3\text{-atm/mol-K})(8.31\ \text{J}/82\ \text{cm}^3\text{-atm}) =$
$150\ \text{J/mol-K} - 15\ \text{J/mol-K} = 135\ \text{J/mol-K}$;

(g) $(\partial A/\partial V)_T = -P = -(1\ \text{atm})(8.31\ \text{J}/82\ \text{cm}^3\text{-atm}) = -0.10\ \text{J/cm}^3$,
where (1.19) and (1.21) were used.

4.9 **(a)** The Gibbs equation $dU = T\,dS - P\,dV$ becomes $dU_T = T\,dS_T - P\,dV_T$ at constant T. Division by dP_T gives $dU_T/dP_T = T\,dS_T/dP_T - P\,dV_T/dP_T$ or $(\partial U/\partial P)_T = T(\partial S/\partial P)_T - P(\partial V/\partial P)_T$. But the Maxwell equation (4.45) gives $(\partial S/\partial P)_T = -(\partial V/\partial T)_P$, so $(\partial U/\partial P)_T = -T(\partial V/\partial T)_P - P(\partial V/\partial P)_T =$
$-TV\alpha + PV\kappa$.

(b) From (1.35) we have $(\partial U/\partial P)_T = (\partial U/\partial V)_T(\partial V/\partial P)_T = -(\partial U/\partial V)_T\kappa V$. Substitution of (4.47) gives the desired result.

4.10 **(a)** At constant P, $dU_P = T\,dS_P - P\,dV_P$. Division by dT_P gives $(\partial U/\partial T)_P = T(\partial S/\partial T)_P - P(\partial V/\partial T)_P = C_P - PV\alpha$, where (4.31) and (4.39) were used.

(b) $C_P = (\partial H/\partial T)_P = [\partial(U + PV)/\partial T]_P = (\partial U/\partial T)_P + P(\partial V/\partial T)_P$, so $(\partial U/\partial T)_P = C_P - P(\partial V/\partial T)_P = C_P - PV\alpha$.

4.11 At constant T, we have $dH_T = T\,dS_T + V\,dP_T$. Division by dV_T gives $(\partial H/\partial V)_T = T(\partial S/\partial V)_T + V(\partial P/\partial V)_T = T(\partial P/\partial T)_V + V(\partial P/\partial V)_T = \alpha T/\kappa - 1/\kappa$, where (4.45), (1.42), (1.44), and (1.32) were used.

4.12 Liquids. Solids.

4.13 Differentiation followed by use of (4.51) gives

$$\left(\frac{\partial(GT^{-1})}{\partial T}\right)_P = -\frac{1}{T^2}G + \frac{1}{T}\left(\frac{\partial G}{\partial T}\right)_P = -\frac{H - TS}{T^2} - \frac{S}{T} = -\frac{H}{T^2}$$

4.14 At constant V, $dU = T\,dS - P\,dV$ becomes $dU_V = T\,dS_V$. Division by dT_V and use of (4.29) gives $C_V = (\partial U/\partial T)_V = T(\partial S/\partial T)_V$. At constant P, (4.34) is $dH_P = T\,dS_P$ and division by dT_P gives $C_P = (\partial H/\partial T)_P = T(\partial S/\partial T)_P$.

4.15 From (2.63), $\mu_J = -(\partial U/\partial V)_T/C_V$. Use of (4.47) gives the desired result.

> Reminder: Don't look up the solution to a problem until you have made a serious effort to solve it.

4.16 $V_m = RT/P + bRT$. $\alpha = (1/V_m)(\partial V_m/\partial T)_P = (1/V_m)(R + bPR)/P = R(1 + bP)/PV_m$. $\kappa = -(1/V_m)(\partial V_m/\partial P)_T = RT/P^2V_m$. $(\partial U/\partial V)_T = \alpha T/\kappa - P = (1 + bP)P - P = bP^2$. $C_{P,m} - C_{V,m} = TV_m\alpha^2/\kappa = (R + bPR)^2/R = R(1 + bP)^2$. $\mu_{JT} = V(\alpha T - 1)/C_P = (V/C_P)[RT(1 + bP)/PV_m - 1] = (V/C_P)(PV_m/PV_m - 1) = 0$.

4.17

$$\left(\frac{\partial C_P}{\partial P}\right)_T = \left[\frac{\partial}{\partial P}\left(\frac{\partial H}{\partial T}\right)_P\right]_T = \left[\frac{\partial}{\partial T}\left(\frac{\partial H}{\partial P}\right)_T\right]_P = \left\{\frac{\partial}{\partial T}\left[-T\left(\frac{\partial V}{\partial T}\right)_P + V\right]\right\}_P$$

$$= -\left(\frac{\partial V}{\partial T}\right)_P - T\left(\frac{\partial^2 V}{\partial T^2}\right)_P + \left(\frac{\partial V}{\partial T}\right)_P = -T\left(\frac{\partial^2 V}{\partial T^2}\right)_P$$

4.18 **(a)** $V = V_0 + aV_0(T - 273\text{ K}) + bV_0(T - 273\text{ K})^2$, $(\partial V/\partial T)_P = aV_0 + 2bV_0(T - 273\text{ K})$ and $(\partial^2 V/\partial T^2) = 2bV_0$. $(\partial C_{P,\text{m}}/\partial P)_T = -T(\partial^2 V_\text{m}/\partial T^2)_P = -2bV_{\text{m},0}T$, so $(\partial C_{P,\text{m}}/\partial P)_T =$ $-2(0.78 \times 10^{-8}\text{ K}^{-2})(298\text{ K})(200.6\text{ g/mol})/(13.595\text{ g/cm}^3) =$ $(-6.8_6 \times 10^{-5}\text{ cm}^3\text{ K}^{-1}\text{ mol}^{-1})(1.987\text{ cal})/(82.06\text{ cm}^3\text{-atm}) =$ -1.66×10^{-6} cal/mol-K-atm.

(b) $\Delta C_{P,\text{m}} \approx (\partial C_{P,\text{m}}/\partial P)_T\,\Delta P = (-1.66 \times 10^{-6}\text{ cal/mol-K-atm})(10^4\text{ atm}) =$ -0.02 cal/mol-K. $C_{P,\text{m}} = 6.66$ cal/mol-K $- 0.02$ cal/mol-K $=$ 6.64 cal/mol-K.

4.19 **(a)** $C_P - C_V = TV\alpha^2/\kappa$. As in Prob. 1.49, $\alpha = (1/V_\text{m})(c_2 + 2c_3T - c_5P)$ and $\kappa = (c_4 + c_5T)/V_\text{m}$. So $C_P - C_V = nT(c_2 + 2c_3T - c_5P)^2/(c_4 + c_5T)$, since $V/V_\text{m} = n$.

(b) $(\partial U/\partial V)_T = \alpha T/\kappa - P = T(c_2 + 2c_3T - c_5P)/(c_4 + c_5T) - P$.

(c) $(\partial S/\partial P)_T = -\alpha V = -n(c_2 + 2c_3T - c_5P)$.

(d) $\mu_{JT} = (V_\text{m}/C_{P,\text{m}})(\alpha T - 1) = (T/C_{P,\text{m}})(c_2 + 2c_3T - c_5P) - V_\text{m}/C_{P,\text{m}}$.

(e) $(\partial S/\partial T)_P = C_P/T$.

(f) $(\partial G/\partial P)_T = V$.

4.20 **(a)** $(\partial V/\partial T)_S = 1/(\partial T/\partial V)_S = -1/(\partial P/\partial S)_V = -(\partial S/\partial P)_V =$ $-(\partial S/\partial T)_V(\partial T/\partial P)_V = -(C_V/T)/(\partial P/\partial T)_V = -C_V\kappa/\alpha T$, where (1.45) was used. Hence $\alpha_S = -C_V\kappa/TV\alpha$.

(b) For a perfect gas, $\alpha = 1/T$ and $\kappa = 1/P$ [Eqs. (1.46) and (1.47)], so $\alpha_S =$ $-C_V/PV = -C_V/nRT = -C_{V,\text{m}}/RT = V^{-1}(\partial V/\partial T)_S$, so $dV/V =$ $-(C_{V,\text{m}}/RT)\,dT$ at constant S, and $\ln(V_2/V_1) = (C_{V,\text{m}}/R)\ln(T_1/T_2) =$ $\ln(T_1/T_2)^{C_{V,\text{m}}/R}$; hence $V_2/V_1 = (T_1/T_2)^{C_{V,\text{m}}/R}$ or $T_2/T_1 = (V_1/V_2)^{R/C_{V,\text{m}}}$.

(c) $(\partial V/\partial P)_S = (\partial V/\partial T)_S(\partial T/\partial P)_S$. From (a), $(\partial V/\partial T)_S = -C_V\kappa/\alpha T$. From (4.44), $(\partial T/\partial P)_S = (\partial V/\partial S)_P = [(\partial S/\partial V)_P]^{-1} = [(\partial S/\partial T)_P(\partial T/\partial V)_P]^{-1} = (C_P/T)^{-1}(\partial V/\partial T)_P = (T/C_P)\alpha V = \alpha TV/C_P$. So $(\partial V/\partial P)_S = (-C_V\kappa/\alpha T)(\alpha TV/C_P) = -C_V\kappa V/C_P$; $\kappa_S = -V^{-1}(\partial V/\partial P)_S = C_V\kappa/C_P$.

4.21 $(\partial H/\partial P)_T = V - T(\partial V/\partial T)_P$. For an ideal gas, $V = nRT/P$ and $(\partial V/\partial T)_P = nR/P$, so $(\partial H/\partial P)_T = nRT/P - T(nR/P) = 0$.

4.22 **(a)** Integration of $dU_T = (\partial U/\partial V)_T\, dV_T$ at constant T gives

$$\int_{\infty}^{V'} (\partial U/\partial V)_T\, dV = \int_{\infty}^{V'} dU = U(T,V') - U(T,\infty) = U_{\text{intermol}}(T,V')$$

(b) Use of (4.57) gives for a van der Waals gas

$$\int_{\infty}^{V'} (\partial U/\partial V)_T\, dV = \int_{\infty}^{V'} (an^2/V^2)\, dV = -an^2/V' = U_{\text{intermol}} \text{ so}$$

$U_{\text{intermol,m}} = -a/V_{\text{m}}$, where the unnecessary prime was dropped.

(c) At 25°C and 1 atm, $V_{\text{m}} = RT/P = 24500\ \text{cm}^3/\text{mol}$ and $U_{\text{intermol,m}} = -a/V_{\text{m}} = -(10^6 \text{ to } 10^7)(\text{cm}^6\text{-atm/mol}^2)/(24500\ \text{cm}^3/\text{mol}) = (-40 \text{ to } -400)(\text{cm}^3\text{-atm/mol})(1.99\ \text{cal}/82\ \text{cm}^3\text{-atm}) = -1 \text{ to } -10\ \text{cal/mol} = -4 \text{ to } -40\ \text{J/mol}$, where (1.19) and (1.21) were used. At 25°C and 40 atm, $V_{\text{m}} \approx RT/P = 610\ \text{cm}^3/\text{mol}$ and $U_{\text{intermol,m}} \approx -40 \text{ to } -400\ \text{cal/mol} = -160 \text{ to } -1600\ \text{J/mol}$.

4.23 **(a)** $a = (1.34 \times 10^6\ \text{cm}^6\ \text{atm/mol}^2)(8.314\ \text{J}/82.06\ \text{cm}^3\text{-atm}) = 1.36 \times 10^5\ \text{J cm}^3/\text{mol}^2$. $U = U_{\text{tr,m}} + U_{\text{intermol,m}} + U_{\text{el,m}} \approx (3/2)RT - a/V_{\text{m}} + \text{const} = (12.5\ \text{J/mol-K})T - (1.36 \times 10^5\ \text{J cm}^3/\text{mol}^2)/V_{\text{m}} + \text{const}$, where $U_{\text{tr,m}}$ is the molar molecular translational energy.

(b) For both the liquid and the gas, $U_{\text{tr,m}} = (3/2)RT = 1090\ \text{J/mol}$. $V_{\text{m,liq}} = M/\rho = (39.95\ \text{g/mol})/(1.38\ \text{g/cm}^3) = 28.9\ \text{cm}^3/\text{mol}$. $V_{\text{m,gas}} = RT/P = 7160\ \text{cm}^3/\text{mol}$. $U_{\text{intermol,m,liq}} \approx -a/V_{\text{m,liq}} = -(1.36 \times 10^5\ \text{J cm}^3/\text{mol}^2)/(28.9\ \text{cm}^3/\text{mol}) = -4710\ \text{J/mol}$. $U_{\text{intermol,m,gas}} \approx -a/V_{\text{m,gas}} = -19\ \text{J/mol}$.

(c) $\Delta U_{\text{m,vap}} = U_{\text{intermol,m,gas}} - U_{\text{intermol,m,liq}} \approx -19\ \text{J/mol} + 4710\ \text{J/mol} = 4.7\ \text{kJ/mol}$.

4.24 **(a)** T [See the paragraph before Eq. (4.64).] **(b)** T [See Eq. (4.56).]

4.25 Since T is constant, we have $\Delta A = \Delta U - T\,\Delta S$ and $\Delta G = \Delta H - T\,\Delta S$. From Prob. 2.50c, $\Delta H = 0$ and $\Delta U = 0$. From Prob. 3.12c, $\Delta S = 8.06$ J/K. So $\Delta A = 0 - (400\text{ K})(8.06\text{ J/K}) = -3220$ J and $\Delta G = -3220$ J.

4.26 Each process is isothermal, so $\Delta A = \Delta U - T\,\Delta\text{S}$ and $\Delta G = \Delta H - T\,\Delta S$. (For q, w, ΔU and ΔH, see the answers to Prob. 2.45.)

(a) $\Delta U = q + w$ and $\Delta S = q/T$, so $\Delta A = w$. Since $w < 0$, we have $\Delta A < 0$. Since $\Delta H = q$, $\Delta G = q - q = 0$.

(b) The same as (a), except that $w > 0$ and $\Delta A > 0$.

(d) $\Delta U = 0 = \Delta H$. $\Delta A = -T\,\Delta S = -T(q/T) = -q = w < 0$. Also, $\Delta G = -T\,\Delta S < 0$.

(e) $\Delta U = 0 = \Delta H$. For this irreversible adiabatic process, ΔS is positive and $\Delta A = -T\,\Delta S$ is negative. Also, $\Delta G = -T\,\Delta S$ is negative.

(f) The same as (e).

4.27 Each is zero, since the process is cyclic.

4.28 **(a)** $\Delta G = \Delta H - T\,\Delta S = \Delta H - T(\Delta H/T) = 0$, as it must be for a reversible (equilibrium) process at constant T and P.

(b) From Prob. 3.14, $\Delta S = -11.4$ J/K $= -2.73$ cal/K. For the reversible path in Fig. 3.7, ΔH for each step is:
$\Delta H_1 = (1.01\text{ cal/g-K})(10.0\text{ g})(10\text{ K}) = 101$ cal,
$\Delta H_2 = -(79.7\text{ cal/g})(10.0\text{ g}) = -797$ cal,
$\Delta H_3 = -(0.50\text{ cal/g-K})(10.0\text{ g})(10\text{ K}) = -50$ cal.
The overall process therefore has $\Delta H = -746$ cal. For the isothermal process at $-10°$C, we have $\Delta G = \Delta H - T\,\Delta S =$
$-746\text{ cal} - (263.15\text{ K})(-2.73\text{ cal/K}) = -27_{.6}\text{ cal} = -115\text{ J}$.

4.29 Since T is constant and U and H of ideal gases depend on T only, $\Delta U = 0$ and $\Delta H = 0$. From (3.33), $\Delta S = -0.200\text{ mol } R \ln 0.400 - 0.300\text{ mol } R \ln 0.600 =$ 2.80 J/K. $\Delta A = \Delta U - T\,\Delta S = -T\,\Delta S = -(300\text{ K})(2.80\text{ J/K}) = -840$ J. $\Delta G = \Delta H - T\,\Delta S = -T\,\Delta S = -840$ J.

4.30 Let the path be

$(27°C, 1 atm) \xrightarrow{a} (100°C, 1 atm) \xrightarrow{b}$ (100°C, 50 atm) where step (a) is isobaric and step (b) is isothermal. For a liquid, V varies slowly with T and P.

(a) Use of (4.63) with C_P, α, and V assumed constant gives
$\Delta H_a = C_P(T_2 - T_1) = (18.0 \text{ cal/K})(73 \text{ K}) = 1310 \text{ cal} = 5.50 \text{ kJ}$
$\Delta H_b = (V - TV\alpha)(P_2 - P_1) =$
$[18.1 \text{ cm}^3 - (373 \text{ K})(18.1 \text{ cm}^3)(3.04 \times 10^{-4} \text{ K}^{-1})](49 \text{ atm}) \times$
$(1.987 \text{ cal})/(82.06 \text{ cm}^3\text{-atm}) = 19 \text{ cal} = 0.080 \text{ kJ}$

$\Delta H = \Delta H_a + \Delta H_b = 1.33 \text{ kcal} = 5.58 \text{ kJ}.$

(b) $\Delta U = \Delta H - \Delta(PV)$. Since V changes only slightly,
$\Delta(PV) \approx V\,\Delta P = (18.1 \text{ cm}^3)(49 \text{ atm}) = 887 \text{ cm}^3 \text{ atm} =$
$(887 \text{ cm}^3 \text{ atm})(1.987 \text{ cal}/82.06 \text{ cm}^3\text{-atm}) = 21 \text{ cal} = 90 \text{ J}.$
$\Delta U = 1330 \text{ cal} - 21 \text{ cal} = 1.31 \text{ kcal} = 5.49 \text{ kJ}.$

(c) Use of (4.60) [or (4.61) and (4.62)] with C_P, α, and V assumed constant gives $\Delta S_a = C_P \ln(T_2/T_1) = (18.0 \text{ cal/K}) \ln(373/300) = 3.92 \text{ cal/K} =$ 16.4 J/K;

$\Delta S_b = -\alpha V(P_2 - P_1) =$

$$-(3.04 \times 10^{-4} \text{ K}^{-1})(18.1 \text{ cm}^3)(49 \text{ atm}) \frac{1.987 \text{ cal}}{82.06 \text{ cm}^3 \text{ atm}} =$$

$-0.0065 \text{ cal/K} = -0.027 \text{ J/K}$

$\Delta S = \Delta S_a + \Delta S_b = 3.91 \text{ cal/K} = 16.4 \text{ J/K}$

4.31 $dG = -S\,dT + V\,dP = V\,dP$ at constant T.

$\Delta G = V\,\Delta P$ at constant T and V.

$$\Delta G = \frac{30.0 \text{ g}}{0.997 \text{ g/cm}^3} 99.0 \text{ atm} \frac{1.987 \text{ cal}}{82.06 \text{ cm}^3 \text{ atm}} = 72.1 \text{ cal} = 302 \text{ J}$$

4.32 $V_m = RT/P + bRT + cRTP$. $\alpha = (1/V_m)(\partial V_m/\partial T)_P = (1/V_m)(R/P + bR + cRP)$. Use the path

$$(P_1, T_1) \xrightarrow{a} (P_1, T_2) \xrightarrow{b} (P_2, T_2)$$

where step (a) is isobaric and step (b) is isothermal.

$V_m - TV_m\alpha = RT/P + bRT + cRTP - (RT/P + bRT + cRPT) = 0.$

Equation (4.63) gives $\Delta H_{m,a} = \int_{T_1}^{T_2} C_{P,m}\, dT = C_{P,m}(T_2 - T_1)$;

$\Delta H_{m,b} = 0$. Then $\Delta H_m = \Delta H_{m,a} + \Delta H_{m,b} = C_{P,m}(T_2 - T_2)$, where $C_{P,m}$ is assumed constant. From (4.60) with $C_{P,m}$ constant: $\Delta S_{m,a} = C_{P,m} \ln(T_2/T_1)$;

$\Delta S_{m,b} = -\int_1^2 (R/P + bR + cRP)\, dP = R \ln(P_1/P_2) + bR(P_1 - P_2) + ½cR(P_1^2 - P_2^2)$;

$\Delta S_m = \Delta S_{m,a} + \Delta S_{m,b} =$ etc.

4.33 From Prob. 4.20c, $(\partial V/\partial P)_S = -V\kappa_S$ and $\Delta V_S \approx -V\kappa_S\, \Delta P_S =$ $-V\kappa(C_V/C_P)\,\Delta P_S$. From (4.54) and preceding data:

$\Delta V_S \approx -(18.1\ \text{cm}^3)(4.52 \times 10^{-5}/\text{atm})(74.2/75.3)(9.00\ \text{atm}) = -7.26 \times 10^{-3}\ \text{cm}^3$

We have $V_{\text{final}} = 18.1\ \text{cm}^3 - 0.007\ \text{cm}^3 = 18.1\ \text{cm}^3$.

From Prob. 4.20a, $(\partial V/\partial T)_S = \alpha_S V$ and $\Delta V_S \approx \alpha_S V\, \Delta T_S$, so

$\Delta T_S \approx \Delta V_S/\alpha_S V = -\Delta V_S(\alpha T/C_V\kappa)$.

$$\Delta T_S \approx 0.00726\ \text{cm}^3 \frac{(3.04 \times 10^{-4}\ \text{K}^{-1})(303.15\ \text{K})}{(74.2\ \text{J/K})(4.52 \times 10^{-5}/\text{atm})} \times \frac{8.314\ \text{J}}{82.06\ \text{cm}^3\ \text{atm}} =$$

0.0202 K, and $T_{\text{final}} = 30.02°\text{C}$.

We have $dU = dq + dw = dw = -P_S\, dV_S = -P_S(\partial V/\partial P)_S\, dP_s = P_S V\kappa_S\, dP_S$, since $(\partial V/\partial P)_S = -V\kappa_S$. Approximating V and κ_S as constants, we have $\Delta U \approx V\kappa_S \int_1^2 P_S\, dP_S = V\kappa_S \frac{1}{2}(P_2^2 - P_1^2)$. Using $\Delta V_S \approx -V\kappa_S\, \Delta P_S$ (given earlier in this problem), we have $\Delta U \approx -(\Delta V_S/\Delta P_S)\frac{1}{2}(P_2^2 - P_1^2) =$

$[(0.00725\ \text{cm}^3)/(9.00\ \text{atm})]0.5(100 - 1)\ \text{atm}^2 = 0.040\ \text{cm}^3\ \text{atm} = 0.00097\ \text{cal} =$ 0.0040 J.

4.34 We have $dU = (\partial U/\partial T)_V\, dT + (\partial U/\partial V)_T\, dV = C_V\, dT + (\partial U/\partial V)_T\, dV$. Integrating and using a path similar to that in Fig. 4.3 but with V as the vertical axis, we get $\Delta U = \int_{T_1}^{T_2} C_V\, dT + \int_{V_1}^{V_2} (\partial U/\partial V)_T\, dV$. Using $(\partial U/\partial V)_T = a/V_m^2 = an^2/V^2$ for a van der Waals gas, we have $\Delta U =$ $\int_{T_1}^{T_2} C_V\, dT - an^2/V_2 + an^2/V_1$. If C_V is approximately constant over the temperature interval, then $\Delta U \approx C_V(T_2 - T_1) + an^2(1/V_1 - 1/V_2)$ for a van der Waals gas.

4.35 **(a)** T; **(b)** T; **(c)** F; **(d)** T; **(e)** F; **(f)** T; **(g)** F.

4.36 Set $dS = 0$, $dP = 0$, and $dn_{j\neq i} = 0$ in Eq. (4.76). Then set $dT = 0$, $dV = 0$, and $dn_{j\neq i} = 0$ in (4.77).

4.37 For a closed system, $dU = dq + dw$. For a mechanically reversible process in a closed system with *P-V* work only, $dw = -P\,dV$ and $dU = dq - P\,dV$. Equating this expression for dU to that in (4.74), we get $dq = T\,ds + \sum_i \mu_i\,dn_i$ under the conditions stated in (4.73).

4.38 **(a)** F; **(b)** F; **(c)** T; **(d)** F.

4.39 Use of (4.88) gives the following results.

(a) $\mu_{H_2O(solid)} = \mu_{H_2O(liquid)}$

(b) $\mu_{sucrose(s)} = \mu_{sucrose(aq)}$, where s and aq denote solid and aqueous solution.

(c) $\mu_{ether(in\ water\ phase)} = \mu_{ether(in\ ether\ phase)}$ and $\mu_{water(in\ ether\ phase)} = \mu_{water(in\ water\ phase)}$

(d) $\mu_{H_2O(s)} = \mu_{H_2O(aq)}$

(e) $\mu_{sucrose(s)} = \mu_{sucrose(aq)}$ and $\mu_{glucose(s)} = \mu_{glucose(aq)}$.

4.40 The more-stable phase at the given T and P has the lower μ.
(a) $H_2O(g)$; **(b)** neither; the two phases are in equilibrium and have equal μ's; **(c)** $H_2O(l)$; **(d)** $C_6H_{12}O_6(s)$; substance *i* flows out of the phase with the higher μ_i; **(e)** neither; **(f)** $C_6H_{12}O_6(aq)$; **(g)** $H_2O(g)$, since $\mu = G_m$ for a pure substance.

4.41 $\mu_{H_2O(s)} = \mu_{H_2O(l)} = G_{m,H_2O(s)} = G_{m,H_2O(l)}$, since $\mu = G_m$ for a pure substance. Multiplication by n gives $G_{H_2O(s)} = G_{H_2O(l)}$ or $\Delta G = 0$.

4.42 $\nu_{C_3H_8} = -1$, $\nu_{O_2} = -5$, $\nu_{CO_2} = 3$, $\nu_{H_2O} = 4$.

4.43 $\mu_{N_2} + 3\mu_{H_2} = 2\mu_{NH_3}$, where (4.98) was used.

4.44 $\xi = \Delta n_{O_3}/\nu_{O_3} = (7.10\text{ mol} - 6.20\text{ mol})/(-2) = -0.45\text{ mol}$.

4.45 **(a)** No; **(b)** no; **(c)** no; **(d)** yes; **(e)** yes; **(f)** no.

4.46 **(a)** Heat is needed to vaporize the liquid, so q is positive. Hence $\Delta H = q_P > 0$. For this reversible isothermal process, $\Delta S = q/T > 0$. Since the process is reversible, $\Delta S_{\text{univ}} = 0$. Also, $\Delta G = \Delta H - T\,\Delta S = q - q = 0$, as it must be for a reversible process at constant T and P.

(b) q is positive. $\Delta U = q + w = q > 0$. For this irreversible isothermal process, Eq. (4.8) gives $dS > dq/T$ and $\Delta S > q/T$. Since q is positive, ΔS is positive. Since the process is irreversible, ΔS_{univ} is positive. Finally, $\Delta A = \Delta U - T\,\Delta S$. Since $\Delta U = q$ and $T\,\Delta S > q$, we have $\Delta A < 0$. [This also follows from (4.22) with $w_{\text{by}} = 0$.]

4.47 Only processes (a) and (d) are reversible [assuming the surroundings in (d) are only infinitesimally warmer than the system] and so (a) and (d) have $\Delta S_{\text{univ}} = 0$. As to ΔU, ΔH, ΔS, ΔA, and ΔG, the following is true:

(a) All are zero, since the process is cyclic.

(b) The process is adiabatic, so $q = 0$. Since V is constant, $w = 0$. The system is closed. Hence $\Delta U = q + w = 0$. There is no reason for any of the others to be zero.

(c) $q = 0$, but $w \neq 0$. Hence $\Delta U \neq 0$. Section 2.7 gives $\Delta H = 0$. There is no reason for any of the others to be zero.

(d) $\Delta G = 0$ for this reversible constant-T-and-P process. None of the others is zero.

4.48 From (4.47), $(\partial U/\partial V)_T = \alpha T/\kappa - P$. Both κ and T are always positive. For liquid water between 0°C and 4°C at 1 atm, α is negative and hence $(\partial U/\partial V)_T$ is negative.

4.49 **(a)** nu, stoichiometric number; **(b)** mu, chemical potential;

(c) xi, extent of reaction; **(d)** alpha, thermal expansivity;

(e) kappa, isothermal compressibility; **(f)** rho, density.

4.50 **(a)** Closed system at rest in the absence of external fields.

(b) Closed system (at rest in the absence of external fields), reversible process, P-V work only.

(c) System (at rest in the absence of fields) in mechanical and thermal equilibrium, P-V work only.

4.51 **(a)** J; **(b)** J mol^{-1} K^{-1}; **(c)** J/K; **(d)** J/mol.

4.52 **(a)** The chemical potential of substance i is the same in every phase in which i is present, and this condition holds for each substance.

(b) $\sum_i \nu_i \mu_i = 0$.

(c) $dG = 0$ is a valid equilibrium condition only for systems held at constant T and P.

4.53 $G = H - TS$. At constant T and P, $dG = dH - T\,dS = dq_P - T\,dS$. Solving for dS, we get $dS = dq_P/T - dG/T$.

4.54 **(a) and (b)**: From $dG = -S\,dT + V\,dP$, we get $S_m = -(\partial G_m/\partial T)_P$ and $V_m = (\partial G_m/\partial P)_T$.

(c) $H_m = G_m + TS_m = G_m - T(\partial G_m/\partial T)_P$.

(d) $U_m = H_m - PV_m = G_m - T(\partial G_m/\partial T)_P - P(\partial G_m/\partial P)_T$.

(e) $C_{P,m} = (\partial H_m/\partial T)_P$ and partial differentiation of (c) gives $C_{P,m} = (\partial G_m/\partial T)_P - (\partial G_m/\partial T)_P - T(\partial^2 G_m/\partial T^2)_P = -T(\partial^2 G_m/\partial T^2)_P$.

(f) $C_{V,m} = C_{P,m} - TV_m\alpha^2/\kappa$ and the results of (b), (e), (g), and (h) give $C_{V,m} = -T(\partial^2 G_m/\partial T^2)_P + T(\partial G_m/\partial P)_T[(\partial G_m/\partial P)_T]^{-2}(\partial^2 G_m/\partial T\,\partial P)^2 \times (\partial G_m/\partial P)_T/(\partial^2 G_m/\partial P^2)_T = -T(\partial^2 G_m/\partial T^2)_P + T(\partial^2 G_m/\partial T\,\partial P)^2/(\partial^2 G_m/\partial P^2)_T$.

(g) $\alpha = (1/V_m)(\partial V_m/\partial T)_P$ and the result of (b) gives $\alpha = (\partial^2 G_m/\partial T\partial P)/(\partial G_m/\partial P)_T$.

(h) $\kappa = -(\partial V_m/\partial P)_T/V_m = -(\partial^2 G_m/\partial P^2)_T/(\partial G_m/\partial P)_T$.

4.55 The same since S is a state function.

4.56 **(a)** T, which is intensive; **(b)** S, which does not have units of energy.

4.57 **(a)** C_P. **(b)** μ_i^α and μ_i^β. **(c)** C_V. **(d)** S. **(e)** S_{univ}. **(f)** G.

4.58 **(a)** False; the equation holds only for ideal gases.

(b) True.

(c) False; the system must be held at constant T and P.

(d) False; the system must be held at constant T and P.

(e) True; $w_{\text{by}} = -w = q - \Delta U$ and if q is positive then $w_{\text{by}} > -\Delta U$.

(f) True.

(g) False; there is no law of conservation of free energy.

(h) False; ΔS_{univ} is positive for an irreversible process, but ΔS of the system can be positive, negative, or zero.

(i) True. **(j)** False. **(k)** True. **(l)** False. **(m)** False.

(n) False; the system must be isolated or adiabatically enclosed.

(o) False.

Chapter 5

5.1 **(a)** F; **(b)** F; **(c)** F (The gas must be ideal.).

5.2 **(a)** F (They are J/mol.); **(b)** T; **(c)** T; **(d)** T.

5.3 $\Delta H_T^\circ = 2H^\circ_{\mathrm{m},T,\mathrm{H_2O}(\ell)} + 2H^\circ_{\mathrm{m},T,\mathrm{SO_2(g)}} - 2H^\circ_{\mathrm{m},T,\mathrm{H_2S(g)}} - 3H^\circ_{\mathrm{m},T,\mathrm{O_2(g)}}$.

5.4 **(a)** The stoichiometric coefficients are doubled, so Eq. (5.3) gives 2(–319 kJ/mol) = –638 kJ/mol.

(b) 4(–319 kJ/mol) = –1276 kJ/mol. **(c)** –1(–319 kJ/mol) = 319 kJ/mol.

5.5 **(a)** F; **(b)** T; **(c)** T.

5.6 **(a)** C(graphite) + $2Cl_2$(g) → $CCl_4(\ell)$;

(b) $\frac{1}{2}N_2$(g) + $\frac{5}{2}H_2$(g) + 2C(graphite) + O_2(g) → NH_2CH_2COOH(s).

(c) $\frac{1}{2}H_2$(g) → H(g). **(d)** N_2(g) → N_2(g).

5.7 The 25°C reference form is the form most stable at 25°C and 1 bar. The elements that are liquid at 25°C and 1 bar are Hg and Br_2. Those that are gaseous at 25°C and 1 bar are He, Ne, Ar, Kr, Xe, Rn, H_2, F_2, Cl_2, N_2, O_2.

5.8 **(a)** C_4H_{10}(g) + $\frac{13}{2}O_2$(g) → $4CO_2$(g) + $5H_2O(\ell)$

(b) $C_2H_5OH(\ell)$ + $3O_2$(g) → $2CO_2$(g) + $3H_2O(\ell)$

5.9 **(a)** T; **(b)** T; **(c)** T; **(d)** T; **(e)** F.

5.10 **(a)** [2(–285.830) + 2(–296.830) – 2(–20.63) – 3(0)] kJ/mol = –1124.06 kJ/mol

(b) [2(–241.818) + 2(–296.830) – 2(–20.63) – 3(0)] kJ/mol = –1036.04 kJ/mol

(c) [–187.78 + 4(0) – 2(294.1) – 2(90.25)] kJ/mol = –956.5 kJ/mol

5.11 **(a)** $C_6H_{12}O_6(c) + 6O_2(g) \rightarrow 6CO_2(g) + 6H_2O(\ell)$. $\Delta H^\circ_{c,298}/(\text{kJ/mol}) =$
$6(-393.509) + 6(-285.830) - (-1274.4) - 6(0) = -2801.6$.
$\Delta H^\circ = \Delta U^\circ + (\Delta n_g/\text{mol})RT = \Delta U^\circ + (6-6)RT = \Delta U^\circ = -2801.6$ kJ/mol.

(b) (0.7805 g)(1 mol/180.158 g) = 0.004332 mol. The heat flowing out of the bomb is (2801.6 kJ/mol)(0.004332 mol) = 12.13_7 kJ. The water mass is $(2500\ \text{cm}^3)(0.9973\ \text{g/cm}^3) = 2493$ g. C_P of the steel bomb plus surrounding water is (14050 g)(0.450 J/g-°C) + (2493 g)(4.180 J/g-°C) = $1.67_4 \times 10^4$ J/°C. So 12137 J = $(1.67_4 \times 10^4\ \text{J/°C})\Delta t$ and $\Delta t = 0.725$°C. $t_{\text{final}} = 24.030°\text{C} + 0.725°\text{C} = 24.755°\text{C}$.

5.12 Initially, $n_{O_2} = PV/RT = (30\ \text{atm})(380\ \text{cm}^3)/R(297.2\ \text{K}) = 0.47$ mol. At the end, $n_{O_2} = 0.47\ \text{mol} - 6(0.004332\ \text{mol}) = 0.44$ mol, $n_{H_2O(\ell)} = 0.026$ mol, $n_{CO_2} = 0.026$ mol. The heat capacity C_{syst} of the system is gotten by adding the heat capacity C_{con} of the bomb contents to the heat capacity of the steel bomb and surrounding water. The gases are heated at constant V, so we use $C_{V,\text{m}}$ of the gases. Appendix data gives C_{con} = (0.44 mol)(29.36 − 8.31)(J/mol-K) + (0.026 mol)(37.11 − 8.31)(J/mol-K) + (0.026 mol)(75.29 J/mol-K) = 12.0 J/K. $C_{\text{syst}} = 12.0\ \text{J/K} + 1.67_4 \times 10^4\ \text{J/K} = 1.67_5 \times 10^4$ J/K. We get $\Delta t = 0.725$°C and $t_{\text{final}} = 24.755$°C.

5.13 **(a)** For the benzoic acid run, $\Delta_r U =$
(−26.434 kJ/g)(0.5742 g) + (0.0121 g)(−6.28 kJ/g) = -15.25_4 kJ
and (5.8) gives $-15.25_4\ \text{kJ} = -C_{K+P}(1.270\ \text{K})$ and $C_{K+P} = 12.01$ kJ/K.

(b) For the naphthalene run, (5.8) gives $\Delta_r U = -(12.01\ \text{kJ/K})(2.035\ \text{K}) =$ −24.44 kJ. The contributions of the combustion wire and the naphthalene to ΔU_r are (0.0142 g)(−6.28 kJ/g) + (0.6018 g)$\Delta_c U_{\text{naph}}$, where $\Delta_c U_{\text{naph}}$ is per gram of naphthalene. Then
−24.44 kJ = −0.089 kJ + (0.6018 g)$\Delta_c U_{\text{naph}}$ and $\Delta_c U_{\text{naph}} = -40.46$ kJ/g.
$\Delta_c U^\circ \cong \Delta_c U_{\text{m,naph}} = (-40.46\ \text{kJ/g})(128.17\ \text{g/mol}) = -5186$ kJ/mol. The combustion reaction $C_{10}H_8(s) + 12O_2(g) \rightarrow 10CO_2(g) + 4H_2O(\ell)$ has $\Delta n_g/\text{mol} = 10 - 12 = -2$ and $\Delta_c H^\circ = \Delta_c U^\circ - 2RT =$
$-5186\ \text{kJ/mol} - 2(8.314 \times 10^{-3}\ \text{kJ/mol-K})(298\ \text{K}) = -5191$ kJ/mol.

5.14 $U_{el} = VIt = (8.412\ \text{V})(0.01262\ \text{A})(812\ \text{s}) = 86.2$ J.
$\Delta_r U_{298} = -U_{el} = -86.2$ J. If we neglect the difference between U of the

standard states and U of the states in the calorimeter, then ΔU°_{298} is $\Delta_r U_{298}$ per mole, where "per mole" means for $\Delta\xi = 1$ mol. The reaction as written involves a coefficient of 3 for B, so $\Delta\xi = 1$ mol corresponds to $\Delta n(\text{B}) = 3$ mol. We have $n(\text{B}) = (1.450\text{ g})/(168.1\text{ g/mol}) = 0.008626$ mol, and

$$\Delta U^\circ_{298} \approx 3\frac{-86.2\text{ J}}{0.008626\text{ mol}} = -29.98\text{ kJ/mol}$$

Then $\Delta n_g/\text{mol} = 6 - 2 = 4$ and $\Delta H^\circ_{298} = \Delta U^\circ_{298} + \Delta n_g RT/\text{mol} \approx$ -29.98 kJ/mol + 4(0.0083145 kJ/mol-K)(298.15 K) = -20.06 kJ/mol.

5.15 **(a)** With V_{liq} neglected, $\Delta H^\circ_{298} - \Delta U^\circ_{298} = (\Delta n_g/\text{mol})RT =$ $(-1.5)(8.3145\text{ J/mol-K})(298.15\text{ K}) = -3718.5$ J/mol.

(b) $\Delta H^\circ_{298} - \Delta U^\circ_{298} = P\,\Delta V^\circ = PV^\circ_{\text{m,H}_2\text{O}(\ell)} + P\,\Delta V^\circ_{\text{gas}} =$ $(1\text{ atm})(18\text{ cm}^3/\text{mol})(8.3\text{ J}/82\text{ cm}^3\text{-atm}) + (\Delta n_g/\text{mol})RT =$ 1.8 J/mol -3718.5 J/mol = -3716.7 J/mol.

5.16 For $(CH_3)_2CO(\ell) + 4O_2(g) \rightarrow 3CO_2(g) + 3H_2O(\ell)$, Eq. (5.6) gives:

$$\Delta H^\circ_{298} = \Sigma_i \nu_i \Delta_f H^\circ_{298,i} =$$

$$3\,\Delta_f H^\circ_{298,\text{CO}_2(g)} + 3\,\Delta_f H^\circ_{298,\text{H}_2\text{O}(\ell)} - \Delta_f H^\circ_{298,(\text{CH}_3)_2\text{CO}(\ell)} - 4\,\Delta_f H^\circ_{298,\text{O}_2(g)}.$$

Use of Appendix data for CO_2 and H_2O gives

-1790 kJ/mol =

$$3(-393.509\text{ kJ/mol}) + 3(-285.830\text{ kJ/mol}) - 4(0) - \Delta_f H^\circ_{298,(\text{CH}_3)_2\text{CO}(\ell)}$$

and $\Delta_f H^\circ_{298,(\text{CH}_3)_2\text{CO}(\ell)} = -248$ kJ/mol. The formation reaction is $3\text{C(graphite)} + 3\text{H}_2(g) + \tfrac{1}{2}\text{O}_2(g) \rightarrow (\text{CH}_3)_2\text{CO}(\ell)$ and has $\Delta n_g/\text{mol} = -3.5$.

Neglecting the volumes of condensed phases, we have

$$\Delta_f U^\circ_{298} = \Delta_f H^\circ_{298} - \Delta_f (PV)^\circ_{298} = \Delta_f H^\circ_{298} - RT\,\Delta n_g/\text{mol} =$$

-248 kJ/mol $-$ (0.008314 kJ/mol-K)(298.1 K)(-3.5) = -239 kJ/mol.

5.17 $NH_2CH(CH_3)COOH(s) + \frac{15}{4}O_{2(g)} \rightarrow 3CO_2(g) + \frac{7}{2}H_2O(\ell) + \frac{1}{2}N_2(g)$.

For this reaction, Eq. (5.6) gives

$$\Delta H^\circ_{298} = 3\,\Delta_f H^\circ_{298,\text{CO}_2(g)} + 3.5\,\Delta_f H^\circ_{298,\text{H}_2\text{O}(\ell)} + \tfrac{1}{2}\Delta_f H^\circ_{298,\text{N}_2(g)} -$$

$3.75\,\Delta_f H^\circ_{O_2(g)} - \Delta_f H^\circ_{298,\text{alanine(s)}}$. Use of Appendix data for CO_2 and H_2O gives -1623 kJ/mol = 3(−393.51 kJ/mol) + 3.5(−285.83 kJ/mol) + $0 - 0 - \Delta_f H^\circ_{298,\text{alanine(s)}}$ and $\Delta_f H^\circ_{298,\text{alanine(s)}} = -558$ kJ/mol.

The formation reaction is

$\frac{1}{2}N_2(g) + \frac{7}{2}H_2(g) + 3C(\text{graphite}) + O_2(g) \rightarrow NH_2CH(CH_3)COOH(s)$

and has $\Delta n_g/\text{mol} = -5$. Then $\Delta_f U^\circ_{298} = \Delta_f H^\circ_{298} - RT\,\Delta n_g/\text{mol} =$ −558000 J/mol − (8.314 J/mol-K)(298.1 K)(−5) = −546 kJ/mol.

5.18 Let the reactions be numbered (1), (2), (3), (4). Taking −(2) + (4) −½(3), we get the desired formation reaction Fe(s) + ½O_2(g) → FeO(s).
Hence $\Delta_f H^\circ_{298}$(FeO) = [−37 − 94 + ½(135)]kcal/mol = −63½ kcal/mol.
Taking −(1) + 3(4) −$\frac{3}{2}$(3), we get 2Fe + $\frac{3}{2}O_2 \rightarrow Fe_2O_3$, so
$\Delta_f H^\circ_{298}(Fe_2O_3)$ = [−117 + 3(−94) − $\frac{3}{2}$(−135)] kcal/mol = −196½ kcal/mol.

5.19 Let the reactions be numbered (1), (2), and (3). If we take ¼(1) + ½(3) + ¾(2), we get the desired reaction. Hence ΔH°/(kJ/mol) = 0.25(−1170) + 0.50(−72) + 0.75(−114) = −414.

5.20 For reaction (1), $\Delta H^\circ_{298} = \sum_i \nu_i\,\Delta_f H^\circ_{298,i} = -1560$ kJ/mol = 2(−393½ kJ/mol) + 3(−286 kJ/mol) − $\Delta_f H^\circ_{298,C_2H_6(g)} - \frac{7}{2}(0)$, where the data of reactions (2) and (3) were used to give $\Delta_f H^\circ_{298}$ of CO_2(g) and $H_2O(\ell)$, and where $\Delta_f H^\circ$ of the stable-form element O_2(g) is zero. Solving, we find $\Delta_f H^\circ_{298,C_2H_6(g)} = -85$ kJ/mol.

5.21 **(a)** $V_m = RT/P + b$. Eq. (5.16) gives $H_{m,id}(T,P) - H_{m,re}(T,P) =$
$\int_0^P [T(\partial V_m/\partial T)_P - V_m]\,dP' = \int_0^P [RT/P' - (RT/P' + b)]\,dP' =$
$-b\int_0^P dP' = -bP$

(b) $-bP = -(45\ \text{cm}^3/\text{mol})(1\ \text{atm})\dfrac{8.314\ \text{J}}{82.06\ \text{cm}^3\ \text{atm}} = -4.6$ J/mol

5.22 **(a)** From Eq. (5.17), $H^\circ_{m,298} = 0$.

(b) $(\partial H/\partial T)_P = C_P$ and with the T dependence of C_P neglected, $\Delta H_m \cong C_{P,m}\,\Delta T$; thus $H^\circ_{m,308} = H^\circ_{m,298} + (28.824 \text{ J/mol-K})(10 \text{ K}) = 288.2$ J/mol.

(c) From Sec. 5.4, $H^\circ_{m,H_2O(\ell)} = \Delta_f H^\circ_{H_2O(\ell)} + H^\circ_{m,H_2(g)} + \frac{1}{2}H^\circ_{m,O_2(g)} =$ -285.83 kJ/mol at 25°C.

(d) Use of $\Delta H_m \cong C_{P,m}\,\Delta T$ gives for $H_2O(\ell)$: $H^\circ_{m,308} = H^\circ_{m,298} + C_{P,m}(10 \text{ K})$ $= -285830 \text{ J/mol} + (75.291 \text{ J/mol-K})(10 \text{ K}) = -285.08$ kJ/mol.

5.23 **(a)** T; **(b)** T; **(c)** F; **(d)** F.

5.24 Eq. (5.19) with ΔC°_P assumed constant gives $\Delta H^\circ_{T_2} - \Delta H^\circ_{T_1} = \Delta C^\circ_P(T_2 - T_1)$. (See the Prob. 5.10 solution for ΔH°_{298}.)

(a) $\Delta C^\circ_P/(\text{J/mol-K}) = 2(75.291) + 2(39.87) - 2(34.23) - 3(29.355) = 73.80$

$\Delta H^\circ_{370} = -1124.06 \text{ kJ/mol} + (0.07380 \text{ kJ/mol-K})(72 \text{ K}) =$

-1118.75 kJ/mol

(b) $\Delta C^\circ_P/(\text{J/mol-K}) = 2(33.577) + 2(39.87) - 2(34.23) - 3(29.355) = -9.63$

$\Delta H^\circ_{370} = -1036.04 \text{ kJ/mol} + (-0.00963 \text{ kJ/mol-K})(72 \text{ K})$

$\Delta H^\circ_{370} = -1036.73$ kJ/mol

(c) $\Delta C^\circ_P = 58.55$ J/mol-K, $\Delta H^\circ_{370} = -952.3$ kJ/mol

5.25 The formation reaction is $\frac{1}{2}H_2(g) + \frac{1}{2}Cl_2(g) \rightarrow HCl(g)$. From Example 5.6, $\Delta H^\circ_{T_2} - \Delta H^\circ_{T_1} = \Delta a(T_2 - T_1) + \frac{1}{2}\Delta b(T_2^2 - T_1^2) + \frac{1}{3}\Delta c(T_2^3 - T_1^3) + \frac{1}{4}\Delta d(T_2^4 - T_1^4)$. We have $\Delta a/(\text{J/mol-K}) = 30.67 - \frac{1}{2}(27.14) - \frac{1}{2}(26.93) = 3.63_5$; $\Delta b/(\text{J/mol-K}^2) = -0.007201 - \frac{1}{2}(0.009274) - \frac{1}{2}(0.03384) = -0.028758$; $\Delta c/(\text{J/mol-K}^3) = 1.3925 \times 10^{-5}$; $\Delta d/(\text{J/mol-K}^4) = -15.455_5 \times 10^{-9}$. From the Appendix, $\Delta_f H^\circ_{298} = -92307$ J/mol. Then $\Delta_f H^\circ_{1000}/(\text{J/mol}) =$ $-92307 + 3.63_5(1000 - 298.15) + \frac{1}{2}(-0.028758)(1000^2 - 298.15^2) + \frac{1}{3}(1.3925 \times 10^{-5})(1000^3 - 298.15^3) - \frac{1}{4}(15.455_5 \times 10^{-9})(1000^4 - 298.15^4)$ and $\Delta_f H^\circ_{1000} = -102.17$ kJ/mol.

5.27 $25.665 + 0.013045(T/\text{K}) - 3.8115 \times 10^{-6}(T/\text{K})^2 - 5.5756 \times 10^{-11}(T/\text{K})^3$.

5.29 $23.272 + 0.013246(T/\text{K}) - 3.5603 \times 10^{-6}(T/\text{K})^2 + 2.0154 \times 10^5/(T^2/\text{K}^2)$ gives a sum of squares of residuals of 0.0140 compared with 0.0920 for the cubic polynomial.

5.30 F, since this is not a reversible process.

5.31 Equation (5.33) is $aT_{\text{low}}^3 = C_{P,\text{m}}^\circ(T_{\text{low}})$, so $a = (0.62 \text{ J/mol-K})/(10.0 \text{ K})^3 = 0.00062 \text{ J/mol-K}^4$. At 10 K, Eq. (5.35) gives $S_\text{m}^\circ(T_{\text{low}}) - S_{\text{m},0}^\circ = S_\text{m}^\circ(T_{\text{low}}) = \int_0^{T_{\text{low}}} (C_{P,\text{m}}^\circ/T)\, dT = C_{P,\text{m}}^\circ(T_{\text{low}})/3 = (0.62 \text{ J/mol-K})/3 = 0.21 \text{ J/mol-K}$. At 6.0 K, $C_{P,\text{m}}^\circ = aT^3 = (0.00062 \text{ J/mol-K}^4)(6.0 \text{ K})^3 = 0.13_4 \text{ J/mol-K}$ and $S_\text{m}^\circ = C_{P,\text{m}}^\circ/3 = (0.13_4 \text{ J/mol-K})/3 = 0.045 \text{ J/mol-K}$.

5.32 **(a)** We use (5.29) and $\int_U^W f(x)\, dx = \int_U^V f(x)\, dx + \int_V^W f(x)\, dx$. Equations (5.31), (5.33), and (5.35) give $S_{\text{m},10\text{K}}^\circ = a(10 \text{ K})^3/3 = C_{P,\text{m},10\text{K}}^\circ/3 = [c(10 \text{ K})^3 + d(10 \text{ K})^4]/3$. Then $S_{\text{m},300\text{K}}^\circ = 333c \text{ K}^3 + 3333d \text{ K}^4 + \int_{10\text{ K}}^{20\text{ K}} (cT^2 + dT^3)\, dT + \int_{20\text{ K}}^{200\text{ K}} (e/T + f + gT + hT^2)\, dT + (1450 \text{ J/mol})/(200 \text{ K}) + \int_{200\text{ K}}^{300\text{ K}} (i/T + j + kT + lT^2)\, dT = (2666 \text{ K}^3)c + (40833 \text{ K}^4)d + 2.3026e + (180 \text{ K})f + (19800 \text{ K}^2)g + (2664000 \text{ K}^3)h + 7.25 \text{ J/mol-K} + 0.4055i + (100 \text{ K})j + (25000 \text{ K}^2)k + (6333333 \text{ K}^3)l$.

(b) From $(\partial H_\text{m}/\partial T)_P = C_{P,\text{m}}$, we have $H_{\text{m},T'}^\circ - H_{\text{m},0}^\circ = \int_0^{T'} C_{P,\text{m}}^\circ\, dT$ for the solid at T'. Use of the Debye T^3 law (5.31) gives
$H_{\text{m},10\text{ K}}^\circ - H_{\text{m},0}^\circ = \int_0^{10\text{ K}} C_{P,\text{m}}^\circ\, dT = \int_0^{10\text{ K}} aT^3\, dT = \frac{1}{4}(10 \text{ K})^4 a$.
As noted in part (a), $a = C_{P,\text{m},10\text{ K}}^\circ/(10 \text{ K})^3$, so
$H_{\text{m},10\text{ K}}^\circ - H_{\text{m},0}^\circ = \frac{1}{4}(10 \text{ K})C_{P,\text{m},10\text{ K}}^\circ = (2.5 \text{ K})[c(10 \text{ K})^3 + d(10 \text{ K})^4]$. Also, $H_{\text{m},200\text{ K}}^\circ - H_{\text{m},10\text{ K}}^\circ = \int_{10\text{ K}}^{20\text{ K}} C_{P,\text{m}}^\circ(\text{s})\, dT + \int_{20\text{ K}}^{200\text{ K}} C_{P,\text{m}}^\circ(\text{s})\, dT$, $\Delta H_{\text{m,fus}}^\circ = 1450 \text{ J/mol}$, and $H_{\text{m},300}^\circ - H_{\text{m},200}^\circ = \int_{200\text{K}}^{300\text{K}} C_{P,\text{m}}^\circ(\ell)\, dT$. Substitution of the expressions for $C_{P,\text{m}}^\circ$ and integration gives $H_{\text{m},300}^\circ - H_{\text{m},0}^\circ = (40000 \text{ K}^4)c + (645000 \text{ K}^5)d + (180 \text{ K})e + (19800 \text{ K}^2)f + (2664000 \text{ K}^3)g +$

$(3.9996 \times 10^8\ K^4)h + 1450$ J/mol + $(100\ K)i + (25000\ K^2)j +$
$(6333333\ K^3)k + (1.625 \times 10^9\ K^4)l$.

5.33 **(a)** For the solid's data, a spreadsheet fit gives the quartic polynomial $C_{P,m}$/(cal/mol-K) = $aT^4 + bT^3 + cT^2 + dT + e$, where $a = -1.64630 \times 10^{-8}\ K^{-4}$, $b = 1.17413 \times 10^{-5}\ K^{-3}$, $c = -0.00288851\ K^{-1}$, $d = 0.343422$, $e = -4.02817$. The fit is good except for the first couple of points. (To achieve greater accuracy, one could fit the first few points with one polynomial and use a different polynomial for the remaining points.) For the liquid, the quadratic polynomial $C_{P,m}$/(cal/mol-K) = $fT^2 + gT + h$ with $f = 3.2724 \times 10^{-6}\ K^{-2}$, $g = -0.0066796\ K^{-1}$, $h = 22.173$ fits well.
For the gas, $C_{P,m}$/(cal/mol-K) = $kT^2 + mT + n$, with $k = 4.0182 \times 10^{-5}\ K^{-2}$, $m = -0.018273\ K^{-1}$, $n = 11.676$.

(b) Equation (5.35) gives
$\int_0^{15\text{ K}} (C_{P,m}/T)\,dT = \frac{1}{3}(0.83 \text{ cal/mol-K}) = 0.28$ cal/mol-K.

Then substitution of the solid's polynomial gives $\int_{15\text{ K}}^{197.64\text{ K}} (C_{P,m}/T)\,dT =$
$\int_{15\text{ K}}^{197.64\text{ K}} (aT^3 + bT^2 + cT + d + e/T)\,dT$ (cal/mol-K) =
$(\frac{1}{4}aT^4 + \frac{1}{3}bT^3 + \frac{1}{2}cT^2 + dT + e\ln T)\,\Big|_{15\text{ K}}^{197.64\text{ K}}$ (cal/mol-K) =
20.17 cal/mol-K. For the liquid,
$\int_{197.64\text{ K}}^{263.1\text{ K}} (C_{P,m}/T)\,dT = \int_{197.64\text{ K}}^{263.1\text{ K}} (fT + g + h/T)\,dT$ (cal/mol-K) =
$(\frac{1}{2}fT^2 + gT + h\ln T)\,\Big|_{197.64\text{ K}}^{263.1\text{ K}}$ (cal/mol-K) = 5.96 cal/mol-K. For the gas
$\int_{263.1\text{ K}}^{298.15\text{ K}} (C_{P,m}/T)\,dT = \int_{263.1\text{ K}}^{298.15\text{ K}} (kT + m + n/T)\,dT$ (cal/mol-K) =
$(\frac{1}{2}kT^2 + mT + n\ln T)\,\Big|_{263.1\text{ K}}^{298.15\text{ K}}$ (cal/mol-K) = 1.21 cal/mol-K. Addition of these contributions gives (0.28 + 20.17 + 5.96 + 1.21) cal/mol-K = 27.62 cal/mol-K. The contributions of the entropies of fusion and vaporization at the normal melting and boiling points are calculated in Example 5.7 as 8.95 cal/mol-K and 22.65 cal/mol-K. We also include the contributions (found in Example 5.7) of 0.07 and 0.03 cal/mol-K for going from the real to the ideal gas and for going from 1 atm to 1 bar. Therefore, $S^\circ_{m,298}$ = (27.62 + 8.95 + 22.65 + 0.10) cal/mol-K = 59.32 cal/mol-K.

5.34 **(a)** $S^\circ_{m,298}$ would be increased by a for graphite, by b for $H_2(g)$, and by c for $O_2(g)$. The formation reactions C(graphite) + $2H_2(g) \rightarrow CH_4(g)$, $H_2(g) + \frac{1}{2}O_2(g) \rightarrow H_2O(\ell)$, and C(graphite) + $O_2(g) \rightarrow CO_2(g)$ and the fact that ΔS° of a reaction is independent of any convention show that $S^\circ_{m,298}$ would be increased by $a + 2b$ for $CH_4(g)$, by $b + \frac{1}{2}c$ for $H_2O(\ell)$, and by $a + c$ for $CO_2(g)$.

(b) No change, as verified by $a + c + 2(b + \frac{1}{2}c) - (a + 2b) - 2c = 0$.

5.35 **(a)** The Appendix gives $S^\circ_{m,298} = 69.91$ J/mol-K.

(b) Equation (4.61) with the T dependence of $C^\circ_{P,m}$ neglected gives $S^\circ_{m,348} \cong S^\circ_{m,298} + C^\circ_{P,m} \ln(T_2/T_1)$, so $S^\circ_{m,348} \cong 69.91$ J/mol-K + (75.291 J/mol-K) ln(348.1/298.1) = 81.58 J/mol-K.

(c) Equation (4.62) with the P dependence of α and V_m neglected gives $S_m(298\text{ K}, 100\text{ bar}) \cong S^\circ_{m,298} - \alpha V_m \Delta P$, so $S_m(298\text{ K}, 100\text{ bar}) \cong$ 69.91 J/mol-K – (0.000304 K^{-1})(18.1 cm^3/mol)(99 bar) × (8.134 J/83.14 cm^3 bar) = 69.86 J/mol-K.

(d) Use the path in Fig. 4.3. The calculations of parts (b) and (c) of this problem give $\Delta S = \Delta S_a + \Delta S_b = 11.67$ J/mol-K – 0.05 J/mol-K = 11.62 J/mol-K. So S_m(348 K, 100 bar) = (69.91 + 11.62) J/mol-K = 81.53 J/mol-K.

5.36 **(a)** [2(69.91) + 2(248.22) – 2(205.79) – 3(205.138)] J/mol-K = –390.73 J/mol-K.

(b) [2(188.825) + as in (a)] J/mol-K = –152.90 J/mol-K.

(c) [109.6 + 4(191.61) – 2(238.97) – 2(210.761)] J/mol-K = –23.4 J/mol K.

5.37 With the temperature dependence of ΔC°_P neglected, Eq. (5.37) gives $\Delta S^\circ_{T_2} = \Delta S^\circ_{298} + \Delta C^\circ_P \ln(T_2/298\text{ K})$. The ΔC°_P's are calculated in the Prob. 5.24 solution and the ΔS°_{298}'s are calculated in Prob. 5.36.

(a) $\Delta S^\circ_{370} = -390.73$ J/mol-K + (73.80 J/mol-K) ln(370/298) = –374.76 J/mol-K.

(b) $\Delta S^\circ_{370} = -152.90$ J/mol-K + (–9.63 J/mol-K) ln(370/298) = –154.98 J/mol-K.

(c) $\Delta S^\circ_{370} = -23.4$ J/mol-K + (58.55 J/mol-K) ln(370/298) = –10.7 J/mol-K.

5.38 Differentiation of (5.36) gives $d\,\Delta S^\circ/dT = \sum_i \nu_i\, dS^\circ_{m,i}/dT = \sum_i \nu_i C^\circ_{P,m,i}/T = \Delta C^\circ_P/T$. Integration gives $\Delta S^\circ(T_2) - \Delta S^\circ(T_1) = \int_1^2 (\Delta C^\circ_P/T)\, dT$.

5.39 **(a)** From Eq. (5.37), $\Delta S^\circ_{1000} = \Delta S^\circ_{298} + \int_{298\,\text{K}}^{1000\,\text{K}} (\Delta C^\circ_P/T)\, dT =$ –173.01 J/mol-K + $\int_{298\,\text{K}}^{1000\,\text{K}} (\Delta a/T + \Delta b + T\,\Delta c + T^2\,\Delta d)\, dT =$ –173.01 J/mol-K + Δa ln(1000/298.1) + Δb(1000 – 298.1)K + (1/2)Δc[(1000^2 – 298.1^2)K^2 + (1/3)Δd[(1000^3 – 298.1^3)K^3 = –173.01 J/mol-K + (–39.87 J/mol-K)(1.21003) + (0.11744 J/mol-K)(701.85) + 0.5(–9.8296 × 10^{-5})(911136)J/mol-K + (1/3)(2.8049 × 10^{-8})(9.7351 × 10^8)J/mol-K = –174.51 J/mol-K.

(b) $\Delta S^\circ_{1000} \approx \Delta S^\circ_{298} + \Delta C^\circ_{P,298}$ ln(1000/298.15) = –173.01 J/mol-K + (–13.367 J/mol-K)(1.21016) = –189.19 J/mol-K. The approximation that ΔC°_P is independent of T is a poor one.

5.40 $V_m = RT/P + RTf(T)$ and $(\partial V_m/\partial T)_P = R/P + Rf + RTf'$. Substitution in (5.30) gives $S_{m,id}(T, P) - S_{m,re}(T, P) = \int_0^P [Rf(T) + RTf'(T)]\, dP = RP[f(T) + Tf'(T)]$.

5.41 $\Delta_f G^\circ_{298} = \Delta_f H^\circ_{298} - T\,\Delta_f S^\circ_{298}$. The formation reaction is C(graphite) + ½O_2(g) + N_2(g) + 2H_2(g) → $CO(NH_2)_2$(*c*). Appendix data gives for this reaction, $\Delta S^\circ = \sum_i \nu_i S^\circ_{m,i}$ = 104.60 J/mol-K – 5.740 J/mol-K – ½(205.138 J/mol-K) – 191.61 J/mol-K – 2(130.684 J/mol-K) = –456.69 J/mol-K = $\Delta_f S^\circ_{298,\text{urea}}$. Then $\Delta_f G^\circ_{298}$ = –333.51 kJ/mol – (298.15 K)(–0.45669 kJ/mol-K) = –197.35 kJ/mol.

5.42 **(a)** $\Delta G^\circ_a = \Delta H^\circ_a - T\,\Delta S^\circ_a =$ –1124.06 kJ/mol – (298.15 K)(–0.39073 kJ/mol-K) = –1007.56 kJ/mol.

$\Delta G_b^\circ = -1036.04$ kJ/mol – (298.15 K)(–0.15290 kJ/mol-K) = –990.45 kJ/mol. $\Delta G_c^\circ = -956.5$ kJ/mol – (298.15 K)(–0.0234 kJ/mol-K) = –949.5 kJ/mol.

(b) ΔG_a°/(kJ/mol) = 2(–237.129) + 2(–300.194) – 2(–33.56) – 3(0) = –1007.53. ΔG_b°/(kJ/mol) = 2(–228.572) + 2(–300.194) – 2(–33.56) –3(0) = –990.41. ΔG_c°/(kJ/mol) = –120.35 + 4(0) – 2(328.1) – 2(86.55) = –949.6.

5.43 **(a)** $\Delta G_{370}^\circ = \Delta H_{370}^\circ - T\,\Delta S_{370}^\circ =$ –1118.75 kJ/mol – (370 K)(–0.037476 kJ/mol-K) = –980.09 kJ/mol.

(b) $\Delta G_{370}^\circ = -1036.73$ kJ/mol – (370 K)(–0.15498 kJ/mol-K) = –979.39 kJ/mol.

(c) $\Delta G_{370}^\circ = -948.3$ kJ/mol.

5.44 **(a)** $G_{\text{m},298}^\circ = {}^\circ H_{\text{m},298} - TS_{\text{m},298}^\circ = 0$ – (298.15 K)(205.138 J/mol-K) = –61.16 kJ/mol.

(b) Using the result of Prob. 5.22c, we get $G_{\text{m},298}^\circ = H_{\text{m},298}^\circ - TS_{\text{m},298}^\circ =$ –285830 J/mol – (298.15 K)(69.91 J/mol-K) = –306.67 kJ/mol.

5.45 Landolt-Börnstein data give 70.7 kJ/mol.

5.46 ΔH_{298}° = 2(–46.11 kJ/mol) – 0 – 3(0) = – 92.22 kJ/mol.
$\Delta H_{2000}^\circ = \Delta H_{298}^\circ + \sum_i \nu_i (H_{\text{m},2000}^\circ - H_{\text{m},298}^\circ)$ = –92.22 kJ/mol + 2(98.18 kJ/mol) – 56.14 kJ/mol – 3(52.93 kJ/mol) = –110.79 kJ/mol.

5.47 The reaction is $\frac{1}{2}N_2(g) + \frac{3}{2}H_2(g) \rightarrow NH_3(g)$. The Appendix gives $\Delta_f H_{298}^\circ =$ –46.11 kJ/mol. Use of (5.43) gives $\Delta_f G_{2000}^\circ = -46110$ J/mol + (2000 K)[–242.08 – $\frac{1}{2}$(–223.74) – $\frac{3}{2}$(–161.94)](J/mol-K) = 179.29 kJ/mol.

5.48 $\Delta H^\circ_{298} + T\sum_i \nu_i[G^\circ_{\text{m},T} - H^\circ_{\text{m},298})/T]_i = \Delta H^\circ_{298} + T\sum_i \nu_i G^\circ_{\text{m},i,T}/T -$
$T\sum_i \nu_i H^\circ_{\text{m},i,298}/T = \Delta H^\circ_{298} + \Delta G^\circ_T - \Delta H^\circ_{298} = \Delta G^\circ_T$.

5.49 **(a)** $\Delta G_T^{\text{bar}} - \Delta G_T^{\text{atm}} = \Delta H_T^{\text{bar}} - T\Delta S_T^{\text{bar}} - (\Delta H_T^{\text{atm}} - T\Delta S_T^{\text{atm}}) =$
$-T(\Delta S_T^{\text{bar}} - \Delta S_T^{\text{atm}})$, since the difference between 1-bar and 1-atm enthalpies of a solid or liquid is negligible and H° of a gas is independent of P. Since the effect of a slight change in P on S of a solid or liquid is negligible, we have $\Delta S_T^{\text{bar}} - \Delta S_T^{\text{atm}} = \Delta S_{g,T}^{\text{bar}} - \Delta S_{g,T}^{\text{atm}} =$
$\sum_{\text{gases}} \nu_i(S_{\text{m},T,i}^{\text{bar}} - S_{\text{m},T,i}^{\text{atm}}) = (0.1094 \text{ J/mol-K}) \sum_{\text{gases}} \nu_i =$
$(0.1094 \text{ J/mol-K})\Delta n_g/\text{mol}$ and
$\Delta G_T^{\text{bar}} - \Delta G_T^{\text{atm}} = -T(0.1094 \text{ J/mol-K})\Delta n_g/\text{mol}$.

(b) $H_2(g) + \frac{1}{2}O_2(g) \rightarrow H_2O(\ell)$. $\Delta_f G_{298}^{\text{bar}} - \Delta_f G_{298}^{\text{atm}} =$
$-(298 \text{ K})(0.1094 \text{ J/mol-K})(-1.5) = 48.9 \text{ J/mol}$.

5.50

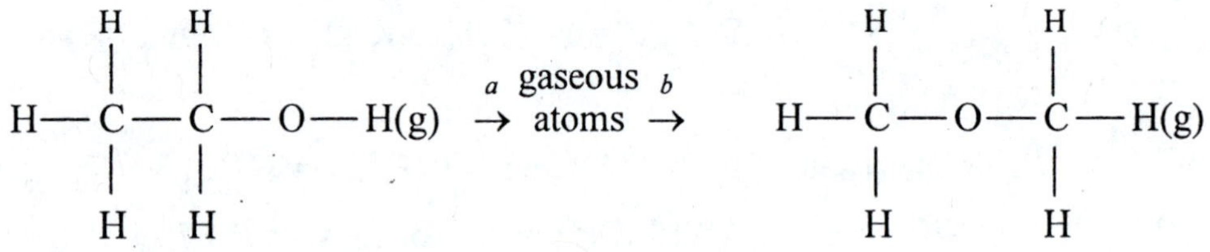

(a) $\Delta H_a/(\text{kJ/mol}) = 5(415) + 344 + 350 + 463 = 3232$
$\Delta H_b/(\text{kJ/mol}) = -[6(415) + 2(350)] = -3190$
$\Delta H = \Delta H_a + \Delta H_b = 42 \text{ kJ/mol}$

(b) $\Delta H/(\text{kcal/mol}) = [6(-3.83) + 2(-12.0)] - [5(-3.83) - 12.0 - 27.0 + 2.73]$
$= 8.44$, so $\Delta H = 8.44 \text{ kcal/mol} = 35.3 \text{ kJ/mol}$.

(c) $\Delta H/(\text{kJ/mol}) = 2(-41.8) - 99.6 - (-41.8 - 33.9 - 158.6) = 51.1 \text{ kJ/mol}$.

5.51 **(a)** $3C(\text{graphite}) + 4H_2(g) + \frac{1}{2}O_2(g) \xrightarrow{a} 3C(g) + 8H(g) + O(g) \xrightarrow{b}$
$CH_3OCH_2CH_3(g)$. $\Delta_f H^\circ = \Delta H_a + \Delta H_b$. Appendix data give
$\Delta H_a/(\text{kJ/mol}) = 3(716.682) + 8(217.965) + 249.170 = 4142.94$. Bond energies give

ΔH_b/(kJ/mol) = –[8(415) + 344 + 2(350)] = –4364. So $\Delta_f H°$/(kJ/mol) = 4143 – 4364 = –221.

(b) $\Delta_f H°$/(kJ/mol) = 4.184[2.73 + 2(–12.0) + 8(–3.83)] = –217.2

(c) $\Delta_f H°$/(kJ/mol) = –41.8 – 99.6 – 33.9 – 41.8 = –217.1

5.52 In the Benson–Buss bond-contribution method, the carbonyl group is treated as a unit and no explicit contribution is made for the C═O bond. The effect of this bond is absorbed into the contributions for bonds to the carbonyl carbon. The bond contribution to $S°_{m,298}$ of the F—CO bond is 31.6 cal/mol-K, so the total bond contribution is 2(31.6 cal/mol-K) = 63.2 cal/mol-K for COF_2(g). In addition, the quantity $R \ln \sigma$ must be subtracted to allow for molecular symmetry. For COF_2, the symmetry number σ is 2, since there are two indistinguishable orientations of the molecule (obtained by 180° rotation about the CO bond). The symmetry correction is $-R \ln 2 = -1.38$ cal/mol-K and the predicted $S°_{m,298}$ is 61.8 cal/mol-K.

5.53 $\Delta H°_{298}$ of vaporization refers to a change from liquid at 1 bar and 25°C to vapor at 1 bar and 25°C. A path to accomplish this is the following 25°C isothermal path:

$$\text{liq(1 bar)} \xrightarrow{1} \text{liq(23.8 torr)} \xrightarrow{2} \text{gas(23.8 torr)} \xrightarrow{3} \text{gas(1 bar)}$$

$\Delta H° = \Delta H_{m,1} + \Delta H_{m,2} + \Delta H_{m,3}$. As noted in Sec. 4.5, a modest change in pressure from 1 bar to 24 torr will have only a *very* slight effect on H and S of a liquid, so we can take $\Delta H_{m,1} = 0$. Also, since the vapor is assumed to behave ideally, its H depends only on T, and $\Delta H_{m,3} = 0$. Thus $\Delta H° = \Delta H_{m,2} = 10.5$ kcal/mol. For comparison, Appendix data give $\Delta H°$ = (–241.818 + 285.830) kJ/mol = 10.519 kcal/mol. Next, $\Delta S° = \Delta S_{m,1} + \Delta S_{m,2} + \Delta S_{m,3}$. To a good approximation, $\Delta S_{m,1} = 0$. Equations (3.25) and (3.30) and Boyle's law give $\Delta S_{m,2} + \Delta S_{m,3} = \Delta H_{m,2}/T + R \ln(P_1/P_2)$ = (10500 cal/mol)/(298.1 K) + (1.987 cal/mol-K) ln(23.8/750) = 28.3_6 cal mol^{-1} K^{-1} = $\Delta S°$. The Appendix data give $\Delta S°$ = (188.825 – 69.91) J/mol-K = 28.42 cal/mol-K. Finally, $\Delta G° = \Delta H° - T\Delta S°$ = 10.5 kcal/mol – (298.1 K)(0.0283_6 kcal/mol-K) = 2.0_4 kcal/mol. The Appendix gives $\Delta G°$ = 2.045 kcal/mol.

5.54 We use the 25°C path (where M is methanol)

$$\mathrm{M}(\ell, 1\text{ bar}) \xrightarrow{1} \mathrm{M}(\ell, 125\text{ torr}) \xrightarrow{2} \mathrm{M}(\mathrm{g}, 125\text{ torr}) \xrightarrow{3} \mathrm{M}(\mathrm{g}, 1\text{ bar}) \xrightarrow{4} \mathrm{M}(\text{ideal gas},$$

1 bar). For this path, $\Delta H = H^\circ_{\mathrm{m},M(\mathrm{g})} - H^\circ_{\mathrm{m},M(\ell)} = H^\circ_{\mathrm{m},M(g)} - H^\circ_{el} - (H^\circ_{\mathrm{m},M(\ell)} - H^\circ_{el})$ $= \Delta_f H^\circ_{M(g)} - \Delta_f H^\circ_{M(\ell)}$, where H°_{el} is the standard-state enthalpy of the elements needed to form methanol. Since a moderate change in P has little effect on thermodynamic properties of a liquid, we have $\Delta H_1 \cong 0$ (and $\Delta S_1 \cong 0$). $\Delta H_2 = 37.9$ kJ/mol. $\Delta H_3 \cong 0$ and $\Delta H_4 \cong 0$, where we neglected nonideality of the gas. Then $\Delta H = \Delta H_1 + \Delta H_2 + \Delta H_3 + \Delta H_4 = 37.9$ kJ/mol and $\Delta_f H^\circ_{M(g)} =$ -238.7 kJ/mol + 37.9 kJ/mol = -200.8 kJ/mol. We have $\Delta S_1 \cong 0$, $\Delta S_2 = \Delta H_2/T$ = (37900 J/mol)/(298 K) = $127._1$ J/mol-K, $\Delta S_3 \cong R \ln(V_2/V_1) = R \ln(P_1/P_2) =$ (8.314 J/mol-K) ln (125/750) = -14.90 J/mol-K [where (3.30) was used], and $\Delta S_4 \cong 0$, where the gas is approximated as ideal. Then $\Delta S = (127._1 - 14.9)$ J/mol-K = $112._2$ J/mol-K. So $S^\circ_{\mathrm{m},M(g)} = 126.8$ J/mol-K + $112._2$ J/mol-K = 239 J/mol-K.

5.55 **(a)** $\Delta_f H^\circ_{298} = (2n+2)b_{\mathrm{CH}} + (n-1)b_{\mathrm{CC}}$

(b) Breaking the formation reaction into two steps, we have $n\mathrm{C(graphite)} + (n+1)\mathrm{H_2(g)} \rightarrow n\mathrm{C(g)} + (2n+2)\mathrm{H(g)} \rightarrow \mathrm{C}_n\mathrm{H}_{2n+2}(\mathrm{g})$. So $\Delta_f H^\circ_{298} = \Delta H_1 + \Delta H_2 = n\,\Delta_f H^\circ_{298}[\mathrm{C(g)}] + (2n+2)\Delta_f H^\circ_{298}[\mathrm{H(g)}] -$ $(n-1)D_{\mathrm{CC}} - (2n+2)D_{\mathrm{CH}}$.

(c) Equating the expressions in parts a and b, we have for $n = 1$: $4b_{\mathrm{CH}} = \Delta_f H^\circ_{298}[\mathrm{C}(g)] + 4\Delta_f H^\circ_{298}[\mathrm{H}(g)] - 4D(\mathrm{CH})$ and $b_{\mathrm{CH}} = \frac{1}{4}\Delta_f H^\circ_{298}[\mathrm{C}(g)] + \Delta_f H^\circ_{298}[\mathrm{H}(g)] - D(\mathrm{CH})$. For $n = 2$, $6b_{\mathrm{CH}} + b_{\mathrm{CC}} = 2\,\Delta_f H^\circ_{298}[\mathrm{C(g)}] + 6\Delta_f H^\circ_{298}[\mathrm{C(g)}] - D_{\mathrm{CC}} - 6D_{\mathrm{CH}}$. Substitution of the result for b_{CH} gives after cancellation: $b_{\mathrm{CC}} = 0.5\Delta_f H^\circ_{298}[\mathrm{C(g)}] - D_{\mathrm{CC}}$

5.56 **(a)** Pa; **(b)** J; **(c)** J/mol-K; **(d)** J; **(e)** $\mathrm{m^3/mol}$; **(f)** K.

5.57 If ΔH° is independent of T, then $d\,\Delta H^\circ/dT = 0$ and (5.18) gives $\Delta C^\circ_P = 0$. Then (5.37) gives $\Delta S^\circ_{T_2} = \Delta S^\circ_{T_2}$. Q.E.D.

5.58 (a) Equation (4.45) gives $(\partial V/\partial T)_P = -(\partial S/\partial P)_T$. The 3rd law gives $\lim_{T\to 0} \Delta S = 0$ for an isothermal pressure change in an equilibrium system. Hence $(\partial S/\partial P)_T \to 0$ as $T \to 0$, and $\alpha = (1/V)(\partial V/\partial T)_P \to 0$ as $T \to 0$.

(b) $\alpha = (1/V)(\partial V/\partial T)_P = 1/T$ [Eq. (1.46)], which goes to ∞ as $T \to 0$.

5.59 (a) Nonzero.

(b) Nonzero. $\Delta_f H^\circ_{298}$ refers to formation of the substance from elements in their stable forms at 298 K. The 298 K stable form of chlorine is $Cl_2(g)$. not Cl(g), and ΔH°_{298} for ½$Cl_2(g) \to Cl(g)$ is not zero.

(c) Zero, since $Cl_2(g)$ is the stable form of an element.

(d) Nonzero. (Entropies are zero at 0 K.)

(e) Zero.

(f) Zero, since 350 K formation of $N_2(g)$ from its stable-form element(s) is a process in which nothing happens.

(g) Zero.

(h) Zero, since heat-capacities go to zero as T goes to zero, as shown by the Debye equation (5.31).

(i) Nonzero.

5.60 The 25°C reaction (step c) is $CH_4(g) + 2O_2(g) \to CO_2(g) + 2H_2O(\ell)$ and has $\Delta H = [-393.509 + 2(-285.83) - (-74.81) - 2(0)]$ kJ $= -890.4$ kJ for burning 1 mole of CH_4. We have $\Delta H_a = \Delta H_c + \Delta H_b = 0 = -890$ kJ $+ \Delta H_b$ and $\Delta H_b = 890$ kJ. From the Appendix, the heat needed to vaporize 2 mol of H_2O is $2(-241.8 + 285.8)$ kJ $= 88$ kJ. In step b, the products are being heated from 25°C to the flame temperature T. The heat required to do this is 88000 J + (1 mol)(54.3 J/mol-K)(T – 298 K) + (2 mol)(41.2 J/mol-K)(T – 298 K) + 2(3.76 mol)(32.7 J/mol-K)(T – 298 K) = 890000 J. So $T - 298$ K $= 2096$ K; $T \cong 2400$ K.

5.61 (a) n-$C_4H_{10}(g)$, since larger molecules have larger entropies.

(b) $H_2O(g)$, since gases have higher entropies than the corresponding liquids.

(c) $H_2(g)$, which has larger molecules than H(g).

(d) $C_{10}H_8(g)$.

5.62 (a) $\Delta H^\circ = q_P$ is positive, since heat is needed to vaporize the liquid. For liquid → gas, $\Delta S^\circ > 0$.

(b) ΔH° is positive since energy is needed to break the bond. ΔS° is positive, since the number of moles of gas is increasing.

(c) q is negative when the vapor condenses, so ΔH° is negative. For gas → solid, $\Delta S^\circ < 0$.

(d) $(COOH)_2(s) + \frac{1}{2}O_2(g) \rightarrow 2CO_2(g) + H_2O(\ell)$. We can expect the reaction to be exothermic, as is generally true for combustion reactions. Hence, $\Delta H^\circ < 0$. Also, $\Delta S^\circ > 0$, since the number of moles of gases is increasing.

(e) $\Delta S^\circ < 0$, since the number of moles of gases is decreasing. All species are gases, and the Table 20.1 bond energies yield $\Delta H^\circ \approx$ (2711 – 2834) kJ/mol = –123 kJ/mol, so $\Delta H^\circ < 0$.

5.63 10^3 MW = 10^9 J/s. Let y = the absolute value of the heat of combustion of the fuel burned per second. Then $0.39 = (10^9 \text{ J/s})/y$ and $y = 2.6 \times 10^9$ J/s. 10^4 Btu/lb = 1.06×10^7 J/lb. Then $(2.6 \times 10^9 \text{ J/s})/(1.06 \times 10^7 \text{ J/lb}) = 245$ lb/s of coal burned, which is 15000 pounds per minute, 21 million pounds per day, and 7.7×10^9 pounds per year.

5.64 (a) True, since $\Delta H = q_P = 0$.

(b) False. There is a thermodynamic standard state at each temperature.

(c) False. $G = H - TS$. Although the conventional H is taken as zero at 25°C, S is zero at 0 K and is not zero at 25°C.

5.65 To find $S^\circ_{\text{m},298}$ from (5.29), we need to measure (a) $C^\circ_{P,\text{m}}$ of the solid from a very low T up to the melting point, (b) $\Delta_{\text{fus}}H^\circ_{\text{m}}$, (c) the melting point, and (d) $C^\circ_{P,\text{m}}$ for the liquid from the melting point to 298 K. If there are any solid–solid phase transitions, we need $\Delta H^\circ_{\text{m}}$ and the temperature of each such transition. To find $\Delta_f H^\circ_{298}$ we need ΔH°_{298} of combustion of the hydrocarbon. $\Delta_f G^\circ$ is found from $\Delta_f H^\circ - T\,\Delta_f S^\circ$.

5.66 The formation reaction is $2C(\text{graphite}) + 3H_2(g) + 0.5O_2(g) \rightarrow C_2H_5OH(l)$, so $\Delta_f S^\circ_{298}/(\text{J/mol-K}) = 160.7 - 2(5.740) - 3(130.684) - 0.5(205.138) = -345.4$.

$\Delta_f G^\circ_{298} = \Delta_f H^\circ_{298} - T\,\Delta_f S^\circ_{298} = -277690$ J/mol – (298.15 K)(–345.4 J/mol-K) = –174.4 kJ/mol.

Chapter 6

6.1 For a pure substance, $G = nG_m = n\mu$ and $\Delta G = n\,\Delta\mu$. For an isothermal process in an ideal gas, $\Delta\mu = \mu°(T) + RT\ln(P_2/P°) - [\mu°(T) + RT\ln(P_1/P°)] = RT\ln(P_2/P_1)$ and $\Delta G = nRT\ln(P_2/P_1) =$ (3.00 mol)(8.314 J/mol-K)(400 K) ln [(1.00 bar)/(2.00 bar)] = –6.92 kJ.

6.2 **(a)** T; **(b)** T; **(c)** T.

6.3 **(a)** $P_i = x_iP$. P_{SO_3} = (0.440)(1767 torr) = 777 torr; P_{SO_2} = (0.310)(1767 torr) = 548 torr; P_{O_2} = 442 torr. $K_P^\circ = \prod_i (P_i/P°)^{\nu_i} =$ $(P_{SO_3}/P°)^2/(P_{SO_2}/P°)^2(P_{O_2}/P°) = (777/750)^2/(548/750)^2(442/750) = 3.41$. $\Delta G° = -RT\ln K_P^\circ$ = –(8.314 J/mol-K)(1000 K) ln 3.41 = –10.2 kJ/mol.

(b) $K_P^\circ = (P_{SO_3}^2/P_{SO_2}^2P_{O_2})P° = K_PP°$ and $K_P = K_P^\circ/P° = 3.41/(1\text{ bar}) =$ 3.41 bar^{-1}.

(c) $K_c^\circ = K_P^\circ(RTc°/P°)^{-\Delta n/\text{mol}} =$ 3.41[(82.06 cm^3-atm/mol-K)(1000 K)(1 mol/1000 cm^3)/(0.987 atm)$^{-(2-3)}$ = 284.

6.4 A + B $\rightleftarrows$ 2C + 3D. If 10.0 mmol of C is formed, then 5.0 mmol of A and 5.0 mmol of B must have reacted and (3/2)(10.0 mmol) = 15.0 mmol of D is formed. At equilibrium, n_A = 10.0 mmol, n_B = 13.0 mmol, n_C = 10.0 mmol, n_D = 15.0 mmol; n_{tot} = 48.0 mmol. $x_A = n_A/n_{tot}$ = 0.208, x_B = 0.271, x_C = 0.208, x_D = 0.312$_5$. $P_A = x_AP$ = (0.208)(1085 torr) = 226 torr, P_B = 294 torr, P_C = 226 torr, P_D = 339 torr. $K_P^\circ = (P_C/P°)^2(P_D/P°)^3/(P_A/P°)(P_B/P°) =$ $(226/750)^2(339/750)^3/(226/750)(294/750) = 0.0710$. $\Delta G° = -RT\ln K_P^\circ =$ –(8.314 J/mol-K)(600 K) ln 0.0710 = 13.2 kJ/mol.

6.5 $n_{tot} = PV/RT$ = [(231.2/760) atm](1055 cm^3)/(82.06 cm^3-atm/mol-K)(323.7 K) = 0.01208 mol. Let x mol of Br_2 react to reach equilibrium. At equilibrium,

n_{NO}/mol = 0.01031 – 2x, n_{Br_2} = 0.00440 – x, n_{NOBr} = 2x, n_{tot} = 0.01471 – x = 0.01208; x = 0.00263. So n_{NO} = 0.00505, n_{Br_2} = 0.00177, n_{NOBr} = 0.00526; $x_{NO} = n_{NO}/n_{tot}$ = 0.418, x_{Br_2} = 0.146$_5$, x_{NOBr} = 0.435.
$P_{NO} = x_{NO}P$ = (0.418)(231.2 torr) = 96.6 torr, P_{Br_2} = 33.9 torr, P_{NOBr} = 100.6 torr, $P°$ = 750 torr. $K_P^\circ = (P_{NOBr}/P°)^2/(P_{NO}/P°)^2(P_{Br_2}/P°) = 24.0$.
$\Delta G° = -RT \ln K_P^\circ$ = –(8.314 J/mol-K)(323.7 K) ln 24.0 = –8.55 kJ/mol.

6.6 $(P_N/P°)^2/(P_{N_2}/P°) = (0.12)^2/(720)(750) = 2.7 \times 10^{-8}$, which is less than K_P°. The system is not at equilibrium; $P_{N(g)}$ and hence $n_{N(g)}$ must increase to reach equilibrium.

6.7 Consider the reaction aA + bB = cC + dD. Use of $P_i = x_iP$ gives

$$K_P^\circ = \frac{(P_C/P^\circ)^c(P_D/P^\circ)^d}{(P_A/P^\circ)^a(P_B/P^\circ)^b} = \frac{x_C^c x_D^d}{x_A^a x_B^b} P^{c+d-a-b}(P^\circ)^{a+b-c-d} = K_x(P/P^\circ)^{\Delta n/\text{mol}}$$

6.8 $\prod_{j=1}^{4} j(j+1) = (1 \cdot 2)(2 \cdot 3)(3 \cdot 4)(4 \cdot 5) = 2880$.

6.9 ΔG_{298}° = 2(231.731 kJ/mol) – 0 = 463.462 kJ/mol = $-RT \ln K_P^\circ$.
So $\ln K_P^\circ$ = –(463462 J/mol)/[(8.3145 J/mol-K)(298.15 K)] = –186.96.
$K_P^\circ = 6.4 \times 10^{-82}$.

6.10 **(a)** T; **(b)** F; **(c)** F; **(d)** F; **(e)** T; **(f)** F; **(g)** T; **(h)** T.

6.11 **(a)** $\Delta G_{300,1}^\circ = -R(300\text{ K})\ln K_{P,300,1}^\circ$ and $\Delta G_{300,2}^\circ = -R(300\text{ K})\ln K_{P,300,2}^\circ$. Since $\Delta G_{300,1}^\circ < \Delta G_{300,2}^\circ$, then $-R(300\text{ K})\ln K_{P,300,1}^\circ < -R(300\text{ K})\ln K_{P,300,2}^\circ$. Division by the positive quantities R and 300 K does not change the direction of the inequality sign, so $-\ln K_{P,300,1}^\circ < -\ln K_{P,300,2}^\circ$. Addition of each log to both sides of the inequality gives $\ln K_{P,300,2}^\circ < \ln K_{P,300,1}^\circ$, so $K_{P,300,2}^\circ < K_{P,300,1}^\circ$. The statement is true.

(b) $\Delta G_{300,1}^\circ = -R(300\text{ K})\ln K_{P,300,1}^\circ$ and $\Delta G_{400,1}^\circ = -R(400\text{ K})\ln K_{P,400,1}^\circ$. Then $-R(300\text{ K})\ln K_{P,300,1}^\circ > -R(400\text{ K})\ln K_{P,400,1}^\circ$ and

$1.33 \ln K^\circ_{P,400,1} > \ln K^\circ_{P,300,1}$. This inequality does not allow us to conclude that $\ln K^\circ_{P,400,1} > \ln K^\circ_{P,300,1}$, and we do not know whether K°_P is greater at 300 K or at 400 K, so the statement is false.

6.12 $\Delta G^\circ = -RT \ln K^\circ_P = -(8.314 \text{ J/mol-K})(298.1 \text{ K}) \ln 0.144 = 4.80$ kJ/mol. If ΔH° is assumed constant over the range 25°C to 35°C, then Eq. (6.39) applies and $\ln(0.321/0.144) = [\Delta H^\circ/(8.314 \text{ J/mol-K})][(298.1 \text{ K})^{-1} - (308.1 \text{ K})^{-1}]$; $\Delta H^\circ = 61.2$ kJ/mol. $\Delta G^\circ = \Delta H^\circ - T\Delta S^\circ =$
4800 J/mol = 61200 J/mol – (298.1 K)ΔS° and $\Delta S^\circ = 189$ J/mol-K.

6.13 We plot $\ln K^\circ_P$ versus $1/T$. The data are

$\ln K^\circ_P$	−1.406	0.688	1.601	2.235
T^{-1}/K^{-1}	0.00206_2	0.00187_3	0.00179_9	0.00174_2

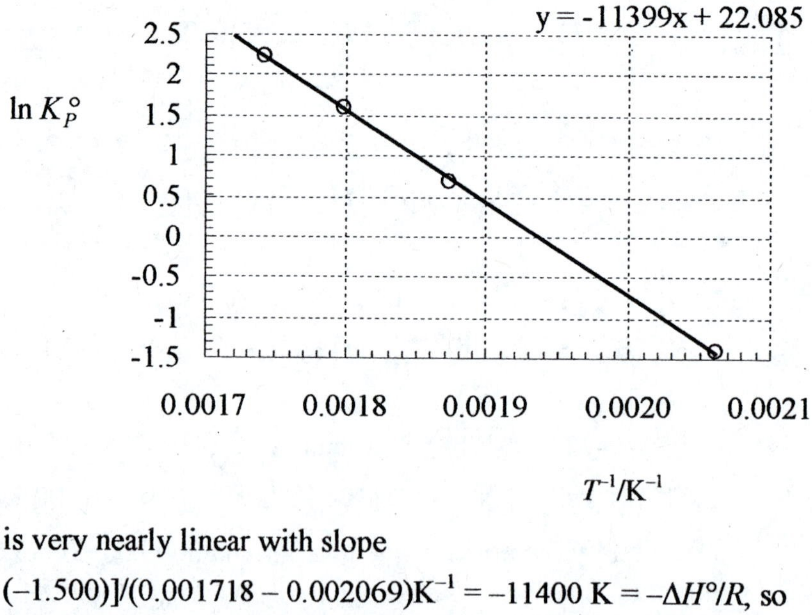

The plot is very nearly linear with slope
$[2.500 - (-1.500)]/(0.001718 - 0.002069)\text{K}^{-1} = -11400 \text{ K} = -\Delta H^\circ/R$, so
$\Delta H^\circ = (11400 \text{ K})(8.314 \text{ J/mol-K}) = 94._8$ kJ/mol = 22.6 kcal/mol in this range of T. Then $\Delta G^\circ_{534} = -RT \ln K^\circ_P = -(8.314 \text{ J/mol-K})(534 \text{ K}) \ln 1.99 = -3.05_5$ kJ/mol = –730 cal/mol. From $\Delta G^\circ = \Delta H^\circ - T\Delta S^\circ$, $\Delta S^\circ_{534} =$

$(\Delta H° - \Delta G°)/T = (94800 + 3055)(\text{J/mol})/(534\text{ K}) = 183$ J/mol-K = 43.7 cal/mol-K.

(b) $\Delta H° = 22.6$ kcal/mol = $94._8$ kJ/mol, as in (a). (The near linearity of the plot shows that $\Delta H°$ is essentially constant over this range of T.)
$\Delta G^\circ_{574} = -RT \ln K^\circ_P = -10.7$ kJ/mol = -2.55 kcal/mol.
$\Delta S^\circ_{574} = (\Delta H^\circ_{574} - \Delta G^\circ_{574})/T = 183$ J/mol-K = 43.8 cal/mol-K.

6.14 Appendix data give $\Delta G^\circ_{298} = 37.2$ kJ/mol. $\Delta G° = -RT \ln K^\circ_P$. 37200 J/mol = $-(8.314\text{ J/mol-K})(298.1\text{ K}) \ln K^\circ_{P,298}$ and $K^\circ_{P,298} = 3.0 \times 10^{-7}$. $d \ln K^\circ_P/dT = \Delta H°/RT^2$; $\ln K^\circ_P(T_2) - \ln K^\circ_P(T_1) \approx -(\Delta H°/R)(1/T_2 - 1/T_1)$. $\ln K^\circ_{P,400} \approx \ln(3.0 \times 10^{-7}) - [(87900\text{ J/mol})/(8.134\text{ J/mol-K})][(400\text{ K})^{-1} - (298.1\text{ K})^{-1}] = -5.99$ and $K^\circ_{P,400} \approx 0.0025$.

6.15 $\Delta G° = -RT \ln K^\circ_P = 72400\text{ J/mol} = -(8.314\text{ J/mol-K})(600\text{ K}) \ln K^\circ_P$ and $K^\circ_P = 4.9_7 \times 10^{-7}$ at 600 K. Integration of the van't Hoff equation (6.36) with $\Delta H°$ assumed constant gives $\ln(K^\circ_{P,2}/K^\circ_{P,1}) = -(\Delta H°/R)(1/T_2 - 1/T_1)$, so $\ln[26/(4.97 \times 10^{-7})] = -[(217900\text{ J/mol})/(8.314\text{ J/mol-K})][1/T_2 - 1/(600\text{ K})]$ and $T_2 = 1010$ K.

6.16 **(a)** $\Delta G° = -RT \ln K^\circ_P = -RT[\ln a + b \ln(T/\text{K}) - c/(T/\text{K})]$. Equation (6.36) gives $\Delta H° = RT^2 d \ln K^\circ_P/dT = RT^2[b/T + (c\text{ K})/T^2] = bRT + (cR\text{ K})$. Then $\Delta S° = (\Delta H° - \Delta G°)/T = bR + R[\ln a + b \ln(T/\text{K})]$ and $\Delta C^\circ_P = d\Delta H°/dT = bR$.

(b) At 300 K, $\Delta H° = (8.314\text{ J/mol-K})[(-1.304)(300\text{ K}) + 7307\text{ K}] = 57.5$ kJ/mol. At 600 K, $\Delta H° = 54.2$ kJ/mol.

6.17 Substitution of $\Delta H°(T) \approx \Delta H°(T_1) + \Delta C^\circ_P(T_1)(T - T_1)$ into (6.37) gives
$$\ln[K^\circ_P(T_2)/K^\circ_P(T_1)] \approx R^{-1} \int_{T_1}^{T_2} \Delta H°(T_1)T^{-2}\, dT + R^{-1} \Delta C^\circ_P(T_1) \int_{T_1}^{T_2} (1/T - T_1/T^2)\, dT$$
$$= R^{-1}\Delta H°(T_1)\,(1/T_1 - 1/T_2) + R^{-1} \Delta C^\circ_P(T_1)[\ln(T_2/T_1) + T_1/T_2 - 1].$$

For $N_2O_4(g) \rightleftarrows 2NO_2(g)$, $\Delta G^\circ_{298} = 4.73$ kJ/mol, $\ln K^\circ_{P,298} = -1.908$, and $K^\circ_{P,298} = 0.148$. $\Delta H^\circ_{298} = 57.2$ kJ/mol; $\Delta C^\circ_{P,298} = -2.88$ J/mol-K.

$\ln K^\circ_{P,600} \approx -1.908 + (57200\text{ J/mol})(8.314\text{ J/mol-K})^{-1}(1/298.1 - 1/600)\text{K}^{-1} + (8.314\text{ J/mol-K})^{-1}(-2.88\text{ J/mol-K})[\ln(600/298) + 298/600 - 1] = 9.632$ and $K^\circ_{P,600} \approx 1.5 \times 10^4$.

6.18 **(a)** Substitute (6.28) into (6.14).

(b) $\ln[K^\circ_P(T_2)/K^\circ_P(T_1)] = \ln K^\circ_P(T_2) - \ln K^\circ_P(T_1) = -\Delta H^\circ(T_2)/RT_2 + \Delta S^\circ(T_2)/R + \Delta H^\circ(T_1)/RT_1 - \Delta S^\circ(T_1)/R \approx \Delta H^\circ(T_1)R^{-1}(1/T_1 - 1/T_2)$ if we take $\Delta H^\circ(T_2) = \Delta H^\circ(T_1)$ and $\Delta S^\circ(T_2) = \Delta S^\circ(T_1)$.

6.19 **(a)** When T_2 in the ΔH° expression in Example 5.6 is replaced by T and this expression is inserted into (6.37), the right side of (6.37) becomes $R^{-1}\int_{T_1}^{T_2}(T^{-1}\,\Delta a + T^{-2}w + \tfrac{1}{2}\Delta b + \tfrac{1}{3}\Delta c T + \tfrac{1}{4}\Delta d\, T^2)\,dT$, where $w \equiv \Delta H^o_{T_1} - T_1\,\Delta a - \tfrac{1}{2}T_1^2\,\Delta b - \tfrac{1}{3}T_1^3\,\Delta c - \tfrac{1}{4}T_1^4\,\Delta d$. After integration, we have $R^{-1}[\Delta a \ln(T_2/T_1) + w(T_1^{-1} - T_2^{-1}) + \tfrac{1}{2}\Delta b(T_2 - T_1) + \tfrac{1}{6}\Delta c(T_2^2 - T_1^2) + \tfrac{1}{12}\Delta d(T_2^3 - T_1^3)] = \ln[K^o_P(T_2)/K^o_P(T_1)]$ [Eq. (1)]. Let $T_1 = 298.15$ K. Then $-RT\ln K^\circ_{P,298} = \Delta G^\circ_{298} = [2(-394.359) - 2(-137.168) - 0]$ kJ/mol $= -514382$ J/mol and $\ln K^\circ_{P,298} = 207.50$. Using data from Example 5.6, we find $w = -558488$ J/mol for $T_1 = 298.15$ K. We can thus find $K^o_P(T_2)$ using the preceding Eq. (1).

(b) Substitution in Eq. (1) of part (a) gives $\ln K^o_P(1000\text{ K}) = 207.50 + (1/8.3145)(-48.25 - 1314.69 + 41.21 - 14.93 + 2.28) = 47.01$ and $K^o_P(1000\text{ K}) = 3 \times 10^{20}$.

6.20 Equation (5.19) gives $\Delta H^\circ_T = \Delta H^\circ_{T_1} + \int_{T_1}^{T}[2e - a + (2f - b)T' + (2g - c)T'^2]\,dT' = \Delta H^\circ_{T_1} + (2e - a)(T - T_1) + (f - \tfrac{1}{2}b)(T^2 - T_1^2) + \tfrac{1}{3}(2g - c)(T^3 - T_1^3)$. Substitution in (6.37) and integration gives $\ln K^\circ_P(T_2) = \ln K^\circ_P(T_1) + R^{-1}[\Delta H^\circ_{T_1} + (a - 2e)T_1 + (\tfrac{1}{2}b - f)\,T_1^2 + \tfrac{1}{3}(c - 2g)T_1^3](T_1^{-1} - T_2^{-1}) + R^{-1}(2e - a)\ln(T_2/T_1) + (f - \tfrac{1}{2}b)(T_2 - T_1)/R + \tfrac{1}{6}(2g - c)(T_2^2 - T_1^2)/R$.

6.21 From (6.25), $\ln K_c^\circ = \ln K_P^\circ - (\Delta n/\text{mol}) \ln T + \text{const.}$ Differentiation and use of (6.36) and (5.10) give

$$\frac{d \ln K_c^\circ}{dT} = \frac{d \ln K_P^\circ}{dT} - \frac{\Delta n/\text{mol}}{T} = \frac{\Delta H^\circ}{RT^2} - \frac{\Delta nRT/\text{mol}}{RT^2} = \frac{\Delta U^\circ}{RT^2}$$

6.22 From (6.27), $\ln K_x = \ln K_P^\circ - (\Delta n/\text{mol}) \ln(P/\text{bar})$. Differentiation with respect to T and use of (6.36) gives $(\partial \ln K_x/\partial T)_P = d \ln K_P^\circ/dT = \Delta H^\circ/RT^2$. Partial differentiation with respect to P gives (since K_P° is independent of P):

$(\partial \ln K_x/\partial P)_T = -(\Delta n/\text{mol})/P$.

6.23 **(a)** T; **(b)** F.

6.24 Since $P_i = n_iRT/V$, the partial pressures are proportional to the moles. Let zV/RT moles of CO react. The equilibrium partial pressures are then $P_{CO} = 342 \text{ torr} - z$, $P_{Cl_2} = 351.4 \text{ torr} - z$, $P_{COCl2} = z$.

Thus 439.5 torr = 342.0 torr $- z +$ 351.4 torr $- z + z$ and $z = 253.9$ torr.

$K_P^\circ = (253.9/750)/(88.1/750)(97.5/750) = 22.2$.

6.25 $\Delta G^\circ/(\text{kJ/mol}) = -394.359 + (-879) - 2(-619.2) = -35$. $\ln K_P^\circ = -\Delta G^\circ/RT =$ (35000 J/mol)/(8.314 J/mol-K)(298 K) $\ln K_P^\circ = 14._1$, $K_P^\circ = 1.3 \times 10^6$.

Let $2z$ moles of COF_2 react. The equilibrium amounts are $n(COF_2) = 1 - 2z$, $n(CO_2) = 1 + z$, $n(CF_4) = z$. Because K_P° is so large, we have $2z \approx 1$ and $z \approx$ ½. Hence, $n(CO_2) = 1.50$ mol, $n(CF_4) = 0.50$ mol, and

$$1._3 \times 10^6 = \frac{(1.5/n_{tot})P(0.5/n_{tot})P}{[n(COF_2)/n_{tot}]^2P^2}, \quad n(COF_2) = 7 \times 10^{-4} \text{ mol}$$

6.26 $\ln K_P^\circ = -\Delta G^\circ/RT = -1.258$. $K_P^\circ = 0.284$.

(a) Let z moles of A react. The equilibrium amounts are $n_A = 1 - z$, $n_B = 1 - z$, $n_C = 2z$, $n_D = 2z$, and $n_{tot} = 2 + 2z$. Use of $P_i = x_iP$ gives

$$0.284 = \frac{\left(\frac{z}{1+z}\right)^2\left(\frac{P}{P°}\right)^2\left(\frac{z}{1+z}\right)^2\left(\frac{P}{P°}\right)^2}{\left(\frac{1-z}{2(1+z)}\right)^2\left(\frac{P}{P°}\right)^2} = \frac{4z^4}{(1-z^2)^2}\left(\frac{1200}{750}\right)^2$$

$$0.533 = \frac{2z^2}{1-z^2}\frac{1200}{750}, \quad 2.333z^2 = 0.333_1, \quad z = 0.378$$

Both times we took the positive square root, since z and $1 - z^2$ are positive. $n_A = 0.622$ mol $= n_B$, $n_C = 0.756$ mol $= n_D$.

(b) The equilibrium amounts are $n_A = 1 - z$, $n_B = 2 - z$, $n_C = 2z = n_D$. We have

$$0.284 = \frac{[2z/(3+2z)]^4(P/P°)^2}{[(1-z)/(3+2z)][(2-z)/(3+2z)]}$$

$$0.00693_1 - z^4/[(1-z)(2-z)(3+2z)^2] = 0$$

z lies between 0 and 1, and trial and error gives the desired root as $z = 0.530$. Hence $n_A = 0.470$, $n_B = 1.470$, $n_C = 1.060 = n_D$.

6.27 Appendix $\Delta_f G°$ data give $\Delta G°_{298} = -2.928$ kJ/mol. $\ln K°_P = -\Delta G°/RT =$ 1.181 and $K°_P = 3.26$. $K°_P = [P(\text{HD})/P°]^2/[P(H_2)/P°][P(D_2)/P°] =$ $[P(\text{HD})]^2/P(H_2)P(D_2) = [n(\text{HD})RT/V]^2/[n(H_2)RT/V][n(D_2)RT/V] =$ $[n(\text{HD})]^2/n(H_2)n(D_2)$. Let x mol of H_2 react to reach equilibrium. The equilibrium amounts are then $n(H_2) = (0.300 - x)$mol, $n(D_2) = (0.100 - x)$mol, $n(\text{HD}) = 2x$ mol. Substitution in the $K°_P$ expression gives after simplification: $0.227x^2 + 0.400x - 0.0300 = 0$. The quadratic formula gives $x = 0.072$ and -1.83. Since we started with no HD, x must be positive. So $x = 0.072$, and $n(H_2) = (0.300 - x)$mol $= 0.228$ mol, $n(D_2) = 0.028$ mol, $n(\text{HD}) = 0.144$ mol.

6.28 **(a)** $K°_P = [P(NH_3)/P°]^2/[P(N_2)/P°][P(H_2)/P°]^3 = P(NH_3)^2(P°)^2/P(N_2)[P(H_2)]^3$
Use of $P_i = n_iRT/V$ gives $K°_P = [n(NH_3)^2/n(N_2)n(H_2)^3](P°V/RT)^2$.
Now $K°_P(RT/P°V)^2 =$
$36[(82.06 \text{ cm}^3\text{-atm/mol-K})(400 \text{ K})/(750/760)\text{atm}(2000 \text{ cm}^3)]^2 = 9960$ mol^{-2}. Let x mol of N_2 react to reach equilibrium. The equilibrium amounts are $(0.100 - x)$ mol of N_2, $(0.300 - 3x)$ mol of H_2, and $2x$ mol of NH_3. So $9960 = (2x)^2/(0.100 - x)(0.300 - 3x)^3 = 4x^2/3^3(0.1 - x)(0.1 - x)^3$

$= 4x^2/27(0.100 - x)^4$. Taking the square root of both sides, we get $\pm 99.8 = 0.385x/(0.100 - x)^2$. Since x must be positive (we started with no NH_3), the negative sign is rejected and $0.01 - 0.2x + x^2 = 0.00386x$. The quadratic formula gives $x = 0.082$ (the other root exceeds 0.1, which is impossible). Then $n(N_2) = 0.018$ mol, $n(H_2) = 0.054$ mol, and $n(NH_3) = 0.164$ mol.

(b) At equilibrium, $n(N_2)/\text{mol} = 0.200 - x$, $n(H_2)/\text{mol} = 0.300 - 3x$, and $n(NH_3)/\text{mol} = 0.100 + 2x$. Then $9960 = (0.100 + 2x)^2/(0.200 - x)(0.300 - 3x)^3 \equiv W$. To make W equal 9960, $3x$ must be close to 0.3. For an initial guess of $x = 0.09$, we find $W = 26400$, which is too big. For $x = 0.08$, $W = 2600$. Repeated trial and error (or use of the Solver in a spreadsheet) gives $x = 0.0865$. Then $n(N_2) = 0.113$ mol, $n(H_2) = 0.040_5$ mol, $n(NH_3) = 0.273$ mol.

6.29 The equilibrium amounts are $n_{N_2} = 4.50 \text{ mol} - \xi$, $n_{H_2} = 4.20 \text{ mol} - 3\xi$, $n_{NH_3} = 1.00 \text{ mol} + 2\xi$. The requirement that each of these amounts be positive gives $\xi < 4.50$ mol, $\xi < 1.4$ mol, $\xi > -0.5$ mol. So $-0.5 \text{ mol} < \xi < 1.4$ mol. Hence n_{N_2} lies between 4.50 mol – (–0.50 mol) and 4.50 mol – 1.4 mol; that is, $3.1 \text{ mol} < n_{N_2} < 5.0$ mol. Also, $0 < n_{H_2} < 5.7$ mol and $0 < n_{NH_3} < 3.8$ mol.

6.30 **(a)** Use of $P_i \equiv x_i P$ gives $K_P^\circ = (P_B/P^\circ)^2/(P_A/P^\circ) = P_B^2/P_A P^\circ = (x_B P)^2/x_A P P^\circ = [x_B^2/(1 - x_B)](P/P^\circ)$.

(b) With this definition of z, the equation in (a) becomes $z = x_B^2/(1 - x_B)$, so $x_B^2 + z x_B - z = 0$. The quadratic formula gives $x_B = [-z + (z^2 + 4z)^{1/2}]/2$ [Eq. (1)], where the negative root was discarded since x_B must be positive.

(c) $z = 49.3(1 \text{ bar})/(1.50 \text{ bar}) = 32.9$. Substitution in Eq. (1) of part (b) gives $x_B = x_O = 0.971$, so $x_{O_2} = 1 - 0.971 = 0.029$.

(d) $\Delta G^\circ_{298} = 4.73 \text{ kJ/mol} = -RT \ln K_P^\circ$ and $K_P^\circ = 0.148$. So $z = 0.148(750 \text{ torr})/(2 \times 760 \text{ torr}) = 0.0730$. Equation (1) of part (b) gives $x_B = x_{NO_2} = 0.236$, so $x_{N_2O_4} = 0.764$

6.31 $K_P^\circ = [P(SO_3)/P^\circ]^2/[P(SO_2)/P^\circ]^2[P(O_2)/P^\circ] =$ $\{[n(SO_3)]^2/[n(SO_2)]^2 n(O_2)\}(P^\circ V/RT)$, where $P_i = n_i RT/V$ was used. We find $P^\circ V/RT = 0.00222$ mol, so $[n(SO_2)]^2/[n(SO_2)]^2 n(O_2) = 1540\ \text{mol}^{-1}$. Calculation with the initial mole numbers shows that to reach equilibrium, more SO_3 must be formed. Let x moles of O_2 react to reach equilibrium. The equilibrium mole numbers are $n(SO_2)/\text{mol} = 0.00265 - 2x$, $n(O_2)/\text{mol} = 0.00310 - x$, and $n(SO_3) = 0.00144 + 2x$. Then $1540 = (0.00144 + 2x)^2/(0.00265 - 2x)^2(0.00310 - x) \equiv W$. An initial guess of $x = 0.001$ gives $W = 13340$, which is too big. Repeated trial and error (or use of the Solver in a spreadsheet) gives $x = 0.000632$. So $n(SO_2) = 0.00139$ mol, $n(O_2) = 0.00247$ mol, and $n(SO_3) = 0.00270$ mol.

6.32 $x_A = (1.000\ \text{mol}/6.000\ \text{mol}) = 0.1667$, $x_B = 0.5000$, $x_C = 0.3333$. $P_A = x_A P = 0.1667(1.000\ \text{bar}) = 0.1667$ bar, $P_B = 0.5000$ bar, $P_C = 0.3333$ bar. $K_P^\circ = 0.3333/(0.1667 \times 0.5000) = 4.000$. Let z moles of A react to reach the new equilibrium position at $P = 2.000$ bar. For this equilibrium, $n_A = 1 - z$, $n_B = 3 - z$, $n_C = 2 + z$ and $x_A = (1 - z)/(6 - z)$, $x_B = (3 - z)/(6 - z)$, $x_C = (2 + z)/(6 - z)$. $P_A = x_A P = [(1 - z)/(6 - z)](2\ \text{bar})$, $P_B =$ etc.

$$K_P^\circ = 4.00 = \frac{[(2 + z)/(6 - z)]2}{[(1 - z)/(6 - z)]2[(3 - z)/(6 - z)]2}$$

$9z^2 - 36z + 12 = 0$, $z = 0.367$ and 3.63.
We have $n_A = 1 - z$, so z cannot exceed 1. Therefore $z = 0.367$.
$n_A = 0.633$ mol, $n_B = 2.633$ mol, $n_C = 2.367$ mol.

6.33 $\Delta G_{298}^\circ/(\text{kJ/mol}) = -267.8 + 0 - (-305.0) = 37.2$. $\Delta G^\circ = -RT \ln K_P^\circ$. $37200\ \text{J/mol} = -(8.314\ \text{J/mol-K})(298\ \text{K}) \ln K_{P,298}^\circ$. $K_{P,298}^\circ = 3.0 \times 10^{-7}$. $\Delta H_{298}^\circ/(\text{kJ/mol}) = -287.0 + 0 - (-374.9) = 87.9$. Eq. (6.39) gives $\ln K_{P,500}^\circ \approx \ln(3.0 \times 10^{-7}) + [(87900\ \text{J/mol})/(8.314\ \text{J/mol-K})][(298\ \text{K})^{-1} - (500\ \text{K})^{-1}] = -0.69_9$. $K_{P,500}^\circ = 0.50$. The mole fractions are $x_{PCl_3} = x_{Cl_2} = x$ and $x_{PCl_5} = 1 - x_{PCl_3} - x_{Cl_2} = 1 - 2x$. We have $K_{P,500}^\circ = 0.50 = (xP/P^\circ)^2/[(1 - 2x)P/P^\circ] = x^2/(1 - 2x)$, since $P = P^\circ = 1$ bar. $x^2 = 0.50 - 1.00x$, and the quadratic formula gives $x = 0.36_6 = x_{PCl_3} = x_{Cl_2}$; $x_{PCl_5} = 1 - 2x = 0.26_8$.

6.34 **(a)** Let $N_2 + 3H_2 \rightleftarrows 2NH_3$ be called reaction I. Since reaction (a) has its coefficients equal to one-half those of reaction I, we have $K^\circ_{P,(a)} = (K^\circ_{P,\text{I}})^{1/2} = (36)^{1/2} = 6$.

(b) $K^\circ_{P,(b)} = 1/K^\circ_{P,\text{I}} = 0.028$.

6.35 For $n \rightarrow$ iso, $\Delta G^\circ_{1000} = -670$ cal/mol and we find $K^\circ_{P,1000} = 1.40 \equiv K_{\text{B/A}}$, where A $\equiv$ *n*-pentane, B $\equiv$ isopentane, C $\equiv$ neopentane. For $n \rightarrow$ neo, $\Delta G^\circ_{1000} = 4900$ cal/mol and $K^\circ_{P,1000} = 0.085 \equiv K_{\text{C/A}}$. Eqs. (6.45) and (6.46) give $x_\text{A} = x_n = 1/(1 + 1.40 + 0.085) = 0.40$, $x_\text{B} = x_{\text{iso}} = 1.40/2.48_5 = 0.56$, $x_\text{C} = x_{\text{neo}} = 0.085/2.48_5 = 0.034$.

6.36 $\Delta G^\circ_{6000}/(\text{kJ/mol}) = 1059.72 + 0 - 71.74 = 987.98$.
$\ln K^\circ_{P,6000} = -(987980 \text{ J/mol-K})/(8.314 \text{ J/mol-K})(6000 \text{ K}) = -19.804$ and $K^\circ_{P,6000} = 2.5 \times 10^{-9}$.

6.37 $\Delta G^\circ = -RT \ln K^\circ_P$; $\Delta G^\circ_{\text{wrong}} = -RT \ln K^\circ_{P,\text{wrong}}$. $\Delta G^\circ - \Delta G^\circ_{\text{wrong}} =$ 2500 J/mol $= -RT \ln (K^\circ_P/K^\circ_{P,\text{wrong}})$. We get $\ln(K^\circ_P/K^\circ_{P,\text{wrong}}) = -1.00$ and $K^\circ_{P,\text{wrong}}/K^\circ_P = 2.7$. The error is a factor of 2.7.

6.38 **(a)** $I_2(g) \rightleftarrows 2I(g)$. Let n^* be the number of I_2 moles before dissociation, and let z moles of I_2 dissociate to give $2z$ moles of I, leaving $n^* - z$ moles of I_2. The total number of moles at equilibrium is $n^* - z + 2z = n^* + z$. Then $x_\text{I} = 2z/(n^* + z)$ and $x_{\text{I}_2} = (n^* - z)/(n^* + z)$. We have $n^* = P^*V/RT$ and $n^* + z = PV/RT$, so $z = PV/RT - n^* = (P - P^*)V/RT$. Use of these expressions for z, $n^* + z$, and n^* gives $x_\text{I} = [2(P - P^*)V/RT]/(PV/RT) = 2(P - P^*)/P$ and $x_{\text{I}_2} = [P^*V/RT - (P - P^*)V/RT]/(PV/RT) = (2P^* - P)/P$.

(b) $K^\circ_P = (x_\text{I}P/P^\circ)^2/(x_{\text{I}_2}P/P^\circ) = [2(P - P^*)/P^\circ]^2/[(2P^* - P)/P^\circ] = 4(P - P^*)^2/(2P^* - P)P^\circ$.

(c) We use the result of (b) to calculate the following K°_P values:
$K^\circ_{P,973} = 4(0.0624 - 0.0576)^2/[2(0.0576) - 0.0624]0.987 = 0.00177$;

$K^\circ_{P,1073} = 0.0112$; $K^\circ_{P,1173} = 0.0493$; $K^\circ_{P,1274} = 0.172_5$.

We plot $\ln K^\circ_P$ vs. $1/T$. The data are

$\ln K^\circ_P$	–6.34	–4.49	–3.01	-1.75_7	
$10^4/T$	10.28	9.32	8.52_5	7.85	K^{-1}

One finds a straight line with slope

$[-1.1_0 - (-6.7_3)]/(0.00075 - 0.00105)K^{-1} = -18800$ K. Eq. (6.40) gives $\Delta H^\circ = (18800 \text{ K})(8.314 \text{ J/mol-K}) = 156$ kJ/mol = 37.4 kcal/mol.

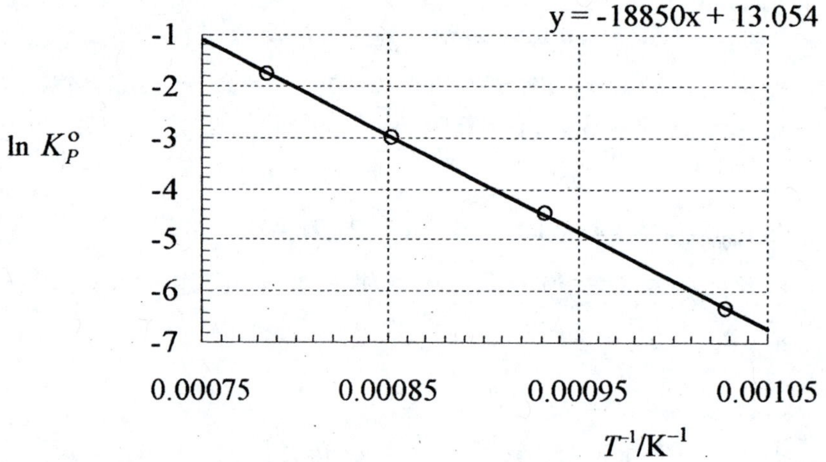

6.39 $K_P^{\text{bar}}/K_P^{\text{atm}} = [\Pi_i (P_i/\text{bar})^{\nu_i}]/[\Pi_i (P_i/\text{atm})^{\nu_i}] = \Pi_i (P_i/\text{bar})^{\nu_i}/(P_i/\text{atm})^{\nu_i} =$ $\Pi_i (\text{atm/bar})^{\nu_i}$. For $a\text{A} + b\text{B} \rightleftarrows c\text{C} + d\text{D}$, $\Pi_i (\text{atm/bar})^{\nu_i} = (\text{atm/bar})^{c+d-a-b} =$ $(\text{atm/bar})^{\Delta n/\text{mol}} = (760 \text{ torr}/750.062 \text{ torr})^{\Delta n/\text{mol}} = (1.01325)^{\Delta n/\text{mol}}$.

6.40 At equilibrium, $n_{N_2}/\text{mol} = 1 - x$, $n_{H_2}/\text{mol} = 3 - 3x$, $n_{NH_3}/\text{mol} = 2x$; $n_{\text{tot}}/\text{mol} = 4 - 2x$. $P_{N_2} = x_{N_2}P = (n_{N_2}/n_{\text{tot}})P = [(1-x)/(4-2x)]P$, $P_{H_2} = [3(1-x)/(4-2x)]P$, $P_{NH_3} = [2x/(4-2x)]P$. $K^\circ_P =$ $(P_{NH_3}/P^\circ)^2/(P_{N_2}/P^\circ)(P_{H_2}/P^\circ)^3 = (P^\circ)^2P^{-2}(4-2x)^2(2x)^2/(1-x)3^3(1-x)^3 =$ $(P^\circ/P)^2(16/27)(2-x)^2x^2/(1-x)^4 = K^\circ_P$. x must be between 0 and 1. Taking the positive square root of each side gives $(P^\circ/P)(4/27^{1/2})(2-x)x/(1-x)^2 = K_P^{\circ 1/2}$ or $4(2x - x^2)/(1-x)^2 = (27K^\circ_P)^{1/2}(P/P^\circ) = s$. Hence $8x - 4x^2 = s - 2sx + x^2s$ and $(4+s)x^2 - (2s+8)x + s = 0$ or $x^2 - 2x + s/(s+4) = 0$. The quadratic formula gives $x = \{2 \pm [4 - 4s/(s+4)]^{1/2}\}/2 = 1 \pm [1 - s/(s+4)]^{1/2}$. Since x is less than 1, we take the minus sign: $x = 1 - [1 - s/(s+4)]^{1/2}$.

6.41 **(a)** No; **(b)** no; **(c)** no; **(d)** no; **(e)** no; **(f)** no; **(g)** yes; **(h)** no; **(i)** yes.

6.42 (b).

6.43 **(a)** See Fig. 6.11. Results for $n(CH_4)$/mol in order of increasing pressure are 0.00061, 0.053, 0.690, 1.355, 1.515, 1.62, 1.71.

(b) The NIST-JANAF tables give at 1200 K: $\Delta G_1^\circ = -77.92$ kJ/mol, $\Delta G_2^\circ = -74.77$ kJ/mol, $K_1 = 2.46 \times 10^3$ and $K_2 = 1.80 \times 10^3$. One changes the values of T, P, K_1 and K_2 in the Fig. 6.9 spreadsheet. With the guessed values of the equilibrium amounts taken as equal to the starting amounts, the Solver might not find a solution. The large values of K_1 and K_2 indicate that one or both of the reactants CH_4 and H_2O are present in small amounts at equilibrium. The 0.01 bar 900 K equilibrium amounts in Fig. 6.10 have small amounts of the reactants, so it is a good idea to use the 0.01 bar 900 K equilibrium amounts as the initial guess for the 1200 K problem.

6.44 Suppose one equilibrium constant K_1 is very small with the value 1×10^{-8}. If the Solver changes the mole numbers and finds a calculated value of 8×10^{-8} for K_1, then the calculated and true values of K_1 differ by 7×10^{-8}, which is less than the Solver's default precision of 1×10^{-6}, and the Solver will declare it has found a solution, even though the calculated K is 8 times the true K and at least some mole numbers must be greatly in error.

6.45 Let ξ_1 additional moles of CO be formed when reaction 1 reaches equilibrium and let ξ_2 additional moles of CO_2 be formed when reaction 2 reaches equilibrium. The equilibrium amounts are then $n(CH_4) = 1 - \xi_1 - \xi_2$, $n(H_2O) = 1 - \xi_1 - 2\xi_2$, $n(CO_2) = 1 + \xi_2$, $n(CO) = 2 + \xi_1$, $n(H_2) = 1 + 3\xi_1 + 4\xi_2$ (where the unit mol is omitted). The conditions that each of these amounts be positive give the inequalities: (I) $1 > \xi_1 + \xi_2$; (II) $1 > \xi_1 + 2\xi_2$; (III) $\xi_2 > -1$; (IV) $\xi_1 > -2$; (V) $3\xi_1 + 4\xi_2 > -1$. Addition of (I) and (III) gives $2 > \xi_1$ (VI). Addition of (II) and (IV) gives $\xi_2 < 1.5$ (VII). From (III), (IV), (VI), and (VII), we have $-2 < \xi_1 < 2$ and $-1 < \xi_2 < 1.5$, which can be used to help set limits on the mole numbers.

6.46 (a) CO occurs only in reaction 1 and CO_2 occurs only in reaction 2. Therefore, we can use the changes in amounts of CO and CO_2 to find the extents of reaction. CO goes from an initial amount of 2 mol to 3.3236 mol at equilibrium and has a stoichiometric coefficient of 1. Therefore $\xi_1 = 1.3236$. CO_2 goes from 1 mol to 0.6758 mol, so $\xi_2 = -0.3242$ mol.

(b) Now CH_4 is the only species that occurs only in reaction 1. CH_4 has stoichiometric coefficient –1 and goes from 1 mol to 0.00061 mol, so $\xi_1 = 0.99939$ mol, which is quite different than in (a).

6.47 (a) (1) $N_2 \rightleftarrows 2N$; (2) $O \rightleftarrows 2O$; (3) $N_2 + O_2 \rightleftarrows 2NO$. The reaction $N + O \rightleftarrows NO$ is $-\frac{1}{2}(1) - \frac{1}{2}(2) + \frac{1}{2}(3)$. The reaction $N + O_2 \rightleftarrows NO + O$ is $-\frac{1}{2}(1) + \frac{1}{2}(2) + \frac{1}{2}(3)$.

(b) Check your results against the mole fractions at 4000 K in Fig. 6.5.

6.48 (a) $d \ln K_P^\circ/dT = \Delta H^\circ/RT^2$. Appendix data give $\Delta H^\circ = 87.9$ kJ/mol > 0, so K_P° decreases as T decreases and the equilibrium shifts to the left.

(b) As V decreases at constant T, the pressure increases and the equilibrium shifts to the side with fewer moles of gas, i.e., it shifts to the left.

(c) This removal decreases P_{PCl_5}, so to restore equilibrium, the reaction must shift to the left.

(d) Constant T and V addition of He does not affect the partial pressures of PCl_5, PCl_3, or Cl_2, and there is no shift.

(e) To keep P constant as He(g) is added, V must be increased. Since $P_i = n_iRT/V$, this volume increase will decrease each partial pressure P_i by the same percentage. Since the reaction has more moles of products than of reactants, the numerator of the reaction quotient Q_P will decrease more than the denominator. Therefore the equilibrium will shift to the right to make Q_P again equal to K_P.

6.49 From Prob. 6.21, $d \ln K_c^\circ/dT = \Delta U^\circ/RT^2$; if ΔU° is positive, then K_c° increases as T increases. Since $c_i = n_i/V$ and V is held fixed, the mole numbers n_i undergo changes proportional to the changes in the concentrations, and the equilibrium shifts to the right if $\Delta U^\circ > 0$.

6.50 **(a)** $\ln Q_x = \ln [\Pi_i (x_i)^{\nu_i}] = \Sigma_i \ln (x_i)^{\nu_i} = \Sigma_i \nu_i \ln x_i = \Sigma_i \nu_i \ln (n_i/n_{tot}) = \Sigma_i \nu_i \ln n_i - (\Sigma_i \nu_i) \ln (n_1 + n_2 + \cdots)$. We have $\partial \ln Q_x/\partial n_j = \nu_j/n_j - \Delta n/(n_1 + n_2 + \cdots)\text{ mol} = (\nu_j - x_j \Delta n/\text{mol})/n_j$.

(b) Case I: suppose substance j is a reactant. Then ν_j is negative. To get a shift to the left, we want $\partial Q_x/\partial n_j > 0$, so that addition of j will make $Q_x > K_x$. We thus want $\nu_j - x_j \Delta n/\text{mol} > 0$ and $\nu_j > x_j \Delta n/\text{mol}$. Since ν_j is negative, Δn must be negative. Division of the inequality by the negative quantity Δn reverses its direction to give $\nu_j/(\Delta n/\text{mol}) < x_j$. The fact that Δn is negative means that the total moles of reactants exceeds the total moles of products, in agreement with condition (1). Case II: suppose substance j is a product. Then ν_j is positive. We want $\partial Q_x/\partial n_j < 0$, so that addition of j will make $Q_x < K_x$ and shift the equilibrium to the right. We thus want $\nu_j - x_j\Delta n/\text{mol} < 0$ or $\nu_j < x_j\Delta n/\text{mol}$. ν_j is positive, so Δn must be positive and division by Δn gives $x_j > \nu_j/(\Delta n/\text{mol})$. The fact that Δn is positive means that the product total moles exceed the reactant total moles.

(c) The left side of the reaction has the greater sum of the coefficients, so constant-T-and-P addition of NH_3 will never produce more NH_3. For N_2, condition (2) is $x_{N_2} > (-1)/(-2) = ½$, so for $x_{N_2} > 0.5$, addition of N_2 at constant T and P produces more N_2. For H_2, condition (2) is $x_{H_2} > (-3)/(-2) = 1.5$, which can never be satisfied.

6.51 **(a)** $K_x = (1/5)^2/(3/5)(1/5)^3 = 8.333$. If x is the extent of reaction to reach the new equilibrium position, the new equilibrium amounts are $n(N_2) = 3.1 - x$, $n(H_2) = 1 - 3x$, $n(NH_3) = 1 + 2x$ and

$$8.333 = \frac{[(1+2x)/(5.1-2x)]^2}{[(3.1-x)/(5.1-2x)][(1-3x)/(5.1-2x)]^3}$$

The condition that the mole numbers be positive gives $-0.5 < x < 0.333$. Use of the Solver gives $x = -0.0005438$, so $n(N_2) = 3.1005438$, $n_{tot} = 5.1010876$, $x(N_2) = 0.607820$.

(b) $K_x = (0.4)^2/(0.2)(0.4)^3 = 12.5$. If x is the extent of reaction to reach the new equilibrium position, the new equilibrium amounts are

$n(N_2) = 14 - x$, $n(H_2) = 4 - 3x$, $n(NH_3) = 4 + 2x$ and

$$12.5 = \frac{[(4+2x)/(20-2x)]^2}{[(12-x)/(20-2x)][(4-3x)/(20-2x)]^3}$$

The condition that the mole numbers be positive gives $-2 < x < 1.333$. Use of the Solver gives $x = 0.12608$, so $n(N_2) = 11.8739$, $n_{tot} = 19.7478$, $x(N_2) = 0.6013$.

6.52 $\Delta\xi_1$ equals Δn_{CO} because CO appears only in reaction (1) and not in reaction (2). Likewise $\Delta\xi_2$ equals Δn_{CO_2} because CO_2 appears only in reaction (2). The results of Prob. 6.43 show that when P is increased from 10 bar to 30 bar at 900 K, $\Delta n_{CO} = 0.553$ mol – 0.886 mol = –0.333 mol = $\Delta\xi_1$, and show that Δn_{CO_2} = 1.931 mol – 1.758 mol = 0.173 mol = $\Delta\xi_2$. The changes in amounts of H_2O due to reactions (1) and (2) are $\Delta n_{H_2O,1} = -\Delta\xi_1 = 0.333$ mol and $\Delta n_{H_2O,2} = -2\Delta\xi_2 = -0.346$ mol. So $\Delta n_{H_2O} = -0.013$ mol. Since the stoichiometry of each of the reactions (1) and (2) has an increase of 2 moles, we have Δn_{tot} = 2(–0.333 mol) + 2(0.173 mol) = –0.320 mol.

6.53 The reaction rate at lower temperatures is too slow to make the reaction economically practical.

6.54 $\Delta G° = -RT \ln K_P° = -2.303RT \log K_P° =$ –2.303(8.314 J/mol-K)(500 K)[7.55 – (4830 K)/(500 K)] = 20.2 kJ/mol.
$d \ln K_P°/dT = \Delta H°/RT^2$. $\Delta H° = RT^2\, d \ln K_P°/dT = 2.303RT^2\, d \log K_P°/dT =$ $2.303RT^2(4830\text{ K})/T^2$ = 2.303(8.314 J/mol-K)(4830 K) = 92.5 kJ/mol.
$\Delta S° = (\Delta H° - \Delta G°)/T$ = (72.3 kJ/mol)/(500 K) = $144._6$ J/mol-K.
$C_P = (\partial H/\partial T)_P$. $\Delta C_P° = (\partial\, \Delta H°/\partial T)_P = 0$.

6.55 $\Delta G°$ refers to a change from pure, separated reactants, each in its standard state, to pure, separated products, each in its standard state. $\Delta G°$ is not the change in G that occurs in the actual reaction mixture. The reacting system does not contain substances in their standard states.

6.56 $K_P°$ is a function of T only. Therefore only (d) will change $K_P°$.

6.57 **(a)** $\Delta H^\circ = \Sigma_i \nu_i H^\circ_{m,i}$. Enthalpies of ideal gases depend on T only and are unaffected by pressure changes or by mixing with other ideal gases; so the observed ΔH per mole of reaction *does* equal the reaction's ΔH°.

(b) Entropies of ideal gases depend strongly on P and the gases in the reaction mixture are not at 1 bar partial pressures. Therefore, the entropies in the mixture do not equal the standard-state entropies and ΔS per mole of reaction differs substantially from ΔS°.

(c) Since $\Delta S_m \neq \Delta S^\circ$, it follows that $\Delta G_m \neq \Delta G^\circ$.

6.58 Let the superscripts 750 and 1000 denote standard states based on 750 torr and on 1000 torr, respectively.

$K_P^{\circ,1000}/K_P^{\circ,750} = [(P_{NO_2}/P^{\circ,1000})^2/(P_{N_2O_4}/P^{\circ,1000})] \div$

$[(P_{NO_2}/P^{\circ,750})^2/(P_{N_2O_4}/P^{\circ,750})] = (P^{\circ,1000})^{-1} \div (P^{\circ,750})^{-1} = 750/1000.$

(The gases are unaware of what choice of standard state has made, so the equilibrium partial-pressure ratio $P^2_{NO_2}/P_{N_2O_4}$ is independent of the standard-state choice and cancels.) The Appendix $\Delta_f G^\circ$ data give $\Delta_f G^{\circ}_{\text{'ɹ20}}$ = 4.73 kJ/mol and use of $\Delta_f G^{\circ}_{\text{'ɹ20}} = -RT \ln K_P^{\circ,750}$ gives $K_P^{\circ,750} = 0.14_8$. Then $K_P^{\circ,1000} = (750/1000)0.14_8 = 0.11_1$.

6.59 **(a)** Since $\Delta H^\circ < 0$, the relation $d \ln K^\circ_P/dT = \Delta H^\circ/RT^2$ shows that K°_P decreases as T increases; thus the equilibrium amount of the cis isomer increases as T increases.

(b) We have $\Delta G^\circ = \Delta H^\circ - T\Delta S^\circ \approx -T\Delta S^\circ$ in the high-T limit. Since $\Delta S^\circ > 0$, we have $\Delta G^\circ < 0$ at high T and $K^\circ_P > 1$ at high T. Thus the high-T limit has more trans isomer than cis isomer.

(c) Even though the percentage of cis isomer continually increases as T increases, it is still possible to have more trans than cis at high T. In the graph at the right, the % cis continually increases with T but always remains below its asymptotic $T = \infty$ limit of 43%.

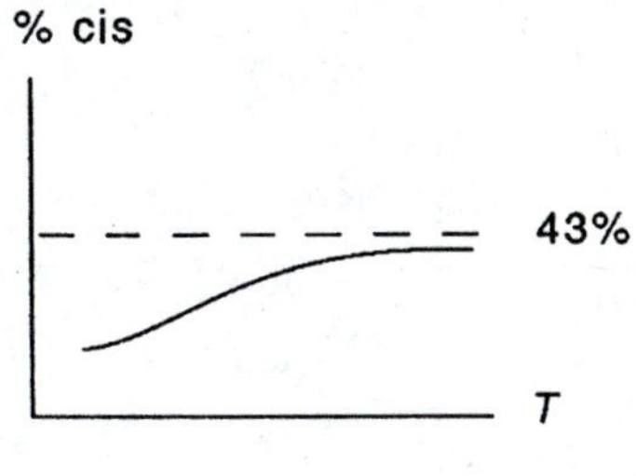

(d) $\Delta G° = \Delta H° - T\Delta S°$, where $\Delta S°$ and $\Delta H°$ are constant. In this equation, the coefficient $-\Delta S°$ of T is negative, so $\Delta G°$ decreases as T increases. As noted in (a), K_P° decreases as T increases. We have $\Delta G°/T = -R \ln K_P^\circ$, and since K_P° is decreasing as T increases, $\Delta G°/T$ is increasing as T increases, which is the opposite behavior as shown by $\Delta G°$. (If this seems puzzling, note that $\Delta G°$ is negative.)

(e) Yes. For the reverse reaction, trans → cis, the results of (d) show that $\Delta G°$ increases and K_P° increases as T increases. (The behavior of K_P° is determined not by $\Delta G°$ but by $\Delta G°/T$.)

6.60 $0 = (\partial/\partial m) \sum_i (y_i - mx_i - b)^2 = \sum_i 2(-x_i)(y_i - mx_i - b) =$ $-2\sum_i x_i y_i + 2m \sum_i x_i^2 + 2b \sum_i x_i$, so $b = (\sum_i x_i y_i - m \sum_i x_i^2)/\sum_i x_i$. (Eq. A) $0 = (\partial/\partial b) \sum_i (y_i - mx_i - b)^2 = -\sum_i (y_i - mx_i - b) = -\sum_i y_i + m \sum_i x_i + nb$, since $\sum_{i=1}^{n} b = nb$; thus $b = (\sum_i y_i - m \sum_i x_i)/n$ (Eq. B). Equating the right sides of Eqs. A and B, and solving the resulting equation for m, we find $m = (n \sum_i x_i y_i - \sum_i x_i \sum_i y_i)/[n \sum_i x_i^2 - (\sum_i x_i)^2]$. Substituting this expression for m into $b = (\sum_i y_i - m \sum_i x_i)/n$ and using a common denominator for the terms on the right, we obtain the equation given in the text.

6.61 **(a)** $n_i - n_{i,0} = \Delta n_i = \nu_i \xi$, so $n_{i,\text{eq}} = n_{i,0} + \nu_i \xi$. $n_{N_2} = 1 \text{ mol} - \xi$, $n_{H_2} = 3 \text{ mol} - 3\xi$, $n_{NH_3} = 2\xi$.

(b) $G = \sum_i n_i G^*_{m,i}(T, P_i) = \sum_i \mu_i^*(T, P_i) = \sum_i n_i[\mu_i^\circ(T) + RT \ln (P_i/P°)]$, where the sum goes over N_2, H_2, and NH_3, the n_i's are given in (a), and $P_{H_2} = [(3 \text{ mol} - 3\xi)/(4 \text{ mol} - 2\xi)]P$, $P_{NH_3} = [2\xi/(4 \text{ mol} - 2\xi)]P$, $P_{N_2} = (n_{N_2}/n_{\text{tot}})P = [(1 \text{ mol} - \xi)/(4 \text{ mol} - 2\xi)]P$. $H = \sum_i n_i H^*_{m,i}(T, P_i)$. Since H of an ideal gas is independent of P, $H^*_{m,i} = H^\circ_{m,i}$ and $H = \sum_i n_i H^\circ_{m,i,T}$, where the n_i's are given in (a).

(c) Results at $\xi = 0, 0.2, 0.3, 0.4, 0.6, 0.8, 1.0$ are: G/(kJ/mol) = –284.73, –288.45, –289.06, –289.20, –288.17, –286.07, –277.21; H/(kJ/mol) = 23.55, 3.60, –6.37, –16.34, –36.29, –56.23, –76.18; TS/(kJ/mol) = 308.28, 292.05, 282.69, 272.86, 251.88, 228.83, 201.03.

```
10 HN = 5.91
12 HH = 5.88
14 HA = –38.09
16 G1 = –97.46
18 G2 = –66.99
20 G3 = –144.37
22 T = 500
24 R = 8.314E–3
26 P0 = 1
28 P = 4
30 FOR X = 0 TO 1 STEP 0.1
32 NN = 1 – X
34 NH = 3 – 3*X
36 NA = 2*X
38 PN = ((1–X)/(4–2*X))*P
40 PH = ((3–3*X)/(4–2*X))*P
42 PA = (2*X/(4–2*X))*P
44 IF PA = 0 THEN PA = 1E–7
46 IF PN = 0 THEN PN = 1E–7
48 IF PH = 0 THEN PH = 1E–7
50 GN = G1 + R*T*LOG(PN/P0)
52 GH = G2 + R*T*LOG(PH/P0)
54 GA = G3 + R*T*LOG(PA/P0)
56 G = NN*GN + NH*GH + NA*GA
58 H = NN*HN + NH*HH + NA*HA
60 TS = H – G
62 PRINT "XI=";X;"G=";G;
   " H=;H;"TS=";TS
64 NEXT X
66 END
```

6.62 **(a)** Any reaction with $\Delta n = 0$; e.g., $H_2(g) + Cl_2(g) \rightleftarrows 2HCl(g)$.

(b) From (6.36), to have K_P° independent of T requires that $\Delta H^\circ = 0$. Two mirror-image species have the same energy, so ΔH° is zero for *d*-CHFClBr(g) $\rightleftarrows$ *l*-CHFClBr(g). [Actually, because of the nonconservation of parity, two mirror-image molecules have an *extremely* tiny, experimentally undetectable energy difference.]

6.63 **(a)** The system discussed near the end of Sec. 6.6 with $n_{N_2} = 3.00$ mol, $n_{H_2} = 1.00$ mol, $n_{NH_3} = 1.00$ mol has $x_{N_2} = 3.00/5.00 = 0.600$. If the reaction $N_2 + 3H_2 \rightleftarrows 2NH_3$ shifts slightly to the right with 0.100 mol of N_2 being consumed, the new amounts are $n_{N_2} = 2.90$ mol, $n_{H_2} = 0.70$ mol, $n_{NH_3} = 1.20$ mol, and $x_{N_2} = 2.90/4.80 = 0.604$. Thus x_{N_2} has increased even though n_{N_2} has decreased.

(b) $x_i = n_i/n_{tot}$; $dx_i = dn_i/n_{tot} - (n_i/n_{tot}^2)\, dn_{tot}$. We have $dn_i = \nu_i\, d\xi$ and $dn_{tot} = d\sum_j n_j = \sum_j dn_j = \sum_j \nu_j\, d\xi = (\sum_j \nu_j)\, d\xi = (\Delta n/\text{mol})\, d\xi$. So $dx_i = (\nu_i/n_{tot})\, d\xi - (x_i/n_{tot})(\Delta n/\text{mol})\, d\xi = n_{tot}^{-1}\,[\nu_i - x_i(\Delta n/\text{mol})]\, d\xi$.

6.64 Rodolfo is right. The equations cited by Mimi show that $(d/dT)(RT \ln K_P^\circ) = \Delta S^\circ$, so that the sign of ΔS° determines whether $T \ln K_P^\circ$ increases or decreases as T increases, but the sign of $(d/dT)(T \ln K_P^\circ)$ can differ from the sign of $(d/dT)(\ln K_P^\circ)$.

6.65 (b).

6.66 Clementine is correct. Joel's statement that if ΔH° is positive, then K_P° keeps increasing as T increases is correct, but this does not guarantee that K_P° will become large at a high-enough T. The following figure shows a function that keeps increasing as x increases, but never exceeds 0.01.

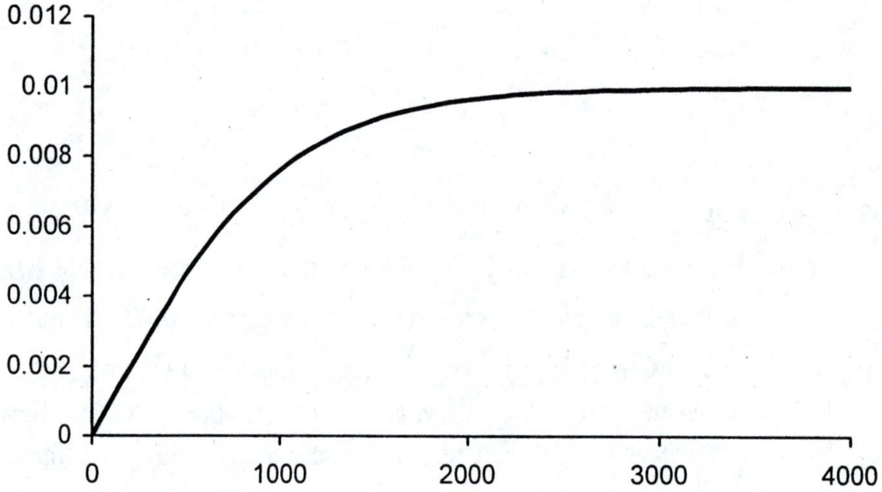

6.67 **(a)** False.

(b) False; G is minimized only if the system is held at constant T and P.

(c) True; see Eq. (6.4). **(d)** False; see Sec. 6.6. **(e)** False; See Fig. 6.8.

(f) True; see Fig. 6.8. **(g)** True. **(h)** False; this is true only if $\Delta n = 0$.

(i) False. **(j)** True.

(k) True; from (6.13) and (6.19), K_P° depends on P° but K_P does not.

(l) False. **(m)** False; **(n)** True.

Reminder: Use the solutions manual as an incentive to work problems, not as a way to avoid working problems.

R6.1 Integration of the van't Hoff equation (6.36) with the approximation that $\Delta H°$ is constant gives $\ln(K^\circ_{P,2}/K^\circ_{P,1}) \approx (\Delta H°/R)(T_1^{-1} - T_2^{-1})$, so

$\Delta H° \approx$ (8.314 J/mol-K)[ln(0.125/0.84)][(298 K)$^{-1}$ – (315 K)$^{-1}$]$^{-1}$ = 87.5 kJ/mol. Then ΔG°_{298} = –(8.314 J/mol-K)(298 K) ln 0.84 = 430 J/mol. Then $\Delta S° = (\Delta H° - \Delta G°)/T$ = (87500 J/mol – 430 J/mol)/(298 K) = 292 J/mol-K at 298 K.

R6.2 **(a)** Eq. (4.31) gives C_P/T.

(b) $dG = -S\,dT + V\,dP$ gives V.

(c) Eq. (4.30) gives C_P.

(d) $dG = -S\,dT + V\,dP$ gives $-S$.

(e) The Euler relation applied to $dG = -S\,dT + V\,dP$ gives $-(\partial V/\partial T)_P$.

(f) Starting with $dH = T\,dS + V\,dP$, the steps given preceding (4.48) lead to $-TV\alpha + V$.

R6.3 **(a)** $\mu_i = (\partial G/\partial n_i)_{T,P,n_{j\neq i}}$; **(b)** $\mu_i^\beta = (\partial G^\beta/\partial n_i^\beta)_{T,P,n^\beta_{j\neq i}}$.

R6.4 $CH_3CH_2CH_2OH(l)$ + 4.5 $O_2(g) \rightarrow 3CO_2(g) + 4H_2O(l)$. Appendix data give –2021.3 kJ/mol =

$$3(-393.509 \text{ kJ/mol}) + 4(-285.830 \text{ kJ/mol}) - 4.5(0) - \Delta_f H^\circ_{298,\text{propanol}}$$

and $\Delta_f H^\circ_{298,\text{propanol}} = -302.4$ kJ/mol. The propanol formation reaction is 3C(graphite) + $4H_2O(l)$ + 0.5 $O_2(g) \rightarrow CH_3CH_2CH_2OH(l)$. For this reaction, Δn_g/mol = 0 – 0.5 = –0.5, so $\Delta_f U^\circ_{298}$ =

–302.4 kJ/mol – (–0.5)(0.008314 kJ/mol-K)(298.1 K) = –301.2 kJ/mol.

R6.5 The correct choice is (c). When the total number of moles of product gases equals the total number of moles of reactant gases, a change in P does not change the equilibrium composition. Otherwise it does.

R6.6 $\Delta G° = 97.89 \text{ kJ/mol} - 2(51.31 \text{ kJ/mol}) = -4.73 \text{ kJ/mol}$.

$-4730 \text{ J/mol} = -(8.314 \text{ J/mol-K})(298.1 \text{ K}) \ln K_P°$. $\ln K_P° = 1.908$; $K_P° = 6.74$.

(a) At equilibrium, $n_{NO_2} = 0.100 - 2z$ and $n_{N_2O_4} = z$. $P_i/P° = n_iRT/VP°$.

$VP°/RT = (3000 \text{ cm}^3)(750/760)\text{atm}/(82.06 \text{ cm}^3 \text{ atm/mol-K})(298.1 \text{ K}) = 0.121$

$$K_P° = \frac{z}{(0.100 - 2z)^2} \frac{RT/VP°}{(RT/VP°)^2} = 6.74 = \frac{z}{0.0100 - 0.400z + 4z^2} 0.121$$

$222.8z^2 - 23.28z + 0.557 = 0$. $z = 0.0674$ or 0.0371. The root 0.0674 makes n_{NO_2} negative and is rejected. So $n_{NO_2} = 0.100 - 2(0.0371) = 0.0258$ mol and $n_{N_2O_4} = 0.0371$ mol.

(b) $n_{tot} = 0.100 - z$, $P_i/P° = x_iP/P°$.

$$K_P° = 6.74 = \frac{[z/(0.100 - z)]}{[(0.100 - 2z)/(0.100 - z)]^2} \frac{P/P°}{(P/P°)^2} = \frac{z(0.100 - z)}{(0.100 - 2z)^2 1.25}$$

$z = 0.0585$ or 0.0415. The first root is rejected, so $n_{NO_2} = 0.100 - 2(0.0415) = 0.017$ and $n_{N_2O_4} = 0.0415$.

R6.7 **(a)** F. **(b)** F. **(c)** T.

R6.8 **(a)** Solid sucrose. **(b)** Equal. **(c)** Water vapor.

R6.9 **(a)** J; **(b)** J/mol; **(c)** no units; **(d)** J/mol; **(e)** J mol^{-1} K^{-1}; **(f)** J/mol.

R6.10 $\Delta H = q_P = (333.6 \text{ J/g})(18.015 \text{ g}) = 6.01 \text{ kJ}$.

$\Delta S = q_{rev}/T = (6.01 \text{ kJ})/(273.15 \text{ K}) = 22.0 \text{ J/K}$.

$\Delta G = \Delta H - T\,\Delta S = \Delta H - T(\Delta H/T) = 0$.

$A = G - PV$ and $\Delta A = \Delta G - P\,\Delta V = -P\,\Delta V =$

$-(1.00 \text{ atm})(18.01 \text{ g})[(1.000 \text{ g/cm}^3)^{-1} - (0.917 \text{ g/cm}^3)^{-1}] =$

$(1.63 \text{ cm}^3 \text{ atm})(8.314 \text{ J}/82.06 \text{ cm}^3 \text{ atm}) = 0.165 \text{ J}$.

R6.11 **(a)** T and P; minimum; **(b)** isolated; decrease; **(c)** system plus surroundings; a maximum; **(d)** $\Delta H°$.

R6.12 $C_{P,\text{m}} = aT^3$; 0.54 J/(mol K) = a(6.0 K)^3; a = 0.0025 J/(mol K^4).

$S_{\text{m},T'} = \int_0^{T'} (C_{P,\text{m}}/T)\,dT = \int_0^{T'} (aT^2)\,dT = aT'^3/3 =$

[0.0025 J/(mol K^4)](6.0 K)^3/3 = 0.18 J/(mol K).

R6.13 **(a)** $\mu_i = \mu_i^\circ + RT\ln(P_i/P^\circ)$; **(b)** $\sum_i \nu_i\mu_i = 0$; **(c)** See Sec. 6.2.

R6.14 **(a)** See Sec. 3.1. **(b)** See Sec. 5.7.

R6.15 Similar to the first figure in Fig. 5.11 but ending with the liquid phase.

R6.16 **(a)** $\mu_{\text{sucrose}(s)} = \mu_{\text{sucrose}(aq)}$; **(b)** $2\mu_{\text{NO}(g)} + \mu_{\text{O}_2(g)} = 2\mu_{\text{NO}_2(g)}$.

R6.17 From $(\partial H/\partial T)_P = C_P$, we get $(\partial\,\Delta H^\circ/\partial T)_P = \Delta C_P^\circ$. Integration gives $\Delta H_2^\circ - \Delta H_1^\circ = \int_1^2 \Delta C_P^\circ\,dT \approx \Delta C_P^\circ(T_2 - T_1)$, where made the approximation that ΔC_P° is constant. $\Delta C_{P,298}^\circ$/(J/mol-K) = 2(37.20) – 2(29.844) –29.355 = –14.643. ΔH_{298}°/(kJ/mol) = 2(33.18) – 2(90.25) – 0 = –114.14.
$\Delta H_{360}^\circ \approx$ –114.14 kJ/mol + (–0.014643 kJ/mol-K)(360 K – 298.1 K) = –115.05 kJ/mol.

Chapter 7

7.1 **(a)** F; **(b)** F.

7.2 **(a)** $c = 2$ (water and sucrose), $p = 1, r = 0, a = 0; f = c - p + 2 - r - a = 3$. The degrees of freedom are T, P, and sucrose mole fraction.

(b) $c = 3, p = 1, r = 0, a = 0;\ f = 4$. The degrees of freedom are T, P, x_{sucrose}, x_{ribose}.

(c) $c = 3$ (sucrose, ribose, water), $p = 2$ (solid sucrose, the solution); $f = 3 - 2 + 2 - 0 - 0 = 3$ (T, P, x_{ribose}). Note that x_{sucrose} in the solution saturated with sucrose is fixed.

(d) $c = 3$ (sucrose, ribose, water), $p = 3$ (solid sucrose, solid ribose, the solution); $f = 3 - 3 + 2 - 0 - 0 = 2$ (T and P).

(e) $c = 1$ (water), $p = 2; f = 1 - 2 + 2 - 0 - 0 = 1$. The degree of freedom can be taken as either T or P. (Once T is fixed, P is fixed at the vapor pressure of water.)

(f) $c = 2$ (water, sucrose), $p = 2;\ f = 2 - 2 + 2 - 0 - 0 = 2$ [T and x_{sucrose} (or P and x_{sucrose})].

(g) $c = 2, p = 3; f = 2 - 3 + 2 - 0 - 0 = 1$ (either T or P).

(h) $c = 2, p = 3; f = 2 - 3 + 2 = 1$ (either T or P).

7.3 **(a)** $f = c_{\text{ind}} - p + 2$ and $p = c_{\text{ind}} + 2 - f$. The smallest possible f is zero, so $p_{\text{max}} = c_{\text{ind}} + 2$.

(b) If $p = 10$, then c_{ind} must be at least 8.

7.4 **(a)** $H_2O \rightleftarrows H^+ + OH^-$, $H_3PO_4 \rightleftarrows H^+ + H_2PO_4^-$, $H_2PO_4^- \rightleftarrows H^+ + HPO_4^{2-}$, $HPO_4^{2-} \rightleftarrows H^+ + PO_4^{3-}$. The equilibrium conditions are:

$\mu_{H_2O} = \mu_{H^+} + \mu_{OH^-}$, $\mu_{H_3PO_4} = \mu_{H^+} + \mu_{H_2PO_4^-}$,

$\mu_{H_2PO_4^{2-}} = \mu_{H^+} + \mu_{HPO_4^{2-}}$, $\mu_{HPO_4^{2-}} = \mu_{H^+} + \mu_{PO_4^{3-}}$. The electroneutrality condition is $x_{H^+} = x_{OH^-} + x_{H_2PO_4^-} + 2x_{HPO_4^{2-}} + 3x_{PO_4^{3-}}$. There are 7 species ($H_2O$, H^+, OH^-, H_3PO_4, $H_2PO_4^-$, HPO_4^{2-}, PO_4^{3-}) and Eq. (7.10) gives $c_{\text{ind}} = 7 - 4 - 1 = 2$. So $f = 2 - 1 + 2 = 3$. T, P, H_3PO_4 mole fraction.

(b) $x_{K^+} = x_{Br^-}$ and $x_{Na^+} = x_{Cl^-}$ (assuming that no solid precipitates out of the solution). We shall neglect the ionization of water. (Its inclusion would not change *f*.) The electroneutrality relation is $x_{K^+} + x_{Na^+} = x_{Br^-} + x_{Cl^-}$ and is not an additional condition since it follows from the two preceding equations. The species are K^+, Br^-, Na^+, Cl^-, and H_2O. Equation (7.10) gives $c_{ind} = 5 - 0 - 2 = 3$. Hence $f = 3 - 1 + 2 = 4$.

7.5 (a) $c_{ind} = 3 - 0 - 0 = 3$ and $f = 3 - 1 + 2 = 4$. A reasonable choice is *T*, *P*, and two of the mole fractions.

(b) $2NH_3 \rightleftarrows N_2 + 3H_2$, so $\mu_{N_2} + 3\mu_{H_2} = 2\mu_{NH_3}$ and $r = 1$. Hence $c_{ind} = 3 - 1 - 0 = 2$ and $f = 2 - 1 + 2 = 3$. *T*, *P*, and one mole fraction. (the other two mole fractions are determined by the reaction equilibrium condition and $\sum_i x_i = 1$.

(c) $2\mu_{NH_3} = \mu_{N_2} + 3\mu_{H_2}$ and $x_{H_2} = 3x_{N_2}$ (since all the N_2 and H_2 come from decomposition of NH_3). Hence $c_{ind} = 3 - 1 - 1 = 1$ and $f = 1 - 1 + 2 = 2$. [In (b) and (c), the catalyst was not counted in finding *c*. If the catalyst is considered to be a species whose amount can be varied, then *c* and c_{ind} are each increased by 1. If the catalyst is a solid, then *p* is increased by 1 and *f* is unchanged; if the catalyst is in the gas phase, then *f* is increased by 1.] *T* and *P*.

(d) We have $r = 1$. The mole fractions satisfy $x_{N_2} + x_N = 1$ and the condition that all the N come from N_2 does not give an additional relation between mole fractions. $f = c - p + 2 - r - a = 2 - 1 + 2 - 1 - 0 = 2$. *T* and *P*.

(e) We have the reaction equilibrium $CaCO_3(s) \rightleftarrows CaO(s) + CO_2(g)$ and the phase equilibria $CaCO_3(s) \rightleftarrows CaCO_3(g)$ and $CaO(s) \rightleftarrows CaO(g)$. There is 1 reaction equilibrium condition. The phase equilibrium conditions have already been taken into account in deriving the phase rule; see Eqs. (7.3)–(7.6). There are no relations between mole fractions. There are three species ($CaCO_3$, CaO, CO_2) and $c_{ind} = 3 - 1 - 0 = 2$. Hence $f = 2 - 3 + 2 = 1$. *T*.

7.6 We have the reactions (1) $HCN \rightleftarrows H^+ + CN^-$ and (2) $H_2O \rightleftarrows H^+ + OH^-$. Let n_{1,H^+} and n_{2,H^+} denote the moles of H coming from reactions 1 and 2, respectively. Stoichiometry gives $n_{1,\text{H}^+} = n_{\text{CN}^-}$ and $n_{2,\text{H}^+} = n_{\text{OH}^-}$, so $n_{\text{CN}^-} + n_{\text{OH}^-} = n_{1,\text{H}^+} + n_{2,\text{H}^+} = n_{\text{H}^+}$.

7.7 **(a)** $c = 2$ (H_2O and NaCl), $p = 2$, $f = 2 - 2 + 2 - 0 - 0 = 2$ (T and P).

(b) $c = 4$ (H_2O, NaCl, Na^+, Cl^-), $p = 2$; $r = 1$, since we have the equilibrium $NaCl(s) \rightleftarrows Na^+(aq) + Cl^-(aq)$; $a = 1$, since we have the electroneutrality condition $x_{Na^+(aq)} = x_{Cl^-(aq)}$. So $f = 4 - 2 + 2 - 1 - 1 = 2$.

7.8 **(a)** $c = 4$, $p = 1$, $r = 2$ [$H_2O \rightleftarrows H^+ + OH^-$ and $2H_2O \rightleftarrows (H_2O)_2$], and $a = 1$ ($x_{H^+} = x_{OH^-}$); $f = c - p + 2 - r - a = 4 - 1 + 2 - 2 - 1 = 2$ (T and P). $c_{ind} = c - r - a = 1$.

(b) $c = 5$, $p = 1$, $r = 3$ [the reaction in (a) and $H_2O + (H_2O)_2 \rightleftarrows (H_2O)_3$], $a = 1$; $f = 2$. $c_{ind} = 1$.

7.9 **(a)** Here, the equilibrium conditions $P^\alpha = P^\beta = P^\gamma = \cdots$ are eliminated. Instead of specifying 1 pressure, we must specify p pressures (where p is the number of phases). This increases f by $p - 1$, so $f = c_{ind} - p + 2 + p - 1 = c_{ind} + 1$.

(b) Here, the equilibrium conditions of Eqs. (7.3) to (7.6) are eliminated. There are $c(p - 1)$ such conditions, so f is increased by $c(p - 1)$ and $f = c - p + 2 - r - a + cp - c = cp - p + 2 - r - a = c_{ind} - p + 2 + cp - c$.

7.10 **(a)** T; **(b)** T; **(c)** F; **(d)** F; **(e)** T; **(f)** T; **(g)** T; **(h)** T; **(i)** T.

7.11 **(a)** Liquid; **(b)** gas; **(c)** liquid; **(d)** liquid; **(e)** gas.

7.12 **(a)** $c = 1$, $p = 2$; $f = 1 - 2 + 2 - 0 - 0 = 1$, as in Prob. 7.2e.

(b) $c = 1$, $p = 1$; $f = 2$. **(c)** $c = 1$, $p = 3$; $f = 1 - 3 + 2 - 0 - 0 = 0$

7.13 **(a)** Treat the vapor as an ideal gas. The liquid's volume is negligible compared to the container's volume. If the equilibrium vapor pressure is reached before all the liquid vaporizes, then the gas will be at 23.76 torr, and

$$n_{\text{gas}} = \frac{PV}{RT} = \frac{(23.76/760)\text{ atm }(10000\text{ cm}^3)}{(82.06\text{ cm}^3\text{-atm/mol-K})(298.15\text{ K})} = 0.01278\text{ mol}$$

$m_{\text{gas}} = (0.01278\text{ mol})(18.015\text{ g/mol}) = 0.230\text{ g}$

There are 0.230 g of vapor and 0.130 g of liquid.

(b) With $V = 20000\text{ cm}^3$, we would get 0.460 g of vapor if the equilibrium vapor pressure of 23.76 torr were reached before all the liquid vaporized. However, there is only 0.360 g of water present initially, so all the liquid vaporizes to give a system with only vapor present.

7.14

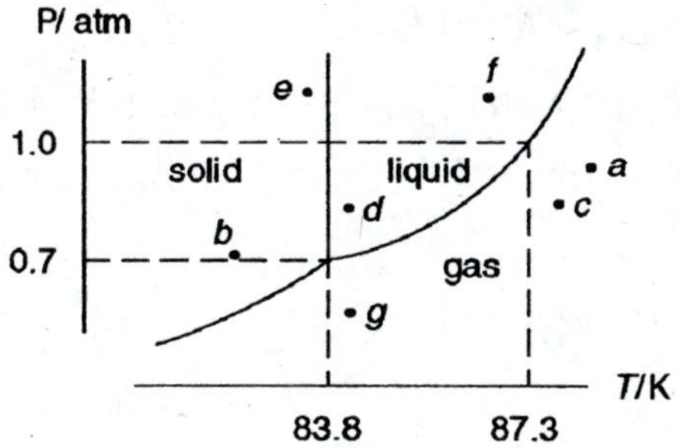

(a) Gas; **(b)** solid; **(c)** gas; **(d)** liquid; **(e)** solid; **(f)** liquid; **(g)** gas.

7.15 Isothermally and reversibly increase the pressure to the pressure at which ice melts at –10°C. Then reversibly freeze the water at constant T and P. Finally, reversibly and isothermally reduce the pressure on the ice to 1 atm.

7.16 **(a)** Ar, which is larger and so has greater intermolecular attractions.

(b) H_2O, due to hydrogen bonding. **(c)** C_3H_8, which is larger.

7.17 **(a)** $\Delta_{\text{vap}}S_{\text{m}} \approx 21$ cal/mol-K = 87 J/mol-K $\approx \Delta_{\text{vap}}H_{\text{m}}/(319.4\text{ K})$, and $\Delta_{\text{vap}}H_{\text{m}} \approx 6.7$ kcal/mol = 28 kJ/mol.

(b) $\Delta_{\text{vap}}S_{\text{m}} \approx 4.5R + R\ln 319.4 = 20.4$ cal/mol-K = 85.4 J/mol-K $\approx \Delta_{\text{vap}}H_{\text{m}}/(319.4\text{ K})$ and $\Delta_{\text{vap}}H_{\text{m}} \approx 6.52$ kcal/mol = 27.3 kJ/mol.

7.18 The hydrogen bonding increases the degree of order in the liquid and so decreases S of the liquid. Therefore, ΔS for the transition liquid → gas is increased.

7.19 **(a)** The T-H-E rule gives $\Delta_{vap}H_{m,nbp,CO}/(81.7\text{ K}) =$ (8.314 J/mol-K)(4.5 + ln 81.7) = 74.0 J/mol-K and $\Delta_{vap}H_{m,,nbp,CO} = 6.05$ kJ/mol. Similarly we get 55.7 kJ/mol for anthracene, 168 kJ/mol for $MgCl_2$, and 295 kJ/mol for Cu.

(b) Experimental values in kJ/mol (from Barin and Knacke, Reed et al., and the AIP handbook) are: CO—6.04, anthracene—56.5, $MgCl_2$—156, and Cu—304.

7.20 Assuming ideal vapor and using (3.30) we have $\Delta S_m = \text{const.} =$ $\Delta S_{m,1} + \Delta S_{m,2} = \Delta_{vap}S_{m,nbp} + R\ln(V^{\dagger}_m/V_{m,i}) = \Delta_{vap}S_{m,nbp} +$ $R\ln V^{\dagger}_m - R\ln(RT_{nbp}/1\text{ atm}) = \Delta_{vap}S_{m,nbp} - R\ln(T_{nbp}/\text{K}) + R\ln V^{\dagger}_m -$ $R\ln(R\text{ K}/1\text{ atm}) = \text{const.}$, so $\Delta_{vap}S_{m,nbp} = R\ln(T_{nbp}/\text{K}) + \text{const.} + R\ln V^{\dagger}_m -$ $R\ln(R\text{ K}/1\text{ atm}) = R\ln(T_{nbp}/\text{K}) + \text{const}'.$, which is the T-H-E rule.

7.21 **(a)** T; **(b)** T; **(c)** F; **(d)** T; **(e)** F; **(f)** F.

7.22 Assuming ideal vapor, neglecting $V_{m,liq}$ in $V_{m,gas} - V_{m,liq}$, and neglecting the T dependence of $\Delta_{vap}H_m$, we use the integrated Clausius–Clapeyron equation (7.21). If state 1 is the normal boiling point 34.5°C and state 2 is 25.0°C, then $\ln(P_2/760\text{ torr}) \approx -[(6380\text{ cal/mol})/(1.987\text{ cal/mol-K})] \times$ $[(298.1\text{ K})^{-1} - (307.6_5\text{ K})^{-1}] = -0.332_5 = \ln(P_2/\text{torr}) - \ln 760$ and $\ln(P_2/\text{torr}) = 6.301$; $P_2/\text{torr} = 545$; $P_2 = 545$ torr.

7.23 **(a)** Integration of $dP/dT = \Delta H/(T\,\Delta V)$ gives $P_2 - P_1 = (\Delta H/\Delta V)\ln(T_2/T_1)$ if ΔH and ΔV are assumed constant. From Prob. 2.49, for one gram $\Delta H =$ 333.6 J and $\Delta V = (1.000)^{-1}\text{ cm}^3 - (0.917)^{-1}\text{ cm}^3 = -0.091\text{ cm}^3$. We have

$$P_2 = 1\text{ atm} + \frac{333.6\text{ J}}{-0.091\text{ cm}^3}\frac{82.06\text{ cm}^3\text{ atm}}{8.3145\text{ J}}\ln\frac{272.15\text{ K}}{273.15\text{ K}} = 133._7\text{ atm}$$

If ΔS and ΔV are assumed constant, then $P_2 - P_1 = (\Delta H/T_1\,\Delta V)(T_2 - T_1)$;

$$P_2 = 1\text{ atm} + \frac{333.6\text{ J}}{(273.15\text{ K})(-0.091\text{ cm}^3)}\frac{82.06\text{ cm}^3\text{ atm}}{8.3145\text{ J}}(-1.00\text{ K}) = 133._5\text{ atm}$$

(b) With 272.15 K replaced by 263.15 K, we get $P_2 = 1350$ atm if ΔH and ΔV are assumed constant or 1325 atm if ΔS and ΔV are assumed constant.

(c) ΔH and ΔV change a lot for this large ΔP.

7.24 Use Eq. (7.24). For 1 g, $\Delta_{fus}V = (1\text{g})/(13.690\text{ g/cm}^3) - (1\text{ g})/(14.193\text{ g/cm}^3) = 0.00259\text{ cm}^3$.

(a) $$(T_2 - T_1) = (234.25\text{ K})\frac{0.00259\text{ cm}^3}{2.82\text{ cal}}99\text{ atm}\frac{1.987\text{ cal}}{82.06\text{ cm}^3\text{ atm}} = 0.52\text{ K}$$

$T_2 = 234.2_5\text{ K} + 0.52\text{ K} = 234.7_7\text{ K} = -38.4°\text{C}$

(b) Replacing 99 atm by 499 atm in the above equation, we get $\Delta T = 2.60$ K and $T_2 = 236.8_5\text{ K} = -36.3°\text{C}$.

7.25 **(a)** $\Delta_{vap}H_{m,ave} = 40.7$ kJ/mol and $\ln(760\text{ torr}/P_1) \approx -[(40700\text{ J/mol})/8.314\text{ J/mol-K}][1/(351.5\text{ K}) - 1/(298.2\text{ K})] = 2.49$. We get $P_1 = 63$ torr, which is pretty close to the true value 59 torr.

(b) $dP/dT = \Delta H/(T\,\Delta V) \approx \Delta H/TV_{gas}$. The true V_{gas} is less than $V_{gas,ideal}$, so the true dP/dT is greater in magnitude than the dP/dT used in the calculation in (a). Therefore the change in P calculated allowing for nonideality will be greater in magnitude and the 25°C calculated vapor pressure will be less than 63 torr, likely bringing it closer to the true value 59 torr.

7.26 Since IT = const. × P, we have $\ln(IT/\text{K}) = \ln\text{ const} + \ln P$. The slope of an $\ln(IT/\text{K})$ vs. T^{-1} curve therefore equals the slope of an $\ln P$ vs. $1/T$ curve, which equals $-\Delta_{sub}H_m/R$, according to the Clausius–Clapeyron equation.

(b) $-2.18 \times 10^4\text{ K} = -\Delta_{sub}H_m/(8.314\text{ J/mol-K})$ and $\Delta_{sub}H_m = 181$ kJ/mol.

7.27 Equation (7.21) gives

$$\ln\frac{760\text{ torr}}{23.76\text{ torr}} = -\frac{\Delta H_m}{1.987\text{ cal/mol-K}}\left(\frac{1}{373.15\text{ K}} - \frac{1}{298.15\text{ K}}\right)$$

$\Delta H_m = 10.2\text{ kcal/mol} = 42.7\text{ kJ/mol}$

7.28 $\Delta H_m = (539.4\text{ cal/g})(18.015\text{ g/mol}) = 9717\text{ cal/mol}$.

(a) Equation (7.21) gives

$$\ln\frac{P}{760\text{ torr}} = -\frac{9717\text{ cal/mol}}{1.987\text{ cal/mol-K}}\left(\frac{1}{393.15\text{ K}} - \frac{1}{373.15\text{ K}}\right)$$

$\ln(P/760 \text{ torr}) = 0.667, \quad P/760 \text{ torr} = 1.950, \quad P = 1480 \text{ torr}$

(b) $\ln \dfrac{446 \text{ torr}}{760 \text{ torr}} = -\dfrac{9717 \text{ cal/mol}}{1.987 \text{ cal/mol - K}}\left(\dfrac{1}{T} - \dfrac{1}{373.15 \text{ K}}\right)$

$1/T = 2.789 \times 10^{-3} \text{ K}^{-1}, \quad T = 358._6 \text{ K} = 85._4°\text{C}$

7.29 **(a)** A plot of $\ln(P/\text{torr})$ vs. $1/T$ has slope $-\Delta H_m/R$.

$\ln(P/\text{torr})$	−2.4214	−1.2986	−0.2934	0.6125	
$10^4/T$	28.317	26.799	25.436	24.204	K^{-1}

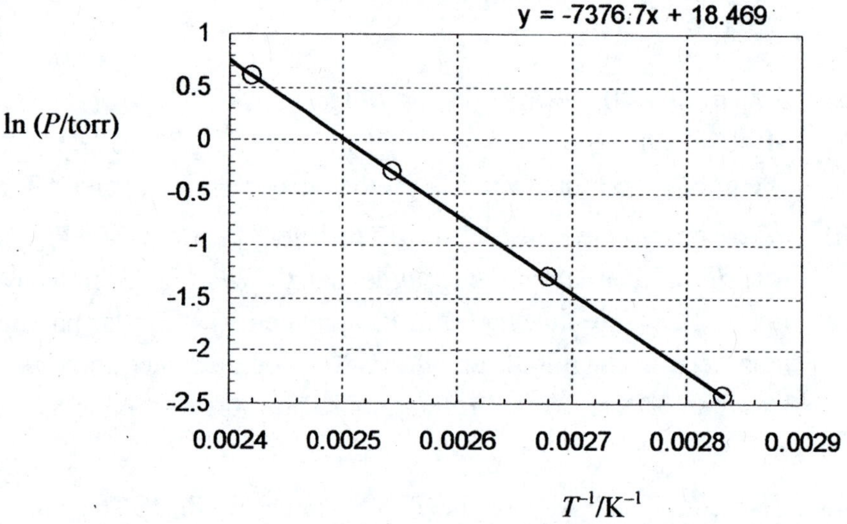

The plot is linear with slope $[0.76 - (-2.50)]/(0.00240 - 0.00284_2)\text{K}^{-1} = -7380 \text{ K} = -\Delta H_m/(1.987 \text{ cal/mol-K})$ and $\Delta H_m = 14.6_6$ kcal/mol = 61.3 kJ/mol.

(b) Equation (7.21) gives

$$\ln \frac{P}{1.845 \text{ torr}} = -\frac{61300 \text{ J/mol}}{8.3145 \text{ J/mol-K}}\left(\frac{1}{433.15 \text{ K}} - \frac{1}{413.15 \text{ K}}\right)$$

$\ln (P/1.845 \text{ torr}) = 0.8240, \quad P = 4.206 \text{ torr}$

(Alternatively, the graph can be extrapolated.)

(c) Use of Eq. (7.21) gives

$$\ln \frac{760 \text{ torr}}{1.845 \text{ torr}} = -\frac{14600 \text{ cal/mol}}{1.987 \text{ cal/mol-K}}\left(\frac{1}{T} - \frac{1}{413.15 \text{ K}}\right)$$

$T = 623 \text{ K} = 350°\text{C}$

(The true normal boiling point is 356.6°C. The error arises because ΔH_m is not constant over the long temperature interval from 140 to 350°C.)

(d) Setting up a spreadsheet like Fig. 7.8, one finds the regression analysis of the ln P versus $1/T$ data gives intercept $b = 18.4706$ and slope -7377.53 (corresponding to $\Delta_{vap}H_m = 61.3_4$ kJ/mol), and these values give a sum of squares of residuals of P experimental versus P calculated of 7×10^{-6}. When the Solver is run to minimize this sum, we get 18.4326 and -7362.4 as the intercept and slope, with the sum reduced to 6×10^{-7}. This gives $\Delta_{vap}H_m = 61.2_1$ kJ/mol. When this value is used in (b), we get $P =$ 4.200 torr.

7.30 **(a)** A plot of $\ln(P/\text{torr})$ vs. $1/T$ has slope $-\Delta H_m/R$.

$\ln(P/\text{torr})$	2.283	3.544_7	4.6521	5.6330	
$10^3/T$	6.5295	6.1293	5.7753	5.4600	K^{-1}

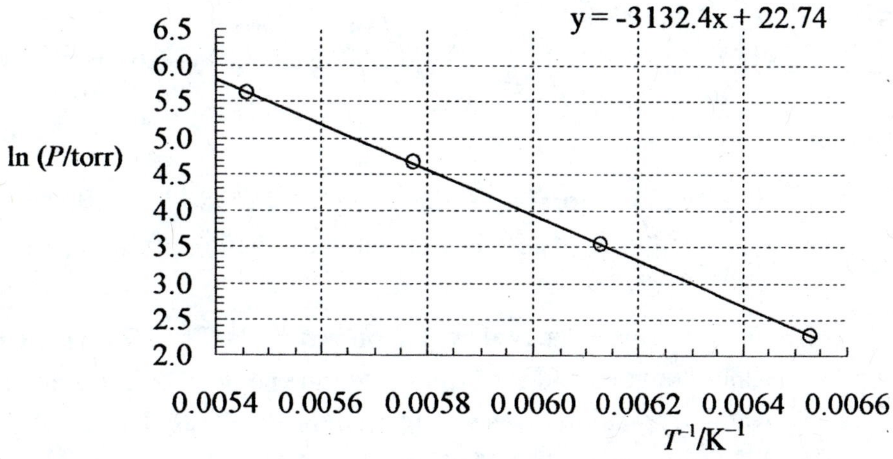

The plot is linear with slope $(5.82 - 2.08)/(0.00540 - 0.00660)\text{K}^{-1} =$ -3120 K $= -\Delta H_m/(8.3145$ J/mol-K) and $\Delta H_m = 25.9$ kJ/mol $=$ 6.20 kcal/mol.

(b) Equation (7.21) gives $\ln(P/279.5 \text{ torr}) =$ $-[(25900 \text{ J/mol})/(8.314 \text{ J/mol-K})][(198 \text{ K})^{-1} - (183 \text{ K})^{-1}]$ and $P = 1015$ torr.

7.31 Trouton's rule is $\Delta H_m/T \approx 10.5\,R$. At the normal boiling point, $\Delta V_m = V_{m,gas} - V_{m,liq} \approx V_{m,gas} = RT/P = RT_b/(1 \text{ atm})$. For a small change in P, we have $dP/dT \approx \Delta P/\Delta T$ and the reciprocal of Eq. (7.18) becomes $\Delta T/\Delta P \approx$ $(\Delta H_m/T)^{-1}\Delta V_m \approx (10.5R)^{-1}RT_b/(1 \text{ atm}) = T_b/(10\tfrac{1}{2} \text{atm})$.

7.32 **(a)** The 0°C path solid → liquid → gas shows that $\Delta_{sub}H_m = \Delta_{fus}H_m + \Delta_{vap}H_m$ = 51.07 kJ/mol. (Pressure changes have no significance since P has little effect on H.)

(b) The Clapeyron equation is $dP/dT = \Delta H_m/(T\,\Delta V_m)$. For the solid–vapor line, $\Delta H_m = \Delta_{sub}H_m$ and $\Delta V_m = V_{m,gas} - V_{m,sol.} = RT/P - M_{sol.}/\rho_{sol.}$ = (82.06 cm^3-atm/mol-K)(273 K)/[(4.585/760)atm] – (18.0 g/mol)/(0.92 g/cm^3) = 3.716×10^6 cm^3/mol – 20 cm^3/mol = 3.716×10^6 cm^3/mol, assuming ideal vapor. $(dP/dT)_{sol.-gas} = \Delta H_m/(T\,\Delta V_m)$ = (51070 J/mol)/(273.16 K)(3.716×10^6 cm^3/mol) = (5.031×10^{-5} J/cm^3-K)(82.06 cm^3-atm/8.3145 J) = 4.966×10^{-4} atm/K = 0.3774 torr/K. For the liquid–vapor line, $\Delta H_m = \Delta_{vap}H_m$ = 45060 J/mol; $\Delta V_m = 3.716 \times 10^6$ cm^3/mol; $(dP/dT)_{liq\text{-}vap}$ = 0.3330 torr/K. For the solid–liquid line, ΔH_m = 6.01 kJ/mol, $\Delta V_m = M/\rho_{liq} - M/\rho_{solid}$ = (18.015 g/mol)(1.000 cm^3/g – 1.0905 cm^3/g) = –1.630 cm^3/mol, and $dP/dT = -1.012 \times 10^5$ torr/K.

7.33 **(a)** At 25°C, ln (P/torr) = 18.3036 – 3816.44/(298.15 – 46.13) = 3.1602 and P = 23.58 torr. At 150°C, the Antoine equation gives ln (P/torr) = 8.1810 and P = 3572 torr.

(b) The Clausius–Clapeyron equation is $d \ln P/dT = \Delta H_m/RT^2$, with ideal vapor assumed and the liquid's volume neglected. Differentiation of the Antoine equation gives at 100°C: $d \ln P/dT = B/(T/\text{K} + C)^2\text{K}$ = $3816.44/(373.15 - 46.13)^2\text{K} = 0.035687/\text{K} = \Delta H_m/RT^2$ = $\Delta H_m/(8.3145\ \text{J/mol-K})(373.15\ \text{K})^2$ and $\Delta H_{m,vap}$ = 41.315 kJ/mol.

7.34 $\ln(T_2/T_1) = \ln[(T_1 + \Delta T)/T_1] = \ln(1 + \Delta T/T_1) = \Delta T/T_1 - \frac{1}{2}(\Delta T/T_1)^2 + \cdots \approx \Delta T/T_1 = (T_2 - T_1)/T_1$.

7.35 **(a)** We make the approximations that the solid and liquid volumes are negligible compared to the vapor volume, that ΔH of vaporization and sublimation are constant, and that the vapor is ideal. Then Eqs. (7.21) and (7.22) apply and show that ln P varies linearly with $1/T$. We plot ln P vs. $1/T$ for the solid and join the two points by a straight line; we do the same for the liquid. At the triple point, the solid and liquid vapor pressures are equal, so the intersection point of the two lines gives the triple point; this is found to be $P = 15._4$ torr, $T = 200$ K.

(b) Equation (7.21) gives ln (10.0 torr/1.00 torr) = $(\Delta_{sub}H_m/1.987 \text{ cal mol}^{-1} \text{ K}^{-1})(1/177.0 \text{ K} - 1/195.8 \text{ K})$ and $\Delta_{sub}H_m =$ 8430 cal/mol. Also ln (100.0 torr/33.4 torr) = $\Delta_{vap}H_m/1.987 \text{ cal mol}^{-1} \text{ K}^{-1})(1/209.6 \text{ K} - 1/225.3 \text{ K})$ and $\Delta_{vap}H_m =$ 6550 cal/mol. $\Delta_{fus}H_m/(\text{cal/mol}) = 8430 - 6550 = 1880$.

7.36 To use the Clausius–Clapeyron equation to find the solid's 1200°C vapor pressure, we need the enthalpy $\Delta_{sub}H$ of sublimation of the solid and we need to know the solid's vapor pressure at some particular temperature T'. We shall take T' as the triple-point temperature T_{tr}, since the solid and liquid vapor pressures are equal at T_{tr}. The triple-point temperature differs only slightly from the normal melting point, so we take $T_{tr} = 1452°\text{C} = 1725$ K. We use Eq. (7.21) to find the liquid's vapor pressure at $T_{tr} = 1725$ K; this equals the solid's vapor pressure at 1725 K. Equation (7.21) gives for the liquid–vapor equilibrium: $\ln (P_2/P_1) = -\Delta H_m(1/T_2 - 1/T_1)$ and

$$\ln \frac{1.00 \text{ torr}}{0.100 \text{ torr}} = -\frac{\Delta_{vap}H_m}{1.987 \text{ cal/mol-K}}\left(\frac{1}{2078 \text{ K}} - \frac{1}{1879 \text{ K}}\right)$$

which gives $\Delta_{vap}H_m = 89.79$ kcal/mol. Use of (7.21) for the liquid between $T_{tr} = 1725$ K and 1606°C gives

$$\ln \frac{0.100 \text{ torr}}{P_{tr}} = -\frac{89790 \text{ cal/mol}}{1.987 \text{ cal/mol - K}}\left(\frac{1}{1879 \text{ K}} - \frac{1}{1725 \text{ K}}\right)$$

which gives $P_{tr} = 0.0117$ torr. The paths solid → gas and solid → liquid → gas at the triple point give $\Delta_{sub}H = \Delta_{fus}H + \Delta_{vap}H$. Thus $\Delta_{sub}H = 4.2_5$ kcal/mol + 89.79 kcal/mol = 94.0_4 kcal/mol. Now we use (7.21) for the solid:

$$\ln \frac{P}{0.0117 \text{ torr}} = -\frac{94040 \text{ cal/mol}}{1.987 \text{ cal/mol-K}}\left(\frac{1}{1473 \text{ K}} - \frac{1}{1725 \text{ K}}\right) = -4.694$$

$P = 1.1 \times 10^{-4}$ torr

7.37 When the two forms are in equilibrium, their chemical potentials are equal. We thus want to make $G_{m,di} = G_{m,gr}$. For each form, $dG_m = -S_m\, dT + V_m\, dP = V_m\, dP$ at constant T. Solids are nearly incompressible, so we neglect the change in V_m with P. For each form, $\Delta G_m = V_m\, \Delta P$ or $G_m(P_2) = G_m(P_1) + V_m\, \Delta P$, where $P_1 = 1$ bar. Setting $G_{m,di}(P_2) = G_{m,gr}(P_2)$, we have $G_{m,di}(P_1) + V_{m,di}\,\Delta P = G_{m,gr}(P_1) + V_{m,gr}\,\Delta P$ and $\Delta P = [G_{m,gr}(P_1) - G_{m,di}(P_1)]/(V_{m,di} - V_{m,gr}) = -\Delta_f G^\circ_{di}/(V_{m,di} - V_{m,gr})$, since $P_1 = 1$ bar $= P^\circ$. Using $V_m = M/\rho$, we get $\Delta P =$ $[(-6070 \text{ J/mol})/(-1.97 \text{ cm}^3\text{/mol})](82.06 \text{ cm}^3 \text{ atm})/(8.314 \text{ J}) = 30400$ atm and $P_2 = 30400$ atm.

7.38 When gray (g) tin is more stable than white (w), we have $\mu_g < \mu_w$ and $G_{m,g} < G_{m,w}$. We have $dG_m = -S_m\,dT + V_m\,dP = -S_m\,dT$ at the constant P of 1 bar. Appendix data show that $S_{m,w} > S_{m,g}$, so $G_{m,w}$ decreases faster than $G_{m,g}$ as T increases, and $G_{m,w}$ increases more rapidly than $G_{m,g}$ as T decreases. At the temperature T_{eq} with $G_{m,g} = G_{m,w}$, the two forms are in equilibrium. Below T_{eq}, we have $G_{m,g} < G_{m,w}$ and gray tin is more stable. Let $T_1 = 25°C$, $T_2 = T_{eq}$, and $\Delta T = T_{eq} - 25°C$. We have $\Delta G_{m,g} = G_{m,g}(T_{eq}) - G_{m,g}(25°C)$ and $\Delta G_{m,w} = G_{m,w}(T_{eq}) - G_{m,w}(25°C)$, so $G_{m,g}(T_{eq}) = G_{m,w}(T_{eq})$ becomes $G_{m,g}(25°C) + \Delta G_{m,g} = G_{m,w}(25°) + \Delta G_{m,w}$ or $\Delta G_{m,g} - \Delta G_{m,w} = G_{m,w}(25°C) - G_{m,g}(25°C) =$ $0 - 0.13$ kJ/mol, where Appendix data were used. Since $\Delta G_m = -\int_1^2 S_m\,dT$ at constant P, we have $\Delta G_{m,g} - \Delta G_{m,w} = -\int_1^2 (S_{m,g} - S_{m,w})\,dT \approx$ $(S_{m,w} - S_{m,g})(T_2 - T_1) = (51.55 - 44.14)(\text{J/mol-K})(T_2 - 25°C)$, if we neglect the T dependence of $S_{m,g} - S_{m,w}$. Then -130 J/mol $\approx (7.41$ J/mol-K$)(T - 25°C)$ and $T_{eq} - 25°C \approx -17.5$ K $= -17.5°C$, so $T_{eq} \approx 7\tfrac{1}{2}$ °C.

7.39 In Example 7.7, $V_{m,gr} - V_{m,di}$ was assumed independent of pressure. At 25°C and 1 atm, $V_{m,gr} = 5.34$ cm^3/mol and $V_{m,di} = 3.41$ cm^3/mol. Since graphite is more compressible than diamond, as P increases, $V_{m,gr}$ will decrease more rapidly than $V_{m,di}$ and the difference $V_{m,gr} - V_{m,di}$ will decrease as P increases. Since this volume difference occurs in the denominator of the expression for $P_2 - P_1$, $P_2 - P_1$ will be greater than calculated in Example 7.7 and P_2 will be greater than 15100 bar.

7.40 $\mu_{B(s)} = \mu_{B(\text{sat sln})}$ and $\mu_{C(s)} = \mu_{C(\text{sat sln})}$, where sat soln denotes the substance in a saturated solution. Since B is the more-stable solid, we have $\mu_{B(s)} < \mu_{C(s)}$. Therefore $\mu_{B(\text{sat sln})} < \mu_{C(\text{sat sln})}$. The crystal structures of the two solids differ, but in a solution we can expect that B and C will have the same structure. The reason that $\mu_{C(\text{sat sln})}$ exceeds $\mu_{B(\text{sat sln})}$ must therefore be due to different concentrations in the saturated solutions in equilibrium with the two solids. From the inequality (4.90), the higher chemical potential of the solute in the solution in equilibrium with C(*s*) must mean that the solute's mole fraction is greater in this solution.

7.41 $C_P = (\partial H/\partial T)_P = T(\partial S/\partial T)_P$. Since $C_P > 0$, both H and S increase as T increases. A first-order transition has $\Delta H > 0$ and $\Delta S > 0$, and so we see a sudden jump in H and in S at the transition temperature T_{trs}. A second-order transition has $\Delta H = 0$ and $\Delta C_P \neq 0$, so there is no sudden jump in H but there is a sudden change in the slope $\partial H/\partial T = C_P$ at T_{trs}. A lambda transition has $\Delta H = 0$ and if $C_P \to \infty$ at T_{trs} the H vs. T curve has a vertical (infinite-slope) inflection point at T_{trs}. Since $(\partial S/\partial T)_P = C_P/T$, the S-vs.-T curves resemble the H-vs.-T curves.

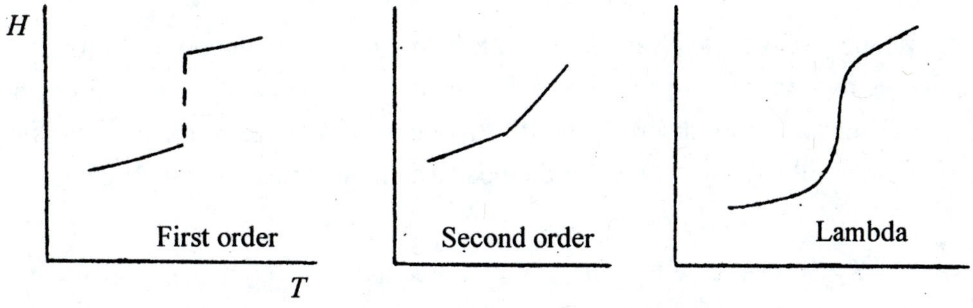

7.42 **(a)** At $T = 0$, $w = 0$ and $\sigma_\ell = r/r = 1$.

(b) If $r = 0$, $\sigma_\ell = -w/w = -1$. This is a highly ordered state; with Cu and Zn atoms interchanged, each Cu is surrounded only by Zn atoms.

(c) If $r = w$, then $\sigma_\ell = 0$.

(d) In the upper ($T = 0$) diagram, $w = 0$ and $\sigma_\ell = 1$. In the lower diagram, comparison with the upper diagram gives $r = 16$, $w = 16$, and $\sigma_\ell = (16 - 16)/(16 + 16) = 0$.

(e) At $T = 0$, $n_{rp} = n_p$ and $\sigma_s = 2 - 1 = 1$.

(f) As $T \to \infty$, $n_{rp} = \frac{1}{2}n_p$ and $\sigma_s = 2(\frac{1}{2}) - 1 = 0$.

(g) In the upper ($T = 0$) figure, $n_{rp} = n_p$ and $\sigma_s = 1$. In the lower figure: $n_p = 4(3) + 2 + 4(3) + 2 + 4(3) + 2 + 7 = 49$, and $n_{rp} = 2 + 2 + 2 + 1 + 2 + 1 + 3 + 1 + 3 + 2 + 3 + 1 + 2 + 1 + 2 + 1 = 29$, where the counting was done by looking at the neighbors of the atoms in rows, 2, 4, 6, and 8. Hence $\sigma_s = 2(29)/49 - 1 = 0.18$.

(h) From Sec. 7.5, both curves have infinite slope at T_λ:

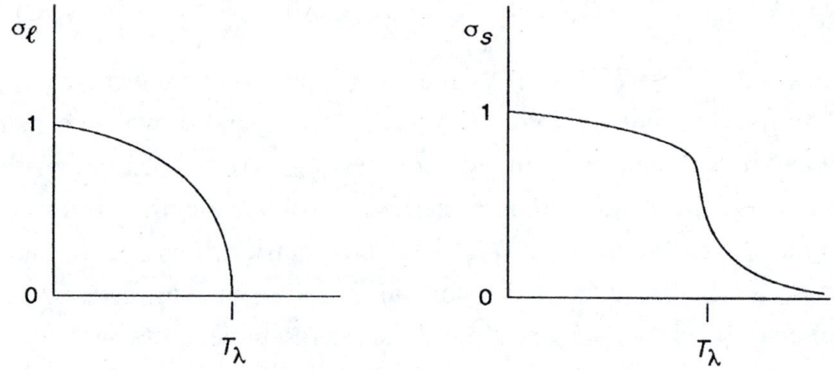

7.43 At 0 K, the most stable arrangement in the crystal has the negative end of each molecule next to the positive end of the adjacent molecule. As T increases, the crystal becomes more disordered until there is a 50-50 chance for the negative end of one molecule to be next to the positive end of the adjacent molecule.

7.44 **(a)** At $T = T_\lambda$ we have $t = 0$. For $T < T_\lambda$, we have $t > 0$. As the positive quantity t goes to zero, $t^{-\alpha} = t^{0.01285}$ goes to zero and C_P/(J/mol-K) goes to B. Likewise, s is positive for $T > T_\lambda$ and $s^{-\alpha} \to 0$ as the positive quantity s goes to 0. Hence $C_P = 456$ J/mol-K at T_λ.

(b) For $T < T_\lambda$, $\partial C_P/\partial T = (\partial C_P/\partial t)(\partial t/\partial T) = -(\partial C_P/\partial t)/T_\lambda =$ $-(A'/\alpha)[-\alpha t^{-\alpha-1} + D(0.5-\alpha)t^{-0.5-\alpha} + E(1-\alpha)t^{-\alpha}]/T_\lambda$. We have $\alpha = -0.013$, so $-\alpha - 1 < 0$, $-0.5 - \alpha < 0$, and $-\alpha > 0$. As the positive quantity t goes to 0, $1/t^{\alpha+1} \to 1/0 = \infty$ and $1/t^{0.5+\alpha} \to \infty$. The quantity $1/t^{\alpha+1} \approx 1/t^{0.99}$ goes to infinity faster than $1/t^{0.5+\alpha} \approx 1/t^{0.49}$ and the coefficient A' is positive, so $\partial C_P/\partial T \to +\infty$ as $t \to 0$ and $T \to T_\lambda$ from below. Similarly for s.

7.45 **(a)** T; **(b)** T.

7.46 From Prob. 2.6, 1 erg = 1 dyn cm and 1 dyn = 10^{-5} N, so 1 erg/cm^2 = 1 dyn/cm = (10^{-5} N)/(10^{-2} m) = 10^{-3} N/m = 10^{-3} J/m^2 = 1 mN/m = 1 mJ/m^2.

7.47 **(a)** $V = (4/3)\pi r^3$ and $r = (3V/4\pi)^{1/3} = [3(1.0\text{ cm}^3)/4\pi]^{1/3} = 0.62$ cm. $\mathscr{A} = 4\pi r^2 = 4\pi(0.62\text{ cm})^2 = 4.8\text{ cm}^2$.

(b) The volume of one particle is $(4/3)\pi(30 \times 10^{-9} \times 10^2\text{ cm})^3 = 1.13 \times 10^{-16}\text{ cm}^3$. The number of particles is $(1.0\text{ cm}^3)/(1.13 \times 10^{-16}\text{ cm}^3) = 8.8_4 \times 10^{15}$. The area of one particle is $4\pi(30 \times 10^{-9} \times 10^2\text{ cm})^2 = 1.13 \times 10^{-10}\text{ cm}^2$. The total area of the particles is $(8.8_4 \times 10^{15})(1.13 \times 10^{-10}\text{ cm}^2) = 1.0 \times 10^6\text{ cm}^2$.

7.48 $dw_{rev} = \gamma\, d\mathscr{A}$ and $w_{rev} = \gamma\, \Delta\mathscr{A} = (73 \times 10^{-3}\text{ N/m})(3.0 \times 10^{-4}\text{ m}^2) = 2.2 \times 10^{-5}$ J.

7.49 $\gamma = \gamma_0(1 - T/T_c)^{11/9}$. At 0°C, $26.5\text{ mN/m} = \gamma_0[1 - (273.1\text{ K})/(523.2\text{ K})]^{11/9} = 0.4057\gamma_0$ and $\gamma_0 = 65.3$ mN/m.
At 50°C, $\gamma = (65.3\text{ mN/m})[1 - (323.1\text{ K})/(523.2\text{ K})]^{11/9} = 20.2$ mN/m.

7.50 $\gamma_0 = (37.8)^{2/3}(523.2)^{1/3}(0.432/0.252 - 0.951)\text{ dyn/cm} = 69.3$ dyn/cm. $\gamma = \gamma_0(1 - T/T_c)^{11/9} = (69.3\text{ dyn/cm})[1 - (273.1\text{ K})/(523.2\text{ K})]^{11/9} = 28.1$ dyn/cm. The percent error is $100(28.1 - 26.5)/26.5 = 6.0\%$.

7.51 $\gamma/l_z = [(50 \times 10^{-3}\text{ J/m}^2)/(0.1\text{ m})](82.06 \times 10^{-6}\text{ m}^3\text{ atm})/(8.314\text{ J}) = 4.9 \times 10^{-6}$ atm.

7.52 **(a)** F; **(b)** T.

7.53 $P^\alpha - P^\beta = 2\gamma/R = 2(73 \times 10^{-3}\text{ N/m})/(0.00040\text{ m}) = 365\text{ J/m}^3 = (365\text{ J/m}^3)(82.06 \times 10^{-6}\text{ m}^3\text{ atm})/(8.314\text{ J}) = 0.0036\text{ atm} = 2.7$ torr.
$P^\alpha = P^\beta + 2.7\text{ torr} = 762.7$ torr.

7.54 $\gamma = \frac{1}{2}(0.7914 - 0.0012)(10^{-3}\text{ kg}/10^{-6}\text{ m}^3)(9.807\text{ m/s}^2)(0.0333\text{ m})(0.000175\text{ m}) = 0.0226\text{ N/m} = 22.6$ mN/m, where (7.37) was used.

7.55 $h = 2\gamma \cos\theta/(\rho_\beta - \rho_\alpha)gr = 2(490\text{ dyn/cm})\cos 140°/(13.59 - 0.001)(\text{g/cm}^3)(980.7\text{ cm/s}^2)(0.0175\text{ cm}) = -3.22$ cm.

7.56 $h = 2\gamma/(\rho_\beta - \rho_\alpha)gr = 2(52.2 \text{ dyn/cm})/(0.3383 \text{ g/cm}^3)(980.7 \text{ cm/s}^2)(0.0175 \text{ cm}) =$ 18.0 cm.

7.57 **(a)** For $\theta = 0$, we have a hemispherical interface (as in Fig. 7.24b). The volume of the liquid above the meniscus is the difference in volume between a cylinder and a hemisphere and equals $(\pi r^2)r - \frac{1}{2}(4/3)\pi r^3 = \pi r^3/3 = (\pi r^2)(r/3)$. Hence, this extra liquid has volume and mass equal to those of a cylindrical column of liquid of height $r/3$, and we must replace h by $h + r/3$ in the equation for γ.

(b) $h + r/3 = 3.33 \text{ cm} + (0.0175 \text{ cm})/3 = 3.33_6 \text{ cm}$. Replacing 0.0333 m by 0.0333_6 m, we get $\gamma = 22.6$ mN/m.

7.58 $\gamma = \frac{1}{2}(\rho_\beta - \rho_\alpha)gh_1r_1$ and $\gamma = \frac{1}{2}(\rho_\beta - \rho_\alpha)gh_2r_2$. $h_1 - h_2 = [2\gamma/(\rho_\beta - \rho_\alpha)g] \times (1/r_1 - 1/r_2)$ and $\gamma = \frac{1}{2}(\rho_\beta - \rho_\alpha)g(h_1 - h_2)r_1r_2/(r_2 - r_1) = \frac{1}{2}(0.900 \text{ g/cm}^3) \times (980.7 \text{ cm/s}^2)(1.00 \text{ cm})(0.0600 \text{ cm})(0.0400 \text{ cm})/(0.0200 \text{ cm}) = 53.0$ dyn/cm.

7.59 **(a)** Taking activity coefficients as 1, we have $K_c = [L_n]/[L]^n$. Conservation of matter gives $c = [L] + n[L_n]$, so $K_c = [L_n]/(c - n[L_n])^n = x/(c - nx)^n$, where $x = [L_n]$; $(c - nx)^n = x/K_c$ and $c = nx + (x/K_c)^{1/n}$.

(b) $f = [L]/c = (c - n[L_n])/c = 1 - nx/c$.

(c) With $K_c^\circ = 10^{200}$, we have $K_c = 10^{200}/(c^\circ)^{49}$, where $c^\circ = 1 \text{ mol/dm}^3$. The equations of (a) and (b) give $c = 50x + c^\circ(x/c^\circ)^{0.02}/10^4$, $f = 1 - 50x/c$, and $[L] = fc$. For various assumed values of $x \equiv [L_n]$, we calculate c, etc., as

x/c°	0	10^{-18}	10^{-10}	10^{-8}	10^{-7}	10^{-6}
$10^5c/c^\circ$	0	4.37	6.31	6.97	7.74	12.6
$10^5n[L_n]/c^\circ$	0	5×10^{-12}	0.00050	0.050	0.500	5.00
$10^5[L]/c^\circ$	0	4.37	6.31	6.92	7.24	7.60
f	1	1.000000	0.99992	0.993	0.935	0.603

x/c°	$10^{-5.5}$	10^{-5}	$10^{-4.5}$	10^{-4}
$10^5c/c^\circ$	23.6	57.9	166	508
$10^5n[L_n]/c^\circ$	15.8	50.0	158	500
$10^5[L]/c^\circ$	7.76	7.94	8.13	8.32
f	0.329	0.137	0.049	0.016

The plots of [L] and n[L_n] have the form of Fig. 7.25b.

(d) From the graph of f, we find that $c = 0.00015$ mol/dm^3 at $f = 0.5$.

7.60 Since ice is the stable phase at –20°C, we have $\mu_{ice} < \mu_{sc\ liq}$, where sc = supercooled. The phase equilibrium condition gives $\mu_{ice} = \mu_{vapor\ above\ ice}$ and $\mu_{sc\ liq} = \mu_{vapor\ above\ sc\ liq}$; therefore $\mu_{vapor\ above\ ice} < \mu_{vapor\ above\ sc\ liq}$. Use of $\mu_i = \mu_i^\circ + RT \ln (P_i/P^\circ)$ [Eq. (6.4)] for the vapors gives

$\mu_i^\circ + RT \ln (P_{i,above\ ice}) < \mu_i^\circ + RT \ln (P_{i,above\ sc\ liq})$, so

$\ln (P_{i,above\ ice}) < \ln (P_{i,above\ sc\ liq})$ and $P_{i,above\ ice} < P_{i,above\ sc\ liq}$.

7.61 $dP/dT = \Delta H/(T\,\Delta V)$. Since V of the solid or liquid is negligible compared to V of the gas, we have to high degree of accuracy, $\Delta V_{solid\to vapor} = \Delta V_{liq\to vapor}$. As shown in Prob. 7.32, $\Delta_{sub}H_m$ is greater than $\Delta_{vap}H_m$ by $\Delta_{fus}H_m$, which is not negligible. Therefore $\Delta H_{solid\to gas} > \Delta H_{liq\to gas}$ and $(dP/dT)_{solid\to gas} > (dP/dT)_{liq\to gas}$. The solid–vapor line has greater slope.

7.62 The pressure due to 10 cm of water is given by (1.9) as $P = \rho gh =$ $[1.00\ (10^{-3}\ \text{kg})/(10^{-6}\ \text{m}^3)](9.80\ \text{m/s}^2)(0.10\ \text{m}) = 980\ \text{N/m}^2 =$ $(980\ \text{J/m}^3)(82.06 \times 10^{-6}\ \text{m}^3\ \text{atm})/(8.314\ \text{J}) = 0.009_7$ atm, where (2.7), (2.12), (1.19) and (1.20) were used. The total pressure 10 cm below the surface is 1.009_7 atm and Eq. (7.24) and data in the example that follows it give

$$0.009_7\ \text{atm} = \frac{333.6\ \text{J}}{-0.091\ \text{cm}^3}\,\frac{82.06\ \text{cm}^3\ \text{atm}}{8.314\ \text{J}}\,\frac{T_2 - 273.15\ \text{K}}{273.15\ \text{K}}$$

$T_2 - 273.15\ \text{K} = -7.3 \times 10^{-5}$ K.

7.63 Neglecting the change in water density with pressure and the change in g with depth, we have as the pressure at 3000 m: $P = \rho gh =$ $(1.0\ \text{g/cm}^3)(1\ \text{kg}/10^3\ \text{g})(10^6\ \text{cm}^3/1\ \text{m}^3)(9.8\ \text{m/s}^2)(3000\ \text{m}) =$ $(2.9 \times 10^7\ \text{Pa})(1\ \text{atm}/1.01 \times 10^5\ \text{Pa}) = 290$ atm. P exceeds the 350°C vapor pressure of water and the stable phase is liquid water.

7.64 Let sc denote supercooled liquid water. All equations in this problem are for –10°C. The phase equilibrium condition and Eq. (6.4) give $\mu_{ice} =$ $\mu_{vapor\ above\ ice} = \mu_i^\circ + RT \ln (1.950\ \text{torr}/750\ \text{torr})$ and $\mu_{sc} = \mu_{vapor\ above\ sc} =$

$\mu_i^\circ + RT \ln (P_i/750 \text{ torr})$. Subtraction gives $\mu_{ice} - \mu_{sc} = RT \ln (1.95 \text{ torr}/P_i)$. But Prob. 4.28*b* gives $\mu_{ice} - \mu_{sc} = G_{m,ice} - G_{m,sc} = (-2.76 \text{ cal/g})(18.0 \text{ g/mol}) = -49.7 \text{ cal/mol}$. (This relation holds at 1 atm, but the effect of pressure on liquid and solid thermodynamic properties is slight and can be ignored.) Therefore $-49.7 \text{ cal/mol} = RT \ln (1.95 \text{ torr}/P_i) =$ $(1.987 \text{ cal/mol-K})(263.1 \text{ K}) \ln (1.95 \text{ torr}/P_i)$. We find $1.95 \text{ torr}/P_i = 0.909_3$ and $P_i = 2.14$ torr. (The experimental value is 2.15 torr.)

7.65 0.01°C is the triple-point temperature, so ice has a vapor pressure of 4.585 torr at 0.01°C.

7.66 **(a)** The initial state (state 1) has partial pressures P_A and P_B above liquid A; the liquid is subject to a pressure $P = P_A + P_B$. The final state (state 2) has partial pressures $P_A + dP_A$ and $P_B + dP_B$ above liquid A, which experiences a pressure $P + dP$, where $dP = dP_A + dP_B$. Let the changes in liquid and vapor A chemical potentials on going from state 1 to 2 be $d\mu(A^\ell)$ and $d\mu(A^g)$. The phase equilibrium conditions for states 1 and 2 are $\mu_1(A^\ell) = \mu_1(A^g)$ and $\mu_2(A^\ell) = \mu_2(A^g) = \mu_1(A^\ell) + d\mu(A^\ell) = \mu_1(A^g) + d\mu(A^g)$. So $d\mu(A^\ell) = d\mu(A^g)$. We have $d\mu(A^\ell) = dG_m(A^\ell) = -S_m(A^\ell)\, dT + V_m(A^\ell)\, dP = V_m(A^\ell)\, dP$, since T is constant. Also, $\mu(A^g) = \mu^\circ(A^g) + RT \ln (P_A/\text{bar})$ and $d\mu(A^g) = (RT/P_A)\, dP_A$ at constant T. Hence $V_m(A^\ell)\, dP = (RT/P_A)\, dP_A = V_m(A^g)\, dP_A$. Q.E.D.

(b) $dP_A/P_A = [V_m(A^\ell)/RT]\, dP$. Integration with $V_m(A^\ell)$ assumed constant gives $\ln (P_{A,2}/P_{A,1}) = [V_m(A^\ell)/RT](P_2 - P_1)$. So $\ln(P_{A,2}/23.76 \text{ torr}) = (18.1 \text{ cm}^3/\text{mol})(1 \text{ atm})/(82.06 \text{ cm}^3\text{-atm/mol-K})(298 \text{ K}) = 0.000740$ and $P_{A,2}/23.76 \text{ torr} = 1.000740$, $P_{A,2} = 23.78$ torr.

7.67 Use the 25°C path:

$$\text{liq(1 bar)} \xrightarrow{a} \text{liq(23.766 torr)} \xrightarrow{b} \text{vap(23.766 torr)} \xrightarrow{c} \text{vap(1 bar)}.$$

We have $dG_m = -S_m\, dT + V_m\, dP = V_m\, dP$ at constant T. For the liquid, $\Delta G_{m,a} = V_m\, \Delta P$, and

$$\Delta G_a = (18.1 \text{ cm}^3/\text{mol})(-0.956 \text{ atm})\frac{8.314 \text{ J}}{82.06 \text{ cm}^3 \text{ atm}} = -1.8 \text{ J/mol}$$

Step b is a constant-T-and-P equilibrium process, so $\Delta G_b = 0$. For step c, $\Delta G_c = \int_1^2 V_m\, dP = \int_1^2 (RT/P)\, dP = RT \ln (P_2/P_1) =$ (8.3145 J/mol-K)(298.15 K) ln (750.062/23.766) = 8557.2 J/mol. $\Delta G_a + \Delta G_b + \Delta G_c = 8555.4$ J/mol. From the Appendix, ΔG/(J/mol) = –228572 + 237129 = 8557.

7.68 **(a)** Trouton's rule is $\Delta H_m/T_b = 10.5R = c$. Substitution into Eq. (7.22) gives $\ln (P/\text{atm}) = -(c/R)(T_b/T) + c/R = -(10.5R/R)T_b/T + 10.5R/R = 10.5(1 - T_b/T)$. $P/\text{atm} = \exp\{10.5[1 - (353.2_5 \text{ K})/T]\}$.

(b) $P/\text{atm} = \exp [10.5(1 - 353.2_5/298.15)] = 0.1436$ and $P = 109$ torr.

(c) $\ln (620/760) = 10.5(1 - 353.2_5/T)$ and $T = 346._5$ K = $73._4$°C.

7.69 **(a)** Use of (7.21) gives at 0°C: ln (4.926/4.258) = $-[\Delta H_m/(1.987/\text{cal/mol-K})][(274.15 \text{ K})^{-1} - (272.15 \text{ K})^{-1}]$ and $\Delta H_{m,273} =$ 10.80 kcal/mol = 45.20 kJ/mol. Similarly, $\Delta H_{m,373} = 9.89$ kcal/mol = 41.37 kJ/mol. Then $\Delta S_{m,273} = \Delta H_{m,273}/T =$ (10800 cal/mol)/(273.1 K) = 39.5 cal/mol-K = 165 J/mol-K; $\Delta S_{m,373} = \Delta H_{m,373}/T = 26.5$ cal/mol-K = 111 J/mol-K. $\Delta G_m = \Delta H_m - T\,\Delta S_m = 0$, as it must be for constant-T-and-P equilibrium processes. The calculated 100°C ΔH_m is slightly in error, since the approximations used to derive Eq. (7.21) are less accurate the higher the vaporization temperature and the closer the temperature is to the critical point; for example, $H_2O(g)$ is denser and hence less ideal at 1 atm and 373 K than at 5 torr and 0°C.

(b) $\Delta H°$ and $\Delta S°$ are for the process liq → ideal gas at 1 bar. Let P be the 0°C vapor pressure. A convenient path is the following 0°C path:

$$\text{liq}(P°) \xrightarrow{1} \text{liq}(P) \xrightarrow{2} \text{vap}(P) \xrightarrow{3} \text{vap}(P°) \xrightarrow{4} \text{ideal vap } (P°)$$

Since moderate pressure changes have little effect on liquid thermodynamic properties, we can take $\Delta H_1 = 0$, $\Delta S_1 = 0$. Gas nonideality is slight at 1 bar, so we take $\Delta H_4 = 0$, $\Delta S_4 = 0$. Since the vapor is nearly ideal and H of an ideal gas depends on T only, we take $\Delta H_3 = 0$. Thus $\Delta H^\circ_{273} = \Delta H_{m,2} = 45.20$ kJ/mol. Use of (3.25), (3.30), and Boyle's law gives $\Delta S^\circ_{273} = \Delta S_{m,2} + \Delta S_{m,3} = \Delta H_{m,2}/T + R \ln (P/P°) =$ (45200 J/mol)/(273 K) + (8.314 J/mol-K) ln (4.58/750) = 123.2 J/mol-K.

7.70 **(a)** $dn_A = dn_A^s - a\,d\xi$, $dn_B = dn_B^s - b\,d\xi$, $dn_E = e\,d\xi$, $dn_F = f\,d\xi$. Then $\sum_i \mu_i\,dn_i = \mu_A(dn_A^s - a\,d\xi) + \mu_B(dn_B^s - b\,d\xi) + \mu_E e\,d\xi + \mu_F f\,d\xi =$ $\mu_A\,dn_A^s + \mu_B\,dn_B^s + (-a\mu_A - b\mu_B + e\mu_E + f\mu_F)\,d\xi = \mu_A\,dn_A^s + \mu_B\,dn_B^s$, since the equilibrium condition for the reaction is $\sum_i \nu_i\mu_i \equiv$ $-a\mu_A - b\mu_B + e\mu_E + f\mu_F = 0$.

(b) $dG = -S\,dT + V\,dP + \sum_i \mu_i\,dn_i = -S\,dT + V\,dP + \mu_A\,dn_A^s + \mu_B\,dn_B^s$; at constant T, P, and n_B^s, $dG = \mu_A\,dn_A^s$ and $\mu_A = (\partial G/\partial n_A^s)_{T,P,n_B^s}$.

7.71 (a) and (c). See Fig. 7.1a.

7.72 **(a)** $dP_A/P_A = (V_{m,A}^\ell/RT)\,dP$ at constant T. Integration gives $\ln(P_{A,2}/P_{A,1}) =$ $(V_{m,A}^\ell/RT)(P_2 - P_1)$, where we neglected the P dependence of $V_{m,A}^\ell$. Let state 1 be the bulk state and state 2 be a drop of liquid with radius r. Equation (7.34) gives the extra pressure experienced by the drop due to its curvature as $P_2 - P_1 = 2\gamma/r$. Hence $\ln(P_r/P_{bulk}) = (V_m^\ell/RT)(2\gamma/r)$ and $P_r = P_{bulk}\exp(2\gamma V_m^\ell/rRT)$.

(b) $V_m^\ell = M/\rho = (18.015\text{ g/mol})/(0.998\text{ g/cm}^3) = 18.05\text{ cm}^3\text{/mol}$. The result of part (a) gives $P_r/(17.535\text{ torr}) =$

$$\exp\left[\frac{2(73\text{ dyn/cm})(18.05\text{ cm}^3\text{/mol})}{(1.00\times10^{-5}\text{ cm})(8.314\times10^7\text{ erg/mol-K})(293.1\text{ K})}\right]$$

and $P_r = 17.726$ torr.

7.73 **(a)** T; **(b)** T; **(c)** F; **(d)** F; **(e)** T; **(f)** F (see Sec. 7.4); **(g)** T. **(h)** F (see the discussion of superheating in Sec. 7.4); **(i)** F [see Eq.(7.3)].

Chapter 8

8.1 **(a)** Pa m^6/mol^2 and m^3/mol; **(b)** Pa $K^{1/2}$ m^6/mol^2 and m^3/mol; **(c)** m^3/mol.

8.2 As $\rho \to 0$, $V_m \to \infty$. In Eq. (8.2), $-b$ can be neglected compared with V_m to give $P = RT/V_m - a/V_m^2 = (1/V_m)(RT - a/V_m)$. As $V_m \to \infty$, $RT - a/V_m$ goes to RT and $P \to RT/V_m$. (Note that it's harder to use (1.39) to get the limiting behavior since both P and a/V_m^2 go to 0.) In (8.4), the terms B/V_m, C/V_m^2, ... each go to 0, giving $PV_m = RT$. When (8.3) is solved for P and b is neglected compared with V_m, we get $P = RT/V_m - a/V_m^2 T^{1/2} = (1/V_m)(RT - a/V_m T^{1/2}) \to RT/V_m$.

8.3 **(a)** $n = (28.8 \text{ g})/(30.07 \text{ g/mol}) = 0.958 \text{ mol}$;
$V_m = V/n = (999 \text{ cm}^3)/(0.958 \text{ mol}) = 1043 \text{ cm}^3/\text{mol}$.

$$P = \frac{(82.06 \text{ cm}^3\text{-atm/mol-K})(298.1 \text{ K})}{1043 \text{ cm}^3/\text{mol}} \times$$

$$\left(1 - \frac{186 \text{ cm}^3/\text{mol}}{1043 \text{ cm}^3/\text{mol}} + \frac{1.06 \times 10^4 \text{ cm}^6/\text{mol}^2}{(1043 \text{ cm}^3/\text{mol})^2}\right) = 19.5 \text{ atm}$$

The ideal-gas P is $P = RT/V_m = 23.5$ atm.

(b) $B^\dagger = B/RT = (-186 \text{ cm}^3/\text{mol})/(82.06 \text{ cm}^3\text{-atm/mol-K})(298.1 \text{ K}) = -0.00760 \text{ atm}^{-1}$;
$C^\dagger = (C - B^2)/R^2T^2 = (1.06 \times 10^4 \text{ cm}^6/\text{mol}^2 - 186^2 \text{ cm}^6/\text{mol}^2)/(82.06 \text{ cm}^3\text{-atm/mol-K})^2(298.1 \text{ K})^2 = -4.01 \times 10^{-5} \text{ atm}^{-2}$.
$V_m = [(82.06 \text{ cm}^3\text{-atm/mol-K})(298.1 \text{ K})/(16.0 \text{ atm})] \times [1 - (0.00760 \text{ atm}^{-1})(16.0 \text{ atm}) - (4.01 \times 10^{-5} \text{ atm}^{-2})(16.0 \text{ atm})^2] = 1327 \text{ cm}^3/\text{mol}$. $V = nV_m = (0.958 \text{ mol})V_m = 1272 \text{ cm}^3$. Also, $V_{m,\text{ideal}} = 1529 \text{ cm}^3/\text{mol}$ and $V_{\text{ideal}} = 1465 \text{ cm}^3$.

8.4 Eq. (8.4) gives $P = RT(1/V_m + B/V_m^2 + C/V_m^3 + \cdots)$. Substitution into the right side of (8.5) gives $PV_m = RT[1 + B^\dagger RT(1/V_m + B/V_m^2 + C/V_m^3 + \cdots) + C^\dagger R^2T^2(1/V_m^2 + 2B/V_m^3 + \cdots) + \cdots]$. We compare this expression for PV_m with (8.4); equating the $1/V_m$ coefficients, we get $B = B^\dagger RT$. Equating the $1/V_m^2$

coefficients, we get $C = B^{\dagger}RTB + C^{\dagger}R^2T^2 = B^{\dagger 2}R^2T^2 + C^{\dagger}R^2T^2 = R^2T^2(B^{\dagger 2} + C^{\dagger})$.

8.5 $V_m = RT/P + B = (82.06\ \text{cm}^3\text{-atm/mol-K})(200\ \text{K})/(1\ \text{atm}) - 47\ \text{cm}^3/\text{mol} = 16.36\ \text{L}$.

8.6 **(a)** $B_{12} \approx (B_1 + B_2)/2 = -387\ \text{cm}^3/\text{mol}$. Let gas 1 be CH_4. Then $x_1 = 0.0300/0.1000 = 0.300$ and $x_2 = 0.700$. The equation at the end of Sec. 8.2 gives $B \approx (0.300)^2(-42\ \text{cm}^3/\text{mol}) + 2(0.300)(0.700)(-387\ \text{cm}^3/\text{mol}) + (0.700)^2(-732\ \text{cm}^3/\text{mol}) = -525\ \text{cm}^3/\text{mol}$. $V/n_{tot} = 10000\ \text{cm}^3/\text{mol}$. $P = [(82.06\ \text{cm}^3\text{-atm/mol-K})(298.1\ \text{K})/(10000\ \text{cm}^3/\text{mol})] \times [1 - (525\ \text{cm}^3/\text{mol})/(1000\ \text{cm}^3/\text{mol})] = 2.32\ \text{atm}$.

(b) With $B_{12} = -180\ \text{cm}^3/\text{mol}$, we get $B = -438\ \text{cm}^3/\text{mol}$ and $P = 2.34$ atm. Also, $P_{ideal} = RT/(V/n_{tot}) = 2.45$ atm.

8.7 $n = (74.8\ \text{g})/(30.07\ \text{g/mol}) = 2.48_8\ \text{mol}$. $V_m = V/n = (200\ \text{cm}^3)/(2.48_8\ \text{mol}) = 80.4\ \text{cm}^3/\text{mol}$.

(a)
$$P = \frac{RT}{V_m} = \frac{(82.06\ \text{cm}^3\text{-atm/mol-K})(310.6\ \text{K})}{80.4\ \text{cm}^3/\text{mol}} = 317\ \text{atm}$$

(b) Equations (8.18) and (8.2) give

$$a = \frac{27(82.06\ \text{cm}^3\text{-atm/mol-K})^2(305.4\ \text{K})^2}{64(48.2\ \text{atm})} = 5.50 \times 10^6\ \text{cm}^6\ \text{atm mol}^{-2}$$

$$b = \frac{(82.06\ \text{cm}^3\text{-atm/mol-K})(305.4\ \text{K})}{8(48.2\ \text{atm})} = 65.0\ \text{cm}^3\ \text{mol}^{-1}$$

$$P = \frac{(82.06\ \text{cm}^3\text{-atm/mol-K})(310.6\ \text{K})}{80.4\ \text{cm}^3/\text{mol} - 65.0\ \text{cm}^3/\text{mol}} - \frac{5.50 \times 10^6\ \text{cm}^6\text{-atm/mol}^2}{(80.4\ \text{cm}^3/\text{mol})^2}$$

$P = 1655\ \text{atm} - 851\ \text{atm} = 804\ \text{atm}$

(c) Equations (8.20), (8.21), and (8.3) give

$$a = \frac{0.4275(82.06\ \text{cm}^3\text{-atm/mol-K})^2(305.4\ \text{K})^{5/2}}{48.2\ \text{atm}}$$

$$= 9.73 \times 10^7\ \text{cm}^6\text{-atm-K}^{1/2}/\text{mol}^2$$

$$b = \frac{0.08664(82.06\ \text{cm}^3\text{-atm/mol-K})(305.4\ \text{K})}{48.2\ \text{atm}} = 45.0\ \text{cm}^3/\text{mol}$$

$P = RT/(V_m - b) - a/V_m(V_m + b)T^{1/2} = 720\ \text{atm} - 548\ \text{atm} = 172\ \text{atm}$

(d) Interpolation gives at 37½°C: $B = -179\ \text{cm}^3/\text{mol} + (7.5/20)(22\ \text{cm}^3/\text{mol}) = -171\ \text{cm}^3/\text{mol}$ and $C = 10119\ \text{cm}^6/\text{mol}^2$.
$P = [(82.06\ \text{cm}^3\text{-atm/mol-K})(310.6\ \text{K})/(80.4\ \text{cm}^3/\text{mol})] \times [1 - (171\ \text{cm}^3/\text{mol})/(80.4\ \text{cm}^3/\text{mol}) + (10119\ \text{cm}^6/\text{mol}^2)/(80.4\ \text{cm}^3/\text{mol})^2] = 139$ atm. Note: The observed pressure is 135 atm.

8.8 $n_{\text{tot}} = 0.2000$ mol, $V/n_{\text{tot}} = 3500\ \text{cm}^3/\text{mol}$.

(a) $P = RT/V_{\text{m}} = (82.06\ \text{cm}^3\text{-atm/mol-K})(313.1\ \text{K})/(3500\ \text{cm}^3/\text{mol}) = 7.34$ atm.

(b) Let C_2H_4 be gas 1. For C_2H_4, $a_1 = 27R^2T_c^2/64P_c = 4.56 \times 10^6\ \text{cm}^6\ \text{atm/mol}^2$, $b_1 = RT_c/8P_c = 58.3\ \text{cm}^3/\text{mol}$. For CO_2, we find $a_2 = 3.61 \times 10^6\ \text{cm}^6\ \text{atm/mol}^2$ and $b_2 = 42.9\ \text{cm}^3/\text{mol}$. Also, $x_1 = 0.0786/0.200 = 0.393$, $x_2 = 0.607$. From the last paragraph of Sec. 8.2,
$a = [(0.393)^2(4.56 \times 10^6) + 2(0.393)(0.607)(4.56 \times 10^6 \times 3.61 \times 10^6)^{1/2} + (0.607)^2(3.61 \times 10^6)](\text{cm}^6\text{-atm/mol}^2) = 3.97 \times 10^6\ \text{cm}^6\text{-atm/mol}^2$;
$b = 0.393(58.3\ \text{cm}^3/\text{mol}) + 0.607(42.9\ \text{cm}^3/\text{mol}) = 49.0\ \text{cm}^3/\text{mol}$.
$P = (82.06\ \text{cm}^3\text{-atm/mol-K})(313.1\ \text{K})/[(3500 - 49)\text{cm}^3/\text{mol}] - (3.97 \times 10^6\ \text{cm}^6\text{-atm/mol}^2)/(3500\ \text{cm}^3/\text{mol})^2 = 7.12$ atm.

(c) $Z = PV_{\text{m}}/RT$; $P = ZRT/V_{\text{m}} = 0.9689(82.06\ \text{cm}^3\text{-atm/mol-K}) \times (313.1\ \text{K})/(3500\ \text{cm}^3/\text{mol}) = 7.11$ atm.

8.9 $P = RT/V_{\text{m}} + RTB(T)/V_{\text{m}}^2 + RTC(T)/V_{\text{m}}^3$. $(\partial P/\partial V_{\text{m}})_T = 0 = -RT/V_{\text{m}}^2 - 2RTB/V_{\text{m}}^3 - 3RTC/V_{\text{m}}^4$. $(\partial^2 P/\partial V_{\text{m}}^2)_T = 0 = 2RT/V_{\text{m}}^3 + 6RTB/V_{\text{m}}^4 + 12RTC/V_{\text{m}}^5$.

$$RT_c V_{\text{m},c}^2 + 2RT_c BV_{\text{m},c} + 3RT_c C = 0 \quad (1)$$

$$2RT_c V_{\text{m},c}^2 + 6RT_c BV_{\text{m},c} + 12RT_c C = 0 \quad (2)$$

Subtract twice equation (1) from equation (2) to get $B = -3C/V_{\text{m},c}$ at $T = T_c$. Substitution of this expression for B into equation (1) gives $C(T_c) = V_{\text{m},c}^2/3$. Also, $B(T_c) = -3C(T_c)/V_{\text{m},c} = -V_{\text{m},c}$. We have $Z_c = P_cV_{\text{m},c}/RT_c = 1 + B(T_c)/V_{\text{m},c} + C(T_c)/V_{\text{m},c}^2 = 1 - 1 + 1/3 = 1/3$.

8.10 **(a)** From (8.18), $b = RT_c/8P_c = (82.06\ \text{cm}^3\text{-atm/mol-K})(150.9\ \text{K})/8(48.3\ \text{atm}) = 32.0\ \text{cm}^3/\text{mol}$; $a = 27R^2T_c^2/64P_c =$

$27(82.06\ \text{cm}^3\text{-atm/mol-K})^2(150.9\ \text{K})^2/64(48.3\ \text{atm}) =$
$1.34 \times 10^6\ \text{cm}^6\ \text{atm/mol}^2.$

(b) Comparison of (8.9) with (8.4) gives as the van der Waals estimate: $B = b - a/RT$. $B_{100\ \text{K}} = 32.0\ \text{cm}^3/\text{mol} - (1.34 \times 10^6\ \text{cm}^6\text{-atm/mol}^2)/(82.06\ \text{cm}^3\text{-atm/mol-K})(100\ \text{K}) = -131\ \text{cm}^3/\text{mol}$. Also, $B_{200\ \text{K}} = -49.6\ \text{cm}^3/\text{mol}$, $B_{300\ \text{K}} = -22.4\ \text{cm}^3/\text{mol}$, $B_{500\ \text{K}} = -0.7\ \text{cm}^3/\text{mol}$, $B_{1000\ \text{K}} = 15.7\ \text{cm}^3/\text{mol}$. Agreement with experiment is fair.

8.11 $\Delta_{\text{vap}}U_{\text{m}} \approx a/V_{\text{m}} = a\rho/M$ and $\Delta_{\text{vap}}H_{\text{m},} \approx a\rho/M + RT$. For N_2 at its normal boiling point, $\Delta_{\text{vap}}H_{\text{m}} \approx (1.35 \times 10^6\ \text{cm}^6\ \text{atm/mol}^2)(0.805\ \text{g/cm}^3)/(28\ \text{g/mol}) + RT = (3.9 \times 10^4\ \text{cm}^3\ \text{atm/mol})(1.987\ \text{cal}/82.06\ \text{cm}^3\ \text{atm}) + (1.987\ \text{cal/mol-K})(77.4\ \text{K}) = 1.1\ \text{kcal/mol}$.
For HCl, $\Delta_{\text{vap}}H_{\text{m}} \approx (3.65 \times 10^6\ \text{cm}^6\ \text{atm/mol}^2)(1.193\ \text{g/cm}^3)/(36.5\ \text{g/mol}) + RT = 2.89\ \text{kcal/mol} + 0.37\ \text{kcal/mol} = 3.3\ \text{kcal/mol}$.
For H_2O, $\Delta_{\text{vap}}H_{\text{m}} \approx (5.46 \times 10^6\ \text{cm}^6\ \text{atm/mol}^2)(0.96\ \text{g/cm}^3)/(18\ \text{g/mol}) + RT = 7.8\ \text{kcal/mol}$. Agreement with experiment is fairly good.

8.12 T in cell D2 is changed. The new graph shows we need to start at a smaller V_{m} value, so the value in A9 is reduced until the graph starts at a substantially positive value for P. An initial V_{m} of 84 cm^3/mol is found to be appropriate. The maximum in P is close to 12 atm, so the vapor pressure is between 0 and 12 atm; averaging these values, we enter an initial guess of 6 atm in C3. The values in columns A and B show that 6 atm corresponds to 2900 cm^3/mol and to a value between 84 and 89 atm, so we use 2900 and 86 cm^3/mol as the initial guesses for V_{m}^{υ} and V_{m}^{l}. The Solver constraints for C3, E3, and G3 need to be appropriately modified. The Solver gives 3.01 atm, 6375 cm^3/mol, and 86.0 cm^3/mol as the vapor pressure, and molar volumes.

8.13 From Table 8.1 and Eqs. (8.20) and (8.21):
$b = 0.08664(82.06\ \text{cm}^3\text{-atm/mol-K})(304.2\ \text{K})/(72.88\ \text{atm}) = 29.68\ \text{cm}^3/\text{mol}$,
$a = 0.42748(82.06\ \text{cm}^3\text{-atm/mol-K})^2(304.2\ \text{K})^{2.5}/(72.88\ \text{atm}) = 6.375 \times 10^7\ \text{K}^{1/2}\ \text{cm}^6\ \text{atm/mol}^2$. These values are entered into C1 and E1 of the Fig. 8.6 spreadsheet and the temperature is changed in D2. The new graph shows no minimum, indicating that we must start at a smaller V_{m} value in A9. One finds 50 or 55 cm^3/mol to be suitable. The local maximum in P is at 48 atm, so the vapor pressure is between 0 and 48 atm, and we use 24 atm as the initial guess.

The column A and B values show 24 atm to correspond to molar volumes of about 60 and 760 cm^3/mol, so we use these as the initial V_m guesses. The Solver constraints for C3, E3, and G3 need to be appropriately modified. The Solver gives 38.7 atm, 57.2 cm^3/mol, and 390 cm^3/mol.

8.14 Table 8.1 and Eq. (8.18) give
$b = (82.06 \text{ cm}^3\text{-atm/mol-K})(369.8 \text{ K})/8(41.9 \text{ atm}) = 90.5 \text{ cm}^3\text{/mol}$,
$a = 27(82.06 \text{ cm}^3\text{-atm/mol-K})^2(369.8 \text{ K})^2/64(41.9 \text{ atm}) = 9.27 \times 10^6$ $\text{cm}^6 \text{ atm/mol}^2$. When the van der Waals expression for P is substituted into (8.24), we get $P(V_m^{\upsilon} - V_m^{l}) = \int_{V_m^l}^{V_m^{\upsilon}} [RT/(V_m - b) - a/V_m^2]\, dV_m$, which gives

$$P = \frac{1}{V_m^{\upsilon} - V_m^{l}}\left[RT \ln\frac{V_m^{\upsilon} - b}{V_m^{l} - b} + a\left(\frac{1}{V_m^{\upsilon}} - \frac{1}{V_m^{l}}\right)\right] \quad (8.25)\text{vdW}$$

The Fig. 8.6 spreadsheet is modified by changing the a and b values, changing (in an efficient way) the formulas in B9, B10,... to the van der Waals expression for P, using the right side of Eq. (8.25)vdW (given above) in C4, and using the right side of the second equation of (8.2) with V_m replaced by V_m^l or V_m^{υ}, in E4, and G4, respectively. The Solver constraints for C3, E3, and G3 need to be appropriately modified. The graph shows P is too high at 95 cm^3/mol, and an appropriate starting V_m is 130 cm^3/mol. The local maximum in P is at 22 atm, and 11 atm is a reasonable initial guess for the vapor pressure. The values in columns A and B show 11 atm corresponds to about 145 and 1885 cm^3/mol, and these are reasonable initial guesses for V_m. The Solver gives 16.6_5 atm , 141.5 cm^3/mol, and 1093 cm^3/mol.

8.15 **(a)** $m = 0.480 + 1.574(0.153) - 0.176(0.153)^2 = 0.717$.

$$a = \frac{0.42748(82.06 \text{ cm}^3\text{-atm/mol-K})^2(369.8 \text{ K})^2}{41.9 \text{ atm}}\left\{1 + 0.717\left[1 - \left(\frac{298.15 \text{ K}}{369.8 \text{ K}}\right)^{1/2}\right]\right\}^2$$

$a = 1.08_2 \times 10^7 \text{ cm}^6 \text{ atm/mol}^2$

(b) We change the a value in C1 and delete the $T^{1/2}$ factors in the denominators of the formulas in cells C4, E4, G4, B9, B10,... (do this in an efficient way). The Solver gives 9.47 atm, 97.6 cm^3/mol, and 2144 cm^3/mol.

8.16 **(a)** Using $\omega = 0.153$, we find $k = 0.604$. From $T_c = 369.8$ K, $P_c = 41.9$ atm, and $k = 0.604$, we get $b = 56.3_5$ cm^3/mol and $a = 1.13_3 \times 10^7$ atm cm^6/mol^2.

(b) When the Peng–Robinson expression for P is substituted into (8.24), we get $P(V_m^\upsilon - V_m^l) = \int_{V_m^l}^{V_m^\upsilon} \{RT/(V_m - b) - a/[V_m(V_m + b) + b(V_m - b)]\}\, dV_m$. Using the given integral with $s = 2b$ and $c = -b^2$, we get

$$P = \frac{1}{V_m^\upsilon - V_m^l}\left[RT\ln\frac{V_m^\upsilon - b}{V_m^l - b} - \frac{a}{b\sqrt{8}}\ln\left(\frac{V_m^\upsilon + (1-\sqrt{2})b}{V_m^\upsilon + (1+\sqrt{2})b}\cdot\frac{V_m^l + (1+\sqrt{2})b}{V_m^l + (1-\sqrt{2})b}\right)\right] \quad \text{(PR)}$$

The Fig. 8.6 spreadsheet is modified by changing the a and b values, changing (in an efficient way) the formulas in B9, B10,... to the Peng–Robinson expression for P, using the right side of Eq. (PR) (given above) in C4, and using the right side of the Peng–Robinson equation with V_m replaced by V_m^l or V_m^υ, in E4, and G4, respectively. An appropriate starting value is 85 cm^3/mol in A9. The maximum in P is at 18 atm, and we take 9 atm as the initial guess for the vapor pressure. This P corresponds to 86 and 2260 cm^3/mol, which are the initial guesses for the molar volumes. The Solver gives 9.38 atm, 2143 and 86.1 cm^3/mol.

8.17 **(a)** For $P = RT/(V_m - b) - a/[V_m(V_m + b)T^{1/2}]$,

$$T\left(\frac{\partial P}{\partial T}\right)_{V_m} - P = T\left(\frac{R}{V_m - b} + \frac{a}{2V_m(V_m + b)T^{3/2}}\right) - \left(\frac{RT}{V_m - b} - \frac{a}{V_m(V_m + b)T^{1/2}}\right)$$

$$= \frac{3a}{2V_m(V_m + b)T^{1/2}}$$

Then, using $\int [\upsilon(\upsilon + b)]^{-1}\, d\upsilon = b^{-1}\ln[\upsilon/(\upsilon + b)]$, we get

$$\Delta_{vap}U_m = \int_{V_m^l}^{V_m^\upsilon} \frac{3a}{2V_m(V_m + b)T^{1/2}}\, dV_m = \frac{3a}{2bT^{1/2}}\ln\frac{V_m^\upsilon(V_m^l + b)}{V_m^l(V_m^\upsilon + b)}$$

and addition of $P\,\Delta V$ gives the desired result.

(b) Example 8.1 gave $V_m^\upsilon = 1823$ cm^3/mol and $V_m^l = 100.3$ cm^3/mol. Then $P(V_m^\upsilon - V_m^l) =$ (10.85 atm)(1723 cm^3/mol) $= 1.869 \times 10^4$ cm^3 atm/mol $=1894$ J/mol. We have

$$\Delta_{vap}U_m = \frac{3(1.807\times 10^8 \text{ cm}^6 \text{ atm K}^{1/2}/\text{mol}^2)}{2(62.7 \text{ cm}^3/\text{mol})(298.15 \text{ K})^{1/2}}\ln\frac{1823(100+63)}{100.3(1823+63)} =$$

1.13×10^5 cm^3 atm/mol $= 1.14_6 \times 10^4$ J/mol.

Then $\Delta_{\text{vap}}H$ = (11460 + 1894) J/mol = 13.3 kJ/mol. (The experimental value is 14.8 kJ/mol.)

8.18 Use of (8.20) and (8.21) gives $a = 3.77 \times 10^8$ cm^6 atm K$^{1/2}$/mol^2 and $b = 92.4$ cm^3/mol. The Fig. 8.6 spreadsheet with a, b, and T changed can be used. One finds a minimum pressure of –14.5 atm at 210 cm^3/mol.

8.19

$$(P + 27R^2T_c^2/64P_cV_m^2)(V_m - RT_c/8P_c) = RT$$

$$[P + (27T_c^2/64P_c V_m^2)(64P_c^2V_{m,c}^2/9T_c^2)](V_m - V_{m,c}/3) = (8P_cV_{m,c}/3T_c)T$$

$$(P + 3P_c/V_r^2)(V_m - V_{m,c}/3) = 8P_cV_{m,c}T_r/3$$

$$(P_r + 3/V_r^2)(V_r - 1/3) = 8T_r/3$$

8.20 (a) $P = RT/(V_m - b) - a/TV_m^2$. The critical-point conditions $(\partial P/\partial V_m)_T = 0$ and $(\partial^2 P/\partial V_m^2)_T = 0$ lead to

$$RT_c/(V_{m,c} - b)^2 = 2a/T_cV_{m,c}^3 \text{ and } RT_c/(V_{m,c} - b)^3 = 3a/T_cV_{m,c}^4$$

Division of the first equation by the second gives $V_{m,c} - b = 2V_{m,c}/3$ and $V_{m,c} = 3b$. Substitution in the first equation gives $RT_c/4b^2 = 2a/27b^3T_c$ and $T_c = (2/3)(2a/3bR)^{1/2}$. Substitution in the Berthelot equation at the critical point gives

$$P_c = \frac{R}{2b}\frac{2}{3}\left(\frac{2a}{3bR}\right)^{1/2} - \frac{a}{9b^2}\frac{3}{2}\left(\frac{3bR}{2a}\right)^{1/2} = \frac{\sqrt{2}}{12\sqrt{3}}\frac{a^{1/2}R^{1/2}}{b^{3/2}}$$

Division of this P_c equation by the preceding T_c expression gives $P_c/T_c = R/8b$, so $b = RT_c/8P_c$. The above P_c expression gives

$$a = \frac{144 \times 3}{2}\frac{P_c^2b^3}{R} = \frac{216P_c^2R^3T_c^3}{8^3P_c^3R} = \frac{27}{64}\frac{R^2T_c^3}{P_c}$$

(b)

$$Z_c = \frac{P_cV_{m,c}}{RT_c} = \frac{\sqrt{2}}{12\sqrt{3}}\frac{a^{1/2}R^{1/2}}{b^{3/2}}\frac{3b}{R}\frac{3}{2}\left(\frac{3bR}{2a}\right)^{1/2} = \frac{3}{8} = 0.375$$

(c) Substitution of $R = 8P_cV_{m,c}/3T_c$ and the above expression for a and b into the Berthelot equation followed by division by $P_cV_{m,c}$ gives

$$\left(P + \frac{27}{64}\frac{R^2T_c^3}{P_c}\frac{1}{TV_m^2}\right)\left(V_m - \frac{RT_c}{8P_c}\right) = RT$$

$$\left(P + \frac{27}{64}\frac{64}{9}\frac{P_c^2 V_{m,c}^2}{T_c^2}\frac{T_c^3}{P_c}\frac{1}{TV_m^2}\right)\left(V_m - \frac{8P_cV_{m,c}}{3T_c}\frac{T_c}{8P_c}\right) = \frac{8P_cV_{m,c}}{3T_c}T$$

$$\left(P_r + \frac{3}{V_r^2 T_r}\right)\left(V_r - \frac{1}{3}\right) = \frac{8}{3}T_r$$

8.21 From (8.29), $P_r = 8T_r/3(V_r - \frac{1}{3}) - 3/V_r^2$. $T_r = T/T_c = 310.6/305.4 = 1.017$; $V_r = V_m/V_{m,c} = V/nV_{m,c} = (200\ \text{cm}^3)/(2.488\ \text{mol})(148\ \text{cm}^3/\text{mol}) = 0.543$. Then $P_r = 8(1.017)/3(0.543 - 0.333) - 3/(0.543)^2 = 2.75 = P/P_c$ and $P = 2.75P_c = 132\frac{1}{2}$ atm.

8.22 $B = (82.06\ \text{cm}^3\text{-atm/mol-K})(150.9\ \text{K})(48.3\ \text{atm})^{-1}[0.597 - 0.462e^{0.7002(150.9\ \text{K})/T}]$ $= [153.1 - 118.4e^{(105.66\ \text{K})/T}]\ \text{cm}^3/\text{mol}$. At $T = 100$, 200, 300, 500, and 1000 K, we get (in cm^3/mol) $-187._5$, $-47._7$, $-15._3$, $6._8$, $21._5$. Agreement with experiment is very good.

8.23 As noted in Sec. 8.8, at T and $P°$, $H_m^{id} - H_m = \int_0^{P°} [T(\partial V_m/\partial T)_P - V_m]\, dP$. Differentiation of (8.5) gives $(\partial V_m/\partial T)_P = RP^{-1}(1 + B^\dagger P + C^\dagger P^2 + \cdots) + RTP^{-1}(PB^{\dagger\prime} + P^2C^{\dagger\prime} + \cdots)$, where the prime means differentiation with respect to T. Use of (8.5) gives $T(\partial V_m/\partial T)_P - V_m = RT^2P^{-1}(PB^{\dagger\prime} + P^2C^{\dagger\prime} + \cdots)$. So $H_m^{id} - H_m = \int_0^{P°} RT^2(B^{\dagger\prime} + PC^{\dagger\prime} + \cdots)\, dP = RT^2[P°B^{\dagger\prime} + \frac{1}{2}(P°)^2C^{\dagger\prime} + \cdots]$.

$S_m^{id} - S_m = \int_0^{P°} [(\partial V_m/\partial T)_P - R/P]\, dP =$

$\int_0^{P°} [R(B^\dagger + C^\dagger P + \cdots) + RT(B^{\dagger\prime} + PC^{\dagger\prime} + \cdots)]\, dP = R(B^\dagger + TB^{\dagger\prime})P° + \frac{1}{2}R(C^\dagger + TC^{\dagger\prime})(P°)^2 + \cdots$. Finally, $G_m^{id} - G_m = (H_m^{id} - H_m) - T(S_m^{id} - S_m)$. Use of the preceding results gives $G_m^{id} - G_m = -RT[B^\dagger P° + \frac{1}{2}C^\dagger(P°)^2 + \cdots]$.

8.24 **(a)** Comparison of (8.9) with (8.4) gives the van der Waals estimates of the virial coefficients as $B = b - a/RT$, $C = b^2, \ldots$. Use of (8.6) gives $B^\dagger = b/RT - a/R^2T^2, \ldots$. Differentiation gives $B^{\dagger\prime} = -b/RT^2 + 2a/R^2T^3, \ldots$. Substitution in the results of Prob. 8.23 gives $H_m^{id} - H_m = RT^2(-b/RT^2 + 2a/R^2T^3)P° + \cdots = (2a/RT - b)P° + \cdots$ and gives $S_m^{id} - S_m = (a/RT^2)P° + \cdots$.

(b) From (8.18), $a = 27R^2T_c^2/64P_c = 5.50 \times 10^6$ cm^6 atm/mol^2 and $b = RT_c/8P_c = 65.0$ cm^3/mol. Then $H_m^{id} - H_m = (0.987 \text{ atm}) \times [2(5.50 \times 10^6 \text{ cm}^6 \text{ atm/mol}^2)/(82.06 \text{ cm}^3\text{-atm/mol-K})(298 \text{ K}) - 65.0 \text{ cm}^3/\text{mol}] = (379 \text{ cm}^3\text{-atm/mol})(1.987 \text{ cal}/82.06 \text{ cm}^3 \text{ atm}) = 9.2$ cal/mol. Also $S_m^{id} - S_m = [(5.50 \times 10^6 \text{ cm}^6 \text{ atm/mol}^2)/(82.06 \text{ cm}^3\text{-atm/mol-K})(298 \text{ K})^2](0.987 \text{ atm}) = (0.74 \text{ cm}^3 \text{ atm/mol-K})(1.987 \text{ cal}/82.06 \text{ cm}^3 \text{ atm}) = 0.018$ cal/mol-K. These are substantially smaller than the experimental values.

8.25 (a) The Berthelot equation when solved for P and multiplied by V_m/RT is $PV_m/RT = V_m/(V_m - b) - a/RT^2V_m = 1/(1 - b/V_m) - a/RT^2V_m$. Use of (8.8) with $x = b/V_m$ gives $PV_m/RT = 1 + (b - a/RT^2)/V_m + B^2/V_m^2 + \cdots$. Comparison with (8.4) gives the Berthelot estimates of the virial coefficients as $B = b - a/RT^2$, $C = b^2, \ldots$. Use of (8.6) gives $B^\dagger = b/RT - a/R^2T^3, \ldots$. Differentiation gives $B^\dagger = -b/RT^2 + 3a/R^2T^4$. Use of Prob. 8.23 results gives $H_m^{id} - H_m = (3a/R^2 - b)P° + \cdots$ and $S_m^{id} - S_m = (2a/RT^3)P° + \cdots$.

(b) Substitution of the a and b expressions of Prob. 8.20a gives $H_m^{id} - H_m = (81RT_c^3/64P_cT^2 - RT_c/8P_c)P° + \cdots$ and $S_m^{id} - S_m = (27RT_c^3/32P_cT^3)P° + \cdots$.

(c) Substitution in the results of (b) gives $S_m^{id} - S_m = (27/32)(1.987 \text{ cal/mol-K})(305.4 \text{ K}/298 \text{ K})^3(0.987/48.2) = 0.037$ cal/mol-K and $H_m^{id} - H_m = 15$ cal/mol. Agreement with experiment is excellent.

(d) $S_m^{id} - S_m = (27/32)(1.987 \text{ cal/mol-K})(430.8/298)^3(1/77.8) = 0.065$ cal/mol-K.

8.26 (a) We use the following isothermal path (rg = real gas, ig = ideal gas):

$$rg(V_m) \xrightarrow{1} rg(V_m = \infty) \xrightarrow{2} ig(V_m = \infty) \xrightarrow{3} ig(V_m) \xrightarrow{4} ig(V_{m,id})$$

From $dA_m = -S_m\,dT - P\,dV_m$, we get $(\partial A_m/\partial V_m)_T = -P$. Then $\Delta A_{m,1} = -\int_{V_m}^{\infty} P\,dV_m' = \int_{\infty}^{V_m} P\,dV_m'$; $\Delta A_{m,2} = 0$ (since $\Delta U_2 = 0$ and $\Delta S_2 = 0$, as noted in Chap. 5); $\Delta A_{m,3} = -\int_{\infty}^{V_m} P_{id}\,dV_m' = -\int_{\infty}^{V_m} (RT/V_m')\,dV_m'$; $\Delta A_{m,4} =$

$-\int_{V_m}^{V_m^{id}} (RT/V'_m)\, dV'_m = -RT \ln (V_m^{id}/V_m)$. $A_{m,id}(T, P) - A_m(T, P) =$
$\Delta A_{m,1} + \Delta A_{m,2} + \Delta A_{m,3} + \Delta A_{m,4} = \int_{\infty}^{V_m} (P - RT/V'_m)\, dV'_m - RT \ln(V_m^{id}/V_m)$.

(b) Solving (8.3) for P and substituting in the result of (a), we get $A_m^{id} - A_m =$
$\int_{\infty}^{V_m} [RT/(V'_m - b) - a/V'_m(V'_m + b)T^{1/2} - RT/V'_m]\, dV'_m - RT \ln (V_m^{id}/V_m) =$
$[RT \ln(V'_m - b) - (a/T^{1/2})(-1/b) \ln(1 + b/V'_m) - RT \ln V'_m]_{\infty}^{V_m} -$
$RT \ln(V_m^{id}/V_m) = [RT \ln(1 - b/V'_m) + (a/bT^{1/2}) \ln(1 + b/V'_m)]_{\infty}^{V_m} -$
$RT \ln(V_m^{id}/V_m) = RT \ln(1 - b/V_m) + (a/bT^{1/2}) \ln(1 + b/V_m) - RT \ln(V_m^{id}/V_m)$

(c) $S_m^{id} - S_m = -(\partial/\partial T)_{V_m} (A_m^{id} - A_m) =$
$-R \ln (1 - b/V_m) + (a/2bT^{3/2}) \ln(1 + b/V_m) + R \ln(V_m^{id}/V_m)$.
$U_m^{id} - U_m = (A_m^{id} - A_m) + T(S_m^{id} - S_m) = (3a/2bT^{1/2}) \ln(1 + b/V_m)$.

8.27 Differentiation gives $dB/dT = (RT_c/P_c)(-0.462)(-0.7002T_c/T^2)e^{0.7002T_c/T} =$
$0.3235(RT_c^2/P_cT^2)e^{0.7002T_c/T}$. From (8.6), $B^\dagger = B/RT$ and $dB^\dagger/dT = (dB/dT)/RT -$
B/RT^2. Substitution of $T_c = 305.4$ K, $P_c = 48.2$ atm, and $T = 298$ K in the equations for B and dB/dT gives $B = -182$ cm^3/mol and $dB/dT = 1.186$ cm^3/mol-K. Then $B^\dagger = B/RT = -0.00744$ atm^{-1} and $dB^\dagger/dT =$
(1.186 cm^3/mol-K)/(82.06 cm^3-atm/mol-K)(298 K) +
(182 cm^3/mol)/(82.06 cm^3-atm/mol-K)(298 K)^2 = 7.35×10^{-5} atm^{-1} K^{-1}.
From Prob. 8.23, $H_m^{id} - H_m = RT^2P° \, dB^\dagger/dT =$
(1.987 cal/mol-K)(298 K)^2(750/760)atm(7.35×10^{-5} atm^{-1} K^{-1}) = 13 cal/mol
and $S_m^{id} - S_m = RP°(B^\dagger + T\, dB^\dagger/dT) =$
(1.987 cal/mol-K)(0.987 atm)[$-0.00744 + 298(7.35 \times 10^{-5})$]atm^{-1} =
0.028 cal/mol-K.

8.28 $f(x) = (1 - x)^{-1}$, $f'(x) = (1 - x)^{-2}$, $f''(x) = 2(1 - x)^{-3}$, $f'''(x) =$
$3 \cdot 2(1 - x)^{-4}, \ldots$. $f(0) = 1$, $f'(0) = 1$, $f''(0) = 2$, $f'''(0) = 3 \cdot 2, \ldots$
$f(x) = 1 + (x - 0) + 2(x - 0)^2/2! + 3 \cdot 2(x - 0)^3/3! + \cdots = 1 + x + x^2 + x^3 + \cdots$

8.29 $f(x) = \ln x$, $f'(x) = 1/x$, $f''(x) = -1/x^2$, $f'''(x) = 2/x^3$, $f^{(iv)}(x) = -3 \cdot 2/x^4, \ldots$
$f(1) = 0$, $f'(1) = 1$, $f''(1) = -1$, $f'''(1) = 2$, $f^{(iv)}(1) = -3 \cdot 2, \ldots$.
$\ln x = 0 + (x - 1) - (x - 1)^2/2! + 2(x - 1)^3/3 \cdot 2 - 3 \cdot 2(x - 1)^4/4 \cdot 3 \cdot 2 + \cdots =$
$(x - 1) - (x - 1)^2/2 + (x - 1)^3/3 - (x - 1)^4/4 + \cdots$

8.30 $f(x) = e^x$, $f'(x) = e^x$, $f''(x) = e^x, \cdots$. $f(0) = 1$, $f'(0) = 1$, $f''(0) = 1, \cdots$
$e^x = 1 + x + x^2/2! + x^3/3! + \cdots$

8.31 $(d/dx)(\sin x) = \cos x = 1 - 3x^2/3! + 5x^4/5! - \cdots = 1 - x^2/2! + x^4/4! - \cdots$.

8.32 x is in radians; 35° = 0.610865 rad. Substitution in (8.35) gives sin 35° = 0.5736.

8.33 The function $1/(x^2 + 4)$ has singularities at $x = 2i$ and $x = -2i$ (where $x^2 + 4 = 0$). The distance between the origin (point a) and either of these singularities is 2, so $b = 2$.

8.34 **(a)** $\Sigma_{n=0}^{5} 1^n/n! = 2.7166667$, $\Sigma_{n=0}^{10} 1^n/n! = 2.7182818$, $\Sigma_{n=0}^{20} 1^n/n! = 2.7182818$. The exact value is $e = 2.7182818$.

(b) For $x = 10$, we find 1477.6667, 12842.305, and 21991.482 for $n = 5$, 10, and 20. The exact value is $e^{10} = 22026.466$. (For $n = 30$, one finds 22026.464.) A BASIC program is

```
10 INPUT "X=";X
20 FOR M = 5 TO 20 STEP 5
30 S = 1
40 NF = 1
50 FOR N = 1 TO M
60 NF = N*NF
70 S = S + X^N/NF
80 NEXT N
90 PRINT "M=";M;" SUM=";S
100 NEXT M
110 END
```

8.35 From Sec. 8.3, $T_c \approx 1.6T_b = 1.6(353 \text{ K}) = 565 \text{ K}$; $V_{m,c} \approx 2.7V_{m,b} = 2.7(78 \text{ g/mol})/(0.81 \text{ g/cm}^3) = 260 \text{ cm}^3\text{/mol}$. From Sec. 8.4, Z_c is usually between 0.25 and 0.30, so we take $Z_c = P_cV_{m,c}/RT_C \approx 0.275$. Thus

$$P_c \approx 0.275\frac{RT_c}{V_{m,c}} \approx \frac{(82.06 \text{ cm}^3\text{-atm/mol-K})(565 \text{ K})0.275}{260 \text{ cm}^3\text{/mol}} = 49 \text{ atm}$$

(The experimental values are 562 K, 259 cm^3/mol, 48 atm.)

8.36 Use the isothermal path: liq(1 bar) $\overset{a}{\rightarrow}$ liq(23.766 torr) $\overset{b}{\rightarrow}$ vap(23.766 torr) $\overset{c}{\rightarrow}$ ideal vap(23.766 torr) $\overset{d}{\rightarrow}$ ideal vap(1 bar). For step (a), Prob. 7.67 gives $\Delta G_{m,a} = -1.8$ J/mol. For step (b), $\Delta G_{m,b} = 0$, since this is a reversible constant-T-and-P process. For step (c), Prob. 8.24a gives $\Delta G_{m,c} = \Delta H_{m,c} - T\,\Delta S_{m,c} \approx (a/RT - b)P =$

$$\left[\frac{5.47\times 10^6\ \text{cm}^6\text{-atm/mol}^2}{(82.06\ \text{cm}^3\text{-atm/mol-K})(298\ \text{K})} - 30.5\frac{\text{cm}^3}{\text{mol}}\right]\frac{23.766}{760}\text{atm}$$

$$= 6.04\frac{\text{cm}^3\ \text{atm}}{\text{mol}}\frac{8.314\ \text{J}}{82.06\ \text{cm}^3\ \text{atm}} = 0.6\ \text{J/mol}$$

For step (d), Prob. 7.67 gives $\Delta G_{m,d} = 8557.2$ J/mol. The net ΔG_m is $8556._0$ J/mol compared with $8555._4$ J/mol in Prob. 7.67 and 8557 J/mol from data in the Appendix.

8.37 **(a)** Use of Eqs. (4.52) and (8.5) gives $\mu_{JT} = (1/C_{P,m})[T(\partial V_m/\partial T)_P - V_m] = C_{P,m}^{-1}\{T[(R/P)(1 + B^\dagger P + C^\dagger P^2 + \cdots) + (RT/P)(P\,dB^\dagger/dT + P^2\,dC^\dagger/dT + \cdots)] - (RT/P)(1 + B^\dagger P + C^\dagger P^2 + \cdots)\} = (RT^2/C_{P,m})(dB^\dagger/dT + P\,dC^\dagger/dT + P^2\,dD^\dagger/dT + \cdots)$

$\lim_{P\to 0} \mu_{JT} = (RT^2/C_{P,m})dB^\dagger/dT$

(b) Equations (4.47) and (8.4) give $(\partial U/\partial V)_T = T(\partial P/\partial T)_V - P =$

$$T\left[\frac{R}{V_m}\left(1 + \frac{B}{V_m} + \cdots\right) + \frac{RT}{V_m}\left(\frac{1}{V_m}\frac{dB}{dT} + \cdots\right)\right] - \frac{RT}{V_m}\left(1 + \frac{B}{V_m} + \cdots\right) =$$

$$\frac{RT^2}{V_m}\left(\frac{1}{V_m}\frac{dB}{dT} + \cdots\right).$$ As $P \to 0$, $V_m \to \infty$ and $(\partial U/\partial V)_T \to 0$.

8.38 From (8.4), $V_m - V_m^{id} = (RT/P)(1 + B/V_m + C/V_m^2 + \cdots) - RT/P = RT(B/PV_m + C/PV_m^2 + \cdots)$. In the limit $P \to 0$, we have $PV_m = RT$ and $V_m - V_m^{id} \to B + C/V_m + \cdots \to B$, since $V_m \to \infty$ as $P \to 0$.

8.39 **(a)** As found in Prob. 8.24a, the van der Waals equation gives $B^\dagger = b/RT - a/R^2T^2$ and $B^{\dagger\prime} = -b/RT^2 + 2a/R^2T^3$. Substitution in the equation of Prob. 8.37a with all terms but the first neglected gives $\mu_{JT} = (2a/RT - b)/C_{P,m}$ as the low-P estimate of μ_{JT}.

(b) At the inversion temperature, the low-P μ_{JT} is zero, so $(2a/RT - b)/C_{P,\mathrm{m}} = 0$ and $T_{i,P\to 0} = 2a/bR = 2(1.35 \times 10^6$ atm cm^6 mol$^{-2})/(38.6$ cm^3/mol$)(82.06$ cm^3-atm/mol-K$) = 852$ K, where N_2 a and b values in Sec. 8.4 were used. The low-P 298-K estimate of μ_{JT} is $\mu_{JT} =$

$$\left[\frac{2(1.35 \times 10^6 \text{ atm cm}^6\text{ mol}^{-2})}{(82.06 \text{ cm}^3\text{-atm/mol-K})(298\text{ K})} - 38.6 \text{ cm}^3\text{/mol}\right]\frac{1}{6.96 \text{ cal/mol-K}} =$$

$$(10.3 \text{ cm}^3\text{ K/cal})\frac{1.987 \text{ cal}}{82.06 \text{ cm}^3\text{ atm}} = 0.250 \text{ K/atm}$$

where C_P is from the Appendix and (1.19) and (1.21) were used. Agreement with the experimental μ_{JT} is good, but T_i is poorly predicted.

8.40 **(a)** The larger size of a neon atom means that Ne has greater intermolecular attractions than He, so Ne has a greater a, a greater T_c, and a greater $\Delta_{\mathrm{vap}}H_{\mathrm{m}}$. Ne atoms are larger so Ne has a greater b value.

(b) C_3H_8 has greater intermolecular attractions and larger size, and so has the greater a, T_c, $\Delta_{\mathrm{vap}}H_{\mathrm{m}}$, and b.

(c) Due to hydrogen bonds, H_2O has greater intermolecular attractions, and so has the greater a, T_c, and $\Delta_{\mathrm{vap}}H_{\mathrm{m}}$. The H_2S molecule is larger than the H_2O molecule, so H_2S has the larger b.

8.41 From (8.18), $b = RT_c/8P_c = (82.06$ cm^3-atm/mol-K$)(190.6$ K$)/8(45.4$ atm$) =$ 43.1 cm^3/mol; $a = 27R^2T_c^2/64P_c = 2.27 \times 10^6$ cm^6 atm mol^{-2}. $V_{\mathrm{m},0} = b + RT/P$ $= 43.1$ cm^3/mol $+ (82.06$ cm^3-atm/mol-K$)(273$ K$)/(100$ atm$) = 267$ cm^3/mol. $V_{\mathrm{m},1} = b + RT/(P + a/V_{\mathrm{m},0}^2) =$

$$43.1 \text{ cm}^3\text{/mol} + \frac{(82.06 \text{ cm}^3\text{-/mol-K})(273\text{ K})}{100 \text{ atm} + (2.27\times 10^6 \text{ cm}^6\text{ atm/mol}^2)/(267 \text{ cm}^3\text{/mol})^2}$$

$V_{\mathrm{m},1} = 213$ cm^3/mol. $V_{\mathrm{m},2} = b + RT/(P + a/V_{\mathrm{m},1}^2) = 192$ cm^3/mol; $V_{\mathrm{m},3} = 182$ cm^3/mol. Successive calculations give (in cm^3/mol): 176, 172, 170, 169, 168, 167, 166.6, 166, and 166. From Fig. 8.1a, for CH_4 at 100 atm and 0°C, $Z = 0.78_5 = PV_{\mathrm{m}}/RT$ and $V_{\mathrm{m}} = 0.78_5RT/P =$ $0.78_5(82.06$ cm^3-atm/mol-K$)(273$ K$)/(100$ atm$) = 176$ cm^3/mol.
A BASIC program is

```
10 DIM V(100)
20 INPUT "A=";A
22 INPUT "B=";B
24 INPUT "P=";P
26 INPUT "T=";T
30 R = 82.06
40 V(1) = B + R*T/P
50 FOR I = 1 TO 99
60 V(I+1) = B + R*T/(P + A/V(I)^2)
70 PRINT"I=";I;" V=";V(I+1)
80 IF ABS(V(I+1) – V(I))/V(I) < 1E–5
   THEN STOP
90 NEXT I
95 END
```

8.42 $T_r = T/T_c = (286 \text{ K})/(190.6 \text{ K}) = 1.50$ and $P_r = P/P_c = (91 \text{ atm})/(45.4 \text{ atm}) = 2.00$. At these T_r and P_r values, Fig. 8.10 gives $Z = 0.83 = PV_m/RT$ and $V_m = ZRT/P = 0.83(82.06 \text{ cm}^3\text{-atm/mol-K})(286 \text{ K})/(91 \text{ atm}) = 214 \text{ cm}^3\text{/mol}$.

8.43 **(a)** As in Prob. 7.33b, $d \ln P/dT = (1/P)(dP/dT) = 0.035687\text{/K}$ and $dP/dT = (0.035687\text{/K})(1 \text{ atm}) = 0.035687 \text{ atm/K}$ at 100°C.

(b) $\Delta V_m = (82.058 \text{ cm}^3\text{-atm/mol-K})(373.15 \text{ K})/(1 \text{ atm}) - 452 \text{ cm}^3\text{/mol} - 19 \text{ cm}^3\text{/mol} = 30149 \text{ cm}^3\text{/mol}$. Then $\Delta H_m + (T\,\Delta V_m)(dP/dT) = (373.15 \text{ K})(30149 \text{ cm}^3\text{/mol})(0.035687 \text{ atm/K}) = (4.0148 \times 10^5 \text{ cm}^3\text{-atm/mol})(8.3145 \text{ J}/82.058 \text{ cm}^3\text{-atm}) = 40.68_0 \text{ kJ/mol}$.

8.44 Cells for B and C are specified. One can set each of these to zero for the initial guess. The experimental P and V_m values are entered into columns A and B. Formulas to calculate P from Eq. (8.4) are entered into column C. In column D, the squares of the deviations between column-A and column-C pressures are calculated. The sum of the squares of the deviations is calculated in a cell, and the Solver is set up to minimize this sum by varying B and C. The result is $B = -83.12 \text{ cm}^3\text{/mol}$ and $C = 3330 \text{ cm}^6\text{/mol}^2$, with an excellent fit to the data.

8.45 **(a)** F; **(b)** F.

Chapter 9

9.1 **(a)** mol/m^3; **(b)** mol/kg; **(c)** no units.

9.2 c_i; c_i.

9.3 **(a)** $n_i = c_iV = (0.800\ mol/L)(0.145\ L) = 0.116\ mol$.

(b) (145 g) × 10.0% = 14.5 g; 0.398 mol.

(c) $m_i = n_i/w_A = n_i/(w - w_i) = n_i/(w - n_iM_i)$, where w is the mass of the solution, w_i is the mass of the HCl, n_i is the number of moles of HCl, and M_i is the HCl molar mass. Solving for n_i, we get $n_i = m_iw/(1 + m_iM_i) =$ (4.85 mol/kg)(0.145 kg)/[1 + (4.85 mol/kg)(0.03646 kg/mol)] = 0.598 mol HCl. Alternatively, a solution with 1000 g of solvent has 4.85 mol HCl, which is 176.8_3 g of HCl. The weight percent of HCl is $[176.8_3/(1000 + 176.8_3)]100\% = 15.03\%$. 15.03% of 145 g is 21.7_9 g of HCl, which is 0.598 mol HCl.

9.4 **(a)** All quantities involved are intensive, so we can use any convenient amount of solution. Let us take 1 dm^3 of solution. This amount of solution has 8.911 mol of CH_3OH, which is 285.5 g of CH_3OH. Since the solution is 30% CH_3OH, the solution's mass is (100/30)(285.5 g) = 951.8 g. Its density is $(951.8\ g)/(1\ dm^3) = 0.9518\ g/cm^3$.

(b) $m_i = n_i/w_A = (8.911\ mol)/(951.8\ g - 285.5\ g) = 0.01337\ mol/g =$ 13.37 mol/kg.

(c) $\rho_{CH_3OH} = m_{CH_3OH}/V = (285.5\ g)/(1\ dm^3) = 285.5\ g/L$.

9.5 Let us take 1000 g of solution. This contains (0.800 %)(1000 g) = 8.00 g of NH_3 and 1000 g – 8 g = 992 g of water. This is 0.469_7 mol of NH_3 and 55.06 mol of water. Then $m_i = n_i/w_A = (0.469_7\ mol)/(992\ g) =$ 0.000474 mol/g = 0.474 mol/kg. Also, $x_i = 0.469_7/(0.469_7 + 55.06) = 0.00846$.

9.6 The solution's mass is $m = \rho V = (1.2885\ g/cm^3)(1000\ cm^3) = 1288.5$ g. The CsCl mass is (2.296 mol) × (168.358 g/mol) = 386.6 g. The solvent mass is

1288.5 g – 386.6 g = 901.9 g. The molality is (2.296 mol)/(0.9019 kg) = 2.546 mol/kg.

9.7 All quantities involved are intensive, so we can take any amount of solution. Take an amount of solution that contains one kg of water and hence contains 1.506 mol of KI. The solution's mass is 1000 g + (1.506 mol)(166.00 g/mol) = 1250.0 g. The solution's volume is $V = m/\rho$ = (1250.0 g)/(1.1659 g/cm^3) = 1072.1 cm^3. The molarity is (1.506 mol)/(1.0721 L) = 1.405 mol/L.

9.8 $c_i = n_i/V = n_i/(w/\rho) = \rho n_i/w$, where w is the solution's mass. Because the solution is very dilute, we have $w = w_i + w_A \approx w_A = n_A M_A$, where A is the solvent. Then $c_i = \rho n_i/n_A M_A$. Also, $x_i = n_i/(n_A + n_i) \approx n_i/n_A$, so $c_i \approx \rho x_i/M_A$. For the molality, we have $m_i = n_i/w_A = n_i/n_A M_A \approx x_i/M_A$. From $c_i \approx \rho x_i/M_A$ and $m_i \approx x_i/M_A$, we get $c_i \approx \rho m_i$.

9.9 As noted after Eq. (1.4), $M_A = M_{r,A} \times 1$ g/mol, so $m_B = n_B/n_A M_A = n_B/(n_A M_{r,A}$ g/mol$) = n_B/(n_A M_{r,A} 10^{-3}$ kg/mol$) = (1000\, n_B/n_A M_{r,A})$ mol/kg.

9.10 **(a)** F **(b)** F; **(c)** F; **(d)** T; **(e)** T; **(f)** F; **(g)** F; **(h)** T; **(i)** T.

9.11 $V = n_1\bar{V}_1 + n_2\bar{V}_2 =$
(0.500 mol)(18.63 cm^3/mol) + (55.51 mol)(18.062 cm^3/mol) = 1011.9 cm^3.

9.12 $C_P = n_1\bar{C}_{P1} + n_2\bar{C}_{P2}$. An amount of 0.1000-mol/kg solution that contains 1000 g of solvent has 0.1000 mol (which is 5.844 g) of NaCl and has an NaCl weight percentage of [5.844/(5.844 + 1000)]100% = 0.581%. Taking 0.581% of 1000 g gives 5.81 g of NaCl in the 1000 g of solution, which is 0.0994 mol of NaCl. The percent water is 100 – 0.581 = 99.419%. The H_2O mass is 994.19 g, which is 55.186 mol H_2O. So C_P = (55.186 mol)(17.992 cal/mol-K) + (0.0994 mol)(–17.00 cal/mol-K) = 991.22 cal/K.

9.13 $V = n_1\bar{V}_1 + n_2\bar{V}_2$ = 307.09 cm^3 = [(72.061/18.0153) mol](16.488 cm^3/mol) + [(192.252/32.0422) mol]$\bar{V}_{CH_3OH}$ and $\bar{V}_{CH_3OH}$ = 40.19 cm^3/mol.

9.14 Take solutions that each contain 1000 g of H_2O (constant n_{H_2O}). Let w be the solution mass. For the 12% solution: (1000 g)/w = 0.88000 and w = 1136.36 g; this solution contains 136.36 g of CH_3OH, which is 4.2557 mol of CH_3OH; the solution has V = (1136.36 g)/(0.97942 g/cm^3) = 1160.24 cm^3. For the 13% solution: (1000 g)/w = 0.87000 and w = 1149.43 g; this solution has 149.43 g of CH_3OH, which is 4.6636 mol of CH_3OH; the solution's volume is V = (1149.43 g)/(0.97799 g/cm^3) = 1175.30 cm^3. We have $\bar{V}_{CH_3OH} \approx (\Delta V/\Delta n_{CH_3OH})_{T,P,n_{H_2O}}$ = (15.06 cm^3)/(0.4079 mol) = 36.9$_2$ cm^3/mol. To find $\bar{V}_{H_2O}$, we now take solutions with 100 g of CH_3OH (constant n_{CH_3OH}) and do the calculations the same way as for the constant n_{H_2O} solutions. For the 12% solution, we find n_{H_2O} = 40.706 mol and V = 850.84 cm^3. For the 13% solution, we find n_{H_2O} = 37.148 mol and V = 786.54 cm^3. Hence $\bar{V}_{H_2O} \approx (\Delta V/\Delta n_{H_2O})_{T,P,n_{CH_3OH}}$ = (–64.30 cm^3)/(–3.558 mol) = 18.07 cm^3/mol.

9.15 **(a)** $\bar{V}$ of $MgSO_4$ at a given composition equals the slope of Fig. 9.3 at that composition. The slope is zero at the minimum, which occurs at n_{MgSO_4} = 0.07 moles in 1000 g of water, which is a molality of 0.07 mol/kg.

(b) Infinite dilution corresponds to $n_{MgSO_4} \to 0$. Drawing the tangent line to the curve at n_{MgSO_4} = 0, one finds its slope to be –3.$_5$ cm^3/mol, which is $\bar{V}^{\infty}$ of $MgSO_4$.

(c) Drawing the tangent line to the curve at n_{MgSO_4} = 0.05 mol, one finds its slope to be –0.5$_4$ cm^3/mol = $\bar{V}_{MgSO_4}$. $V = n_{H_2O}\bar{V}_{H_2O} + n_{MgSO_4}\bar{V}_{MgSO_4}$ = 1001.69$_7$ cm^3 = (55.509 mol)$\bar{V}_{H_2O}$ + (0.05 mol)(–0.5$_4$ cm^3/mol) and $\bar{V}_{H_2O}$ = 18.04$_6$ cm^3/mol.

9.16 $\bar{V}^{\infty}_{NaCl} = \bar{V}^{\infty}(Na^+) + \bar{V}^{\infty}(Cl^-)$, $\bar{V}^{\infty}_{KNO_3} = \bar{V}^{\infty}(K^+) + \bar{V}^{\infty}(NO_3^-)$, and $\bar{V}^{\infty}_{NaNO_3} = \bar{V}^{\infty}(Na^+) + \bar{V}^{\infty}(NO_3^-)$. Then $\bar{V}^{\infty}_{KCl} = \bar{V}^{\infty}(K^+) + \bar{V}^{\infty}(Cl^-) = \bar{V}^{\infty}_{KNO_3} + \bar{V}^{\infty}_{NaCl} - \bar{V}^{\infty}_{NaNO_3}$ = (38.0 + 16.6 – 27.8) cm^3/mol = 26.8 cm^3/mol.

9.17 Substitution of $G = \sum_i n_i\mu_i$ into $G \equiv U + PV - TS$ gives the desired result.

9.18 $\bar{A}_i = (\partial A/\partial n_i)_{T,P,n_{j\neq i}}$, where A is the Helmholtz energy of the solution, n_i is the number of moles of i in the solution, T and P are temperature and pressure, and $n_{j\neq i}$ indicates that all mole numbers except i are held fixed.

9.19 $H \equiv U + PV$. Partial differentiation gives $(\partial H/\partial n_i)_{T,P,n_{j\neq i}} = (\partial U/\partial n_i)_{T,P,n_{j\neq i}} + P(\partial V/\partial n_i)_{T,P,n_{j\neq i}}$, so $\bar{H}_i = \bar{U}_i + P\bar{V}_i$.

9.20 **(a)** $G = \sum_i n_i\mu_i = \sum_i n_i[\mu_i^\circ + RT\ln(P_i/P^\circ)]$. Taking $P_1 = P^\circ$ and $P_2 = P_i$ in (4.65), we have for n_i moles of pure gas i: $G_i^*(T, P_i, n_i) - G_i^*(T, P^\circ, n_i) = n_iRT\int_{P^\circ}^{P_i} P^{-1}dP = n_iRT\ln(P_i/P^\circ)$. So $G_i^*(T, P_i, n_i) = G_i^*(T, P^\circ, n_i) + n_iRT\ln(P_i/P^\circ) = n_iG_{m,i}^*(T, P^\circ) + n_iRT\ln(P_i/P^\circ) = n_i[\mu_i^\circ + RT\ln(P_i/P^\circ)]$, since $G_{m,i}^* = \mu_i$ for a pure substance. Hence the above sum for G of the mixture becomes $G = \sum_i G_i^*(T, P_i, n_i)$.

(b) $S = -(\partial/\partial T)_{P,n_i} G = \sum_i [-(\partial G_i^*/\partial T)_{P,n_i}] = \sum_i S_i^*(T, P_i, n_i)$, where (4.51) was used.

(c) $H = G + TS = \sum_i G_i^* + T\sum_i S_i^* = \sum_i (G_i^* + TS_i^*) = \sum_i H_i^*(T, n_i)$, where P_i is absent since H is independent of P for an ideal gas.

(d) $C_P = (\partial H/\partial T)_{P,n_i} = (\partial/\partial T)_{P,n_i} \sum_i H_i^* = \sum_i (\partial H_i^*/\partial T)_{n_i} = \sum_i C_{P,i}^*(T, n_i)$. $U = H - PV = \sum_i H_i^* - P(\sum_i n_iRT/P) = \sum_i (H_i^* - n_iRT) = \sum_i (H_i^* - P_iV) = \sum_i U_i^*(T, n_i)$.

(e) Assuming the mixture is ideal, we have $C_P = \sum_i C_{P,i}^*(T, n_i) = C_{P,O_2}^* + C_{P,CO_2}^* = n_{O_2}C_{P,m,O_2}^* + n_{CO_2}C_{P,m,CO_2}^* =$ (0.100 mol)(29.355 J/mol-K) + (0.300 mol)(37.11 J/mol-K) = 14.07 J/K.

9.21 Taking $\partial/\partial T$ of (9.32), we get $(\partial\Delta_{mix}G/\partial T)_{P,n_i} = (\partial/\partial T)_{P,n_i}\sum_i n_i(\bar{G}_i - G_{m,i}^*) = \sum_i n_i[(\partial\bar{G}_i/\partial T)_{P,n_i} - (\partial G_{m,i}^*/\partial T)_{P,n_i}] = \sum_i n_i(-\bar{S}_i + S_{m,i}^*) = \sum_i n_iS_{m,i}^* - \sum_i n_i\bar{S}_i = S^* - S = -\Delta_{mix}S$, where (9.30) was used.

9.22 $n(H_2O) = 1.11_0$ mol and $n(C_2H_5OH) = 0.977$ mol. $x(C_2H_5OH) = 0.977/2.087 = 0.468$. For this composition, Fig. 9.9 gives $\bar{V}(H_2O) =$

16.8 cm^3/mol, $\overline{V}(C_2H_5OH)$ = 57.0 cm^3/mol. $V = \overline{V}_1 n_1 + \overline{V}_2 n_2$ = (16.8 cm^3/mol)(1.11 mol) + (57.0 cm^3/mol)(0.977 mol) = 74.3 cm^3.

9.23 We draw the tangent line at $x_{\text{ethanol}} = 0.4$. This line intersects the x_{ethanol} = 0 axis at –1.1 cm^3/mol = $\overline{V}_{H_2O} - V^*_{m,H_2O} = \overline{V}_{H_2O}$ – 18.05 cm^3/mol, and $\overline{V}_{H_2O}$ = 16.9$_5$ cm^3/mol. The tangent line intersects the x_{ethanol} = 1 axis at –1.1$_5$ cm^3/mol = $\overline{V}_{\text{ethanol}} - V^*_{m,\text{ethanol}} = \overline{V}_{\text{ethanol}}$ – 58.4 cm^3/mol, and $\overline{V}_{\text{ethanol}}$ = 57.3 cm^3/mol.

9.24 The pure molar volumes are $V^*_{m,H_2O} = M/\rho$ = (18.015 g/mol)/(0.99705 g/cm^3) = 18.07 cm^3/mol; V^*_{m,CH_3OH} = 40.71 cm^3/mol. We plot $\Delta_{\text{mix}}V/n$ vs. x_{H_2O} (similar to Fig. 9.7).

(a) Drawing the tangent line at $x_{H_2O} = 0$, we find it intersects the $x_{H_2O} = 1$ line at –3.6 cm^3/mol. (With a reasonable choice of scale, the intersection occurs off the paper and can be calculated by extrapolation.) Hence, at $x_{H_2O} = 0$, $\overline{V}_{H_2O} - V^*_{m,H_2O}$ = –3.6 cm^3/mol = $\overline{V}_{H_2O}$ – 18.1 cm^3/mol and $\overline{V}_{H_2O}$ = 14.5 cm^3/mol. Of course, $\overline{V}_{CH_3OH} = V^*_{m,CH_3OH}$ = 40.7 cm^3/mol at $x_{H_2O} = 0$.

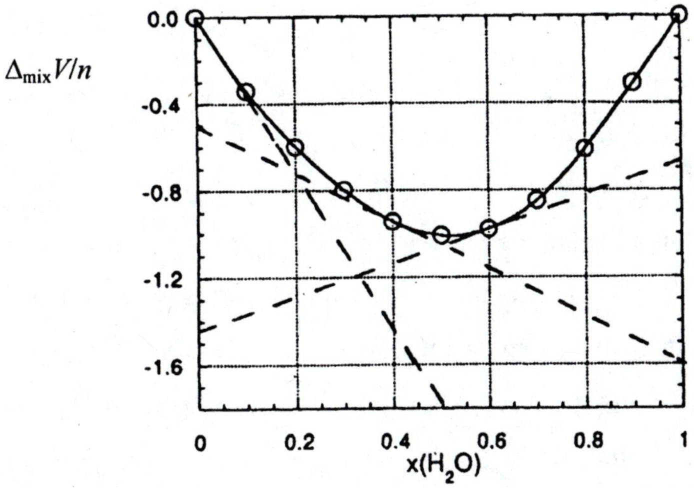

(b) The tangent line at x_{H_2O} = 0.4 intersects the x_{H_2O} = 0 line at –0.5 cm^3/mol and intersects x_{H_2O} = 1 at –1.6 cm^3/mol, so

$\overline{V}_{CH_3OH} - V^*_{m,CH_3OH} = -0.5\ cm^3/mol$ and $\overline{V}_{CH_3OH} = 40.2\ cm^3/mol$; also, $\overline{V}_{H_2O} = 16.5\ cm^3/mol$.

(c) $\overline{V}_{CH_3OH} - V^*_{m,CH_3OH} = -1.4_5\ cm^3/mol$ and $\overline{V}_{CH_3OH} = 39.2_5\ cm^3/mol$; $\overline{V}_{H_2O} - V^*_{m,H_2O} = -0.7\ cm^3/mol$ and $\overline{V}_{H_2O} = 17.4\ cm^3/mol$.

9.25 **(a)** $\overline{V}_B = (\partial V/\partial n_B)_{T,P,n_A} = b + (3/2)cn_B^{1/2} + 2kn_B$ for $n_A M_A = 1$ kg.

(b) For $m_B = 1$ mol/kg, we have $n_B = 1$ mol in a solution with $n_A M_A = 1$ kg. $\overline{V}_B = 16.6253\ cm^3/mol + 1.5(1.7738\ cm^3/mol^{3/2})(1.0000\ mol)^{1/2} + 2(0.1194\ cm^3/mol^2)(1.0000\ mol) = 19.5248\ cm^3/mol$.

(c) $V = n_A\overline{V}_A + n_B\overline{V}_B$ and $n_A\overline{V}_A = V - n_B\overline{V}_B = a + bn_B + cn_B^{3/2} + kn_B^2 - n_B b - 1.5\,cn_B^{3/2} - 2\,kn_B^2 = a - \frac{1}{2}\,cn_B^{3/2} - kn_B^2$. Since $n_A = (1\ kg)/M_A$, we get $\overline{V}_A = (M_A/1000\ g)(a - \frac{1}{2}\,cn_B^{3/2} - kn_B^2)$ for $n_A M_A = 1$ kg.

(d) $m_B = n_B/n_A M_A = n_B/(1\ kg) = n_B/kg$ and $n_B = m_B$ kg. Substitution in the results of (a) and (c) gives the desired equations.

(e) $\overline{V}_{H_2O} = (18.0152\ g/mol)[1002.96\ cm^3 - \frac{1}{2}(1.7738\ cm^3/mol^{3/2}) \times (1\ mol/kg)^{3/2}\ kg^{3/2} - (0.1194\ cm^3/mol^2)(1\ mol/kg)^2\ kg^2]/(1000\ g) = 18.050\ cm^3/mol$.

(f) In the infinite-dilution limit, m_B goes to zero and $\overline{V}_B$ in (d) becomes $\overline{V}_B^\infty = b = 16.6253\ cm^3/mol$.

9.26 **(a)** $z \equiv \Delta_{mix}V/n = (V - V^*)/(n_A + n_B)$, so $V - V^* = (n_A + n_B)z$ and $V = (n_A + n_B)z + V^* = (n_A + n_B)z + n_A V^*_{m,A} \quad n_B V^*_{m,B}$.

(b) Taking $(\partial/\partial n_A)_{n_B,T,P}$ of the result of (a) gives $(\partial V/\partial n_A)_{n_B,T,P} \equiv \overline{V}_A$ for the left side and $z + (n_A + n_B)(\partial z/\partial n_A)_{n_B} + V^*_{m,A}$ for the right side. (Note that $V^*_{m,A}$ is constant at fixed T and P and is independent of n_A.) Equating these expressions, we get $\overline{V}_A = z + n(\partial z/\partial n_A)_{n_B} + V^*_{m,A}$.

(c) Equation (1.35) gives $(\partial z/\partial n_A)_{n_B} = (dz/dx_B)(\partial x_B/\partial n_A)_{n_B}$. Since $x_B = n_B/(n_A + n_B)$, we have $(\partial x_B/\partial n_A)_{n_B} = -n_B/(n_A + n_B)^2 = -x_B/(n_A + n_B) = -x_B/n$. The result of (b) gives $(\partial z/\partial n_A)_{n_B} = (\overline{V}_A - V^*_{m,A} - z)/n$. Substitution of these expressions for $\partial x_B/\partial n_A$ and $\partial z/\partial n_A$ into

$\partial z/\partial n_A = (dz/dx_B)(\partial x_B/\partial n_A)$ gives $dz/dx_B = (V^*_{m,A} - \overline{V}_A + z)/x_B$. The n_B subscript can be omitted from $(\partial z/\partial x_B)_{n_B}$ because $z \equiv (V - V^*)/n$ is an intensive quantity and so is a function of x_B only and is independent of the size of the system and hence of n_B.

(d) The plotted quantity $\Delta_{mix}V/n$ is symmetric in A and B. Since the intercept at $x_B = 0$ is $\overline{V}'_A - V^*_{m,A}$, the A-B symmetry means that the intercept at $x_A = 0$ is $\overline{V}'_B - V^*_{m,B}$. But $x_A = 0$ is $x_B = 1$.

9.27 The tangent to the $\Delta_{mix}H/n$ versus $x_{H_2SO_4}$ curve at $x_{H_2SO_4} = 0.4$ intersects $x_{H_2SO_4} = 0$ at $-13._0$ kJ/mol = $\Delta H_{diff,H_2O}$ and intersects $x_{H_2SO_4} = 1$ at $-16._4$ kJ/mol = $\Delta H_{diff,H_2SO_4}$. Drawing the tangent at $x_{H_2SO_4} = 0.333$, we find $\Delta H_{diff,H_2O} = -9._5$ kJ/mol and $\Delta H_{diff,H_2SO_4} = -22._6$ kJ/mol.

9.28 $\Delta_{mix}H = H - H^*$. $(\partial\, \Delta_{mix}H/\partial n_B)_{T,P,n_A} = (\partial H/\partial n_B)_{T,P,n_A} - (\partial H^*/\partial n_B)_{T,P,n_A}$. But by definition, $\overline{H}_B = (\partial H/\partial n_B)_{T,P,n_A}$. Also, $H^* = n_A H^*_{m,A}(T, P) + n_B H^*_{m,B}(T, P)$, so $(\partial H^*/\partial n_B)_{T,P,n_A} = H^*_{m,B}$. Therefore $(\partial\, \Delta_{mix}H/\partial n_B)_{T,P,n_A} = \overline{H}_B - H^*_{m,B} = \Delta H_{diff,B}$, where (9.38) was used.

9.29 Subtraction of $\Delta_f H^\circ_{NaCl(s)}$ from the apparent $\Delta_f H^\circ$ values give the following integral heats of solution per mole of NaCl vs. x_{NaCl}:

$\Delta H_{int,NaCl}$/(kJ/mol)	1.874	2.347	3.016	3.711
x_{NaCl}	0.1	0.0625	0.03846	0.019608

The $\Delta H_{int,NaCl}$ values are $\Delta_{mix}H/n_{NaCl}$ [Eq. (9.37)]. Hence multiplication of $\Delta H_{int,NaCl} = \Delta_{mix}H/n_{NaCl}$ by $n_{NaCl}/n = x_{NaCl}$ gives the following $\Delta_{mix}H/n$ values:

$(\Delta_{mix}H/n)$/(J/mol)	187.4	146.7	116.0	72.7_7	0
x_{NaCl}	0.1	0.0625	0.03846	0.019608	0

We plot $\Delta_{mix}H/n$ vs. x_{NaCl} and draw the tangent at $x_{NaCl} = 0.05$. The tangent line intersects $x_{NaCl} = 0$ at 69 J/mol = $\overline{H}_{H_2O} - H^*_{m,H_2O}$ and intersects $x_{NaCl} = 0.1$ at 198 J/mol. Extrapolation gives the intersection at $x_{NaCl} = 1$ as
[69 + 10(198 – 69)]J/mol = 1.36 kJ/mol = $\overline{H}_{NaCl} - H^*_{m,NaCl}$.

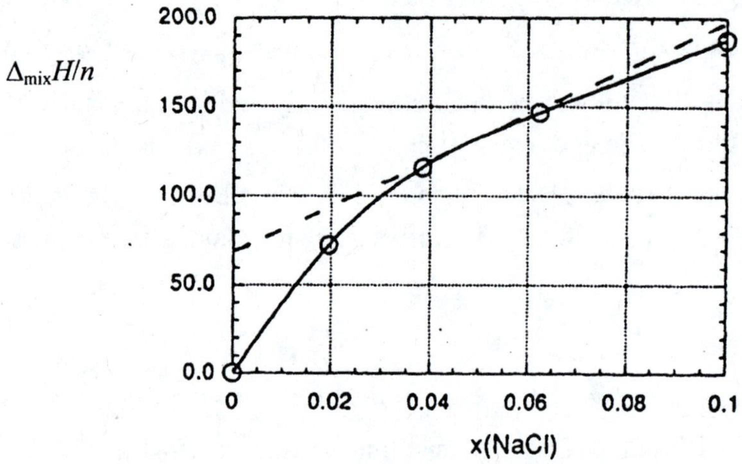

9.30 The NBS tables give $\Delta_f H^\circ_{298} = -92.307$ kJ/mol for HCl(g) and $\Delta_f H^\circ_{298} =$ -121.55 kJ/mol for HCl in 1 mol of H_2O. Thus $\Delta H_{int,HCl} = -29.24$ kJ/mol for a solution with $x_{HCl} = 0.500$. From (9.37), $\Delta_{mix}H = n_{HCl}\,\Delta H_{int,HCl}$ and $\Delta_{mix}H/n_{tot} =$ $x_{HCl}\,\Delta H_{int,HCl} = -14.62$ kJ/mol at $x_{HCl} = 0.5$. Use of further data gives (values in kJ/mol)

x_{HCl}	0.500	0.400	0.333	0.286	0.250	0.200
$\Delta H_{int,\,HCl}$	−29.24	−40.36	−48.65	−53.17	−56.18	−60.61
$\Delta_{mix}H/n_{tot}$	−14.62	−16.15	−16.22	−15.19	−14.05	−12.12

We plot $\Delta_{mix}H/n_{tot}$ vs. x_{HCl} and draw the tangent line at $x_{HCl} = 0.30$. The intercepts at $x_{HCl} = 0$ and $x_{HCl} = 1$ give $\Delta H_{diff,H_2O} = -8._3$ kJ/mol and $\Delta H_{diff,HCl} = -32._6$ kJ/mol.

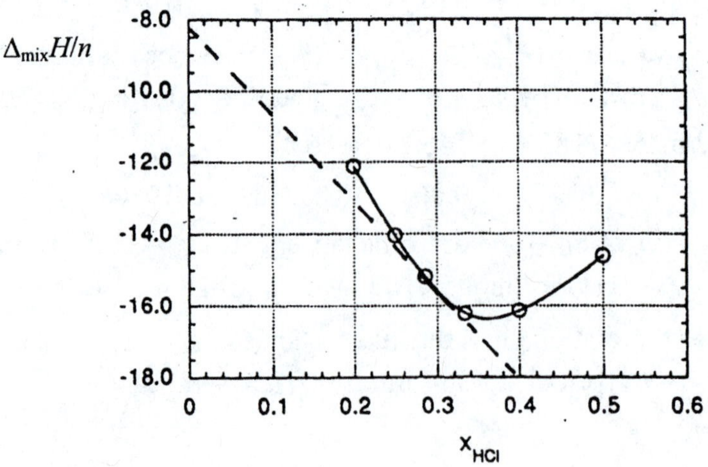

9.31 Use of (9.38) in (9.36) gives the desired result.

9.32 **(a)** F; **(b)** T; **(c)** T; **(d)** F;

9.33 No. Models show that the A-B intermolecular forces differ from the A-A and B-B intermolecular forces.

9.34 **(a)** T; **(b)** T; **(c)** F; **(d)** F; **(e)** F; **(f)** T.

9.35 The vapor is ideal and the pressure dependence of μ of the liquid can be neglected.

9.36 $n_{\text{ben}} = 1.280$ mol and $n_{\text{tol}} = 1.085$ mol. $x_{\text{ben}} = 0.5412$ and $x_{\text{tol}} = 0.4588$.
Equation (9.44) gives $\Delta_{\text{mix}}G = (8.314 \text{ J/mol-K})(293.1 \text{ K}) \times$
$[(1.280 \text{ mol}) \ln 0.5412 + (1.085 \text{ mol}) \ln 0.4588] = -3976 \text{ J} = -950 \text{ cal}$.
$\Delta_{\text{mix}}V = 0$, $\Delta_{\text{mix}}H = 0$. $\Delta_{\text{mix}}G = \Delta_{\text{mix}}H - T\,\Delta_{\text{mix}}S$,
so $\Delta_{\text{mix}}S = (3976 \text{ J})/(293.1 \text{ K}) = 13.56 \text{ J/K} = 3.24 \text{ cal/K}$.

9.37 **(a)** $n_{\text{ben}} = (100.0 \text{ g})/(78.11 \text{ g/mol}) = 1.280$ mol. $n_{\text{tol}} = 1.085$ mol.
$x_{\text{ben}} = 1.280/(1.280 + 1.085) = 0.5412$. $x_{\text{tol}} = 0.4588$.
$P_{\text{ben}} = 0.5412(74.7 \text{ torr}) = 40.4$ torr. $P_{\text{tol}} = 0.4588(22.3 \text{ torr}) = 10.2$ torr.

(b) $P_{\text{tot}} = (40.4 + 10.2)$ torr $= 50.6$ torr. Use of $P_i = x_i^{\upsilon}P$ gives
$x_{\text{ben}}^{\upsilon} = (40.4 \text{ torr})/(50.6 \text{ torr}) = 0.798$. $x_{\text{tol}}^{\upsilon} = 0.202$.

9.38 $P = P_{\text{hex}} + P_{\text{oct}} = x_{\text{hex},\ell}P^*_{\text{hex}} + (1 - x_{\text{hex},\ell})P^*_{\text{oct}}$. $x_{\text{hex},\ell} = (P - P^*_{\text{oct}})/(P^*_{\text{hex}} - P^*_{\text{oct}}) =$
$(666 \text{ torr} - 354 \text{ torr})/(1836 \text{ torr} - 354 \text{ torr}) = 0.211$. $x_{\text{oct},\ell} = 1 - 0.211 = 0.789$.
$P_{\text{hex}} = x_{\text{hex},\ell}P^*_{\text{hex}} = 0.211(1836 \text{ torr}) = 387$ torr. $x_{\text{hex},\upsilon} = P_{\text{hex}}/P =$
$(387 \text{ torr})/(666 \text{ torr}) = 0.581$; $x_{\text{oct},\upsilon} = 0.419$.

9.39 $P_i = x_i^{\upsilon}P = x_i^{l}P_i^*$. $0.555(95.0 \text{ torr}) = 0.305P^*_{\text{hex}}$ and $P^*_{\text{hex}} = 173$ torr.
$P = x_1^{l}P_1^* + x_2^{l}P_2^* = 95.0 \text{ torr} = 0.305P^*_{\text{hex}} + 0.695P^*_{\text{hept}} =$

$0.305(173 \text{ torr}) + 0.695 P^*_{\text{hept}}$ and $P^*_{\text{hept}} = 60.8 \text{ torr}$. We assumed an ideal solution, ideal vapor, and the pressure independence of μ_ℓ^*.

9.40 **(a)** $P_B = x_B^\upsilon P = x_B^l P_B^*$, so $x_B^\upsilon = x_B^l P_B^*/P$. Also, $P = P_B + P_C = x_B^l P_B^* + x_C^l P_C^* = x_B^l P_B^* + (1 - x_B^l) P_C^*$, so $x_B^\upsilon = x_B^l P_B^*/(x_B^l P_B^* + P_C^* - x_B^l P_C^*) = (x_B^l P_B^*/P_C^*)/(x_B^l P_B^*/P_C^* + 1 - x_B^l)$.

(b) If B is benzene, then $P_B^*/P_C^* = 74.7/22.3 = 3.35$ and the equation in (a) gives $x_B^\upsilon = 3.35 x_B^l/(1 + 2.35 x_B^l)$. We get

x_B^υ	0	0.456	0.691	0.834	0.931	1
x_B^l	0	0.2	0.4	0.6	0.8	1

The plot resembles the upper curve in Fig. 9.18b. For toluene (t), we find $x_t^\upsilon = 0.299 x_t^l/(1 - 0.701 x_t^l)$ and

x_t^υ	0	0.070	0.166	0.310	0.545	1
x_t^l	0	0.2	0.4	0.6	0.8	1

9.41 For an ideal solution, $V = V^* = n_1 V_{m1}^* + n_2 V_{m2}^* = n_1 M_1/\rho_1^* + n_2 M_2/\rho_2^* = m_1/\rho_1^* + m_2/\rho_2^* = (33.33 \text{ g})/(0.8790 \text{ g/cm}^3) + (33.33 \text{ g})/(0.8668 \text{ g/cm}^3) = 76.37 \text{ cm}^3$. Then $\rho = m/V = (66.66 \text{ g})/(76.37 \text{ cm}^3) = 0.8729 \text{ g/cm}^3$.

9.42 **(a)** $\Delta_{\text{mix}}C_P = C_P - C_P^* = (\partial H/\partial T)_{P,n_i} - (\partial H^*/\partial T)_{P,n_i} = (\partial/\partial T)_{P,n_i}(H - H^*) = (\partial \Delta_{\text{mix}} H/\partial T)_{P,n_i} = 0$, since $\Delta_{\text{mix}}H = 0$ for an ideal solution.

(b) $C_P = C_P^* = n_1 C_{P,m,1}^* + n_2 C_{P,m,2}^* = [(100/78.11) \text{ mol}](136 \text{ J/mol-K}) + (100/92.14)\text{mol}(156 \text{ J/mol-K}) = 343 \text{ J/mol-K}$.

9.43 At $x_B = 0.50$, the tangent is horizontal and intersects the $x_B = 0$ and $x_B = 1$ axes at -0.41_5 kcal/mol $= \mu_A - \mu_A^* = \mu_B - \mu_B^*$; the calculated result is $\mu_i - \mu_i^* = RT \ln 0.5 = (1.987 \text{ cal/mol-K})(298.1 \text{ K}) \ln 0.5 = -0.411$ kcal/mol. At $x_B = 0.25$, the tangent line's intercept at $x_B = 0$ is -0.18_0 kcal/mol $= \mu_A - \mu_A^*$ and that at $x_B = 1$ is -0.81_5 kcal/mol $= \mu_B - \mu_B^*$; the calculated values are $\mu_A - \mu_A^* = RT \ln 0.75 = -0.170$ kcal/mol and $\mu_B - \mu_B^* = RT \ln 0.25 = -0.821$ kcal/mol.

9.44 From (9.30), (9.31), (9.42), and (9.28):

$$\bar{S}_i = -(\partial\mu_i/\partial T)_{P,n_j} = -(\partial\mu_i^*/\partial T)_{P,n_j} - R\ln x_i = S_{m,i}^* - R\ln x_i.$$

$$\bar{V}_i = -(\partial\mu_i/\partial P)_{T,n_j} = (\partial\mu_i^*/\partial P)_T = V_{m,i}^*.$$

$$\bar{H}_i = \mu_i + T\bar{S}_i = \mu_i^* + RT\ln x_i + TS_{m,i}^* - RT\ln x_i = H_{m,i}^* \text{ (since } \mu_i^* = G_{m,i}^*).$$

9.45 $\mu_i = \mu_i^\circ + RT\ln(P_i/P^\circ) = \mu_i^\circ + RT\ln(x_iP/P^\circ) = \mu_i^\circ + RT\ln(P/P^\circ) + RT\ln x_i.$
For $x_i = 1$, this equation becomes $\mu_i^* = \mu_i^\circ + RT\ln(P/P^\circ)$, so
$\mu_i = \mu_i^* + RT\ln x_i.$

9.46 Substitution of Equation (9.42) into the phase-equilibrium condition $\mu_1^\alpha = \mu_1^\beta$
gives $\mu_1^* + RT\ln x_1^\alpha = \mu_1^* + RT\ln x_1^\beta$, so $x_1^\alpha = x_1^\beta$. Similarly, $x_2^\alpha = x_2^\beta$.

9.47 **(a)** The use of $G = n_i\mu_i + n_A\mu_A$ for the solution gives G of the final state in
the dilution process as $G_2 = n_i\mu_{i,2} + n_{A,2}\mu_{A,2}$. For the initial state of the
dilution process, $G_1 = n_i\mu_{i,1} + n_{A,1}\mu_{A,1} + (n_{A,2} - n_{A,1})\mu_A^*$. ΔG of the
dilution process is $G_2 - G_1$ and use of these expressions for G_2 and G_1
gives the desired result.

(b) Equating the right sides of (9.66) and (9.67) and using (9.68), we get
(9.70).

9.48 **(a)** The solvent A is ethanol and Raoult's law gives $P_A = x_A^l P_A^* =$
$P_{eth} = 0.9900(172.76 \text{ torr}) = 171.03 \text{ torr}$. $P_{chl} = P - P_{eth} =$
$(177.95 - 171.03) \text{ torr} = 6.92 \text{ torr}.$

(b) In a gas mixture, $P_i = x_iP$, so $x_{eth}^\upsilon = 171.03/177.95 = 0.9611_1$ and
$x_{chl}^\upsilon = 6.92/177.95 = 0.0388_9.$

(c) $P_i = K_i x_i^l$, so $K_{chl} = P_{chl}/x_{chl}^l = (6.92 \text{ torr})/0.0100 = 692 \text{ torr}.$

(d) $P_A = x_A^l P_A^* = P_{eth} = 0.9800(172.76 \text{ torr}) = 169.30 \text{ torr}$; $P_{chl} = K_{chl} x_{chl}^l =$
$(692 \text{ torr})(0.0200) = 13.84 \text{ torr}$. $P = (169.30 + 13.84) \text{ torr} = 183.14 \text{ torr}.$
$x_{eth}^\upsilon = P_{eth}/P = 169.30/183.14 = 0.9244$ and $x_{chl}^\upsilon = P_{chl}/P = 13.84/183.14$
$= 0.0756.$

9.49 (a) $P_{chl} = x^{v}_{chl} P = 0.9794(438.59 \text{ torr}) = 429.56$ torr; $P_{eth} =$ $(1 - 0.9794)(438.59 \text{ torr}) = 9.03$ torr.

(b) Raoult's law for the solvent chloroform gives $P_{chl} = x^{l}_{chl} P^*_{chl}$ and $P^*_{chl} = (429.56 \text{ torr})/0.9900 = 433.90$ torr.

(c) $P_{eth} = K_{eth} x^{l}_{eth}$ and $K_{eth} = (9.03 \text{ torr})/0.0100 = 903$ torr.

9.50 (a) At $x^{l}_{CS_2} = 1$ in Fig. 9.21b, $P = 424$ torr.

(b) At $x^{l}_{chl} = 0.40$ in Fig. 9.21a, the curves give $P = 266$ torr and $P_{chl} =$ 82 torr, so $x^{v}_{chl} = P_{chl}/P = 0.31$.

9.51 We have $K_i = P_i / x^{l}_i$ for very dilute solutions of i. For CS_2 as solute, we draw the line tangent to the $P(CS_2)$ curve at the point $x^{l}_{CS_2} = 0$. Since $P_i = K_i x^{l}_i$, the slope of this tangent line equals K_{CS_2} ; the slope equals the intercept of the tangent line with the $x^{l}_{CS_2} = 1$ line. We find $K_{CS_2} = 1.2_5 \times 10^3$ torr in acetone as solvent. Drawing the tangent line to the $P(ac)$ curve at the point $x^{l}_{CS_2} = 1$ and finding its intercept with $x^{l}_{CS_2} = 0$, we get $K_{ac} = 2.0 \times 10^3$ torr in the solvent CS_2.

9.52 From (9.62), $\mu^{\circ}_i - \mu^{\circ,v}_i = RT \ln (K_i/P^{\circ}) = (8.314 \text{ J/mol-K})(308 \text{ K}) \ln (145/750)$ $= -4.21$ kJ/mol.

9.53 (a) $n_{H_2} = 8.14 \times 10^{-5}$ mol, $n_{H_2O} = 5.55$ mol; $x^{l}_{H_2} = 1.47 \times 10^{-5}$.
$K_{H_2} = P_{H_2} / x^{l}_{H_2} = (1.00 \text{ atm})/(1.47 \times 10^{-5}) = 6.82 \times 10^4$ atm.

(b) $x^{l}_{H_2} = P_{H_2} / K_{H_2} = (10.00 \text{ atm})/(68200 \text{ atm}) = 1.47 \times 10^{-4}$.
$n_{H_2} = 8.14 \times 10^{-4}$ mol and $m_{H_2} = 1.64$ mg.

9.54 $x^{l}_{N_2} = P_{N_2}/K_{N_2} = (0.780 \times 760 \text{ torr})/(5.75 \times 10^7 \text{ torr}) = 1.03 \times 10^{-5}$, $x^{\ell}_{O_2} = (0.210 \times 760 \text{ torr})/(2.95 \times 10^7 \text{ torr}) = 5.41 \times 10^{-6}$. n_{O_2} and n_{N_2} are negligible compared to $n_{H_2O} = 5.55$ mol, so $n_{N_2} = (1.03 \times 10^{-5})(5.55 \text{ mol}) =$

5.72×10^{-5} mol and $n_{O_2} = (5.41 \times 10^{-6})(5.55 \text{ mol}) = 3.00 \times 10^{-5}$ mol.
$m_{N_2} = 1.60$ mg and $m_{O_2} = 0.960$ mg.

9.55 For the solvent in an ideally dilute solution, μ_A obeys the same equation as the chemical potential (9.42) and (9.43) for the chemical potential of a component of an ideal solution, so the solvent's partial molar properties in an ideally dilute solution obey the equations in Prob. 9.44 for an ideal-solution component.

9.56 **(a)** $\bar{V}_i = (\partial \mu_i / \partial P)_{T,n_j} = (\partial / \partial P)_{T,n_j} (\mu_i^\circ + RT \ln x_i) = (\partial \mu_i^\circ / \partial P)_{T,n_j}$. But (9.31) shows that $(\partial \mu_i^\circ / \partial P)_{T,n_j} = \bar{V}_i^\circ$, so $\bar{V}_i = \bar{V}_i^\circ$. Since μ_i° for an ideally dilute solution is a function of T and P only, its derivative $(\partial \mu_i^\circ / \partial P)_{T,n_j}$ is a function of T and P only, and $(\partial \mu_i^\circ / \partial P)_{T,n_j}$ and $\bar{V}_i^\circ$ (which equals $(\partial \mu_i^\circ / \partial P)_{T,n_j}$) are independent of concentration for concentrations in the ideally dilute range. Therefore $\bar{V}_i$ (which equals $\bar{V}_i^\circ$) equals its limiting infinite-dilution value $\bar{V}_i^\infty$.

(b) Using the intercept method of Fig. 9.7, we draw the tangent line to the curve at $x_{\text{ethanol}} = 1$. The tangent line intersects the $x_{\text{ethanol}} = 0$ vertical axis at -4.7 cm^3/mol $= \bar{V}_{H_2O}^\infty - V_{m,H_2O}^*$. Since $V_{m,H_2O}^* = 18.0$ cm^3/mol at 20°C, we get $\bar{V}_{H_2O}^\infty = 13.3$ cm^3/mol in ethanol, in rough agreement with Fig. 9.9.

9.57 Equations (9.30), (9.28), and (9.58) give: $\bar{S}_i = -(\partial \mu_i / \partial T)_{P,n_j} = -(\partial / \partial T)_{P,n_j} (\mu_i^\circ + RT \ln x_i) = \bar{S}_i^\circ - R \ln x_i$. [The infinite-dilution limit of this equation gives $\bar{S}_i^\circ = (\bar{S}_i + R \ln x_i)^\infty$.] We have $\bar{H}_i = \bar{G}_i + T\bar{S}_i = \mu_i^\circ + RT \ln x_i + T\bar{S}_i^\circ - RT \ln x_i = \bar{H}_i^\circ$. The infinite-dilution limit of this equation gives $\bar{H}_i^\infty = \bar{H}_i^\circ$.

9.58 Equation (9.17) and equations in Probs. 9.55 and 9.56 give $\Delta_{\text{mix}} V = \sum_{i \neq A} n_i(\bar{V}_i - V_{m,i}^*) + n_A(\bar{V}_A - V_{m,A}^*) = \sum_{i \neq A} n_i(\bar{V}_i^\circ - V_{m,i}^*)$. Equations in Probs.

9.55 and 9.57 give

$\Delta_{\text{mix}}H = \Sigma_{i\neq A}\, n_i(\overline{H}_i - H^*_{m,i}) + n_A(\overline{H}_A - H^*_{m,A}) = \Sigma_{i\neq A}\, n_i(\overline{H}^\circ_i - H^*_{m,i})$.

9.59 $0 = \Sigma_i\, \nu_i\mu_i = \Sigma_i\, \nu_i(\mu^o_i + RT\ln x_{i,\text{eq}}) = \Sigma_i\, \nu_i\mu^o_i + RT\,\Sigma_i \ln(x_{i,\text{eq}})^{\nu_i} =$ $\Delta G^\circ + RT\ln \Pi_i\,(x_{i,\text{eq}})^{\nu_i} = \Delta G^\circ + RT\ln K_x$, and $\Delta G^\circ = -RT\ln K_x$, where (1.70) and (1.69) were used.

9.60 **(a)** $x^\ell_i = P_i/K_i$, so as K_i increases, x^ℓ_i decreases. The solubility of O_2 in water decreases as T increases.

(b) $(\overline{H}^{\circ,\upsilon}_i - \overline{H}^{\circ,l}_i)/RT^2 = (\partial \ln K_i/\partial T)_P \approx (\Delta \ln K_i/\Delta T)_P$.
$\Delta \ln K_i = \ln(3.52\times 10^7) - \ln(2.95\times 10^7) = 0.177$.
$\overline{H}^{\circ,l}_i - \overline{H}^{\circ,\upsilon}_i \approx -(8.314\text{ J/mol-K})(298\text{ K})^2 0.177/(10\text{ K}) = -13.1\text{ kJ/mol}$.

(c) The log of (9.62) gives $RT\ln K_i/\text{bar} = \mu^{\circ,l}_i - \mu^{\circ,\upsilon}_i = \overline{G}^{\circ,l}_i - \overline{G}^{\circ,\upsilon}_i$ and we have $\overline{G}^{\circ,l}_i - \overline{G}^{\circ,\upsilon}_i \approx (8.314\text{ J/mol-K})(298\text{ K})\ln(44100\text{ bar}/1\text{ bar}) =$ 26.5 kJ/mol.

(d) $\Delta S = (\Delta H - \Delta G)/T$, so $\overline{S}^{\circ,l}_i - \overline{S}^{\circ,\upsilon}_i \approx [(-13100 - 26500)/298]\text{ J/mol-K} =$ -133 J/mol-K.

9.61 The log of Eq. (9.62) reads $\ln K_i = \ln P^\circ + (\mu^{\circ,l}_i - \mu^{\circ,\upsilon}_i)/RT$. Partial differentiation with respect to T gives $(\partial \ln K_i/\partial T)_P = -(\mu^{\circ,l}_i - \mu^{\circ,\upsilon}_i)/RT^2 +$ $[(\partial \mu^{\circ,l}_i/\partial T)_P - d\mu^{\circ,\upsilon}_i/dT]/RT = (\overline{G}^{\circ,\upsilon}_i - \overline{G}^{\circ,l}_i)/RT^2 + (-\overline{S}^{\circ,l}_i + \overline{S}^{\circ,\upsilon}_i)/RT =$ $(\overline{G}^{\circ,\upsilon}_i + T\,\overline{S}^{\circ,\upsilon}_i - \overline{G}^{\circ,l}_i - T\,\overline{S}^{\circ,l}_i)/RT^2 = (\overline{H}^{\circ,\upsilon}_i - \overline{H}^{\circ,l}_i)/RT^2$. Partial differentiation of $\ln K_i$ with respect to P gives $(\partial \ln K_i/\partial P)_T = (\partial \mu^{\circ,l}_i/\partial P)_T =$ $\overline{V}^{\circ,l}_i/RT = \overline{V}^{\infty,l}_i/RT$ (since $\mu^{\circ,\upsilon}_i$ depends only on T).

9.62 Trouton's rule is $\Delta_{\text{vap}}H_{\text{m,nbp}}/T_b \cong 21$ cal/mol-K. We find $\Delta_{\text{vap}}H_{\text{m,nbp,ben}} \cong$ 7420 cal/mol; $\Delta_{\text{vap}}H_{\text{m,nbp,tol}} \cong 8060$ cal/mol. Equation (7.21) gives for the vapor pressure of pure benzene at 120°C: $\ln[P^*_{\text{ben}}/(1\text{ atm})] =$ $(7420\text{ cal/mol})(1/353.2\text{ K} - 1/393.1\text{ K})/(1.987\text{ cal/mol-K}) = 1.073$ and $P^*_{\text{ben}} = 2.92$ atm. Similarly, we find $P^*_{\text{tol}} = 1.29$ atm. $P_{\text{ben}} = x^\ell_{\text{ben}}P^*_{\text{ben}} =$

$0.68(2.92 \text{ atm}) = 1.99$ atm. $P_{tol} = 0.32(1.29 \text{ atm}) = 0.41$ atm. $P = 2.40$ atm. $x^{v}_{ben} = P_{ben}/P = (1.99 \text{ atm})/(2.40 \text{ atm}) = 0.83$. We assumed the accuracy of Trouton's rule, the T independence of ΔH of vaporization, ideal gases, an ideal solution, and the pressure independence of μ^*.

9.63 $\Delta_{mix}G = G - G^* = G - n_A\mu_A^* - n_B\mu_B^*$. Then $\mu_A = (\partial G/\partial n_A)_{T,P,n_B} = (\partial/\partial n_A)_{T,P,n_B}(n_A\mu_A^* + n_B\mu_B^* + \Delta_{mix}G) = \mu_A^* + RT\ln x_A + n_A RT(\partial \ln x_A/\partial n_A)_{n_B} + n_B RT(\partial \ln x_B/\partial n_A)_{n_B} = \mu_A^* + RT\ln x_A + n_A RT[1/n_A - 1/(n_A + n_B)] + n_B RT[-1/(n_A + n_B)] = \mu_A^* + RT\ln x_A + RT - x_A RT - x_B RT = \mu_A^* + RT\ln x_A$, since $x_A + x_B = 1$. (The partial derivatives of the logs are found as in Prob. 9.66b.)

9.64 **(a)** Use of $(\partial G/\partial P)_T = V$ gives $\Delta G_1 = \int_P^{P_A^*} V_A^*\, dP' + \int_P^{P_B^*} V_B^*\, dP'$. We have $\Delta G_2 = 0$ for this constant-T-and-P equilibrium process. $\Delta G_3 = \Delta H_3 - T\Delta S_3 = -T\Delta S_3 = -n_A RT\ln(P_A^*/P_A) - n_B RT\ln(P_B^*/P_B)$, where (3.30) and Boyle's law were used. $\Delta G_4 = 0$ (see the end of Sec. 6.1). $\Delta G_5 = 0$ (constant-T-and-P equilibrium process). $\Delta G_6 = \int_{P_A+P_B}^{P} V_{A+B}\, dP'$.

(b) They are small because G of a liquid varies only slowly with P.

(c) With ΔG_1 and ΔG_6 neglected,
$\Delta G \equiv \Delta_{mix}G \cong \Delta G_3 = n_A RT\ln(P_A/P_A^*) + n_B RT\ln(P_B/P_B^*)$.

(d) Use (9.51).

9.65 Eq. (9.74) becomes $\Delta_{mix}G/(n_A + n_B) = RT[x_A \ln(x_A^{v}P/P_A^*) + x_B\ln(x_B^{v}P/P_B^*)]$. At $x_{eth} = 0.200$, $\Delta_{mix}G/(n_A + n_B) = (8.314 \text{ J/mol-K})(318 \text{ K}) \times \{0.200\ln[(0.1552)454.53/172.76] + 0.800\ln[(1 - 0.1552)454.53/433.54]\} = -731$ J/mol. At $x_{eth} = 0.400, 0.600, 0.800$, we get $\Delta_{mix}G/n_{tot} = -1034$ J/mol, -1554 J/mol, -997 J/mol, respectively. At $x_{eth} = 0$ and 1, $\Delta_{mix}G/n_{tot} = 0$.

9.66 **(a)** From (9.35), $\Delta_{mix}S = -(\partial\, \Delta_{mix}G/\partial T)_{P,n_j} = -n_A R\ln x_A - n_B R\ln x_B - (n_A + n_B)x_A x_B(\partial W/\partial T)_P$. From (9.33), $\Delta H_{mix} = \Delta_{mix}G + T\Delta_{mix}S = (n_A + n_B)x_A x_B[W - T(\partial W/\partial T)_P]$. From (9.34), $\Delta_{mix}V = (\partial\, \Delta_{mix}G/\partial P)_{T,n_j} = (n_A + n_B)x_A x_B(\partial W/\partial P)_T$.

(b) $\Delta_{mix}G = G - G^*$, so $G = G^* + \Delta_{mix}G = n_A\mu_A^*(T, P) + n_B\mu_B^*(T, P) + \Delta_{mix}G$. Then $\mu_A = (\partial G/\partial n_A)_{T,P,n_B} = \mu_A^* + (\partial \Delta_{mix} G/\partial n_A)_{T,P,n_B}$. We have $\ln x_A = \ln n_A - \ln(n_A + n_B)$ and $\ln x_B = \ln n_B - \ln(n_A + n_B)$, so $\partial \ln x_A/\partial n_A = 1/n_A - 1/(n_A + n_B)$ and $\partial \ln x_B/\partial n_A = -1/(n_A + n_B)$. Also, $x_A x_B(n_A + n_B) = n_A n_B/(n_A + n_B)$. Using these equations, we find $\mu_A = \mu_A^* + RT \ln x_A + n_A RT[1/n_A - 1/(n_A + n_B)] - n_B RT/(n_A + n_B) + W(T, P)[n_B/(n_A + n_B) - n_A n_B/(n_A + n_B)^2] = \mu_A^* + RT \ln x_A + W(T, P)x_B^2$. Since $\Delta_{mix}G$ is symmetric in A and B, by analogy to μ_A we have $\mu_B = \mu_B^* + RT \ln x_B + W(T, P)x_A^2$.

(c) Equating μ_A in the solution to μ_A in the vapor above the solution, we get $\mu_A^* + RT \ln x_A + Wx_B^2 = \mu^\circ_{A,gas} + RT \ln(P_A/P^\circ)$. Also, for pure liquid A in equilibrium with its vapor at T: $\mu_A^* = \mu^\circ_{A,gas} + RT \ln(P_A^*/P^\circ)$. Subtraction of this equation from the preceding one gives $RT \ln x_A + Wx_B^2 = RT \ln(P_A/P_A^*)$, so $\ln(P_A/P_A^*) = \ln x_A + Wx_B^2/RT$; so $P_A/P_A^* = \exp(\ln x_A)\exp(Wx_B^2/RT)$, and $P_A = x_A P_A^* \exp(Wx_B^2/RT)$. By symmetry, $P_B = x_B P_B^* \exp(Wx_A^2/RT)$.

9.67 $\Delta_{vap}G^\circ$ involves isothermal conversion of liquid at 1 bar to vapor at 1 bar. We use this isothermal path:

$$\text{liq}(P^\circ) \xrightarrow{1} \text{liq}(P_i) \xrightarrow{2} \text{vap}(P_i) \xrightarrow{3} \text{vap}(P^\circ)$$

$\Delta G_1 \approx 0$, since moderate pressure changes have little effect on liquid thermodynamic properties. $\Delta G_2 = 0$, since this is an equilibrium process at constant T and P. Assuming ideal vapor, we have $\Delta G_3 = \Delta H_3 - T\Delta S_3 = -T\Delta S_3 = -nRT \ln(P_i/P^\circ)$, where (3.30) and Boyle's law were used. Therefore $\Delta_{vap}G^\circ = \Delta G_{m,1} + \Delta G_{m,2} + \Delta G_{m,3} = -RT \ln(P_i/P^\circ)$ and $P_i = P^\circ \exp(-\Delta_{vap}G^\circ/RT)$.

9.68 **(a)** All; **(b)** ideal; **(c)** ideal and ideally dilute; **(d)** all; **(e)** ideal; **(f)** ideal; **(g)** ideal.

9.69 **(a)** Neither; neither.

(b) CCl_4 — Raoult's law; CH_3OH — Henry's law.

(c) CH_3OH — Raoult's law; CCl_4 — Henry's law.

(d) Both — Raoult's law.

9.70 **(a)** True, since the equilibrium condition at constant T and P is minimization of G. **(b)** False; see Fig. 9.5. **(c)** False. In a liquid solution, intermolecular interactions are large. **(d)** T. **(e)** F. **(f)** T.

R9.1 **(a)** $f = c - p + 2 - r - a = 1 - 3 + 2 = 0.$

(b) $f = 1 - 2 + 2 = 1.$

(c) $f = 1 - 1 + 2 = 2.$

R9.2 From $dG_m = -S_m\, dT + V_m\, dP$, we have $(\partial G_m/\partial P)_T = V_m = RT/P + B(T)$. So $\Delta G_m = \int_1^2 (RT/P + B)\, dP = RT\ln(P_2/P_1) + B(T)(P_2 - P_1)$.

R9.3 Integration of $d\ln P/dT = \Delta H_m/RT^2$ with ΔH_m assumed constant, gives $\ln(P_2/P_1) = (\Delta H_m/R)(T_1^{-1} - T_2^{-1})$. So $\ln(0.74/0.57) = [\Delta H_m/(8.314\text{ J/mol-K})][(105.0\text{ K})^{-1} - (108.0\text{ K})^{-1}]$ and $\Delta H_m = 8200$ J/mol. The other approximations are neglecting the volume of the liquid compared with the vapor and treating the vapor as an ideal gas.

R9.4 **(a)** F. **(b)** T. **(c)** F (see Fig. 9.5). **(d)** T. **(e)** T. **(f)** T. We have $P = P_B + P_C = x_B^l P_B^* + (1 - x_B^l)P_B^* = P_B^*$, as is obvious from Fig. 9.18a with $P_B^* = P_C^*$. Then $P_B \equiv x_B^v P = x_B^l P_B^*$, and setting $P = P_B^*$, we get $x_B^v = x_B^l$.

R9.5 **(a)** See p. 262. **(b)** See p. 264. **(c)** See p. 268. **(d)** See p. 269.

R9.6 $P_B = x_B^v P = x_B^l P_B^*$, so $P_B^* = x_B^v P/x_B^l = 0.650(139\text{ torr})/0.400 = 226$ torr. $P_C^* = x_C^v P/x_C^l = 0.350(139\text{ torr})/0.600 = 81.1$ torr.

R9.7 **(a)** Since the molecules do not closely resemble each other and since the solution is not dilute neither component obeys Raoult's law or Henry's law.

(b) The solution is very dilute, so water approximately obeys Henry's law and acetone approximately obeys Raoult's law.

(c) The two components have very similar molecules, so each component approximately obeys Raoult's law.

R9.8 **(a)** See the derivation of (9.28).

(b) See the lines preceding (9.31).

R9.9 See Example 9.4.

R9.10 **(a)** $f = c - p + 2 - r - a = 2 - 2 + 2 = 2$. T and x^{l}_{benzene}.

(b) $f = 3 - 1 + 2 - 1 - 1 = 2$, where the additional restriction is that $x_{\text{H}_2} = 3x_{\text{NH}_3}$. T and P.

(c) With the ionization of water included, we have 5 chemical species, 2 reactions, and the electroneutrality condition, so $f = 5 - 1 + 2 - 2 - 1 = 3$. T, P, and x_{HF}.

R9.11 Integration of $d \ln P/dT = \Delta H_{\text{m}}/RT^2$ with ΔH_{m} assumed constant, gives $\ln(P_2/P_1) = (\Delta H_{\text{m}}/R)(T_1^{-1} - T_2^{-1})$. We have $\ln[P_2/(760 \text{ torr})] =$ $[(2256.7 \text{ J/g})(18.015 \text{ g/mol})/(8.3145 \text{ J/mol-K})][(373.15 \text{ K})^{-1} - (363.15 \text{ K})^{-1}] =$ -0.3608; so $P_2/(760 \text{ torr}) = 0.697$ and $P_2 = 530$ torr. Figure 14.17 shows that T is not constant as the altitude changes, so there is some uncertainty as to what value of T to use. A reasonable choice might be 270 K. Then 530 torr = $(760 \text{ torr})\exp[-(0.0290 \text{ kg/mol})(9.81 \text{ m/s}^2)z/(8.314 \text{ J/mol-K})(270 \text{ K})]$ and $0.697 = \exp(-0.000126_7\, z/\text{m})$. We get $z = 2850 \text{ m} = 2.85$ km.

R9.12 The equation given is for fixed T, P, and n_{A}, so we have $\bar{V}_{\text{B}} \equiv (\partial V/\partial n_{\text{B}})_{T,P,n_{\text{A}}} = b + 2cn_{\text{B}} + 3en_{\text{B}}^2$. Also, $V = n_{\text{A}}\bar{V}_{\text{A}} + n_{\text{B}}\bar{V}_{\text{B}}$, so $\bar{V}_{\text{A}} = (V - n_{\text{B}}\bar{V}_{\text{B}})/n_{\text{A}} = (V - bn_{\text{B}} - 2cn_{\text{B}}^2 - 3en_{\text{B}}^3)/n_{\text{A}}$.

R9.13 Integration of $dP/dT = \Delta H/(T\,\Delta V)$ with ΔH and ΔV assumed constant gives $P_2 - P_1 = (\Delta H/\Delta V)\ln(T_2/T_1) =$

$$\frac{333.6 \text{ J}}{(1 \text{ g})[(1.000 \text{ g/cm}^3)^{-1} - (0.917 \text{ g/cm}^3)^{-1}]}\,\frac{82.06 \text{ cm}^3 \text{ atm}}{8.3145 \text{ J}}\ln\frac{272.35}{273.15} = 107 \text{ atm}$$

and $P_2 = 108$ atm. The same answer is found by assuming ΔS and ΔV constant.

Chapter 10

10.1 **(a)** T; **(b)** T; **(c)** T; **(d)** T; **(e)** F.

10.2 **(a)** No. **(b)** Yes. **(c)** Yes. $\gamma_i = a_i/x_i$ and a_i in (10.3) depends on μ_i°.
(d) Yes.

10.3 From (10.4) and (10.7), $\mu_i = \mu_i^*(T, P) + RT \ln a_{\mathrm{I},i}$. As $x_i \to 1$, $\gamma_{\mathrm{I},i} \to 1$ and $a_{\mathrm{I},i} \to x_i \to 1$. As x_i decreases (at constant T and P) from 1, Eq. (4.90) shows that μ_i decreases. The equation $\mu_i = \mu_i^* + RT \ln a_{\mathrm{I},i}$ then shows that this decrease in μ_i means that $a_{\mathrm{I},i}$ decreases from its limiting value of 1 as x_i decreases. Hence $a_{\mathrm{I},i}$ can never be greater than 1. The definition (10.3) shows that $a_{\mathrm{I},i}$ can never be negative. Hence $0 \le a_{\mathrm{I},i} \le 1$.

10.4 The Convention II solute standard state is the same as that used for solutes in an ideally dilute solution (Sec. 9.8 and Fig. 9.20), so the Convention II standard-state thermodynamic properties are the same as for solutes in an ideally dilute solution. The equations for these properties are given in Probs. 9.56 and 9.57. The equation for $\bar{S}^\circ_{\mathrm{II},i}$ is valid because in the infinite-dilution limit, the solution becomes ideally dilute and $\bar{S}_i = \bar{S}_i^\circ - R \ln x_i$ (Prob. 9.57) then holds.

10.5 **(a)** $\partial G^E/\partial n_i = (\partial/\partial n_i)(G - G^{\mathrm{id}}) = \partial G/\partial n_i - \partial G^{\mathrm{id}}/\partial n_i = \mu_i - \mu_i^{\mathrm{id}}$, where all derivatives are at constant T, P, $n_{j\ne i}$.

(b) $\mu_i = \mu_i^\circ + RT \ln \gamma_i x_i = (\mu_i^\circ + RT \ln x_i) + RT \ln \gamma_i = \mu_i^{\mathrm{id}} + RT \ln \gamma_i$, so $\mu_i - \mu_i^{\mathrm{id}} = RT \ln \gamma_i$ and use of the result of (a) gives the desired equation.

(c) $(\partial G^E/\partial n_{\mathrm{B}})_{n_{\mathrm{C}}} = (\partial/\partial n_{\mathrm{B}})[(n_{\mathrm{B}} + n_{\mathrm{C}})G^E_{\mathrm{m}}] = G^E_{\mathrm{m}} + (n_{\mathrm{B}} + n_{\mathrm{C}})\partial G^E_{\mathrm{m}}/\partial n_{\mathrm{B}}$ [Eq. (a)]. We have $(\partial G^E_{\mathrm{m}}/\partial n_{\mathrm{B}})_{n_{\mathrm{C}}} = (\partial G^E_{\mathrm{m}}/\partial x_{\mathrm{B}})(\partial x_{\mathrm{B}}/\partial n_{\mathrm{B}})_{n_{\mathrm{C}}}$. Also, $\partial x_{\mathrm{B}}/\partial n_{\mathrm{B}} = (\partial/\partial n_{\mathrm{B}})[n_{\mathrm{B}}(n_{\mathrm{B}} + n_{\mathrm{C}})^{-1}] = 1/(n_{\mathrm{B}} + n_{\mathrm{C}}) - n_{\mathrm{B}}/(n_{\mathrm{B}} + n_{\mathrm{C}})^2 = (n_{\mathrm{B}} + n_{\mathrm{C}} - n_{\mathrm{B}})/(n_{\mathrm{B}} + n_{\mathrm{C}})^2 = n_{\mathrm{C}}/(n_{\mathrm{B}} + n_{\mathrm{C}})^2 = x_{\mathrm{C}}/(n_{\mathrm{B}} + n_{\mathrm{C}})$, so

$(\partial G_{\mathrm{m}}^{E}/\partial n_{\mathrm{B}})_{n_{\mathrm{C}}} = (\partial G_{\mathrm{m}}^{E}/\partial x_{\mathrm{B}})x_{\mathrm{C}}/(n_{\mathrm{B}}+n_{\mathrm{C}})$ [Eq. (b)]. Use of Eqs. (b) and (10.106) in Eq. (a) gives $RT\ln\gamma_{\mathrm{I,B}} = G_{\mathrm{m}}^{E} + x_{\mathrm{C}}(\partial G_{\mathrm{m}}^{E}/\partial x_{\mathrm{B}})_{T,P}$.

10.6 **(a)** T; **(b)** T.

10.7 **(a)** $\gamma_{\mathrm{I},i} = x_i^{\upsilon} P/x_i^{l} P_i^{*}$.

$\gamma_{\mathrm{I,chl}} = (1-0.138)(304.2 \text{ torr})/(1-0.200)(295.1 \text{ torr}) = 1.11$.
$\gamma_{\mathrm{I,eth}} = 0.138(304.2 \text{ torr})/0.200(102.8 \text{ torr}) = 2.04$.
$a_{\mathrm{I,chl}} = \gamma_{\mathrm{I,chl}}x_{\mathrm{chl}} = 1.11(0.800) = 0.889$; $\quad a_{\mathrm{I,eth}} = 2.04(0.200) = 0.408$.

(b) $\mu_i = \mu_i^{\circ} + RT\ln a_i = \mu_i^{*} + RT\ln a_{\mathrm{I},i}$, so $\mu_i - \mu_i^{*} = RT\ln a_{\mathrm{I},i}$.

$\mu_{\mathrm{eth}} - \mu_{\mathrm{eth}}^{*} = (8.3145 \text{ J/mol-K})(308.1 \text{ K})\ln 0.408 = -2300 \text{ J/mol}$.
$\mu_{\mathrm{chl}} - \mu_{\mathrm{chl}}^{*} = RT\ln 0.889 = -301 \text{ J/mol}$.

(c) Eqs. (9.32), (10.7), and (10.4) give $\Delta_{\mathrm{mix}}G = \sum_i n_i(\mu_i - \mu_i^{*}) =$
$\sum_i n_i RT\ln a_{\mathrm{I},i} = (8.3145 \text{ J/mol-K})(308.1 \text{ K})[(0.200 \text{ mol})\ln 0.408 + (0.800 \text{ mol})\ln 0.889] = -700 \text{ J}$. Alternatively, $\Delta_{\mathrm{mix}}G = $
$(0.200 \text{ mol})(-2300 \text{ J/mol}) + (0.800 \text{ mol})(-301 \text{ J/mol}) = -701 \text{ J}$.

(d) For an ideal solution, $\gamma_{\mathrm{I},i} = 1$ and $a_{\mathrm{I},i} = x_i$, so
$\Delta_{\mathrm{mix}}G = (8.3145 \text{ J/mol-K})(308.1 \text{ K}) \times$
$[(0.200 \text{ mol})\ln 0.200 + (0.800)\ln 0.800] = -1280 \text{ J}$.

10.8 **(a)** $a_{\mathrm{I},i} = P_i/P_i^{*} = x_i^{\upsilon} P/P_i^{*}$. $\quad a_{\mathrm{I,w}} = 0.696(5.03 \text{ kPa})/(19.92 \text{ kPa}) = 0.176$.
$\gamma_{\mathrm{I,w}} = a_{\mathrm{I,w}}/x_{\mathrm{w}} = 0.176/0.300 = 0.586$.
$a_{\mathrm{I,hp}} = (1-0.696)(5.03 \text{ kPa})/(2.35 \text{ kPa}) = 0.651$;
$\gamma_{\mathrm{I,hp}} = 0.651/(1-0.300) = 0.930$.

(b) Since water is the solvent, $\gamma_{\mathrm{II,w}} = \gamma_{\mathrm{I,w}} = 0.586$. Example 10.2 in Sec. 10.3 gives $\gamma_{\mathrm{II,hp}} = (P_{\mathrm{hp}}^{*}/K_{\mathrm{hp}})\gamma_{\mathrm{I,hp}}$, where $K_{\mathrm{hp}} = (P_{\mathrm{hp}}/x_{\mathrm{hp}}^{l})^{\infty}$. We plot $x_{\mathrm{hp}}^{\upsilon} P/x_{\mathrm{hp}}^{l}$ versus x_{hp}^{l} and extrapolate to $x_{\mathrm{hp}}^{l} = 0$. We find $K_{\mathrm{hp}} = 0.62 \text{ kPa}$. (Use of a spreadsheet shows that a cubic or quartic polynomial gives an excellent fit to the data; the fitted equation has an intercept of 0.61 for the cubic polynomial and 0.62 for the quartic.) So $\gamma_{\mathrm{II,hp}} = (2.35 \text{ kPa}/0.62 \text{ kPa})0.930 = 3.5$. $\quad a_{\mathrm{II,hp}} = 3.5(0.700) = 2.5$. $\quad a_{\mathrm{II,w}} = a_{\mathrm{I,w}} = 0.176$.

(c) We have $\Delta_{mix}G = G - G^* = \sum_i n_i(\mu_i - \mu_i^*) = \sum_i n_i(\mu_i - \mu^\circ_{I,i}) =$ $\sum_i n_i RT \ln a_{I,i}$. An amount of solution with $n_{tot} = 1$ mol has 0.300 mol of water and 0.700 mol of H_2O_2, which is 5.405 g of water and 23.81 g of H_2O_2. The water wt. % is [5.405/(5.405 + 23.81)]100% = 18.50%; so 125 g of solution has 0.1850(125 g) = 23.1 g of water and 101.9 g of H_2O_2; thus $n_w = 1.284$ mol and $n_{hp} = 3.00$ mol. Then $\Delta_{mix}G =$ (8.314 J/mol-K)(333.1 K)[(1.284 mol) ln 0.176 + (3.00 mol) ln 0.651] = –9.75 kJ.

10.9 (a) Let $b = 3.92$. Equation (10.23) with B = Hg and A = Zn and state 1 being pure Hg gives $\ln \gamma_{Hg} = -\int_1^2 [x_{Zn}/(1 - x_{Zn})]\, d \ln (1 - bx_{Zn}) =$ $b\int_0^{x_{Zn}} [x/(1-x)(1-bx)]\, dx = [b/(1-b)][-\ln(1-x) + b^{-1} \ln (1-bx)]\big|_0^{x_{Zn}} =$ $[b/(b-1)] \ln (1 - x_{Zn}) - [1/(b-1)] \ln (1 - bx_{Zn}) =$ $(3.92/2.92) \ln (1 - x_{Zn}) - (1/2.92) \ln (1 - 3.92x_{Zn})$.

(b) $\gamma_{II,Zn} = 1 - 3.92(0.0400) = 0.843$; $a_{II,Zn} = \gamma_{II,Zn}x_{Zn} = (0.843)(0.0400) =$ 0.0337. $\ln \gamma_{Hg} = (1/2.92)\{3.92 \ln 0.960 - \ln [1 - (3.92)(0.0400)]\} =$ 0.00360; $\gamma_{Hg} = 1.0036$. $a_{Hg} = 1.0036(0.9600) = 0.9635$.

10.10 From Eq. (10.12), $G^E = RT\sum_i n_i \ln \gamma_{I,i}$. So $G^E/n = RT\sum_i x_i \ln \gamma_{I,i} =$ $RT(x_{ac} \ln \gamma_{I,ac} + x_{chl} \ln \gamma_{I,chl})$. Values of γ_I listed in Sec. 10.3 give at 35.2°C:

x_{ac}	0	0.082	0.200	0.336	0.506	0.709	
G^E/n_{tot}	0	–42.8	–96.1	–133.1	–138.3	–102.6	cal/mol

x_{ac}	0.815	0.940	1	
G^E/n_{tot}	–69.8	–24.7	0	cal/mol

10.11 (a) In the second example in Sec. 10.3, it was shown that $\boxed{\gamma_{II,i}/\gamma_{I,i} = P_i^*/K_i}$ for $i \neq$ A. In the infinite-dilution ($x_A \to 1$) limit, we have $\gamma_{II,i} \to 1$ [Eq. (10.10)] and $\gamma_{I,i} \to \gamma_{I,i}^\infty$; in this limit, the boxed equation becomes $1/\gamma_{I,i}^\infty = P_i^*/K$. Equating the two expressions for P_i^*/K_i, we get $\gamma_{II,i} = \gamma_{I,i}/\gamma_{I,i}^\infty$.

(b) From Fig. 10.3a, $\gamma_{I,chl} \to 0.50$ as $x_{ac} \to 1$. Therefore $\gamma_{II,chl} = 2.0\gamma_{I,chl}$ in acetone.

10.12 **(a)** Eq. (10.13) gives $P_{H_2O} = a_{H_2O}P^*_{H_2O}$ and a_{H_2O} = (23.34 torr)/(23.76 torr) = 0.9823.

(b) $a_{H_2O} = 22.75/23.76 = 0.9575$. An amount of solution with 1 kg of solvent has 2.00 mol of sucrose and 55.51 mol of H_2O, so x_{H_2O} = 55.51/(55.51 + 2.00) = 0.9652. Then $a_{H_2O} = \gamma_{H_2O}x_{H_2O} = 0.9575 = \gamma_{H_2O}(0.9652)$ and $\gamma_{H_2O} = 0.9920$.

10.13 **(a)** We use Eq. (10.23) written for Convention I and with A changed to B and B changed to C. Let state 1 be pure C, so $\gamma_{I,C,1} = 1$ and $x_{C,1} = 1$. Then (10.23) is $\ln \gamma_{I,C,2} = -\int_1^2 [x_B/(1 - x_B)]\, d\ln \gamma_{I,B}$ [Eq. (a)] at constant T and P. We have $\ln \gamma_{I,B} = (W/RT)x_C^2$, so $d\ln\gamma_{I,B} = 2(W/RT)x_C\, dx_C$ at constant T and P. Also, the integrand in Eq. (a) is $x_B/(1-x_B) = (1-x_C)/x_C$. Hence Eq. (a) becomes $\ln \gamma_{I,C,2} = -2(W/RT)\int_1^2 (1 - x_C)\, dx_C = -2(W/RT)(x_C - \frac{1}{2}x_C^2)\,\big|_1^{x_{C,2}} = -2(W/RT)(x_{C,2} - \frac{1}{2}x_{C,2}^2 - 1 + \frac{1}{2}) = -2(W/RT)[1 - x_{B,2} - \frac{1}{2}(1 - x_{B,2})^2 - \frac{1}{2}] = (W/RT)x_{B,2}^2$, and $RT\ln\gamma_{I,C} = Wx_B^2$, where the unneeded subscript 2 was dropped

(b) Equation (10.12) gives

$$G^E = n_B RT\ln\gamma_{I,B} + n_C RT\ln\gamma_{I,C} = n_B W x_C^2 + n_C W x_B^2 =$$

$$W\left[n_B\frac{n_C^2}{(n_B+n_C)^2} + n_C\frac{n_B^2}{(n_B+n_C)^2}\right] = Wn_Bn_C\left[\frac{n_C+n_B}{(n_B+n_C)^2}\right] = \frac{Wn_Bn_C}{n_B+n_C}$$

10.14 In the limit $x_{chl} = 0$ and $x_{ac} = 1$, Fig. 10.3 gives $\gamma_{I,chl} = 0.50$. Hence $\ln\gamma_{I,chl} = (W/RT)x_{ac}^2$ becomes ln 0.50 = W/RT and $W/RT = -0.693$. For x_{chl} = 0.494, the simple-solution model gives $\ln\gamma_{I,ac} = (-0.693)(0.494)^2 = -0.169$ and $\gamma_{I,ac} = 0.844$. Also, $\ln\gamma_{I,chl} = (-0.693)(0.506)^2 = -0.177$ and $\gamma_{I,chl} = 0.837$. The true values are 0.824 and 0.772.

10.15 **(a)** With one term taken, we have $G_m^E \equiv G^E/(n_B + n_C) = [n_Bn_C/(n_B + n_C)^2]RTA_1(T, P)$, which is G^E in Prob. 10.13(b) with $W(T, P) = RTA_1$. See also part (c) of Prob. 10.15.

(b) Comparing

$$G_m^E = x_B x_C RT[A_1 + A_2(x_B - x_C) + A_3(x_B - x_C)^2 + A_4(x_B - x_C)^3 + \cdots]$$

with

$$G_m^E = x_C x_B RT[A_1' + A_2'(x_C - x_B) + A_3'(x_C - x_B)^2 + A_4'(x_C - x_B)^3 + \cdots],$$

we see that $A_1' = A_1$, $A_2' = -A_2$, $A_3' = A_3$, $A_4' = -A_4$, etc.

(c) x_B and x_C are not independent. Substitution of $x_C = 1 - x_B$ in the Redlich–Kister equation gives

$G_m^E = (x_B - x_B^2)RT[A_1 + A_2(2x_B - 1) + A_3(2x_B - 1)^2]$, so

$(\partial G_m^E / \partial x_B) = (1 - 2x_B)RT[A_1 + A_2(2x_B - 1) + A_3(2x_B - 1)^2] +$

$(x_B - x_B^2)RT[2A_2 + 4A_3(2x_B - 1)]$. Using $x_B = 1 - x_C$, we have

$\ln \gamma_{I,B} = (RT)^{-1}[G_m^E + x_C(\partial G_m^E / \partial x_B)] =$

$x_C(1 - x_C)[A_1 + A_2(1 - 2x_C) + A_3(1 - 2x_C)^2] +$

$x_C(2x_C - 1)[A_1 + A_2(1 - 2x_C) + A_3(1 - 2x_C)^2] +$

$(1 - x_C)x_C^2[2A_2 + 4A_3(1 - 2x_C)] =$

$(A_1 + 3A_2 + 5A_3)x_C^2 - (4A_2 + 16A_3)x_C^3 + 12A_3x_C^4$. (The tedious algebra can be done by certain calculators or by symbolic algebra programs such as Maple or Mathematica. If you don't have access to such resources, you are excused from doing the algebra.) Interchanging B and C in the final expression for $\ln \gamma_{I,B}$ requires that the sign of A_2 be changed, as shown in part (b), and this gives $\ln \gamma_{I,C}$. Note that if we set $A_2 = 0$ and $A_3 = 0$ in these equations, we get $\ln \gamma_{I,B} = A_1(T, P)x_C^2$ and $\ln \gamma_{I,C} = A_1(T, P)x_B^2$, which have the same form as the simple-solution equations in Prob. 10.13.

10.16 **(a)** At infinite dilution of B, we have $x_C = 1$ and the first equation of Prob. 10.15(c) with $A_3 = 0$ becomes $\ln \gamma_{I,B}^\infty = A_1 - A_2$. With $x_B = 1$ and $A_3 = 0$, the second equation of Prob. 10.15(c) becomes $\ln \gamma_{I,C}^\infty = A_1 + A_2$. Adding and subtracting these equations, we get the desired results.

(b) Let B be CCl_4. From the results of (a), $A_1 = 0.5(1.129 + 1.140) = 1.134$ and $A_2 = 0.5(1.140 - 1.129) = 0.005$. The equations of Prob. 10.15 then give

$$\ln \gamma_{\text{I,CCl}_4} = (A_1 + 3A_2)x^2_{\text{SiCl}_4} - 4A_2x^3_{\text{SiCl}_4} = 0.1826 \text{ and } \gamma_{\text{I,CCl4}} = 1.20;$$
$$\ln \gamma_{\text{I,SiCl}_4} = (A_1 - 3A_2)x^2_{\text{CCl}_4} + 4A_2x^3_{\text{CCl}_4} = 0.407 \text{ and } \gamma_{\text{I,SiCl4}} = 1.50.$$

10.17 **(a)** From (10.12) and the definition of G^E_{m} in Prob. 10.15, we have $G^E_{\text{m}} = RT(x_{\text{chl}} \ln \gamma_{\text{I,chl}} + x_{\text{hep}} \ln \gamma_{\text{I,hep}})$ and use of the data gives G^E_{m}/(J/mol) = 90.7, 197.9, 224, 187.9, and 82.4 at $x_{\text{chl}} = 0.1, 0.3, 0.5, 0.7$, and 0.9.

(b) We designate three cells for A_1, A_2, and A_3. A suitable initial guess is zero for each of these, corresponding to an ideal solution. The x_{chl} values are put in column A, the G^E_{m} values in column B, the Redlich–Kister values in column C, and the squares of the deviations of the column C values from the column B values in column D. We sum these squared deviations in a cell and use the Solver to minimize this sum by varying A_1, A_2, and A_3. The Solver gives $A_1 = 0.334656$, $A_2 = -0.021853$, and $A_3 = 0.038433$. The fit is quite good, with deviations of typically 0.5 J/mol.

(c) At $x_B = x_{\text{chl}} = 0$, the equations of Prob. 10.15c give $\ln \gamma_{\text{I,chl}} = A_1 - A_2 + A_3 = 0.39494$ and $\gamma_{\text{I,chl}} = 1.48$; also, $\ln \gamma_{\text{I,hep}} = 0$ and $\gamma_{\text{I,hep}} = 1$. At $x_B = x_{\text{chl}} = 0.4$, we get $\ln \gamma_{\text{I,chl}} = 0.11188$ and $\gamma_{\text{I,chl}} = 1.118$; $\ln \gamma_{\text{I,hep}} = 0.06163$ and $\gamma_{\text{I,hep}} = 1.064$.

10.18 For the two-parameter fit, the spreadsheet of Prob. 10.17b can be used by setting $A_3 = 0$ and omitting A_3 from the By Changing Cells box of the Solver. One finds $A_1 = 0.340293$ and $A_2 = -0.021853$. The fit is much poorer than for the 3-parameter function, with the sum of squares of the deviations equal to 58 J^2/mol^2 as compared with 1.44 J^2/mol^2 for the three-parameter function. For the four-parameter function, we modify the spreadsheet by adding a cell for A_4, modifying the formulas in column C, and including A_4 in the By Changing Cells box. The sum of squares of deviations is reduced to 1.42 J^2/mol^2, which is not a significant improvement over the 3-parameter function.

10.19 From (10.12) and the definition of G^E_{m} in Prob. 10.15, we have $G^E_{\text{m}} = RT(x_{\text{chl}} \ln \gamma_{\text{I,chl}} + x_{\text{ac}} \ln \gamma_{\text{I,ac}})$ and use of the data gives G^E_{m}/(J/mol) = –179, –402, –557, –579, –429, –292, –103 at $x_{\text{chl}} = 0.918, 0.8, 0.664, 0.494, 0.291, 0.185, 0.060$. The spreadsheet of Prob. 10.17 can be used with the data revised. We take B as chloroform. The two-parameter fit is poor with a sum of squares of deviations equal to 904 J^2/mol^2 and deviations of 3 to 22 J/mol. The three-J^2/mol^2parameter fit is fair, with a sum of squares of deviations equal to 258 J^2/mol^2 and deviations of 1 to 10 J/mol. The four-parameter fit is very good, with a sum of squares of deviations equal to 32 J^2/mol^2 and deviations of 1 to 3 J/mol. The four parameters are $A_1 = -0.869763$, $A_2 = -0.22272$, $A_3 = 0.10898$, $A_4 = 0.14581$. The activity coefficients can be found from equations like those in Prob. 10.15, but since the four-parameter version of these equations was not given, this calculation is omitted.

10.20 All.

10.21 For an amount of solution containing 1 kg of water, $n_{\text{sucrose}} = 1.50$ mol and $n_{H_2O} = 55.51$ mol, so $x_{H_2O} = 55.51/57.01 = 0.9737$ and $x_{\text{suc}} = 0.0263$. From Eq. (10.24), $\gamma_{\text{II,suc}} = \gamma_{\text{m,suc}}/x_A = 1.292/0.9737 = 1.327$. Then $a_{\text{II,suc}} = \gamma_{\text{II,suc}}x_{\text{suc}} = 1.327(0.0263) = 0.0349$. Also, $a_{\text{m,suc}} = \gamma_{\text{m,suc}}(m_{\text{suc}}/m°) = 1.292(1.50) = 1.94$.

10.22 Equations (9.31) and (10.27) give $\bar{V}^\circ_{\text{m},i} = (\partial\mu^\circ_{\text{m},i}/\partial P)_T = (\partial\mu^\circ_{\text{II},i}/\partial P)_T = \bar{V}^\circ_{\text{II},i} = \bar{V}_i^\infty$, where an equation in Prob. 10.4 was used. From (9.30) and (10.27), $\bar{S}^\circ_{\text{m},i} = -(\partial\mu^\circ_{\text{m},i}/\partial T)_P = -(\partial\mu^\circ_{\text{II},i}/\partial T)_P - R\ln M_Am° = \bar{S}^\circ_{\text{II},i} - R\ln M_Am° = (\bar{S}_i + R\ln x_i)^\infty - R\ln M_Am° = [\bar{S}_i + R\ln(x_i/M_Am°)]^\infty$, where an equation in Prob. 10.4 was used. We have $x_i/M_A = n_i/n_{\text{tot}}M_A$; in the infinite-dilution limit, $n_{\text{tot}} = n_A$ and $x_i/M_A = n_i/n_AM_A = n_i/w_A = m_i$, where w_A is the solvent mass and m_i is the solute molality. So $\bar{S}^\circ_{\text{m},i} = [\bar{S}_i + R\ln(m_i/m°)]^\infty$. Finally, (9.28), (10.27), the above relation between $\bar{S}^\circ_{\text{m},i}$ and $\bar{S}^\circ_{\text{II},i}$, and the relation $\bar{H}^\circ_{\text{II},i} = \bar{H}_i^\infty$ (Prob. 10.4) give $\bar{H}^\circ_{\text{m},i} = \mu^\circ_{\text{m},i} + T\bar{S}^\circ_{\text{m},i} = \mu^\circ_{\text{II},i} + RT\ln M_Am° + T(\bar{S}^\circ_{\text{II},i} - R\ln M_Am°) = \mu^\circ_{\text{II},i} + T\bar{S}^\circ_{\text{II},i} = \bar{H}^\circ_{\text{II},i} = \bar{H}_i^\infty$.

10.23 $\mu^\circ_{c,i} + RT\ln(\gamma_{c,i}c_i/c°) = \mu^\circ_{\text{II},i} + RT\ln\gamma_{\text{II},i}x_i$. In the limit $x_A \to 1$, Eqs. (10.31) and (10.10) show that the activity coefficients go to 1; so this limit gives $\mu^\circ_{c,i} = \mu^\circ_{\text{II},i} + RT\ln(x_ic°/c_i)^\infty$. We have $x_i = n_i/n_{\text{tot}} \approx n_i/n_A$ and $c_i = n_i/V \approx n_i/V^*_A$ for x_A near 1. Hence, $(x_ic°/c_i)^\infty = c°\,n_iV^*_A/n_in_A = c°V^*_{\text{m},A}$; $\mu^\circ_{c,i} = \mu^\circ_{\text{II},i} + RT\ln V^*_{\text{m},A}c°$. Substitution of this result in (10.31) gives

$\mu_i = \mu^{\circ}_{\text{II},i} + RT \ln V^*_{\text{m},A}c^{\circ} + RT \ln (\gamma_{c,i}c_i/c^{\circ}) = \mu^{\circ}_{\text{II},i} + RT \ln (V^*_{\text{m},A}\gamma_{c,i}c_i)$.
Comparison with (10.24) shows that $V^*_{\text{m},A}\gamma_{c,i}c_i = \gamma_{\text{II},i}x_i$ and $\gamma_{c,i} = (x_i/V^*_{\text{m},A}c_i)\gamma_{II,i}$. We have $x_i = n_i/n_{\text{tot}} = w_A n_i/w_A n_{\text{tot}} = w_A m_i/n_{\text{tot}} = \rho_A V^*_A m_i/n_{\text{tot}}$ (where w is mass) and $V^*_{\text{m},A} = V^*_A/n_A$, so $x_i/V^*_{\text{m},A} = \rho_A m_i n_A/n_{\text{tot}} = \rho_A m_i x_A$. Then $\gamma_{c,i} = (x_i/V^*_{\text{m},A}c_i)\gamma_{\text{II},i} = (\rho_A m_i x_A/c_i)(\gamma_{\text{m},i}/x_A) = (\rho_A m_i/c_i)\gamma_{\text{m},i}$. Also, $\gamma_{c,i}c_i = \rho_A m_i\gamma_{\text{m},i}$ and $(\gamma_{c,i}c_i/c^{\circ})c^{\circ} = \rho_A(m_i\gamma_{\text{m},i}/m^{\circ})m^{\circ}$, so $a_{c,i}c^{\circ} = \rho_A a_{\text{m},i}m^{\circ}$ and $a_{c,i} = (\rho_A m^{\circ}/c^{\circ})a_{\text{m},i}$.

10.24 **(a)** $KCl \rightarrow K^+ + Cl^-$; $\nu_+ = 1$, $\nu_- = 1$, $z_+ = 1$, $z_- = -1$.
(b) $MgCl_2 \rightarrow Mg^{2+} + 2Cl^-$; $\nu_+ = 1$, $\nu_- = 2$, $z_+ = 2$, $z_- = -1$.
(c) $MgSO_4 \rightarrow Mg^{2+} + SO_4^{2-}$; $\nu_+ = 1$, $\nu_- = 1$, $z_+ = 2$, $z_- = -2$.
(d) $Ca_3(PO_4)_2 \rightarrow 3Ca^{2+} + 2PO_4^{3-}$; $\nu_+ = 3$, $\nu_- = 2$, $z_+ = 2$, $z_- = -3$.
(e) For a 1:1 electrolyte, $z_+ = |z_-| = 1$. KCl is a 1:1 electrolyte.

10.25 $\gamma_{\pm} = [(\gamma_+)^{\nu_+}(\gamma_-)^{\nu_-}]^{1/(\nu_+ + \nu_-)}$.
(a) $\gamma_{\pm} = \gamma_+^{1/2}\gamma_-^{1/2}$. **(b)** $\gamma_{\pm} = (\gamma_+)^{1/3}(\gamma_-)^{2/3}$.
(c) $\gamma_{\pm} = \gamma_+^{1/2}\gamma_-^{1/2}$. **(d)** $\gamma_{\pm} = (\gamma_+)^{3/5}(\gamma_-)^{2/5}$.

10.26 $\mu_i = \nu_+\mu_+ + \nu_-\mu_- = \mu_+ + 2\,\mu_-$.

10.27 $(\nu_{\pm})^{\nu} = (\nu_+)^{\nu_+}(\nu_-)^{\nu_-}$.
(a) $(\nu_{\pm})^2 = 1^1 \cdot 1^1$ and $\nu_{\pm} = 1$.
(b) $(\nu_{\pm})^3 = 1^1 \cdot 2^2$ and $\nu_{\pm} = 1.587$.
(c) $\nu_{\pm} = 1$.
(d) $(\nu_{\pm})^5 = 3^3 \cdot 2^2$ and $\nu_{\pm} = 2.551$.

10.28 With $\nu_+ = \nu_-$, we have $(\nu_{\pm})^{(\nu_+ + \nu_-)} = (\nu_{\pm})^{2\nu_+} = (\nu_+)^{\nu_+}(\nu_-)^{\nu_-} =$ ` $(\nu_+)^{\nu_+}(\nu_+)^{\nu_+} = (\nu_+)^{2\nu_+}$. From $(\nu_{\pm})^{2\nu_+} = (\nu_+)^{2\nu_+}$, we have $\nu_{\pm} = \nu_+$.

10.29 From Prob. 10.27(b), $(\nu_{\pm})^{\nu} = 4$ and (10.52) gives $a_{MgCl_2} = 4\gamma_{\pm}^3(m_i/m^{\circ})^3$.

10.30 **(a)** The second equation in Prob. 10.31 in the textbook gives

$$\phi = \frac{1}{(0.018015\text{ kg/mol})2(4.800\text{ mol/kg})} \ln \frac{23.76\text{ torr}}{20.02\text{ torr}} = 0.990$$

(b) From Eq. (10.56), $a_A = P_A/P_A^* = (20.02\text{ torr})/(23.76\text{ torr}) = 0.843$. One kg of water contains 55.51 mol of H_2O. There is no significant ion pairing in the KCl solution, so $x_A = 55.51/(55.51 + 4.80 + 4.80) = 0.853$. Equation (10.5) gives $\gamma_A = a_A/x_A = 0.843/0.853 = 0.988$.

(c) $a_A = 0.843$. $x_A = 55.51/(55.51 + 4.80) = 0.920$. $\gamma_A = 0.843/0.920 = 0.916$.

10.31 Substitution of $a_A = P_A/P_A^*$ [Eq. (10.56)] into (10.107) gives the desired result.

10.32 **(a)** At constant T and P, $0 = n_A\, d\mu_A + n_i\, d\mu_i =$ $n_A(-\phi RTM_A \nu\, dm_i - RTM_A \nu m_i\, d\phi) + n_i[(\nu RT/m_i)\, dm_i + \nu RT\, d\ln\gamma_i]$ and $0 = (n_i/m_i - \phi n_A M_A)\, dm_i - n_A M_A m_i\, d\phi + n_i\, d\ln\gamma_i$. We have $m_i = n_i/n_A M_A$. Substitution of $n_A M_A = n_i/m_i$ into the preceding equation and division by n_i gives $d\ln\gamma_i = d\phi + [(\phi - 1)/m_i]\, dm_i$ at constant T and P.

(b) $\phi = (\mu_A^* - \mu_A)/RTM_A \nu m_i = -RT\ln\gamma_{x,A}x_A/RTM_A \nu m_i =$ $-\ln(\gamma_{x,A}x_A)/M_A \nu m_i$. At high dilution, $\gamma_{x,A} \to 1$ and $\ln(\gamma_{x,A}x_A) \to \ln x_A$. Equation (8.36) gives for x_A near 1: $\ln x_A \approx x_A - 1$. At high dilution, there is no ion pairing and the electrolyte gives νn_i moles of ions. Hence $x_A = 1 - \nu n_i/n_{tot} \approx 1 - \nu n_i/n_A$. We have $\ln x_A \approx -\nu n_i/n_A$ and $\phi \to (\nu n_i/n_A)/M_A \nu(n_i/n_A M_A) = 1$.

(c) Integration from the infinite-dilution state (where $\gamma_i = 1$ and $\phi = 1$) to a solution with molality m gives $\ln\gamma_i(m) = \phi(m) - 1 + \int_0^m [(\phi - 1)/m_i]\, dm_i$.

10.33 Use of $w_i \equiv m_i^{1/2}$ and $dw_i = (1/2)m_i^{-1/2}\, dm_i$ in the integral in (10.108) gives this integral as $\int_0^{w^2} [(\phi - 1)/m_i]2m_i^{1/2}\, dw_i = 2\int_0^{w^2} [(\phi - 1)/w_i]dw_i$.

10.34 **(a)** $\ln\gamma_m = am/m^\circ + b(m/m^\circ)^2 + c(m/m^\circ)^3 + d(m/m^\circ)^4 +$ $\int_0^m (a/m^\circ + bm'/m^{\circ 2} + cm'^2/m^{\circ 3} + dm'^3/m^{\circ 4})\, dm' =$ $2am/m^\circ + (3/2)b(m/m^\circ)^2 + (4/3)c(m/m^\circ)^3 + (5/4)d(m/m^\circ)^4$.

(b) Substitution of $m/m° = 6.00$ gives $\log \gamma_m = 0.45415$ and $\gamma_m = 2.85$. One kg of H_2O has 55.51 mol H_2O, and $x(H_2O) = 55.51/(55.51 + 6.00) = 0.902_5$. Equation (10.27) gives $\gamma_{II,i} = \gamma_{m,i}/x_A = 2.85/0.902_5 = 3.16$.

10.35 $m(Cl^-) = [0.0100 + 2(0.0050)]\text{mol}/(0.100 \text{ kg}) = 0.200$ mol/kg.
$m(K^+) = 0.100$ mol/kg. $m(Mg^{2+}) = 0.070$ mol/kg. $m(SO_4^{2-}) = 0.020$ mol/kg.
$I_m = \frac{1}{2}\sum_i z_i^2 m_i = \frac{1}{2}[0.200 + 0.100 + 2^2(0.070) + (-2)^2 0.020]$ mol/kg = 0.330 mol/kg.

10.36 For the electrolyte $M_{\nu_+}X_{\nu_-}$ with stoichiometric molality m_i (and no ion pairing), we have $m_+ = \nu_+ m_i$ and $m_- = \nu_- m_i$. Hence $I_m = \frac{1}{2}(z_+^2\nu_+ m_i + z_-^2\nu_- m_i) = \frac{1}{2}m_i(\nu_+ z_+^2 + \nu_- z_-^2) = \frac{1}{2}z_+|z_-|(\nu_+ + \nu_-)m_i = \frac{1}{2}z_+|z_-|\nu m_i$, where (10.62) and (10.45) were used.

10.37 **(a)** $I_m = \frac{1}{2}[2^2(0.02) + 1^2(0.04)]$ mol/kg = 0.06 mol/kg. Then $\log \gamma_\pm \cong -0.51(2)(1)\{(0.06)^{1/2}/[1 + (0.06)^{1/2}] - 0.30(0.06)\} = -0.182$; $\gamma_\pm \cong 0.657$.

(b) $I_m = \frac{1}{2}[2^2(0.02) + 1^2(0.04) + 2^2(0.01) + 2^2(0.01) + 3^2(0.005) + 1^2(0.015)]$ mol/kg = 0.13 mol/kg. $\log \gamma_\pm \cong -0.51(2)(1)\{(0.13)^{1/2}/[1 + (0.13)^{1/2}] - 0.30(0.13)\} = -0.230$; $\gamma_\pm \cong 0.588$.

(c) $\log \gamma_+ \cong -0.51(2)^2\{\ldots\} = 2(-0.182) = -0.364$; $\gamma_+ \cong 0.432$;
$\log \gamma_- \cong -0.51(1)^2\{\ldots\} = \frac{1}{2}(-0.182) = -0.091$ and $\gamma_- \cong 0.811$.

10.38 **(a)** $I_m/m° = \frac{1}{2}[2^2(0.001) + 1^2(0.002)] = 0.003$.
$\log 0.888 = -0.51(2)(1)(0.003)^{1/2}/[1 + 0.328(a/\text{Å})(0.003)^{1/2}]$;
$1.08298 = 1 + 0.01797(a/\text{Å})$ and $a = 4.62$ Å.

(b) $I_m/m° = \frac{1}{2}[2^2(0.01) + 1^2(0.02)] = 0.03$; $\log \gamma_\pm = -0.51(2)(0.03)^{1/2}/[1 + 0.328(4.62)(0.03)^{1/2}] = -0.140$ and $\gamma_\pm = 0.725$.

10.39 From (10.58), we have at 25°C: $A_{CH_3OH}/A_{H_2O} = (\rho_{CH_3OH}/\rho_{H_2O})^{1/2}(\varepsilon_{r,H_2O}/\varepsilon_{r,CH_3OH})^{3/2} = (0.787/0.997)^{1/2}(78.4/32.6)^{3/2} = 3.31$ and $B_{CH_3OH}/B_{H_2O} = (0.787/0.997)^{1/2}(78.4/32.6)^{1/2} = 1.37_8$, where data following (10.66) were used. Using the A_{H_2O} and B_{H_2O} values after (10.66), we get

$A_{CH_3OH} = 3.89\ (kg/mol)^{1/2}$ and $B_{CH_3OH} = 4.52 \times 10^9\ (kg/mol)^{1/2}\ m^{-1}$.

Substitution in (10.63) gives for CH_3OH at 25°C:

$\ln \gamma_\pm = -3.89 z_+ |z_-| (I_m/m°)^{1/2}/[1 + 0.452(a/Å)(I_m/m°)^{1/2}]$. Substitution of $I_m/m° = 0.0200$ and $a/Å = 3$ gives $\ln \gamma_\pm = -0.462$ and $\gamma_\pm = 0.630$.

10.40 $b = 0.76235$. At molality 0.1 mol/kg, $I = (1/2)[1^2(0.2) + (-2)^2(0.1)] = 0.3$ and $c = 0.9896$. At molality 1 mol/kg, $I = 3$ and $c = 0.9944$. At 0.1 mol/kg, we find $\log_{10} \gamma_\pm = -0.3665$ and $\gamma_\pm = 0.430$. At 1 mol/kg, $\log_{10} \gamma_\pm = -0.6826$ and $\gamma_\pm = 0.208$.

10.41 **(a)** Cells are designated for q and b. The values of I, and $\gamma_\pm$ are entered into columns A and B and c is calculated in column C. The Meissner equations are used to calculate $\log \gamma_\pm$ and $\gamma_\pm$ in columns D and E, and the squares of the deviations from experimental values are calculated in column F. The Solver is set up to make the deviation for the $I = 0.3$ value equal to zero by varying q. We get $q = 3.77$. This q gives good values of $\gamma_\pm$ at the lower molalities but very poor values at high molalities (42 and 1229 at 5 and 10 mol/kg).

(b) The Solver is changed to minimize the sum of the squares of the deviations for the first five values. We get $q = 2.123$. The predicted $\gamma_\pm$ values at 5 and 10 mol/kg are 3.36 and 19.7, better than in (a), but still pretty poor.

10.42 With $\nu_+ = \nu_-$, we have $\nu = \nu_+ + \nu_- = 2\nu_+$ and $\nu_+/\nu = \nu_-/\nu = \frac{1}{2}$, so (10.77) becomes $\gamma_\pm^\dagger = \alpha^{1/2}[1 - (1 - \alpha)]^{1/2}\gamma_\pm = \alpha\gamma_\pm$.

10.43 Substituting the m_+ and m_- expressions preceding (10.76) into (10.46) and using (1.70), we find

$\mu_i = \mu_i^\circ + RT \ln \{(\gamma_\pm)^\nu (\alpha\nu_+ m_i/m°)^{\nu_+} [\nu_- - (1-\alpha)\nu_+]^{\nu_-} (m_i/m°)^{\nu_-}\} =$

$\mu_i^\circ + RT \ln\{(\gamma_\pm)^\nu \alpha^{\nu_+} (m_i/m°)^{\nu_+ + \nu_-} (\nu_+)^{\nu_+} (\nu_-)^{\nu_-} [1 - (1-\alpha)\nu_+/\nu_-]^{\nu_-}\} =$

$\mu_i^\circ + \nu RT \ln\{\gamma_\pm \alpha^{\nu_+/\nu} (\nu_+)^{\nu_+/\nu} (\nu_-)^{\nu_-/\nu} [1 - (1-\alpha)\nu_+/\nu_-]^{\nu_-/\nu} m_i/m°\} =$

$\mu_i^\circ + \nu RT \ln\{\nu_\pm \gamma_\pm \alpha^{\nu_+/\nu} [1 - (1-\alpha)\nu_+/\nu_-]^{\nu_-/\nu} m_i/m°\}$, which is (10.77) and (10.76).

10.44 From the m_+ and m_- equations preceding (10.76) and from $m_+^\infty = \nu_+ m_i$ and $m_-^\infty = \nu_- m_i$, we have $m_+/m_+^\infty = \alpha$ and $m_-/m_-^\infty = 1 - (1-\alpha)\nu_+/\nu_-$. Then $(m_+/m_+^\infty)^{\nu_+/\nu}(m_-/m_-^\infty)^{\nu_-/\nu}\gamma_\pm = \alpha^{\nu_+/\nu}[1-(1-\alpha)\nu_+/\nu_-]^{\nu_-/\nu}\gamma_\pm$, which equals $\gamma_\pm^\dagger$ [Eq. (10.77)].

10.45 Use of the reaction-equilibrium condition $\mu_{IP} = \mu_+ + \mu_-$, Eq. (10.71), and Eq. (10.38) gives $G = n_A\mu_A + (\nu_+ n_i - n_{IP})\mu_+ + (\nu_- n_i - n_{IP})\mu_- + n_{IP}(\mu_+ + \mu_-) = n_A\mu_A + (\nu_+\mu_+ + \nu_-\mu_-)n_i = n_A\mu_A + n_i\mu_i$.

10.46 **(a)** $m(Pb^{2+}) = 0.100$ mol/kg – 0.43(0.100 mol/kg) = 0.057 mol/kg; $m(NO_3^-)$ = 0.200 mol/kg – 0.043 mol/kg = 0.157 mol/kg;
$m(PbNO_3^+)$ = 0.043 mol/kg.
$I_m = \frac{1}{2}[4(0.057) + 1(0.157) + 1(0.043)]$ mol/kg = $0.214m^\circ$.

(b) $\log\gamma_\pm = -0.257$ and $\gamma_\pm = 0.553$. Equation (10.77) gives
$\gamma_\pm^\dagger = (0.57)^{1/3}[1 - 0.43(1/2)]^{2/3}0.553 = 0.390$.

10.47 As T increases, the water dielectric constant decreases, which increases the interaction forces between ions, thereby increasing ion pairing.

10.48 For sucrose(aq), $\Delta_f G^\circ = -1551$ kJ/mol = $\Delta_f H^\circ - T\Delta_f S^\circ$ = –2215 kJ/mol – (298.1 K)$\Delta_f S^\circ$ and $\Delta_f S^\circ = -2227$ J/mol-K. $\Delta_f S^\circ = \bar{S}^\circ - S^\circ_{\text{elem}}$, where the entropy of the elements is S°_{elem} = $12S^\circ_m$[C(graphite)] + $11S^\circ_m[H_2(g)]$ + $5.5S^\circ_m[O_2(g)]$ = 2635 J/mol-K, where Appendix data were used. Then $\bar{S}^\circ = -2227$ J/mol-K + 2635 J/mol-K = 408 J/mol-K.

10.49 **(a)** ΔG°_{298}/(kJ/mol) = –237.129 – 0 – (–157.244) = –79.885.
ΔH°_{298}/(kJ/mol) = –285.830 – 0 – (–229.994) = –55.836.
ΔS°_{298}/(J/mol-K) = 69.91 – 0 – (–10.75) = 80.66.

(b) ΔG°_{298}/(kJ/mol) = –237.129 – 394.359 – 2(0) – (–527.81) = –103.68.

ΔH°_{298}/(kJ/mol) = –285.830 – 393.509 – 2(0) – (–677.14) = –2.20.

ΔS°_{298}/(J/mol-K) = 69.90 + 213.74 – 2(0) – (–56.9) = 340.5

10.50 **(a)** Use of (10.91), (10.92), and (10.93) gives

$\Delta_f G^\circ$/(kJ/mol) = 65.49 + 2(–111.25) = –157.01.

$\Delta_f H^\circ$/(kJ/mol) = 64.77 + 2(–207.36) = –349.95.

$\bar{S}^\circ$/(J/mol-K) = –99.6 + 2(146.4) = 193.2.

(b) ΔH°/(kJ/mol) = –240.12 – 167.159 – (–411.153) = 3.87.

10.51 For this reaction, ΔG°_{298}/(kJ/mol) = –1010.61 – (–261.905 – 744.53) = –4.18 (which corresponds to K° = 5.4.)

10.52 From (10.91) and (10.85),

$\Delta_f G^\circ[\text{HNO}_3(ai)] = \Delta_f G^\circ[\text{H}^+(ao)] + \Delta_f G^\circ[\text{NO}_3^-(ao)] = \Delta_f G^\circ[\text{NO}_3^-(ao)]$.

The NBS tables do not satisfy this equality and so contain an error.

10.53 Since the two ways of writing the reaction refer to the same process, ΔS° must be the same for each. Hence $\bar{S}^\circ(\text{H}^+) + \bar{S}^\circ(\text{OH}^-) - \bar{S}^\circ_{\text{m}}(\text{H}_2\text{O}) =$ $\bar{S}^\circ(\text{H}_3\text{O}^+) + \bar{S}^\circ(\text{OH}^-) - 2\bar{S}^\circ_{\text{m}}(\text{H}_2\text{O})$ and $\bar{S}^\circ[\text{H}_3\text{O}^+(aq)] =$ $\bar{S}^\circ(\text{H}^+) + S^\circ_{\text{m}}(\text{H}_2\text{O})$ = 0 + 69.91 J/mol-K = 69.91 J/mol-K at 25°C.

10.54 $\bar{S}^\circ_i = -(\partial\mu^\circ_i/\partial T)_P = -(\partial/\partial T)_P(\nu_+\mu^\circ_+ + \nu_-\mu^\circ_-) = \nu_+(-\partial\mu^\circ_+/\partial T)_P +$ $\nu_-(-\partial\mu^\circ_-/\partial T)_P = \nu_+\bar{S}^\circ_+ + \nu_-\bar{S}^\circ_-$. Subtraction of S°_{el} from each side gives $\Delta_f S^\circ_i = \nu_+ \Delta_f S^\circ_+ + \nu_- \Delta_f S^\circ_-$. Then $\Delta_f H^\circ_i = \Delta_f G^\circ_i + T\,\Delta_f S^\circ_i =$ $\nu_+(\Delta_f G^\circ_+ + T\,\Delta_f S^\circ_+) + \nu_-(\Delta_f G^\circ_- + T\,\Delta_f S^\circ_-) = \nu_+ \Delta_f H^\circ_+ + \nu_- \Delta_f H^\circ_-$.

10.55 **(a)** In Eq. (10.81), $\gamma_{m,i,\text{sat}}$ and $m_{i,\text{sat}}$ are for $P = P^\circ$ = 1 bar. Since the solution is very dilute and nonionic, we can approximate $\gamma_{m,i,\text{sat}}$ as 1 and (10.81) gives $\Delta_f G^\circ_{298}[\text{O}_2(aq)] = 0 - RT \ln 0.00115$ = 16.8 kJ/mol.

(b) $x_i = P_i/K_i = (1 \text{ bar})/(30300 \text{ bar}) = 3.30 \times 10^{-5} = n_i/n_{tot} \cong n_i/n_{H_2O}$ and $n_{C_2H_6} = n_{H_2O}(3.30 \times 10^{-5}) = (55.51 \text{ mol})(3.30 \times 10^{-5}) = 0.00183$ mol in 1000 kg of water, so $m_{C_2H_6} = 0.00183$ mol/kg. Then (10.81) gives $\Delta_f G^\circ_{298}[C_2H_6(aq)] = -32.82 \text{ kJ/mol} - RT \ln(0.00183) = -17.2$ kJ/mol.

10.56 Use of Eq. (10.28) for μ_i gives the following results.

$\bar{S}_i = -(\partial\mu_i/\partial T)_{P,n_j} = -(\partial\mu^\circ_{m,i}/\partial T)_P - (\partial/\partial T)_{P,n_j}[RT\ln(\gamma_{m,i}m_i/m^\circ)] = \bar{S}^\circ_{m,i} - R\ln(\gamma_{m,i}m_i/m^\circ) - RT(\partial \ln\gamma_{m,i}/\partial T)_{P,n_j}$.

$\bar{V}_i = (\partial\mu_i/\partial P)_{T,n_j} = (\partial\mu^\circ_i/\partial P)_{T,n_j} + RT(\partial\ln\gamma_{m,i}/\partial P)_{T,n_j} = \bar{V}^\circ_{m,i} + RT(\partial\ln\gamma_{m,i}/\partial P)_{T,n_j}$.

$\bar{H}_i = \mu_i + T\bar{S}_i = \mu^\circ_{m,i} + T\bar{S}^\circ_{m,i} - RT^2(\partial\ln\gamma_{m,i}/\partial T)_{P,n_j} = \bar{H}^\circ_{m,i} - RT^2(\partial\ln\gamma_{m,i}/\partial T)_{P,n_j}$.

10.57 From (10.91) with i = HCl, $-131.23 \text{ kJ/mol} = \Delta_f G^\circ[\text{HCl}(aq)] = \Delta_f G^\circ[\text{H}^+(aq)] + \Delta_f G^\circ[\text{Cl}^-(aq)] = 0 + \Delta_f G^\circ[\text{Cl}^-(aq)]$ and $\Delta_f G^\circ[\text{Cl}^-(aq)] = -131.23$ kJ/mol. Similarly, (10.93) gives $\Delta_f H^\circ[\text{Cl}^-(aq)] = -167.16$ kJ/mol. Use of $T\Delta_f S^\circ[\text{HCl}(aq)] = \Delta_f H^\circ[\text{HCl}(aq)] - \Delta_f G^\circ[\text{HCl}(aq)]$ gives $\Delta_f S^\circ[\text{HCl}(aq)] = -120.5_1 \text{ J/mol-K} = \bar{S}^\circ[\text{HCl}(aq)] - \frac{1}{2}S^\circ_m[H_2(g)] - \frac{1}{2}S^\circ_m[Cl_2(g)] = \bar{S}^\circ[\text{HCl}(aq)] - 176.88$ J/mol-K, so $\bar{S}^\circ[\text{HCl}(aq)] = 56.3_6 \text{ J/mol-K} = \bar{S}^\circ[\text{H}^+(aq)] + \bar{S}^\circ[\text{Cl}^-(aq)] = \bar{S}^\circ[\text{Cl}^-(aq)]$.

10.58 From (10.84), $\Delta_f G^\circ[i(A)] = \Delta_f G^\circ(i^*) - \nu RT\ln(\nu_\pm\gamma_{m,\pm,sat}m_{i,sat}/m^\circ)$, so $\Delta_f G^\circ_{298}[\text{KCl}(aq)] = -409140 \text{ J/mol} - 2RT\ln[1(0.588)4.82] = -414.30$ kJ/mol. Equation (10.91) gives $-414.30 \text{ kJ/mol} = \Delta_f G^\circ_{298}[\text{K}^+(aq)] - 131.23$ kJ/mol and $\Delta_f G^\circ_{298}[\text{K}^+(aq)] = -283.07$ kJ/mol. Equation (9.38) at 1 bar gives $\Delta H^\infty_{\text{diff,KCl}(aq)} = 17.22 \text{ kJ/mol} = \bar{H}^\infty[\text{KCl}(aq)] - H^\circ_m[\text{KCl}(s)] = \bar{H}^\circ[\text{KCl}(aq)] - H^\circ_m[\text{KCl}(s)] = \bar{H}^\circ[\text{K}^+(aq)] + \bar{H}^\circ[\text{Cl}^-(aq)] - H^\circ_m[\text{KCl}(s)]$. Subtraction and addition of the standard enthalpies of K, $\frac{1}{2}Cl_2$, and e^- on the right side of the last equation gives $17.22 \text{ kJ/mol} = \Delta_f H^\circ_{298}[\text{K}^+(aq)] +$

$\Delta_f H^\circ_{298}[\mathrm{Cl}^-(aq)] - \Delta_f H^\circ_{298}[\mathrm{KCl}(s)] = \Delta_f H^\circ_{298}[\mathrm{K}^+(aq)] - 167.16$ kJ/mol + 436.75 kJ/mol and $\Delta H^\circ_{f,298}[\mathrm{K}^+(aq)] = -252.37$ kJ/mol. To find $\bar{S}^\circ[\mathrm{K}^+(aq)]$, we use the fact that ΔG°_{298} for dissolving KCl in water equals $\Delta_f G^\circ_{298}[\mathrm{KCl}(aq)] - \Delta_f G^\circ_{298}[\mathrm{KCl}(s)] = (-283.07 - 131.23 + 409.14)$ kJ/mol = -5.16 kJ/mol. Since ΔH°_{298} for the solution process is 17.22 kJ/mol, we have $\Delta S^\circ = (\Delta H^\circ - \Delta G^\circ)/T = 75.06$ J/mol-K for dissolving KCl in water.
Then 75.06 J/mol-K = $\bar{S}^\circ[\mathrm{KCl}(aq)] - S^\circ_{\mathrm{m}}[\mathrm{KCl}(s)] =$
$\bar{S}^\circ[\mathrm{K}^+(aq)] + \bar{S}^\circ[\mathrm{Cl}^-(aq)] - S^\circ_{\mathrm{m}}[\mathrm{KCl}(s)] =$
$\bar{S}^\circ[\mathrm{K}^+(aq)]$ + 53.36 J/mol-K – 82.59 J/mol-K and $\bar{S}^\circ[\mathrm{K}^+(aq)] = 104.3$ J/mol-K.

10.59 As the charge increases, the ion binds more H_2O molecules to itself, thereby increasing the degree of order in the solution and decreasing S of the solution and $\bar{S}^\circ_i$.

10.60 **(a)** From (10.104), $\ln \phi_2 = \int_0^{P_2} (V_{\mathrm{m}}/RT - 1/P)\, dP =$
$\int_0^{P_2} (1/P + B^\dagger + C^\dagger P + D^\dagger P^2 + \cdots - 1/P)\, dP =$
$\int_0^{P_2} (B^\dagger + C^\dagger P + D^\dagger P^2 + \cdots)\, dP = B^\dagger P_2 + C^\dagger P_2^2/2 + D^\dagger P_2^3/3 + \cdots$.
Dropping the unnecessary subscript 2 on ϕ and P, we get the desired equation.

(b) Comparison of (8.9) with (8.4) gives $B = b - a/RT$, $C = b^2, \ldots$; use of (8.6) gives $B^\dagger = (bRT - a)/R^2T^2$, $C^\dagger = (2abRT - a^2)/R^4T^4, \ldots$ and substitution into the result of (a) gives the desired result.

10.61 **(a)** From Eq. (8.18), $b = RT_c/8P_c = 42.9$ cm^3/mol and $a = 27R^2T_c^2/64P_c$ = 3.61×10^6 cm^6 atm mol^{-2}. At 75°C, $RT = 28570$ cm^3-atm/mol and Prob. 10.60b gives $\ln \phi = -[(2.38 \times 10^6)/(8.16 \times 10^8 \text{ atm})](1 \text{ atm}) - [(4.18 \times 10^{12})/(1.33 \times 10^{18} \text{ atm}^2)](1 \text{ atm})^2 = -0.00292$ and $\phi = 0.997$ (compared with $\phi_{\text{exper}} = 0.997$). Replacing 1 atm by 25 atm in the preceding equation, we get $\ln \phi = -0.0749$ and $\phi = 0.928$ (compared with $\phi_{\text{exper}} = 0.92$).

(b) $\phi_i = \phi_i^*(T, P) = \phi_i^*(75°\text{C}, 25 \text{ atm}) = 0.928$, where the result of (a) was used. $f_i = \phi_i P_i = \phi_i x_i P = 0.928(0.100)(25.0 \text{ atm}) = 2.32$ atm.

10.62 From (10.102) and (10.103), $G_m = \mu = \mu° + RT \ln (f/P°) = \mu° + RT \ln (\phi P/P°) = \mu° + RT \ln (P/P°) + RT \ln \phi$. For the corresponding ideal gas, $\phi = 1$ and $G_m^{id} = \mu^{id} = \mu° + RT \ln (P/P°)$. So $G_m = G_m^{id} + RT \ln \phi$ and $\ln \phi = (G_m - G_m^{id})/RT$.

10.63 **(a)** Equation (4.65) gives $\Delta G = \int_1^2 V\,dP = \int_1^2 (nRT/P)\,dP = nRT \ln (P_2/P_1) = (1.000 \text{ mol})(8.314 \text{ J/mol-K})(273.15 \text{ K}) \ln 1000 = 15.69 \text{ kJ} = 3.75$ kcal.

(b) $\Delta G = n\,\Delta\mu = n[\mu° + RT \ln (f_2/P°) - \mu° - RT \ln (f_i/P°)] = nRT \ln (f_2/f_1) = nRT \ln (\phi_2 P_2/\phi_1 P_1) = (1.000 \text{ mol})(8.314 \text{ J/mol-K})(273.15 \text{ K}) \times \ln [1.84(1000)/0.9996(1.000)] = 17.1 \text{ kJ} = 4.08$ kcal.

10.64 **(a)** From (10.104), $\ln \phi$ equals the area under the $V_m/RT - 1/P$ vs. P curve from 0 to 120 atm. The data are:

10^3 atm $(V_m/RT - 1/P)$	-4.65_6	-4.70	-4.78	-5.07	-5.42
P/atm	5	10	20	40	60

10^3 atm $(V_m/RT - 1/P)$	-5.56	-5.08	-4.22
P/atm	80	100	120

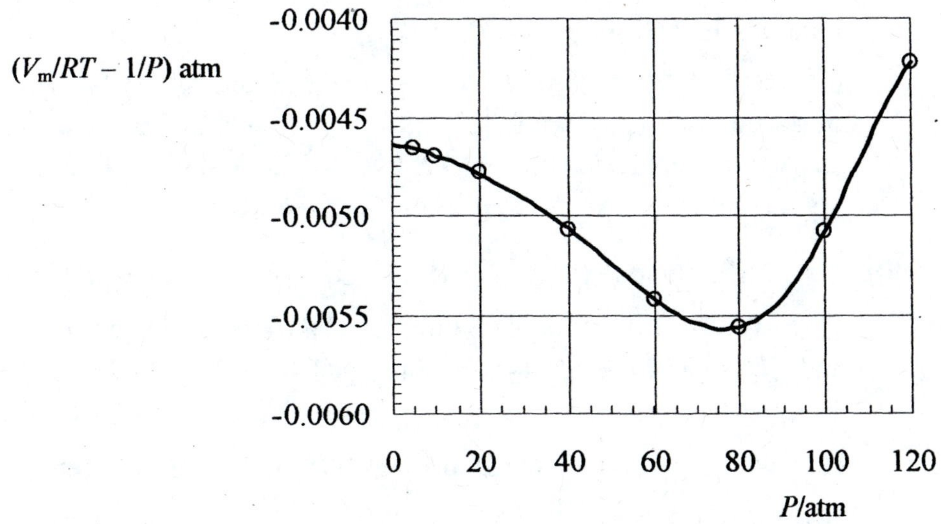

Plotting the graph, cutting it out and weighing it, we find that the weight of the area between the -0.0040 horizontal line and the curve is 0.53_4 times the weight of the rectangular area shown in the figure. The rectangular area shown equals $(0.0020 \text{ atm}^{-1})(120 \text{ atm}) = 0.240$, so the area between the -0.0040 line and the curve is $(0.53_4)(0.240) = 0.128$. Adding in the area $(0.0040)(120) = 0.480$ between the $y = 0$ axis and the -0.0040 line, we get an area of -0.608 as the value of $\ln \phi = \int_0^{120 \text{ atm}} (V_m/RT - 1/P)\, dP$. So $\ln \phi = -0.608$ and $\phi = 0.544$. Also, $f = \phi P = 0.544 \times 120 \text{ atm} = 65.3 \text{ atm}$.

(b) Problem 8.38 gives $B = \lim_{P\to 0} (V_m - V_m^{id}) = \lim_{P\to 0} (V_m - RT/P)$. The data in part (a) give $\lim_{P\to 0} (V_m/RT - 1/P) = -0.0046_4 \text{ atm}^{-1}$. Hence $B(-50°\text{C}) = (-0.0046_4 \text{ atm}^{-1})(82.06 \text{ cm}^3\text{-atm/mol-K})(223 \text{ K}) = -85 \text{ cm}^3\text{/mol}$.

10.65 (a) $B = [(82.06 \text{ cm}^3\text{-atm/mol-K})(126.2 \text{ K})/(33.5 \text{ atm})] \times [0.597 - 0.462e^{0.7002(126.2)/273.15}] = -12._8 \text{ cm}^3\text{/mol}$.

(b) $B^\dagger = B/RT = -5.7 \times 10^{-4} \text{ atm}^{-1}$. At 1 atm, $\ln \phi = B^\dagger P = (-5.7 \times 10^{-4} \text{ atm}^{-1})(1 \text{ atm}) = -5.7 \times 10^{-4}$ and $\phi = 0.99943$. At 25 atm, $\ln \phi = -0.014_3$ and $\phi = 0.9859$.

10.66 (a) $B^\dagger_{chl} = B_{chl}/RT = (-1040 \text{ cm}^3\text{/mol})/RT = -0.04048 \text{ atm}^{-1} = -5.326 \times 10^{-5} \text{ torr}^{-1}$. $B^\dagger_{car} = -0.0569_8 \text{ atm}^{-1} = -7.497 \times 10^{-5} \text{ torr}^{-1}$. In the mixture: $\ln \phi_{chl} \cong \ln \phi^*_{chl}(T, P) \cong B^\dagger_{chl} P = (-5.326 \times 10^{-5} \text{ torr}^{-1})(301.84 \text{ torr}) = -0.01608$ and $\phi_{chl} = 0.9841$; $\ln \phi_{car} = (-7.497 \times 10^{-5} \text{ torr}^{-1})(301.84 \text{ torr}) = -0.02263$ and $\phi_{car} = 0.9776$. In the pure vapors: $\ln \phi^*_{chl} = (-5.326 \times 10^{-5})(360.51) = -0.01920$ and $\phi^*_{chl} = 0.9810$; $\ln \phi^*_{car} = -0.01599$; $\phi^*_{car} = 0.9841$.

(b) $f_i = \gamma_{I,i} x_i^\ell f_i^*$ and $\phi_i x_i^v P = \gamma_{I,i} x_i^\ell \phi_i^* P_i^*$, so $\gamma_{I,i} = \phi_i x_i^v P / x_i^\ell \phi_i^* P_i^*$. Then $\gamma_{I,chl} = (0.9841)(0.6456)(301.84 \text{ torr})/0.5242(0.9810)(360.51 \text{ torr}) = 1.034$ and $\gamma_{I,car} = (0.9776)(0.3544)(301.84 \text{ torr})/0.4758(0.9841)(213.34 \text{ torr}) = 1.047$.

(c) Here we take ϕ_i and ϕ_i^* equal to 1 and $\gamma_{I,i} = x_i^v P / x_i^\ell P_i^*$. We get $\gamma_{I,chl} = 1.031$ and $\gamma_{I,car} = 1.054$.

10.67 At constant T and P, we have $d\mu_i = d(\mu_i^* + RT \ln x_i) = RT\, d \ln x_i = (RT/x_i)\, dx_i$. Therefore $\sum_i n_i\, d\mu_i = RT \sum_i (n_i/x_i)\, dx_i = RT \sum_i (n_i n_{tot}/n_i)\, dx_i = RTn_{tot} \sum_i dx_i$. From $\sum_i x_i = 1$, it follows that $\sum_i dx_i = 0$. Hence $\sum_i n_i\, d\mu_i = 0$ at constant T and P. This completes the proof.

10.68 There are $2N_A = 12 \times 10^{23}$ ions in 1000 cm^3 of solution. With uniform distribution, each ion is in the center of a cube of volume $(1000\text{ cm}^3)/(12 \times 10^{23}) = 8.3 \times 10^{-22}\text{ cm}^3$ and edge length $(8.3 \times 10^{-22}\text{ cm}^3)^{1/3} = 9 \times 10^{-8}\text{ cm} = 9$ Å, and this is the nearest-neighbor distance.

10.69 **(a)** Because A-B attractions are weaker than A-A and B-B attractions, U and H of the solution will be higher than U and H of the corresponding ideal solution. Hence $\Delta_{mix}H > 0 = \Delta_{mix}H^{id}$.

(b) Because A-A and B-B attractions are stronger than A-B attractions and the molecules have similar sizes and shapes, A molecules in the solution will tend to surround themselves preferentially with other A molecules (and similarly for B molecules). Hence the degree of order in the solution is greater than in an ideal solution (where there is complete randomness in distribution of A and B molecules), and S of the solution is less than S^{id}; $S < S^{id}$; $S - S^* < S^{id} - S^*$; $\Delta_{mix}S < \Delta_{mix}S^{id}$.

(c) $\Delta_{mix}G = \Delta_{mix}H - T\,\Delta_{mix}S$. Since $\Delta_{mix}H > \Delta_{mix}H^{id}$ and $\Delta_{mix}S < \Delta_{mix}S^{id}$, we have $\Delta_{mix}G > \Delta_{mix}G^{id}$.

10.70 **(a)** An increase in z_+ increases the attractions between the positive ion and the negative atmosphere that surrounds it, thereby stabilizing the solution and hence lowering μ_+ and lowering γ_+.

(b) An increase in ionic diameter decreases the closest distance of approach between positive and negative ions, decreasing the attractions between them and hence increasing μ_+ and increasing γ_+.

(c) An increase in ionic strength means an increase in the attractions between each cation and the anions in its atmosphere, which lowers μ_+ and hence lowers γ_+.

(d) An increase in solvent dielectric constant decreases the interactions between positive and negative ions, increasing μ_+ and increasing γ_+.

(e) As T increases, the kinetic energy of random ionic motion increases, which tends to distribute the ions more randomly and to thereby break up the ordered ionic atmospheres. This destabilizes the solution and increases μ_+ and γ_+.

10.71 (a) $\mu_{i,(\text{sln})} = \mu_{i,(\upsilon)}$ and use of (10.51) gives $\mu^\circ_{\text{m},i} + \nu RT \ln (\nu_\pm \gamma_i m_i/m^\circ) = \mu_i^{\circ,\upsilon} + RT \ln (P_i/P^\circ)$, so $\ln (P_i/P^\circ) = (\mu^\circ_{\text{m},i} - \mu_i^{\circ,\upsilon})/RT + \ln (\nu_\pm \gamma_i m_i/m^\circ)^\nu$ and $P_i = P^\circ \exp [(\mu^\circ_{m,i} - \mu_i^{\circ,\upsilon})/RT](\nu_\pm \gamma_i m_i/m^\circ)^\nu = K_i(\nu_\pm \gamma_i m_i/m^\circ)^\nu$. For HCl, $\nu_+ = 1 = \nu_-$, $\nu = 2$, and $\nu_\pm = 1$, so $P_i = K_i(\gamma_i m_i/m^\circ)^2$.

(b) $K_i = P^\circ \exp[(\mu^\circ_{\text{m},i} - \mu_i^{\circ,\upsilon})/RT]$. $\mu_i^{\circ,\upsilon} = \mu^\circ_{\text{HCl(g)}}$. $\mu^\circ_{\text{m},i} = \mu^\circ[\text{H}^+(aq)] + \mu^\circ[\text{Cl}^-(aq)]$. $\mu^\circ_{\text{m},i} - \mu_i^{\circ,\upsilon} = \Delta_f G^\circ [\text{H}^+(aq)] + \Delta_f G^\circ [\text{Cl}^-(aq)] - \Delta_f G^\circ [\text{HCl(g)}] = [0 - 131.228 - (-95.299)]$ kJ/mol $= -35.929$ kJ/mol.

$K_i = (1 \text{ bar}) \exp [(-35929 \text{ J/mol})/RT] = 5.08 \times 10^{-7}$ bar.

$P_i = (5.08 \times 10^{-7} \text{ bar})[0.80(0.100 \text{ mol/kg})/(1 \text{ mol/kg})]^2 = 3.2_5 \times 10^{-7}$ bar $= 0.00024$ torr.

10.72 (a) The following become 1 in the limit $x_W \to 1$: $\gamma_{\text{I},W}$, $\gamma_{\text{II},W}$, $\gamma_{\text{II},E}$, $\gamma_{\text{m},E}$.

(b) Only $\gamma_{\text{I},E}$ becomes 1.

10.73 (a) True. When $\mu_i = \mu_i^\circ$ in (10.3), then $a_i = 1$.

(b) False. For example, for Convention I, with $x_i < 1$, γ_i might be such that $\gamma_i x_i = 1 = a_i$.

(c) True. See Eq. (10.3).

(d) True. See Eq. (10.5).

(e) False.

(f) False.

Chapter 11

11.1 **(a)** T: **(b)** T.

11.2 **(a)** F; **(b)** F; **(c)** F; **(d)** F. The H^+ from H_2O ionization can make $m(H^+)$ exceed m if m is extremely small.

11.3 (b).

11.4 **(a)** $m_i = n_i/w_A$ = (4.603 g)(1 mol/46.026 g)/(0.5000 kg) = 0.2000 mol/kg. We neglect H^+ from H_2O. Let $x = m(H^+) = m(HCOO^-)$. Then (11.15) gives K_a = 1.80×10^{-4} mol/kg = $\gamma_\pm^2 x^2/(0.2000 \text{ mol/kg} - x)$. With the initial approximation $\gamma_\pm \cong 1$, we find $x = 0.0059_1$ mol/kg. Hence $I_m \cong$ 0.0059_1 mol/kg and the Davies equation gives $\gamma_\pm = 0.92_2$. Use of this $\gamma_\pm$ in the K_a equation gives $x = 0.0064_0$ mol/kg. Then $I_m = 0.0064_0$ mol/kg, $\gamma_\pm$ = 0.91_9, $x = 0.0064_2$ mol/kg.

(b) m(KCl) = (0.1000 mol)/(0.500 kg) = 0.200 mol/kg. The KCl contribution to I_m is 0.200 mol/kg and the H^+ and $HCOO^-$ contribution to I_m is (from part a) about 0.006 mol/kg, so $I_m \cong 0.206$ mol/kg. The Davies equation gives $\gamma_\pm = 0.74_5$. Use of this $\gamma_\pm$ in the K_a equation gives $x = 0.0079$ mol/kg. This gives a revised I_m of 0.2079 mol/kg, which gives $\gamma_\pm = 0.74_5$. We then obtain $x = 0.0079$ mol/kg.

(c) If x is the moles of HCOOH that ionize per kilogram, then $m(H^+) = x$ and $m(HCOO^-) = x + 0.400$ mol/kg, since the KHCOO contributes 0.400 mol/kg of $HCOO^-$. So 1.80×10^{-4} mol/kg = $\gamma_\pm^2 x(0.400 \text{ mol/kg} + x)/(0.200 \text{ mol/kg} - x)$. With the initial approximation $I_m \cong 0.400$ mol/kg (due to the salt), the Davies equation gives $\gamma_\pm = 0.73_0$. With x in the numerator and denominator neglected compared with 0.400 and 0.200 mol/kg, the K_a equation gives x = 1.69×10^{-4} mol/kg. This x is small enough to neglect its contribution to I_m and to neglect it compared with 0.200 mol/kg.

11.5 **(a)** A BASIC program is:

```
10 G=1
15 INPUT "KA";KA
20 INPUT "M1";M1
25 INPUT "M2";M2
30 INPUT "DM";DM
35 PRINT "KA="KA
40 FOR M=M1 TO M2 STEP DM
45 GOSUB 300
50 XX=X
60 GOSUB 400
65 GOSUB 300
70 IF ABS(X–XX)/X<0.0002
   THEN 85
75 XX=X
80 GOTO 60
85 PRINT "M=";M;" M(H+)=";X
90 NEXT M
95 STOP
300 X=(–KA+SQR(KA*KA+4*KA*M*
       G*G))/(2*G*G)
310 RETURN
400 I=X
410 S=SQR(I)
420 LG = –0.51*(S/(1+S)–0.3*I)
430 G=10^LG
440 RETURN
```

(b) At extremely low molalities, the H^+ from the ionization of water cannot be neglected. At high molalities, the Davies equation is not accurate.

11.6 For $H_2S(aq) \rightleftarrows H^+(aq) + HS^-(aq)$, $\Delta G^\circ_{298} = [12.08 + 0 - (-27.83)]$ kJ/mol = 39910 J/mol = $-(8.3145$ J/mol-K$)(298.15$ K$) \ln K^\circ$, so $K^\circ = 1.02 \times 10^{-7}$.
For $HS^-(aq) \rightleftarrows H^+(aq) + S^{2-}(aq)$, $\Delta G^\circ_{298} = (0 + 85.8 - 12.08)$ kJ/mol = 73.72 kJ/mol = $-RT \ln K^\circ$ and $K^\circ = 1.2 \times 10^{-13}$.

11.7 For this very dilute solution of an extremely weak acid, we cannot neglect H^+ from the H_2O ionization. Example 11.3 in Sec. 11.3 gives $m(H^+) = (K_w + mK_a)^{1/2} = [1.00 \times 10^{-14} + (1.00 \times 10^{-5})(6.2 \times 10^{-10})]^{1/2}$ mol/kg = 1.27×10^{-7} mol/kg.

11.8 $2H_2O \rightleftarrows H_3O^+ + OH^-$. The only significant contribution to I_m is from the NaCl, so $I_m = 0.20$ mol/kg. The Davies equation gives $\log \gamma_\pm = -0.127$ and $\gamma_\pm = 0.746$ for H_3O^+ and OH^-. Equation (11.12) gives $1.00 \times 10^{-14} = (0.746)^2[m(H_3O^+)/m^\circ]^2$ and $m(H_3O^+) = 1.34 \times 10^{-7}$ mol/kg.

11.9 **(a)** Setting T/K equal to 310.1₅ in the equation of Prob. 11.38 gives log K_w° $= -13.618$ and $K_w^\circ = 2.41 \times 10^{-14}$. We can take $I_m \cong 0$ and $\gamma_\pm \cong 1$. Equation (11.12) gives $2.41 \times 10^{-14} = [m(H_3O^+)m^\circ]^2$ and $m(H_3O^+) = 1.55 \times 10^{-7}$ mol/kg.

(b) ΔG_{298}°/(kJ/mol) $= 0 - 157.244 + 237.129 = 79.885$. ln K_w°(25°C) $= -\Delta G_{298}^\circ/RT = -32.225$; K_w°(25°C) $= 1.01 \times 10^{-14}$. ΔH_{298}°/(kJ/mol) $= 0 - 229.994 + 285.830 = 55.836$. Assuming ΔH° to be independent of T and using (6.39), we get ln K_w°(37°C) $= -32.225 +$ [(55836 J/mol)/(8.314 J/mol-K)][1/(298.1 K) – 1/(310.1 K)] $= -31.354$ and K_w°(37°C) $= 2.42 \times 10^{-14}$.

11.10 In this extremely dilute HCl solution, the ionization of H_2O cannot be neglected. $2H_2O \rightleftarrows H_3O^+ + OH^-$. $K_w^\circ = m(H_3O^+)m(OH^-)/m^{\circ 2}$, since $\gamma_\pm \cong 1$ in this very dilute solution. If y moles of H_2O ionize per kilogram, then $m(OH^-) = y$ mol/kg and, since HCl is a strong acid, $m(H_3O^+) = (y + 1.00 \times 10^{-8})$ mol/kg. Therefore $K_w^\circ = 1.00 \times 10^{-14} = (y + 1.00 \times 10^{-8})y$ and $y^2 + (1.00 \times 10^{-8})y - 1.00 \times 10^{-14} = 0$. The quadratic formula gives the positive root as $y = 9.51 \times 10^{-8}$. So $m(H_3O^+) = (y + 1.00 \times 10^{-8})$ mol/kg $= 1.05 \times 10^{-7}$ mol/kg.

11.11 $HX + H_2O \rightleftarrows H_3O^+ + X^-$. $K_a = \gamma_\pm^2 m(H_3O^+)m(X^-)/m(HX)$. $m(H_3O^+) = m(X^-) = 0.0100$ mol/kg, since the H_3O^+ from the ionization of water is negligible. $I_m = 0.0100$ mol/kg. The Davies equation gives log $\gamma_\pm = -0.0448$ and $\gamma_\pm = 0.902$. Hence $K_a = (0.902)^2(0.0100 \text{ mol/kg})^2/[(0.200 - 0.010) \text{ mol/kg}] = 4.28 \times 10^{-4}$ mol/kg.

11.12 $a(H_2O) = \gamma(H_2O)x(H_2O) \approx x(H_2O)$, since H_2O is an uncharged species and its γ should be close to 1 in this fairly dilute solution. In 1 kg of H_2O there are 55.5 mol of H_2O. There is no significant ion pairing, so $n(Na^+) = n(Cl^-) = 0.50$ mol in 1 kg of water. Therefore, $x(H_2O) =$ (55.5 mol)/(55.5 mol + 0.50 mol + 0.50 mol) $= 0.982 \approx a(H_2O)$.

11.13 $C_2H_3O_2^- + H_2O \rightleftarrows HC_2H_3O_2 + OH^-$

$$K_b^\circ = \frac{a(HC_2H_3O_2)a(OH^-)}{a(C_2H_3O_2^-)a(H_2O)} = \frac{K_w^\circ}{K_a^\circ} = \frac{1.00 \times 10^{-14}}{1.75 \times 10^{-5}} = 5.7_1 \times 10^{-10}$$

where $K_a^\circ = a(H_3O^+)a(C_2H_3O_2^-)/a(HC_2H_3O_2)a(H_2O)$ and Eq. (11.11) were used. If we neglect the OH^- from the ionization of water, then $m(HC_2H_3O_2) = m(OH^-)$. The solution is reasonably dilute and the OH^- and $C_2H_3O_2^-$ ions have the same charge, so the Debye–Hückel (or Davies) equation gives $\gamma(OH^-) = \gamma(C_2H_3O_2^-)$; these activity coefficients cancel in K_b°. We take $a(H_2O) = 1$ and $\gamma(HC_2H_3O_2) = 1$. Let $y = m(OH^-)/m^\circ$. Then $5.7_1 \times 10^{-10} = y^2/(0.10 - y) \approx y^2/0.10$ and $y = 7.6 \times 10^{-6}$, so $m(OH^-) = 7.6 \times 10^{-6}$ mol/kg. Equation (11.12) gives $1.00 \times 10^{-14} = \gamma_\pm^2 m(H_3O^+)m(OH^-)/m^{\circ 2}$. We have $I_m \approx 0.10$ mol/kg and the Davies equation gives $\log \gamma_\pm = -0.107$ and $\gamma_\pm = 0.78$. We find $m(H_3O^+) = 2.2 \times 10^{-9}$ mol/kg.

11.14 **(a)** $H_2S + H_2O \rightleftarrows H_3O + HS^-$; $HS^- + H_2O \rightleftarrows H_3O + S^{2-}$. We neglect the H_3O^+ from water. Because K_a of HS^- is much less than K_a of H_2S, the H_3O^+ from the second ionization step is much less than that from the first step. Because of the smallness of K_a of HS^-, the ionization of HS^- does not change the HS^- molality to any significant extent. Hence we can take $m(H_3O^+) = m(HS^-)$. Let $y = m(H_3O^+)/m^\circ$. Neglecting activity coefficients, we have $1.0 \times 10^{-7} = y^2/(0.10 - y) \approx y^2/0.10$ and $y = 1.0 \times 10^{-4}$; hence $m(H_3O^+) = 1.0 \times 10^{-4}$ mol/kg $= m(HS^-)$. Let $z = m(S^{2-})/m^\circ$. For the ionization of HS^-: $0.8 \times 10^{-17} = (1.0 \times 10^{-4})z/(0.00010 - z) \approx (0.00010)z/0.00010 = z$ and $m(S^{2-}) = 0.8 \times 10^{-17}$ mol/kg.

(b) From part (a), $m(H_3O^+) \approx 1 \times 10^{-4}$ mol/kg $\approx m(HS^-)$, so $I_m = 1 \times 10^{-4}$ mol/kg. The Davies equation gives $\log \gamma_\pm = -0.0050$ and $\gamma_\pm = 0.988$ for H_3O^+ and HS^-. Hence $1.0 \times 10^{-7} = (0.988)^2 y^2/0.10$ and $y = 1.0_1 \times 10^{-4}$; hence $m(H_3O^+) = m(HS^-) = 1.0_1 \times 10^{-4}$ mol/kg. The Davies equation gives $\log \gamma(H_3O^+) = -0.0050$ and $\gamma(H_3O^+) = 0.988$; also $\gamma(HS^-) = 0.988$; $\gamma(S^{2-}) = 0.954$. So
$0.8 \times 10^{-17} = (0.988)(1.01 \times 10^{-4})(0.954)z/0.988(1.01 \times 10^{-4})$ and $m(S^{2-}) = 0.8 \times 10^{-17}$ mol/kg.

11.15 $\Delta G^\circ/(\text{kJ/mol}) = -587.0 + 454.8 + 128.0 = -4.2$. $\ln K^\circ = -\Delta G^\circ/RT = 1.694$; $K^\circ = 5.44$.

11.16 $Cu^{2+} + SO_4^{2-} \rightleftarrows CuSO_4(aq)$. Neglecting ionic association, $I_m \approx 0.20$ mol/kg and the Davies equation gives $\log \gamma_\pm \approx -0.5080$ and $\gamma_\pm \approx 0.310$. Let $m[CuSO_4(aq)]/m^\circ = z$. Use of the initial estimate of $\gamma_\pm$ in K_m for the association reaction gives $230 \approx z/(0.310)^2(0.0500 - z)^2$ and $z^2 - 0.1452z + 0.0025 = 0$. The roots are found to be $z = 0.0200$ and 0.125. Since z cannot exceed 0.05, we have $m[CuSO_4(aq)] \approx 0.0200$ mol/kg and $m(Cu^{2+}) \approx 0.0300$ mol/kg. This gives an improved estimate of $I_m \approx 0.12$ mol/kg; the Davies equation gives an improved estimate of $\gamma_\pm \approx 0.354$. Then $230 \approx z/(0.354)^2(0.0500 - z)^2$ and $z^2 - 0.1347z + 0.0025 = 0$; we find $z = 0.0222$. Hence $m(Cu^{2+}) \approx 0.0278$ mol/kg and $I_m \approx 0.1112$ mol/kg. The Davies equation gives $\gamma_\pm \approx 0.361$ and this leads to $z = 0.0226$. Another repetition gives $\gamma_\pm \approx 0.363$ and $z = 0.0227$. Further repetition is clearly unnecessary and $\gamma_\pm = 0.363$ and $m(Cu^{2+}) = 0.0273$ mol/kg. We have $\alpha = 0.0273/0.0500 = 0.546$. Since $\nu_+ = \nu_- = 1$, Eq. (10.77) reduces to (see Prob. 10.42) $\gamma_\pm^\dagger = \alpha\gamma_\pm = 0.546(0.363) = 0.198$. (The experimental $\gamma_\pm^\dagger$ is 0.217.)

11.17 A BASIC program is

```
2 INPUT "M0"; M0
5 INPUT "K"; K
10 INPUT "NUPLUS"; NP
15 INPUT "NUMINUS"; NM
20 INPUT "ZPLUS"; ZP
25 INPUT "ZMINUS"; ZM
30 NU=NP+NM
333 X=0:MP=NP*M0
335 IM=.5*(ZP^2*(NP*M0–X)+
    ZM^2*(NM*M0–X)+(ZP+ZM)*
    (ZP+ZM)*X)
340 GOSUB 600
345 B= –GG/(K*GP*GM)–NU*M0
350 C=NP*NM*M0*M0
355 RT=SQR(B*B–4*C)
360 X1=(–B+RT)/2
365 X2=(–B–RT)/2
370 IF X1>0 THEN 410 ELSE
    X=X2:GOTO 440
410 IF X1<M0 THEN X=X1 ELSE X=X2
440 MM=M0*NP–X
450 IF ABS(MM–MP)/MP<0.0002
    THEN 490
455 MP=MM
460 GOTO 335
490 PRINT "ION PAIR MOLALITY";X
500 STOP
600 S=SQR(IM)
610 CC= –(S/(S+1)–.3*IM)*.51
620 LP=ZP*ZP*CC
```

```
630 LM=ZM*ZM*CC                    660 GM=10^LM
640 LI=(ZP+ZM)*(ZP+ZM)*CC          670 GG=10^LI
650 GP=10^LP                       680 RETURN
```

11.18 The true values of K_w and K_a are entered into designated cells. The initial values $[H^+] = 1 \times 10^{-7}$, $[OH^-] = 1 \times 10^{-7}$, $[OI^-] = 0$, $[HOI] = 1 \times 10^{-4}$ are entered into designated cells. The K_w and K_a values calculated from $[H^+][OH^-]$ and $[H^+][OI^-]/[HOI]$, respectively, are entered into designated cells. The total IO molarity calculated from $[HOI] + [OI^-]$ is entered into a cell. The relative (not absolute) errors for the two equilibrium constants are entered into designated cells. 10^{11} times $[H^+] - [OH^-] - [OI^-]$ is entered into a cell. (Since the ionic concentrations are all very small, the Excel Solver will consider the electroneutrality condition to be satisfied within its designated precision even when the charge is significantly unbalanced. Hence the 10^{11} factor.) The Solver is set up to make the relative error in K_a equal to zero by changing the four concentrations subject to the constraints that the K_w relative error be zero, the total [IO] be 0.0001, that 10^{11} times $[H^+] - [OH^-] - [OI^-]$ be zero, that all concentrations be nonnegative, that $[H^+]$ exceed $[OH^-]$, and that [HOI] and $[OI^-]$ each not exceed 0.0001. We click Options in the Solver Parameters box and check Use Automatic Scaling; this helps the Solver solve problems involving quantities of greatly different magnitudes. The Excel Solver gives $[H^+] = 1.11 \times 10^{-7}$, $[OH^-] = 9.02 \times 10^{-8}$, $[OI^-] = 2.07 \times 10^{-8}$.

11.19 **(a)** The equilibria are $HX + H_2O \rightleftarrows H_3O^+ + X^-$ and $2H_2O \rightleftarrows H_3O + OH^-$. With activity coefficients taken as 1, we have:
(1) $K_a = m(X^-)m(H_3O^+)/m(HX)$ and (2) $K_w = m(H_3O^+)m(OH^-)$.
Electroneutrality gives (3) $m(H_3O^+) = m(OH^-) + m(X^-)$.

(b) Conservation of X gives (4) $m = m(HX) + m(X^-)$.

(c) Using Eq. (2) to eliminate $m(OH^-)$ and Eq. (4) to eliminate $m(X^-)$, we find that Eqs. (1) and (3) become: (1)′ $K_a = m(H_3O^+)m(X^-)/[m - m(X^-)]$ and (3)′ $m(H_3O^+) = K_w/m(H_3O^+) + m(X^-)$. Let $y = m(H_3O^+)$. Solving (3)′ for $m(X^-)$ and substituting into (1)′, we get

$$y^3 + K_a y^2 - (K_w + mK_a)y - K_a K_w = 0.$$

11.20 In the cubic equation of Prob. 11.19, we set $K_a = 2.3 \times 10^{-11}$, $K_w = 1.0 \times 10^{-14}$, and $m = 1.0 \times 10^{-4}$ (where units are omitted). We designate a cell for y and enter an initial guess into this cell. In another cell, we enter the formula for the left side of the cubic equation and set up the Solver to make this cell equal to zero subject to the constraint that y (the H^+ concentration) lie between 10^{-7} and 10^{-4}. We find that for any initial guess for y in this range, the Solver declares it has found a solution without changing the initial guess. This is because the left side of the cubic equation is extremely close to zero for y values in this range. We therefore multiply the expression for the left side of the cubic equation by 10^{22} and require that this quantity be zero. With an initial guess of 1×10^{-7}, the Solver gives $y = 1.11 \times 10^{-7}$.

11.21 (a) For $HX + H_2O \rightleftarrows H_3O^+ + X^-$, $K_{m,a}$ is given by (11.14) with the $m°$'s omitted and where the γ's are γ_m's. Letting + and – indicate the H_3O^+ and X^- ions, we have $\dfrac{K_{c,a}}{K_{m,a}} = \dfrac{(\gamma_{c,+}c_+/\gamma_{m,+}m_+)(\gamma_{c,-}c_-/\gamma_{m,-}m_-)}{\gamma_{c,\mathrm{HX}}c_{\mathrm{HX}}/\gamma_{m,\mathrm{HX}}m_{\mathrm{HX}}}$.

Substitution of $\gamma_{c,i}c_i/\gamma_{m,i}m_i = \rho_A$ for $i = +$, $i = -$, and $i = \mathrm{HX}$ gives $K_{c,a}/K_{m,a} = (\rho_A)^2/\rho_A = \rho_A$.

(b) $c_i = n_i/V$ and $m_i = n_i/w_A$, so $c_i/m_i = w_A/V$. In a dilute solution, the solution's volume is approximately equal to the volume of pure solvent, so $c_i/m_i \approx w_A/V_A^* = \rho_A$.

(c) Substituting $c_i/m_i = \rho_A$ into $\gamma_{c,i}c_i = \rho_A\gamma_{m,i}m_i$ gives $\gamma_{c,i} \approx \gamma_{m,i}$.

11.22 (a) From Sec. 10.7, $\varepsilon_{r,A} = 78.4$ for water at 25°C. For a 1:1 electrolyte, $b =$

$$\frac{(1)(1)(1.602 \times 10^{-19}\,\mathrm{C})^2}{4\pi(8.854 \times 10^{-12}\,\mathrm{C^2/N\text{-}m^2})78.4(4.5 \times 10^{-10}\ \mathrm{m})(1.381 \times 10^{-23}\ \mathrm{J/K})(298.1\ \mathrm{K})}$$

$b = 1.588$

$K_c = (4/3)\pi(4.5 \times 10^{-10}\ \mathrm{m})^3(6.022 \times 10^{23}\ \mathrm{mol^{-1}})e^{1.588} = 0.00112(10\ \mathrm{dm})^3/\mathrm{mol} = 1.12\ \mathrm{dm^3/mol}$

(b) $z_+ = 2$ and $|z_-| = 1$, so $b = 2(1)(1.588) = 3.176$ and we find $K_c = 5.51\ \mathrm{dm^3/mol}$.

(c) $b = 6.352$ and $K_c = 132\ \mathrm{dm^3/mol}$.

(d) $b = 9.528$ and $K_c = 3160\ \mathrm{dm^3/mol}$. The K_c's have the correct magnitudes.

11.23 $V_m = M/\rho = (58.44 \text{ g/mol})/(2.16 \text{ g/cm}^3) = 27.0_6 \text{ cm}^3\text{/mol}$. From (11.23), $\ln a_i \cong (P - P_0)V_{m,i}/RT = (P - 1 \text{ bar})(27.0_6 \text{ cm}^3\text{/mol})/(83.14 \text{ cm}^3\text{-bar/mol-K})(298.1 \text{ K}) = 0.00109_2 \text{ bar}^{-1}(P - 1 \text{ bar})$. At 1, 10, 100, and 1000 bar, we get $\ln a_i = 0$, 0.0098, 0.108, and 1.09, and we get $a_i = 1$, 1.01, 1.11, and 2.98, respectively,

11.24 $K_{sp} = 5.38 \times 10^{-5} \text{ mol}^2\text{/kg}^2 = \gamma_\pm^2 m(\text{Ag}^+)^2$. With the initial approximation $\gamma_\pm \cong 1$, we get $m(\text{Ag}^+) \cong 0.0073_3$ mol/kg, which gives $I_m = 0.0073$ mol/kg. The Davies equation gives $\gamma_\pm \cong 0.91_4$. This $\gamma_\pm$ gives $m(\text{Ag}^+) =$ 0.00802 mol/kg, which gives $I_m = 0.00802$ mol/kg and $\gamma_\pm = 0.91_1$. This $\gamma_\pm$ gives $m(\text{Ag}^+) = 0.00805$ mol/kg as the solubility.

11.25 $CaF_2(s) \rightleftarrows Ca^{2+}(aq) + 2F^-(aq)$. $m(F^-) = 2m(Ca^{2+})$ and $K_{sp} = 3.2 \times 10^{-11} \text{ mol}^3\text{/kg}^3 = \gamma_+ m(\text{Ca}^{2+})\gamma_-^2 [m(\text{F}^-)]^2 = 4\gamma_\pm^3[m(\text{Ca}^{2+})]^3$. With $\gamma_\pm \cong 1$, we get $m(\text{Ca}^{2+}) = 2.0 \times 10^{-4}$ mol/kg. This gives $I_m = \frac{1}{2}[4(2.0 \times 10^{-4} \text{ mol/kg}) + 4.0 \times 10^{-4} \text{ mol/kg}] = 6.0 \times 10^{-4}$ mol/kg. The Davies equation gives $\gamma_\pm = 0.94_6$. With this $\gamma_\pm$ we get $m(\text{Ca}^{2+}) = 2.1 \times 10^{-4}$ mol/kg.

11.26 For $BaF_2(s) \rightleftarrows Ba^{2+}(aq) + 2F^-(aq)$, $\Delta G^\circ_{298} = [-560.77 + 2(-278.79) - (-1156.8)] \text{ kJ/mol} = 38.4_5 \text{ kJ/mol} = -RT \ln K^\circ_{sp}$, and $K_{sp} = 1.8_4 \times 10^{-7} \text{ mol}^3\text{/kg}^3$.

11.27 **(a)** $KCl(s) \rightleftarrows K^+(aq) + Cl^-(aq)$. $\Delta G^\circ_{298} = (-283.27 - 131.228 + 409.14) \text{ kJ/mol} = -5.36 \text{ kJ/mol} = -RT \ln K^\circ_{sp}$. $\ln K^\circ_{sp} = (5360 \text{ J/mol})/(8.3145 \text{ J/mol-K})(298 \text{ K}) = 2.16$. $K^\circ_{sp} = 8.7$ and $K_{sp} = 8.7 \text{ mol}^2\text{/kg}^2$.

(b) $K_{sp} = \gamma_\pm^2 m(\text{K}^+)m(\text{Cl}^-) = \gamma_\pm^2(\alpha m_i)(\alpha m_i) = \gamma_\pm^2\alpha^2 m_i^2 = (\gamma_\pm^\dagger)^2 m_i^2$. $(\gamma_\pm^\dagger)^2 = K_{sp}/m_i^2 = 8.7/(4.82)^2$ and $\gamma_\pm^\dagger = 0.61$.

11.28 $Ca^{2+} + SO_4^{2-} \rightleftarrows CaSO_4(aq)$. The 2.08 g is 0.0152_8 mol of $CaSO_4$. Let z mol/kg of of $CaSO_4$ ion pairs be formed. Then $190 = z/\gamma_\pm^2(0.0152_8 - z)^2$ or $z^2 - (0.0305_6 + 1/190\gamma_\pm^2)z + 0.000233_5 = 0$. The quadratic formula gives

$z = 0.0152_8 + 1/380\gamma_\pm^2 \pm \frac{1}{2}(0.000322/\gamma_\pm^2 + 0.0000277/\gamma_\pm^4)^{1/2}$. Initially we take $\gamma_\pm = 1$ and get $z \approx 0.00856$ (where the minus sign is used, since z can't exceed 0.015). Thus, our initial estimates are $m(Ca^{2+}) \approx 0.0067_2 \approx m(SO_4^{2-})$; these give $I_m \approx 0.0269$ mol/kg. The Davies equation then gives $\log \gamma_\pm \approx -0.0271$ and $\gamma_\pm \approx 0.536$. Use of this $\gamma_\pm$ in the above equation for z gives $z \approx 0.00536$ and $m(Ca^{2+}) \approx 0.0099_2$ mol/kg. Then $I_m \approx 0.0397$ mol/kg, $\log \gamma_\pm \approx -0.315$, $\gamma_\pm \approx 0.485$. This $\gamma_\pm$ gives $z \approx$ gives $z \approx 0.00485$ and $m(Ca^{2+}) \approx 0.0104_3$ mol/kg. Then $I_m \approx 0.0417$ mol/kg, $\log \gamma_\pm \approx -0.320$, $\gamma_\pm \approx 0.478$. This gives $z \approx 0.00478$, $I_m \approx 0.0420$ mol/kg, $\gamma_\pm \approx 0.477$. This gives $z = 0.00477$, which is the converged value. Thus $m(Ca^{2+}) = m(SO_4^{2-}) = 0.0105_1$ mol/kg and $K_{sp} = \gamma_\pm^2 m(Ca^{2+})m(SO_4^{2-}) = (0.477)^2(0.0105_1 \text{ mol/kg})^2 = 2.5_1 \times 10^{-5} \text{ mol}^2/\text{kg}^2$.

11.29 $CaCO_3$(calcite) $\rightleftarrows$ $CaO(s) + CO_2(g)$.
$\Delta G^\circ_{298} = (-604.03 - 394.359 + 1128.79)$ kJ/mol $= 130.40$ kJ/mol $= -RT \ln K^\circ$.
$\ln K^\circ = -(130400 \text{ J/mol})/(8.3145 \text{ J/mol-K})(298.15 \text{ K}) = -52.60$.
$K^\circ = 1.4 \times 10^{-23} = P(CO_2)/P^\circ$, where we took the solid's activities to be 1 and assumed ideal vapor. Therefore $P(CO_2) = 1.4 \times 10^{-23}$ bar.

11.30 Taking the activities of the solids as 1 and assuming ideal gases, we have $K^\circ = P(CO_2)/P(CO) = 1.15 = n(CO_2)/n(CO)$. Since the initial $n(CO_2)/n(CO)$ value exceeds 1.15, the equilibrium position lies to the left. Let z moles of CO_2 react to reach equilibrium. Then $1.15 = (5 - z)/(3 + z)$ and $z = 0.72$ mol. The equilibrium amounts are $n(Fe_3O_4) = 2.72$ mol, $n(CO) = 3.72$ mol, $n(FeO) = 1.84$ mol, $n(CO_2) = 4.28$ mol.

11.31 $K^\circ = P(CO_2)/P^\circ = (183 \text{ torr})/(750 \text{ torr}) = 0.244$.

(a) (5.0 g)/(100.1 g/mol) = 0.050 mol $CaCO_3$. We have $n_{CO_2} =$ $(0.244 \text{ bar})(4000 \text{ cm}^3)/(83.14 \text{ cm}^3\text{-bar/mol-K})(1073 \text{ K}) = 0.0109$ mol. Hence the equilibrium composition is 0.039 mol of $CaCO_3$, 0.0109 mol of CaO, and 0.0109 mol of CO_2.

(b) The 0.0050 mol of $CaCO_3$ is not enough to give 0.0109 mol of CO_2; therefore all the $CaCO_3$ dissociates to give 0 mol of $CaCO_3$, 0.0050 mol of CaO, and 0.0050 mol of CO_2.

11.32 At the pressures involved, we can assume ideal-gas behavior. Initially, $n(\text{CaO}) = 0.00892$ mol. $K° = 0.244 \cong P_{eq}(\text{CO}_2)/(1 \text{ bar})$ and $P_{eq}(\text{CO}_2) =$ 0.244 bar = 183 torr.

(a) The pressure of 125 torr is less than the equilibrium pressure 183 torr when all species are present, so no reaction occurs and the final amounts are $n(\text{CaCO}_3) = 0$, $n(\text{CaO}) = 0.00892$ mol, and $n(\text{CO}_2) =$ (125/760) atm (4000 cm^3)/RT = 0.00747 mol.

(b) $P_{eq}(\text{CO}_2) = 183$ torr and $n_{eq}(\text{CO}_2) = (183/760)$ atm (4000 cm^3)/RT = 0.0109 mol. The initial number of moles of CO_2 is (235/760) atm (V/RT) = 0.0140 mol, so (0.0140 – 0.0109) mol = 0.0031 mol of CO_2 react. Hence $n_{eq}(\text{CaCO}_3) = 0.0031$ mol; $n_{eq}(\text{CaO}) = (0.00892 - 0.0031)$ mol = 0.0058 mol.

(c) The initial number of CO_2 moles is $n(\text{CO}_2) = (825/760)$ atm (4000 cm^3)/ RT = 0.0493 mol. If the pressure $P_{eq}(\text{CO}_2) = 183$ torr were reached, then (0.0493 – 0.0109) mol = 0.0384 mol of CO_2 would have reacted. But only 0.00892 mol of CaO is present initially, so only 0.00892 mol of CO_2 react. The final amounts are thus $n(\text{CaO}) = 0$, $n(\text{CaCO}_3) = 0.00892$ mol, $n(\text{CO}_2) = (0.0493 - 0.00892)$ mol = 0.0404 mol.

11.33 Eq. (11.4) holds: $\Delta G° = -RT \ln K° = 0 + 0 - 2(-95.299 \text{ kJ/mol}) =$ 190.598 kJ/mol. So $K°_{298} = 4.1 \times 10^{-34}$.

11.34 The Lewis–Randall rule gives $\phi_i = \phi_i^*(450°\text{C}, 300 \text{ bar})$. $N_2 + 3H_2 \rightleftharpoons 2NH_3$. Let z mol of N_2 react to reach equilibrium. At equilibrium, $n_{N_2} = 1 - z$, $n_{H_2} = 3 - 3z$, and $n_{NH_3} = 2z$. $P_i = x_i P$ and $P_{N_2} = [(1 - z)/(4 - 2z)]P$, $P_{H_2} = [(3 - 3z)/(4 - 2z)]P$, $P_{NH_3} = [2z/(4 - 2z)]P$. The left side of (11.30) is $(4.6 \times 10^{-5})[(0.91)^2/(1.14)(1.09)^3]^{-1} = 8.2 \times 10^{-5}$. The right side is $(P_{NH_3}/P°)^2/(P_{N_2}/P°)(P_{H_2}/P°)^3 = (P/P°)^{-2}(2z)^2(4 - 2z)^2/(3 - 3z)^3(1 - z) =$ 8.2×10^{-5}. Since $P/P° = 300$, we must solve $4z^2(4 - 2z)^2/3^3(1 - z)^4 = 7.38$. Taking the square root gives $2z(4 - 2z)/(1 - z)^2 = 14.12$, where the negative sign in front of 14.12 is rejected, since z and $4 - 2z$ must be positive. Use of the quadratic formula to solve $18.12z^2 - 36.24z + 14.12 = 0$ gives $z = 1.47$ and 0.530. Since z must be less than 1 (to keep $n_{N_2} > 0$), we have $z = 0.530$ and $n_{N_2} = 0.470$ mol, $n_{N_2} = 1.41$ mol, $n_{NH_3} = 1.06$ mol.

11.35 $N_2 + 3H_2 \rightleftarrows 2NH_3$. $\Delta G^\circ_{700} = 2(6.49\text{ kcal/mol}) = 12.98\text{ kcal/mol} = -RT\ln K^\circ$.

$\ln K^\circ = -(12980\text{ cal/mol})/(1.987\text{ cal/mol-K})(700\text{ K}) = -9.332$.

$K^\circ = 8.85 \times 10^{-5}$. The Lewis rule is $\phi_i \approx \phi_i^*(T, P)$. For NH_3, $P_r =$ (500 atm)/(111.3 atm) = 4.49 and T_r = (700 K)/(405.6 K) = 1.73. For N_2, P_r = 500/33.5 = 14.9 and T_r = 700/126.2 = 5.55. For H_2, P_r = 500/(12.8 + 8) = 24.0 and T_r = 700/(33.8 + 8) = 16.7. The Newton graphs of ϕ (the full reference is in Sec. 10.10) give $\phi_{NH_3} \approx 0.86$, $\phi_{N_2} \approx 1.26$, and $\phi_{H_2} \approx 1.16$. The left side of Eq. (11.30) equals $(8.85 \times 10^{-5})/[(0.86)^2(1.26)^{-1}(1.16)^{-3}] = 2.35 \times 10^{-4}$. Let $2w$ moles of NH_3 decompose. The equilibrium amounts are $n(NH_3)/\text{mol} = 1 - 2w$, $n(N_2)/\text{mol} = w$, and $n(H_2)/\text{mol} = 3w$. $n_{tot}/\text{mol} = 1 + 2w$. Use of $P_i = x_iP$ gives for Eq. (11.30):

$$2.35 \times 10^{-4} = \frac{[(1-2w)/(1+2w)]^2(507)^2}{[w/(1+2w)][3w/(1+2w)]^3(507)^4}$$

$1631w^4 = (1-2w)^2(1+2w)^2 = [(1-2w)(1+2w)]^2 = (1-4w^2)^2$. Let $z = w^2$. Then $1631z^2 = (1-4z)^2 = 1 - 8z + 16z^2$ and $1615z^2 + 8z - 1 = 0$. We find z = 0.0225. Then $w = z^{1/2} = 0.150$ and $n(NH_3)$ = 0.700 mol, $n(N_2)$ = 0.150 mol, and $n(H_2)$ = 0.450 mol.

11.36 In Example 11.1, we found $I_m = 1.96 \times 10^{-3}$ mol/kg = $m(H^+)$. For this I_m, $\log_{10}\gamma(HX) = 0.1(1.96 \times 10^{-3})$ and $\gamma(HX) = 1.00045$, which when used in Eq. (11.14) makes no significant change in $m(H^+)$. In Example 11.2, $I_m = 0.200$ mol/kg, so $\log_{10}\gamma(HC_2H_3O_2) = 0.0200$ and $\gamma(HC_2H_3O_2) = 1.047$. Equation (11.14) becomes

$$1.75 \times 10^{-5}\text{ mol/kg} = \frac{(0.746)^2 m(H_3O^+)(0.200\text{ mol/kg})}{1.047(0.100\text{ mol/kg})}$$

and $m(H^+) = 1.6_5 \times 10^5$ mol/kg, 5% larger than the answer in Example 11.2.

11.37 **(a)** Assuming ideal vapor and taking the solid's activities as 1, we have $K^\circ = P(CO_2)/P^\circ$. Equation (11.32) gives $[\partial \ln K^\circ/\partial(1/T)]_P = -\Delta H^\circ/R$. We plot $\ln K^\circ$ vs. $1/T$. The data are:

$\ln K^\circ$	−3.485	−2.372	−1.411	−0.677	−0.046	
$10^4/T$	10.27	9.79	9.32	8.89	8.57	K^{-1}

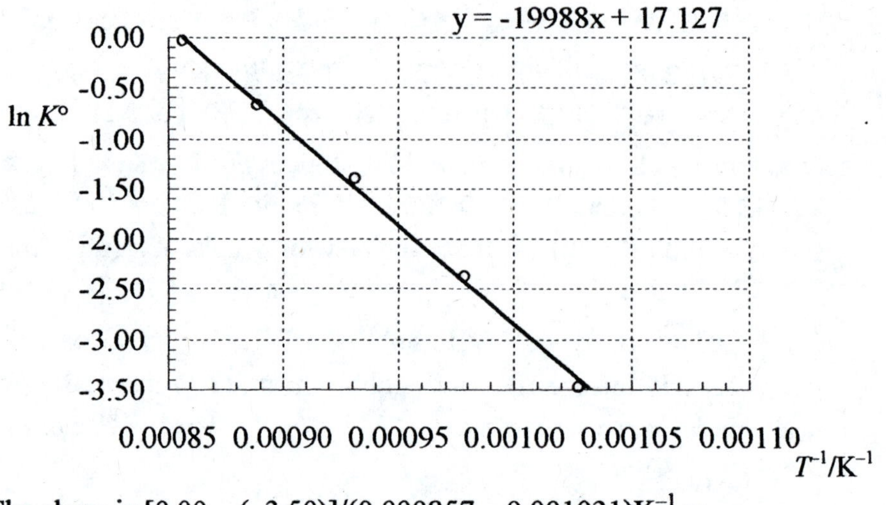

The slope is $[0.00 - (-3.50)]/(0.000857 - 0.001031)\text{K}^{-1}$ = -2.01×10^4 K = $-\Delta H°/R$ and $\Delta H°$ = 167 kJ/mol. At 1073 K, $\Delta G° = -RT \ln K°$ = $-(8.314$ J/mol-K)(1073 K) ln (183/750) = $12._6$ kJ/mol. Then $\Delta G° = \Delta H° - T\Delta S°$ and $\Delta S°_{1073}$ = [(167000 – 12600) J/mol]/(1073 K) = 144 J/mol-K.

(b) From the extrapolated graph, we read ln $K°$ = 1.5 at $10^4/T$ = 7.855, so $K°$ = 4.5 and $P(CO_2) = 4.5P°$ = 4.5 bar.

11.38 Substitution of T/K = 298.15 into the equation gives $\log K_w^\circ = -13.998$ and $K_w^\circ = 1.003 \times 10^{-14}$. We have $\Delta G° = -RT \ln K°$ = $-(8.3145$ J/mol-K)(298.15 K) ln (1.003×10^{-14}) = 79.91 kJ/mol. (Cf. Prob. 10.49.) We have $(\partial \ln K°/\partial T)_P = \Delta H°/RT^2$ = $2.302585(\partial \log K°/\partial T)_P$. Differentiation of the K_w° equation in the text (which is for P = 1 bar) gives $(\partial \log K_w^\circ/\partial T)_P$ = (24746.26 K)/T^2 – $405.8639/2.302585T + 0.48796/\text{K} - 0.0002371(2T/\text{K}^2) = 0.03376\ \text{K}^{-1}$ at 25°C. Hence $\Delta H° = 2.302585RT^2(0.03376\ \text{K}^{-1})$ = 57.45 kJ/mol at 25°C. $\Delta S°_{298} = (\Delta H° - \Delta G°)/T$ = [(57450 – 79910) J/mol]/(298 K) = –75.3 J/mol-K.

11.39 $\Delta V°$ = –5.4 cm³/mol + 1.4 cm³/mol – (18.015 g/mol)/(0.997 g/cm³) = –22.1 cm³/mol. $(\partial \ln K°/\partial P)_T = -\Delta V°/RT$ and $\int_1^2 d\ln K° = -\int_1^2 (\Delta V°/RT)\, dP$ at constant T. Neglecting the pressure dependence of $\Delta V°$, we have $\ln (K_2^\circ/K_1^\circ) = -(\Delta V°/RT)(P_2 - P_1)$ and $\ln [K^\circ_{200\ \text{bar}}/(1.00 \times 10^{-14})]$ =

$(22.1\ \text{cm}^3/\text{mol})(83.14\ \text{cm}^3\text{-bar/mol-K})^{-1}(298\ \text{K})^{-1}(199\ \text{bar}) = 0.178 =$
$\ln K^\circ_{200\,\text{bar}} + 32.236$ and $K^\circ_{200\,\text{bar}} = 1.19 \times 10^{-14}$.

11.40 **(a)** As in Prob. 11.39, with the P dependence of ΔV° neglected, we have $-(\Delta V^\circ/RT)(P_2 - P_1) = \ln (K^\circ_2/K^\circ_1) = \ln 1.191 = 0.1748$ and $\Delta V^\circ =$ $-0.1748(83.14\ \text{cm}^3\text{-bar/mol-K})(298.1\ \text{K})/(399\ \text{bar}) = -10.9\ \text{cm}^3/\text{mol}$.

(b) $\ln 2 = 0.693 = (10.9\ \text{cm}^3/\text{mol})(83.14\ \text{cm}^3\text{-bar/mol-K})^{-1}(298\ \text{K})^{-1}(P_2 - P_1)$ and $P_2 - P_1 = 1.58$ kbar, so $P_2 = 1.58$ kbar.

11.41 **(a)** Since $\mu^\circ_\text{A} = \mu^*_\text{A}$, we have $\bar{H}^\circ_\text{A} = H^*_{m,\text{A}}$. Prob. 10.22 gives $\bar{H}^\circ_{m,i} = \bar{H}^\infty_i$. Hence $\Delta H^\circ_m = \nu_\text{A} H^*_{m,\text{A}} + \sum_{i\neq\text{A}} \nu_i \bar{H}^\infty_i$ and substitution in (11.32) gives the desired result.

(b) $K_c/K_m = (\gamma_{x,\text{A}} x_\text{A})^{\nu_\text{A}} \Pi_{i\neq\text{A}} (\gamma_{c,i} c_i)^{\nu_i} / (\gamma_{x,\text{A}} x_\text{A})^{\nu_\text{A}} \Pi_{i\neq\text{A}} (\gamma_{m,i} m_i)^{\nu_i} =$ $\Pi_{i\neq\text{A}} (\gamma_{c,i} c_i/\gamma_{m,i} m_i)^{\nu_i} = \Pi_{i\neq\text{A}} (\rho_\text{A})^{\nu_i} = \rho^b_\text{A}$, where $b = \sum_{i\neq\text{A}} \nu_i$.

(c) From (b), $K_m = K_c \rho_\text{A}^{-b}$ and $\ln K^\circ_m = \ln K^\circ_c - b \ln \rho_\text{A} + \ln(\text{const})$, where const involves c° and m°. Hence $(\partial \ln K^\circ_m/\partial T)_P = (\partial \ln K^\circ_c/\partial T)_P -$ $b(\partial \ln \rho_\text{A}/\partial T)_P = (\partial \ln K^\circ_c/\partial T)_P + b\alpha_\text{A} = \Delta H^\infty/RT^2$, where we used the result of part (a) and used $(\partial \ln \rho/\partial T)_P = (\partial/\partial T)_P(\ln m - \ln V) =$ $-(1/V)(\partial V/\partial T)_P = -\alpha$. Thus $(\partial \ln K^\circ_c/\partial T)_P = \Delta H^\infty/RT^2 - b\alpha_\text{A} =$ $\Delta H^\infty/RT^2 - \alpha_\text{A} \sum_{i\neq\text{A}} \nu_i$.

(d) From (11.32) with the concentration scale used, $\Delta H^\circ_c/RT^2 = \Delta H^\infty/RT^2 -$ $\alpha_\text{A} \sum_{i\neq\text{A}} \nu_i = \nu_A H^*_{\text{m},A}/RT^2 + \sum_{i\neq\text{A}} (\nu_i \bar{H}^\infty_i/RT^2 - \nu_i \alpha_\text{A})$. Since $\Delta H^\circ_c =$ $\nu_A H^*_{\text{m},A} + \sum_{i\neq\text{A}} \nu_i \bar{H}^\circ_{c,i}$, it follows that $\bar{H}^\circ_{c,i} = \bar{H}^\infty_i - RT^2\alpha_\text{A}$ for $i \neq \text{A}$.

11.42 (c).

11.43 $\Delta G^{\circ\prime} - \Delta G^\circ = \Sigma_{i\neq\text{H}^+} \nu_i \mu^\circ_i + \nu(\text{H}^+)\mu[\text{H}^+ \text{ at } a(\text{H}^+) = 10^{-7}] -$ $[\Sigma_{i\neq\text{H}^+} \nu_i \mu^\circ_i + \nu(\text{H}^+)\mu^\circ(\text{H}^+)] = \nu(\text{H}^+)\{\mu[\text{H}^+ \text{ at } a(\text{H}^+) = 10^{-7}] - \mu^\circ(\text{H}^+)\}$. We have $\mu_i = \mu^\circ_i + RT \ln a_i$, so $\mu[\text{H}^+ \text{ at } a(\text{H}^+) = 10^{-7}] = \mu^\circ(\text{H}^+) + RT \ln 10^{-7} =$ $\mu^\circ(\text{H}^+) - 16.118RT$. $\Delta G^{\circ\prime} - \Delta G^\circ = -16.118\nu(\text{H}^+)RT$.

11.44 $\Delta G° = 2(4.83 \text{ kJ/mol}) = 9.66 \text{ kJ/mol}$. $Q = (P_{NH_3}/P°)^2/(P_{N_2}/P°)(P_{H_2}/P°)^3$ and use of $P_i = x_iP$ gives $Q = [(1/7)3]^2/[(2/7)3][(4/7)3]^3 = 0.0425$. $(\partial G/\partial \xi)_{T,P} = \Delta G° + RT \ln Q = 9660 \text{ J/mol} + (8.3145 \text{ J/mol-K})(500 \text{ K}) \ln 0.0425 = -3.47 \text{ kJ/mol}$. Since $(\partial G/\partial \xi)_{T,P} < 0$, the reaction proceeds to the right.

11.45 As found in Prob. 6.50, constant T and P addition of j will produce more j when $x_j > \nu_j/\Delta|\nu|$ where ν_j and $\Delta|\nu|$ have the same sign. ν_j and $\Delta|\nu|$ are small integers and typical values of $\nu_j/\Delta|\nu|$ are 1/2, 2/1, 1/3, 2/3, etc. Since the solution is dilute and j is a solute, we have $x_j << 1$, and x_j will not exceed $\nu_j/\Delta|\nu|$. Hence the answer is no.

11.46 **(a)** The log of Eq. (9.62) gives $\ln(K_i/P°) = (\mu^\circ_{i,\ell} - \mu^\circ_{i,\upsilon})/RT = -\Delta G°/RT$, which is the desired equation.

(b) For $O_2(aq) \rightarrow O_2(g)$, $\Delta G^\circ_{298}/(\text{kJ/mol}) = 0 - 16.4 = -16.4$, and the equation in (a) gives $\ln(K_{i,m}/P°) = -\Delta G°/RT = (16400 \text{ J/mol})/RT = 6.62$; $K_{i,m}/P° = 747$ and $K_{i,m} = 747$ bar. For $CH_4(aq) \rightarrow CH_4(g)$, $\Delta G^\circ_{298}/(\text{kJ/mol}) = -50.72 + 34.33 = -16.39$; $\ln(K_{i,m}/P°) = -\Delta G°/RT = 6.61$; $K_{i,m} = 744$ bar.

11.47 **(a)** False. Intermolecular interactions between He and the gases in the reaction may change the gases' fugacity coefficients and so shift the equilibrium.

(b) True. See Fig. 4.7.

(c) False. See Eq. (11.19).

(d) False. $\Delta G° \neq \Delta G$ in the reaction mixture.

(e) False. The molality-scale standard state is a state in solution.

(f) False. See (e).

(g) False.

Chapter 12

12.1 **(a)** T; **(b)** T.

12.2 **(a)** T; **(b)** F.

12.3 The contribution of sucrose to the vapor pressure can be neglected. $P = P_A = a_A P_A^*$. Since the solution is reasonably dilute, we can take $a_A \approx x_A$ and $P = x_A P_A^*$. In 100 g of solution, there are (98.00 g)/(18.015 g/mol) = 5.440 mol of water and (2.00 g)/(342.3 g/mol) = 0.00584 mol of sucrose. Then x_A = 0.99893 and P_A = 0.99893(1074.6 torr) = 1073.4 torr.

12.4 **(a)** T; **(b)** F; **(c)** F; **(d)** T; **(e)** T; **(f)** T; **(g)** T.

12.5 $n_{C_5H_{12}}$ = (0.226 g)(1 mol/72.15 g) = 0.00313$_2$ mol. $m_{C_5H_{12}}$ = (0.00313$_2$ mol)/(0.01645 kg) = 0.190$_4$ mol/kg. $k_f = M_A RT_f^{*2}/\Delta_{fus}H_{m,A}$ = (84.16 g/mol)(8.3145 J/mol-K)(279.62 K)^2/(31.3 J/g)(84.16 g/mol) = 2.08 × 10^4 K g/mol = 20.8 K kg/mol. $\Delta T_f = -k_f m$ = –(20.8 K kg/mol) × (0.190$_4$ mol/kg) = –3.96 K. T_f = 6.47°C – 3.96°C = 2.51°C. We assumed an ideally dilute solution and that only pure cyclohexane freezes out.

12.6 $m = -\Delta T_f / k_f$ = (0.112 K)/(1.860 K kg/mol) = 0.0602 mol/kg = n_i/w_A = n_i/(0.0980 kg), so n_i = 0.00590 mol of maltose. $M_i = w_i/n_i$ = (2.00 g)/(0.00590 mol) = 339 g/mol and $M_{r,i}$ = 339.

12.7 **(a)** $\Delta T_f = -k_f m_B = -k_f n_B/w_A = -k_f w_B/M_B w_A$ and $M_B = -k_f w_B/\Delta T_f w_A$.

(b) For 100 g of solution, the 3% solution has w_B = 3.00 g and w_A = 97.00 g; M_B = –(1.86 K kg/mol)(3.00 g)/(–0.169 K)(97.0 g) = 0.340$_4$ kg/mol = 340.$_4$ g/mol. For the 6, 9, 12, and 15% solutions, we find M_B = 337.$_3$, 334.$_5$, 331.$_6$, and 328.$_2$ g/mol. Plotting M_B vs. wt. % maltose, we get a nearly linear graph that extrapolates to M_B = 343 g/mol at 0 wt. %.

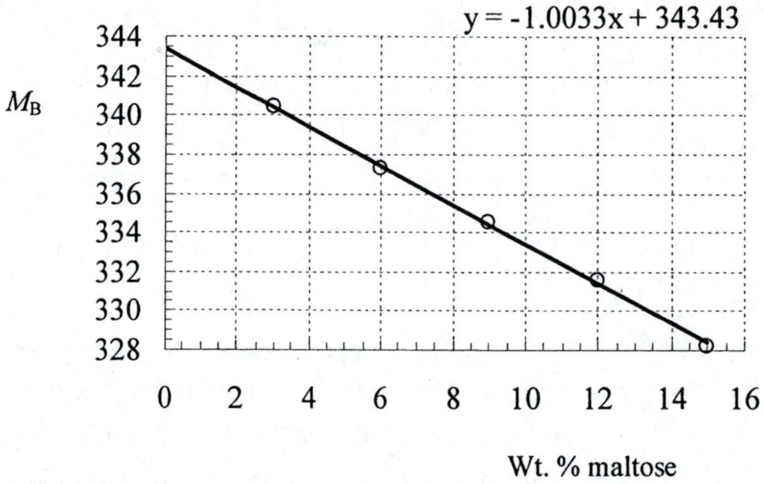

12.8 **(a)** $m_{\text{urea}} = [(1.00\text{ g})/(60.06\text{ g/mol})]/(0.200\text{ kg}) = 0.0833\text{ mol/kg}$;
$k_f = -\Delta T_f/m_{\text{urea}} = (0.250\text{ K})/(0.0833\text{ mol/kg}) = 3.00\text{ K kg/mol}$.
$m_Y = -\Delta T_f/k_f = (0.200\text{ K})/(3.00\text{ K kg/mol}) = 0.0666_7\text{ mol/kg} = n_Y/(0.125\text{ kg})$ and $n_Y = 0.00833$ mol. $M_Y = w_Y/n_Y = (1.50\text{ g})/(0.00833\text{ mol}) = 180\text{ g/mol}$. $M_{r,Y} = 180$.

(b) $\Delta_{\text{fus}}H_{\text{m,A}} = M_A RT_f^{*2}/k_f = (200\text{ g/mol})(1.987\text{ cal/mol-K})(285\text{ K})^2/(3000\text{ g K/mol}) = 10.8\text{ kcal/mol} = 45.0\text{ kJ/mol}$.

12.9 Let U denote $CO(NH_2)_2$. $\Delta T_f(\text{A + Z}) = -k_f m_z$; $\Delta T_f(\text{A + U}) = -k_f m_U$; so $1.65 = \Delta T_f(\text{A + Z})/\Delta T_f(\text{A + U}) = m_z/m_U = (n_Z/w_A)/(n_U/w_A) = n_Z/n_U$ and $n_Z = 1.65 n_U = 1.65(0.679\text{ g})/(60.05\text{ g/mol}) = 0.0186_6\text{ mol}$. $M_Z = (0.542\text{ g})/(0.0186_6\text{ mol}) = 29.0\text{ g/mol}$; $M_{r,Z} = 29.0$.

12.10 $1/T_f^*T_f = (1/T_f^*)[1/(T_f^* + \Delta T_f)] = (1/T_f^*)(1/T_f^*)[1/(1 + \Delta T_f/T_f^*)] = (1/T_f^*)^2[1 - \Delta T_f/T_f^* + (\Delta T_f/T_f^*)^2 + \cdots]$.

12.11 $k_b = M_A RT_b^{*2}/\Delta_{\text{vap}}H_{\text{m,A}} = (18.0153\text{ g/mol})(8.3145\text{ J/mol-K}) \times (373.15\text{ K})^2/(40660\text{ J/mol}) = 512.9_5\text{ K g/mol} = 0.5129_5\text{ K kg/mol}$.

12.12 We have 0.00313_6 mol of $C_{10}H_8$ and $m_B = (0.00313_6\text{ mol})/(26.6\text{ g}) = 1.18 \times 10^{-4}\text{ mol/g}$. Hence $k_b = (0.455\text{ K})/(1.18 \times 10^{-4}\text{ mol/g}) = 3860\text{ K g/mol}$. Equation (12.19) gives $\Delta_{\text{vap}}H_m = (119.4\text{ g/mol})(8.314\text{ J/mol-K}) \times (334.8\text{ K})^2/(3860\text{ K g/mol}) = 28.8\text{ kJ/mol} = 6.89\text{ kcal/mol}$.

12.13 **(a)** We have $\gamma_A x_A = a_A$ and $\Delta T_f = [R(T_f^*)^2/\Delta_{fus}H_{m,A}] \ln a_A =$ $(k_f/M_A)(-\phi M_A \nu m_i) = -k_f \phi \nu m_i$.

(b) In 96.0 g of water we have 4.00 g (or 0.0229_5 mol) of K_2SO_4. Hence $m_i = (0.0229_5 \text{ mol})/(0.0960 \text{ kg}) = 0.239$ mol/kg. Then $\phi = -\Delta T_f/k_f \nu m_i =$ (0.950 K)/(1.86 K kg/mol)3(0.239 mol/kg) = 0.712.

12.14 Equation (12.17) gives $\sum_{i\neq A} m_i$ = (2.37 K)/(14.1 K kg/mol) = 0.168 mol/kg; in 100 g of bromoform, we have 0.0168 total moles of solutes. Let P and P_2 be phenol and its dimer. Then $2P \rightleftarrows P_2$. The 2.58 g is 0.0274 mol of phenol. Let $2z$ mol of phenol react. At equilibrium, n_P/mol = $0.0274 - 2z$, $n_{P_2} = z$, and $n_{tot} =$ $0.0274 - z$. Then $0.0168 = 0.0274 - z$ and $z = 0.0106$. So $n_P = 0.0062$ mol and $n_{P_2} = 0.0106$ mol. The molalities are m_P = (0.0062 mol)/(0.100 kg) = 0.062 mol/kg and $m_{P_2} = 0.106$ mol/kg. Thus K_m = (0.106 mol/kg)/(0.062 mol/kg)^2 = $27._6$ kg/mol.

12.15 $\sum_{i\neq A} m_i = -\Delta T_f/k_f$ = (0.70 K)/(5.1 K kg/mol) = 0.137 mol/kg = $(n_{nap} + n_{an})$/ (0.300 kg); so $n_{nap} + n_{an}$ = 0.0412 mol. We have 6.0 g = n_{nap}(128.2 g/mol) + n_{an}(178.2 g/mol) = n_{nap}(128.2 g/mol) + (0.0412 mol − n_{nap})(178.2 g/mol). We get n_{nap} = 0.027 mol and n_{an} = 0.014 mol.

12.16 **(a)** In 100 g of solution, n_B = (8.00 g)/(342.3 g/mol) = 0.02337 mol and m_B = (0.02337 mol)/(0.092 kg) = 0.254 mol/kg.
ΔT_f = −(1.860 K kg/mol)(0.254 mol/kg) = -0.472_5 K.

(b) $\ln a_A = (\Delta_{fus}H_{m,A}/R)(1/T_f^* - 1/T_f) =$
[(6007 J/mol)/(8.3145 J/mol-K)](1/273.15 − $1/272.66_5$)/K = −0.00470 and $a_A = 0.9953_1$. 100 g of solution has n_A = (92.00 g)/(18.0153 g/mol) = 5.1068 mol of H_2O, so x_A = 5.1068/(5.1068 + 0.02337) = 0.99544 and $\gamma_A = a_A/x_A = 0.9953_1/0.99544 = 0.99987$.

(c) $\phi = -(\ln a_A)/M_A m_i$ = 0.00470/(0.018015 kg/mol)(0.254 mol/kg) = 1.027.

12.17 False.

12.18 **(a)** $\Pi \cong cRT$ = (0.282 mol/L)(1 L/10^3 cm^3)(82.06 cm^3-atm/mol-K)(293.1 K) = 6.78 atm.

(b) $V^*_{m,H_2O} = (18.015 \text{ g/mol})/(0.998 \text{ g/cm}^3) = 18.05 \text{ cm}^3/\text{mol}$.
$\ln a_{H_2O} = -\Pi V^*_{m,A}/RT = -[(7.61 \text{ atm})(18.05 \text{ cm}^3/\text{mol})]/$
$(82.06 \text{ cm}^3\text{-atm/mol-K})(293.1 \text{ K})] = -0.00571$; $a_{H_2O} = 0.99431$. $x_{H_2O} =$
$55.508/(55.508 + 0.300) = 0.99462$. $\gamma_{H_2O} = 0.99431/0.99462 = 0.99969$.

12.19 $c_B = \Pi/RT = (6.1/760)\text{atm}/(82.06 \text{ cm}^3\text{-atm/mol-K})(273 \text{ K}) =$ $3.5_8 \times 10^{-7} \text{ mol/cm}^3$. One cm^3 of solution contains 3.58×10^{-7} mol of the protein and contains 0.0200 g of protein; so $M_B = (0.0200 \text{ g})/(3.5_8 \times 10^{-7} \text{ mol}) = 56000 \text{ g/mol}$; $M_{r,B} = 56000$. This is approximate because the solution is not actually ideally dilute.

12.20 We have 5.55 moles of water and 0.00292 moles of sucrose. Equation (12.26) gives $\Pi = [(82.06 \text{ cm}^3\text{-atm/mol-K})(298.1 \text{ K})/(18.07 \text{ cm}^3/\text{mol})] \times [(0.00292 \text{ mol})/(5.55 \text{ mol})] = 0.712$ atm. We have $\Pi = \rho gh$ and $h = \Pi/\rho g =$

$$\frac{0.712 \text{ atm}}{(1.00 \times 10^{-3} \text{ kg/cm}^3)(9.807 \text{ m/s}^2)} \frac{8.314 \text{ J}}{82.06 \text{ cm}^3 \text{ atm}} = 7.36 \text{ m} = 736 \text{ cm}$$

12.21 $\Pi = \rho gh$. In the infinite-dilution limit, ρ of the solution goes to ρ of water, and we shall use ρ of water in the following calculations. For the first solution, $\Pi = (0.996 \text{ g/cm}^3)(980.7 \text{ cm/s}^2)(2.18 \text{ cm}) = 2129 \text{ erg/cm}^3$ and $\Pi/\rho_B = (2129 \text{ erg/cm}^3)/(0.00371 \text{ g/cm}^3) = 5.74 \times 10^5$ erg/g. The other solutions give Π/ρ_B values of 6.29×10^5, 7.18×10^5, and 8.10×10^5 erg/g. A plot of Π/ρ_B vs. ρ_B is nearly linear and extrapolates to $4.5_4 \times 10^5$ erg/g at $\rho_B = 0$. This intercept equals RT/M_B and $M_B = (8.314 \times 10^7 \text{ erg/mol-K})(303.1 \text{ K})/(4.5_4 \times 10^5 \text{ erg/g}) = 5.5_5 \times 10^4$ g/mol; the number average molecular weight is 55500.

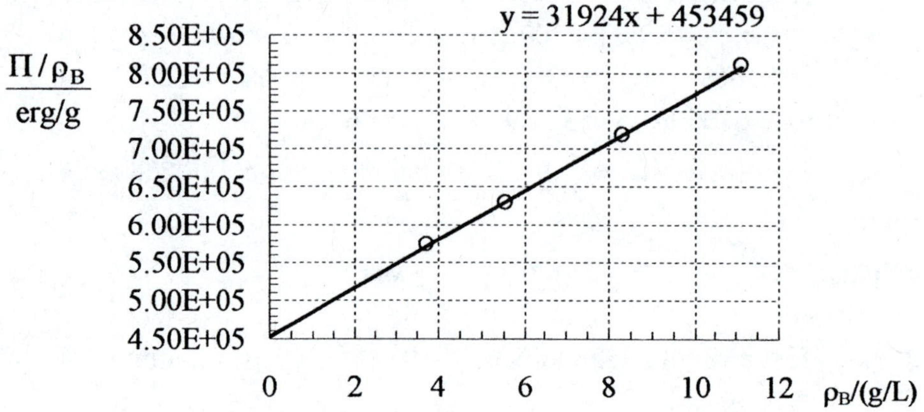

12.22 $\Pi/\rho_B \approx RT/M_B + A_2RT\rho_B + A_3 RT\rho_B^2 = RT/M_B + A_2RT\rho_B + A_2^2 M_B RT\rho_B^2/4 = [(RT/M_B)^{1/2} + A_2(M_BRT)^{1/2}\rho_B/2]^2$ and taking the square root of both sides gives the desired result.

12.23 **(a)** $\sum_{i\neq A} c_i \approx \Pi/RT = (7\text{ atm})/(82.06\text{ cm}^3\text{-atm/mol-K})(310\text{ K}) = 0.000275\text{ mol/cm}^3 = 0.275\text{ mol/dm}^3 \approx c_{Na^+} + c_{Cl^-}$. Therefore $c_{Na^+} \approx 0.138\text{ mol/dm}^3$ as compared with 0.15 mol/dm^3.

(b) We shall approximate the concentration by the molalities. Then $\sum_{i\neq A} c_i \approx [2(0.460) + 3(0.034) + 2(0.019) + 2(0.009)]\text{ mol/dm}^3 = 1.078\text{ mol/dm}^3 = 0.001078\text{ mol/cm}^3$. Then $\Pi \approx (82.06\text{ cm}^3\text{-atm/mol-K})(293\text{ K})(0.001078\text{ mol/cm}^3) = 26\text{ atm}$.

12.24 We have $V_i = n_iRT/P$, so V_i is proportional to n_i and to x_i. Then $M_n = \sum_i x_iM_i$ and $M_{n,r} = (0.78)(28.01) + 0.21(32.00) + 0.01(39.95) = 29.0$.

12.25 $\Pi = -(RT/V^*_{m,A}) \ln a_A = -(RT/V^*_{m,A})(-\phi M_A \nu m_i) = (\phi RT/V^*_{m,A})M_A\nu n_i/n_AM_A = \phi RT\nu n_i/n_A V^*_{m,A}$.

12.26 **(a)** $\mu_{A,1} = \mu_{A,2}$. Assuming ideally dilute solutions, we have $\mu^*_A(P_1, T) + RT \ln x_{A,1} = \mu^*_A(P_1 + \Pi, T) + RT \ln x_{A,2}$ and $\mu^*_A(P_1 + \Pi, T) - \mu^*_A(P_1, T) = RT \ln x_{A,1} - RT \ln x_{A,2} \approx -RTx_{B,1} + RTx_{B,2}$, where (12.11) was used. From (12.22), $\mu^*_A(P_1 + \Pi, T) - \mu^*_A(P_1, T) = V^*_{m,A}\,\Pi$, where the pressure dependence of $V^*_{m,A}$ is neglected. Hence $V^*_{m,A}\,\Pi = RT(x_{B,2} - x_{B,1})$.

(b) $x_{B,2} = 0.100/(0.100 + 55.51) = 0.00180$ and $x_{B,1} = 0.0200/(0.0200 + 55.51) = 0.000360$. $\Pi = (82.06\text{ cm}^3\text{-atm/mol-K})(298.1\text{ K})(0.00144)/(18.07\text{ cm}^3\text{/mol}) = 1.95\text{ atm}$.

12.27 At equilibrium, the equality sign holds in (4.12) and $-P^\alpha\, dV^\alpha - P^\beta\, dV^\beta = dA^\alpha + dA^\beta = -P^\alpha\, dV^\alpha + \sum_i \mu_i^\alpha\, dn_i^\alpha - P^\beta\, dV^\beta + \sum_i \mu_i^\beta\, dn_i^\beta$, so $\sum_i \mu_i^\alpha\, dn_i^\alpha + \sum_i \mu_i^\beta\, dn_i^\beta = 0$. Let dn moles of substance j move from phase α to phase β. Then $dn_j^\beta = dn$ and $dn_j^\alpha = -dn$; we have $-\mu_j^\alpha\, dn + \mu_j^\beta\, dn = 0$ and $\mu_j^\alpha = \mu_j^\beta$.

12.28 $f = c_{ind} - p + 2 = 2 - p + 2 = 4 - p$ and $p = 4 - f$. The minimum f is 0, so the maximum p is 4 in a binary system.

12.29 **(a)** Yes. The horizontal tie line at T_1 intersects the boundaries of the two-phase region at $x_B^l = 0.23$ (point L) and at $x_B^v = 0.68$ (point Q).

(b) No. The overall mole fraction is different for different points on the tie line.

12.30 As noted in Sec. 12.6, the upper curve is the P-vs.-x_B^l curve and the lower curve is the P-vs.-x_B^v curve. We take B as acetone. Table 10.1 gives P and the x_B^l and x_B^v values that correspond to each P value. Plotting these points, we get a phase diagram with a minimum in P at $x_{acetone} = 0.4$.

12.31 The width between the vertical axes is 60½ mm and $x_B = 0.3$ lies at 18 mm. We draw a vertical line at 18 mm, and then a horizontal line that goes from the liquid line at 18 mm to the vapor line; the intersection with the vapor line is at 45½ mm (corresponding to $x_B = 0.75$) and gives the result of the first distillation step; drawing a horizontal line from the liquid line at 45½ mm, we find it intersects the vapor line at 58½ mm, corresponding to $x_B = 0.97$.

12.32 **(a)** The width of the figure is 60 mm and $x = 0.72$ corresponds to $43._2$ mm. Drawing a vertical line at 43 mm and then a horizontal line from the intersection of this vertical line with the liquid line, we find the horizontal line intersects the vapor line at 52½ mm, which corresponds to $x_B^v = 52½/60 = 0.87_5$.

(b) x_B^l is given by point D as (36 mm)/(60 mm) = 0.60. Point G gives x_B^v = (48½ mm)/(60 mm) = 0.81.

12.33 The point giving the system's state lies on the horizontal line at T_1 and cannot lie to the left of point Q, since all points to the left of Q have some liquid present. Point Q corresponds to x_B = (41 mm)/(60½ mm) = 0.68, so $x_B = n_B/(n_B + n_C) \geq 0.68$; $n_B \geq 0.68n_B + 0.68n_C$; $n_C \leq 0.47n_B = 0.47(2.00 \text{ mol}) =$ 0.94 mol. $n_C \leq 0.94$ mol.

12.34 $x_{B,overall} = 4.00/7.00 = 0.571$. Point L gives $x_B^l = (14 \text{ mm})/(60\frac{1}{2} \text{ mm}) = 0.23$. Point Q gives $x_B^v = (41 \text{ mm})/(60\frac{1}{2} \text{ mm}) = 0.68$. The lever rule gives $n^l(0.57_1 - 0.23) = n^v(0.68 - 0.57_1)$ and $n^v = 3.1n^\ell = 3.1(7.00 \text{ mol} - n^v)$ and $n^v = 5.3$ mol; $n^l = 7.00 \text{ mol} - 5.3 \text{ mol} = 1.7$ mol. $n_B^l = 0.23(1.7 \text{ mol}) = 0.39$ mol; $n_C^l = (1.7 - 0.39) \text{ mol} = 1.3$ mol. $n_B^v = 0.68(5.3 \text{ mol}) = 3.6$ mol; $n_C^v = 1.7$ mol.

12.35 $P = P_{ben} + P_{tol} = x_{ben}P^*_{ben} + x_{tol}P^*_{tol} = x_{ben}P^*_{ben} + (1 - x_{ben})P^*_{tol} = P^*_{tol} + (P^*_{ben} - P^*_{tol})x_{ben}$. The graph is linear, so we need only two points to plot it: $P = P^*_{tol} = 22.3$ torr at $x_{ben} = 0$ and $P = P^*_{ben} = 74.7$ torr at $x_{ben} = 1$.

12.36 **(a)** As found in the Prob. 12.31 solution, the first vapor has $x_B^v = 0.75$.

(b) A horizontal line starting from $x_B^v = 0.30$ intersects the liquid line at $x_B^l = 0.03$.

(c) When $n^l = n^v$, the horizontal tie line is bisected by the vertical line at $x_B = 0.30$. By trial and error, we find that the tie line with equal halves is the one that joins $x_B^l = 0.11$ with $x_B^v = 0.50$.

12.37 Regina is correct. Going horizontally in the two-phase region varies the overall mole fraction (whose value depends on the sizes of the two phases present), but does not change the mole fractions in either of the phases present.

12.38 A horizontal line at 80°C in Fig. 12.17b intersects the curve at w_{nic} values of 0.07_6 for phase α and 0.69 for phase β. We have $10 \text{ g} = 0.07_6 m^\alpha + 0.69(20 \text{ g} - m^\alpha)$ and $m^\alpha = 6.2$ g; then $m^\beta = 20 \text{ g} - 6.2 \text{ g} = 13.8$ g. (Alternatively, the lever rule could be used.) The α phase therefore has $0.07_6(6.2 \text{ g}) = 0.5$ g of nicotine and 5.7 g of water. The β phase has $0.69(13.8 \text{ g}) = 9.5$ g of nicotine and 4.3 g of water.

12.39 Drawing a horizontal tie line at 80°C, we find it intersects the water-poor portion of the curve at $w_{nic} = 0.69$. Drawing the vertical line at weight fraction $w_{nic} = 0.5$ and using the lever rule (12.41), we find

$(20 \text{ g})(17\frac{1}{2} \text{ mm}) = m^{wp}(7\frac{3}{4} \text{ mm})$ and the mass of the water-poor phase is $m^{wp} = 45$ g. The mass of nicotine in the water-poor phase is 0.69(45 g) = 31 g and the mass of water in this phase is 45 g – 31 g = 14 g.

12.40 **(a)** Let α be the water-poor phase and β the water-rich phase. $(0.400 - 0.375)m^\alpha = (0.89 - 0.40)m^\beta = (0.89 - 0.40)(10.0 \text{ g} - m^\alpha)$; $0.025m^\alpha = 4.90 \text{ g} - 0.49m^\alpha$ and $m^\alpha = 9.5_1$g, $m^\beta = 0.49$ g. Then $m^\alpha_{H_2O} = 0.375(9.5_1 \text{ g}) = 3.6$ g, $m^\alpha_{\text{phenol}} = (9.5 - 3.6) \text{ g} = 5.9$ g; $m^\beta_{H_2O} = 0.89(0.4_9 \text{ g}) = 0.4_4$ g; $m^\beta_{\text{phenol}} = 0.05$ g.

(b) $0.375 = m^\alpha_{H_2O}/m^\alpha$, $0.89 = m^\beta_{H_2O}/m^\beta = (4.00 \text{ g} - m^\alpha_{H_2O})/(10.0 \text{ g} - m^\alpha)$. The first equation gives $m^\alpha_{H_2O} = 0.375m^\alpha$ and substitution in the second equation gives $0.89 = (4.00 \text{ g} - 0.375m^\alpha)/(10.00 \text{ g} - m^\alpha)$. Solving, we get $m^\alpha = 9.5$ g, and the problem is completed as in (a).

12.41 Let α be the octanol-rich phase. We assume the small amount of DDT does not affect the water (w) and octanol (oc) mole fractions in the phases. We use the mole fractions given on p. 372 of the text. We have a total of 4.44 mol of water and 0.076_8 mol of octanol ($C_8H_{17}OH$, molecular weight 130.2), for a total number of moles of 4.51_8 mol. The total moles of water equals the sum of the water moles in each phase, so $4.44 \text{ mol} = (1 - 0.793)n^\alpha + 0.993(4.51_8 - n^\alpha)$ and $n^\alpha = 0.057_7$ mol. So $n^\beta = 4.46$ mol. Then $n^\alpha_w = 0.207(0.0577 \text{ mol}) = 0.0119$ mol, and $m^\alpha_w = 0.215$ g. We are assuming no volume change on mixing, so we shall take the volume of each phase as the sum of the volumes of the water and octanol that went into that phase (neglecting the small contribution of DDT). The contribution of water (density 1.00 g/cm^3) to the volume of α is thus 0.22 cm^3. We have $n^\alpha_{oc} = (0.793)(0.0577 \text{ mol}) = 0.045_8$ mol, and $m^\alpha_{oc} = 5.96$ g; from the density, the volume of this octanol is 7.1_8 cm^3. The volume of phase α is then $V^\alpha = (0.22 + 7.18) \text{ cm}^3 = 7.4 \text{ cm}^3$. For phase β, $n^\beta_w = (0.993)(4.46 \text{ mol}) = 4.43$ mol; $m^\beta_w = 79.8$ g; $V^\beta_w = 79.8 \text{ cm}^3$. Also, $n^\beta_{oc} = (0.007)(4.46 \text{ mol}) = 0.031$ mol; $m^\beta_{oc} = 4.0_6$ g; $V^\beta_{oc} = 4.9 \text{ cm}^3$. Then $V^\beta = 84.7$ cm^3. The 0.100 g of DDT is 0.000282 mol. We have $K_{ow,DDT} = 8.1 \times 10^6 =$

$$\frac{c^{\alpha}_{\text{DDT}}}{c^{\beta}_{\text{DDT}}} = \frac{n^{\alpha}_{\text{DDT}}/V^{\alpha}}{n^{\beta}_{\text{DDT}}/V^{\beta}} = \frac{(0.000282 - n^{\beta}_{\text{DDT}})/(7.4\text{ cm}^3)}{n^{\beta}_{\text{DDT}}/(84.7\text{ cm}^3)} \text{ and } n^{\beta}_{\text{DDT}} = 4.0 \times 10^{-10}\text{ mol,}$$

which is 1.4×10^{-7} g. $n^{\alpha}_{\text{DDT}} = 0.000282$ mol, which is 100 mg.

12.42 We have $K_{\text{ow,napth}} = 2.0 \times 10^3$. In Eq. (12.45), α corresponds to the octanol-rich phase. The quantity in parentheses in the equation preceding (12.45) is the desired $\Delta G°$. If we make the approximation that the activity coefficient ratio is approximately equal to 1 (which may not be accurate), we have $\Delta G^{\circ}_{c} \approx -RT \ln K_{\text{ow,napth}} = -(8.314\text{ J/mol-K})(298\text{ K})3.30 = -8.2\text{ kJ/mol}$.

12.43 The diagram should bear some resemblance to Fig. 12.19. The phases present in each area are given in Fig. 12.19. On the horizontal line, three phases are present: pure solid B, pure solid C, and liquid solution. On this line, f is zero. In the two-phase areas, f is 1 (the temperature). In the one-phase liquid-solution area, f is 2.

12.44 Let B = benzene and A = cyclohexane. Equation (12.46) becomes $T_{\ell} \approx T^{*}_{B}/[1 - RT^{*}_{B}(\Delta_{\text{fus}}H_{\text{m,B}})^{-1}\ln x_{\text{B}}] = (278.6\text{ K})/(1 - 0.233 \ln x_{\text{B}})$ and $T_r \approx (279.7\text{ K})/(1 - 0.884 \ln x_{\text{A}})$, where T_{ℓ} and T_r are the left- and right-hand curves in Fig. 12.19. We get

x_{A}	0	0.1	0.2	0.3	0.4	0.5	0.6
T_{ℓ}	278.6	271.9	264.8	257.2	249.0	239.9	229.6
T_r							192.7

x_{A}	0.7	0.8	0.9	1
T_{ℓ}	217.6	202.6	181.3	
T_r	212.7	233.6	255.9	279.7

Plotting the curves, we find they intersect at 71½ mole percent cyclohexane and –58°C.

12.45 The halt at 84.3 wt. % Zn must correspond to a compound with melting point 595°C. (It could not be a eutectic halt since its temperature is too high for this.) The compound's empirical formula is found as follows: (84.3 g)/(65.38 g/mol)

= 1.289 mol Zn; (15.7 g)/(24.305 g/mol) = 0.646 mol Mg. The Zn:Mg mole ratio is 2:1 and the compound's empirical formula is $MgZn_2$. The phase diagram is of the type shown in Fig. 12.25. The eutectic temperatures are 345°C and 368°C. One eutectic composition is 97 wt. % Zn (corresponding to 368°C). It is difficult to tell which curve the reading at 50% belongs to, so all that can be said is that the second eutectic composition is close to 50 wt. % Zn.

12.46 **(a)** The phase diagram is the same type as Fig. 12.26c with line PM lying at 0.1°C. The Liquid (ℓ) is a solution of salt in water, B is H_2O, the compound is $NaCl \cdot 2H_2O$, and A is NaCl. Since the diagram goes up to only 100°C, the right-hand melting-point curve (which is almost vertical below 100°C) does not reach the NaCl axis. The compound lies at 62 wt. % NaCl.

(b) 20°C lies above the peritectic temperature of 0.1°C. As the solution evaporates, we move horizontally to the right on the diagram, eventually reaching the ℓ + A (solution + solid NaCl) region; eventually, pure NaCl is obtained.

(c) This system lies initially in the ℓ + A (solution + solid NaCl) region. When the peritectic temperature of 0.1°C is reached, solid $NaCl \cdot 2H_2O$ begins to form. The 80 wt. % composition lies to the right of MN in Fig. 12.26c, and the system stays at 0.1°C until all the liquid disappears, leaving a mixture of the two solids NaCl and $NaCl \cdot 2H_2O$. Then this mixture is cooled to –10°C. No solid H_2O (ice) is present at –10°C.

12.47 A cooling curve that corresponds to l.s. → l.s. + α → α → $\alpha + \beta$ shows three breaks. A cooling curve that corresponds to l.s. → l.s. + α → l.s. + $\alpha + \beta$ → $\alpha + \beta$ shows a break followed by a eutectic halt. A cooling curve at the eutectic composition shows no break and one halt. Similarly for curves to the right of the eutectic composition.

12.48 For $\ell \to \ell + B \to B + A_2B + \ell \to B + A_2B$, a break followed by a (eutectic) halt. For cooling a liquid with the eutectic composition, a halt. For $\ell \to \ell + A_2B \to \ell + B + A_2B \to B + A_2B$, a break and then a halt. For $\ell \to \ell + A \to \ell + A + A_2B \to \ell + A_2B \to \ell + B + A_2B \to B + A_2B$, break, halt, halt. For $\ell \to \ell + A \to \ell + A_2B + A \to A_2B$, break, halt. For $\ell \to \ell + A \to \ell + A + A_2B \to A + A_2B$, break, halt.

12.49 Just as Fig. 12.25 resembles two Fig. 12.19 diagrams placed side by side, the Bi-Te diagram resembles two Fig. 12.22 diagrams placed side by side:

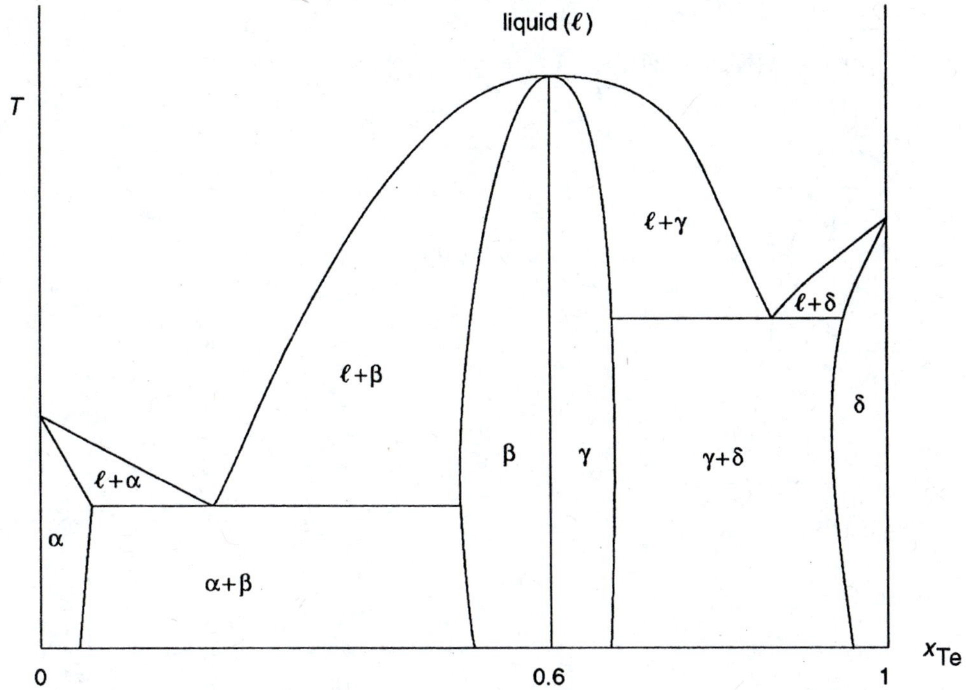

α is a solid solution of Bi_2Te_3 in Bi. β is a solid solution of Bi in Bi_2Te_3. γ is a solid solution of Te in Bi_2Te_3. δ is a solid solution of Bi_2Te_3 in Te.

12.50 The intersection of the miscibility gap with the phase transition loop does not include the composition corresponding to the minimum, so we get a

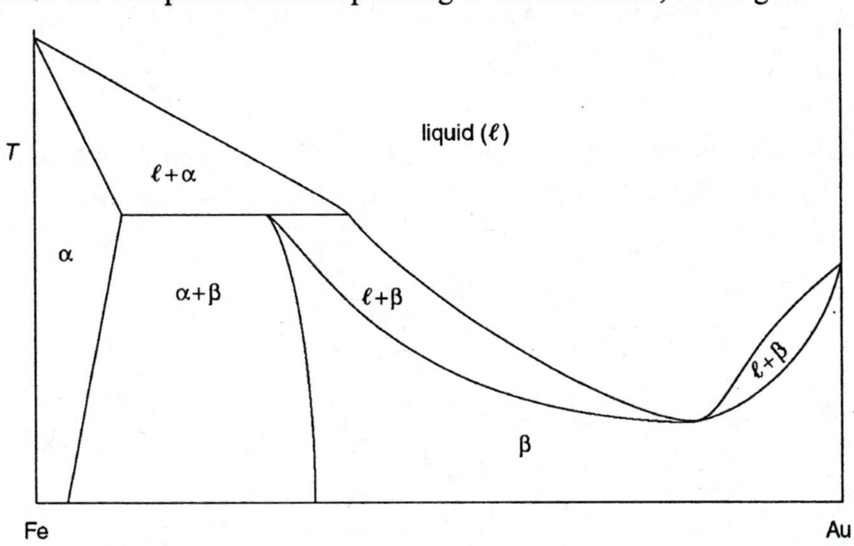

combination of Figs. 12.24 and 12.21. Phase α is a solution of Au in Fe. Phase β is a solution of Fe in Au.

12.51 Let A = $HNO_3 \cdot 3H_2O$ and B = $HNO_3 \cdot H_2O$.

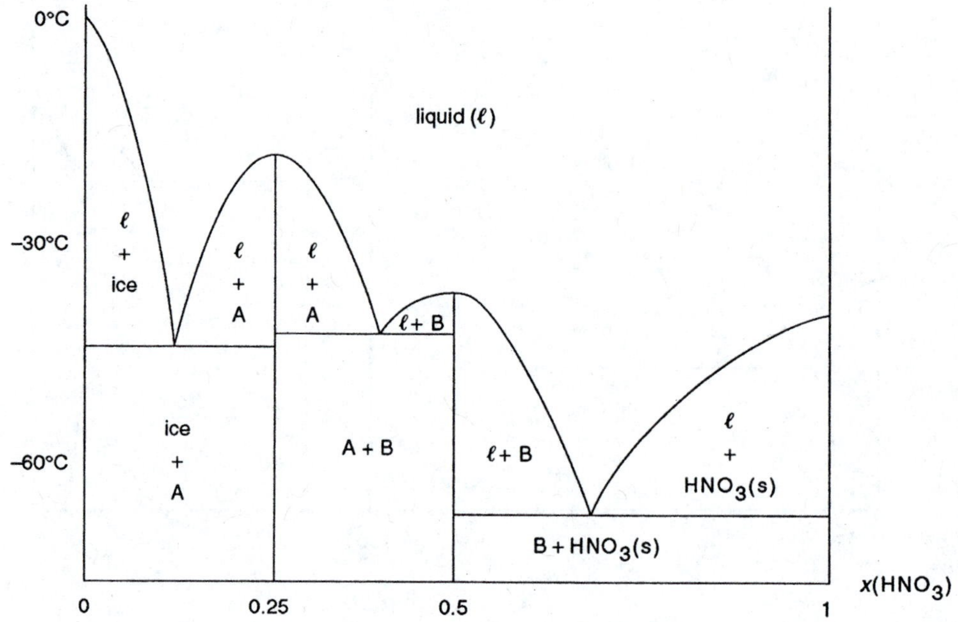

12.52 **(a)** Equation (12.46) gives $\ln x_B = [(147 \text{ J/g})(128.2 \text{ g/mol})/(8.314 \text{ J/mol-K})][1/(353 \text{ K}) - 1/(298 \text{ K})] = -1.185$ and $x_B = 0.306$.

(b) The same as (a); $x_B = 0.306$.

(c) $\ln x_B = [(162 \text{ J/g})(178.2 \text{ g/mol})/(8.314 \text{ J/mol-K})][1/(489 \text{ K}) - 1/(333 \text{ K})] = -3.32_6$ and $x_B = 0.036$.

12.53 (1000 g)(1 mol/78.1 g) = 12.8 mol. The 52½°C point on the solubility curve CE is at $x_{ben} = 0.4_7 = (12.8 \text{ mol})/(12.8 \text{ mol} + n_{naph})$. We get n_{naph} = 14 mol.

12.54

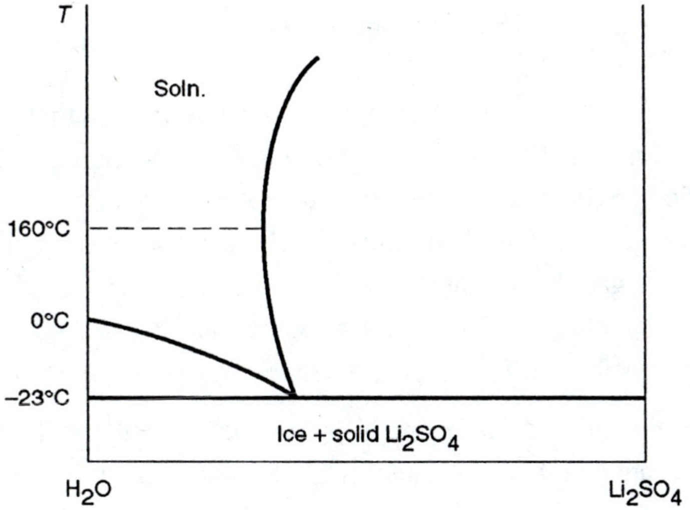

12.55 We change the formula in cell H1, choose a temperature and initial guesses for cells D2 and D3, and run the Solver. Repeating at several temperatures, one finds that at each temperature, the two phases have equal mole fractions, indicating complete miscibility at all temperatures.

12.56 We change the formula in cell H1, choose a temperature and initial guesses for cells D2 and D3, and run the Solver. Repeating at several temperatures, one finds that there is partial miscibility at 240 K and at temperatures below 240 K, and there is complete miscibility at 260 K. Trying various temperatures between 240 K and 260 K, we get the critical solution temperature as 254 K.

12.57 **(a)** We get a *lower* critical solution temperature of about 200 K. (Note that as *T* decreases, *W* becomes more negative, favoring miscibility (Prob. 12.55).

(b) We get a lower critical solution temperature of about 265 K.

12.58 The relation is quite simple. In (b), we get a lower critical solution temperature of about 174 K, and this satisfies the relation.

12.59 When (12.48) and the three other similar equations for the other three chemical potentials are substituted into (12.47), the terms that involve μ_D^* and μ_E^* are

$n_D^{\alpha}\mu_D^* + n_E^{\alpha}\mu_E^* + n_D^{\beta}\mu_D^* + n_E^{\beta}\mu_E^* = (n_D^{\alpha} + n_D^{\beta})\mu_D^* + (n_E^{\alpha} + n_E^{\beta})\mu_E^*$, and the total number of moles of D and of E in the system are constant.

12.60 **(a)** Higher acetone concentrations correspond to tie lines that are closer to point K in Fig. 12.33. Consider moving from the tie line below FH to the FH tie line. In doing so, the left end (water-rich phase) of the tie line moves up a smaller amount than the right end (ether-rich phase) does, so there is a greater enrichment of acetone in the ether-rich phase than in the water-rich phase as the acetone concentration increases. Hence $K_{ew,ac}$ increases, since $c_{ac}^{\text{ether-rich}}$ is in the numerator of $K_{ew,ac}$.

(b) The right end of each tie line is higher than the left end, meaning a greater concentration of acetone in the ether-rich phase, so $K_{ew,ac} > 1$.

12.61 Pure water (wat) lies at the point $x_{wat} = 1$, $x_{eth} = 0$; pure ether (eth) lies at $x_{wat} = 0$, $x_{eth} = 1$; pure acetone (ac) lies at $x_{wat} = 0$, $x_{eth} = 0$. States with $x_{ac} = 0$ correspond to the line $x_{wat} + x_{eth} = 1$ or $x_{eth} = 1 - x_{wat}$. This line has intercept 1 and slope -1; the two-phase region is bounded on one side by part of this line. The plait point K is richer in water than in ether. The phase diagram is

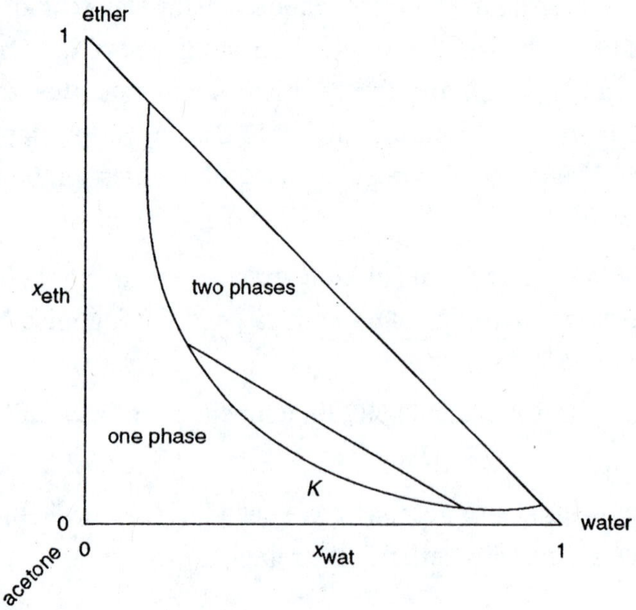

12.62 Draw $\overline{DA}$, $\overline{DB}$, and $\overline{DC}$ in Fig. 12.32a. Let Ar(ABC) be the area of triangle ABC and let h be the altitude of this triangle. We have $\frac{1}{2}h\overline{BC}$ = Ar(ABC) = Ar(ABD) + Ar(BDC) + Ar(CDA) = $\frac{1}{2}(\overline{AB}\cdot\overline{DG}) + \frac{1}{2}(\overline{BC}\cdot\overline{DE}) + \frac{1}{2}(\overline{CA}\cdot\overline{DF}) = \frac{1}{2}\overline{BC}(\overline{DG} + \overline{DE} + \overline{DF})$. Hence $h = \overline{DG} + \overline{DE} + \overline{DF}$.

12.63 **(a)** We drop perpendiculars from point F to each of the three sides of the triangle and take the ratio of each perpendicular distance to the triangle's height. We get for phase F: $x_{ac} = 0.06$, $x_{wat} = 0.92$, $x_{eth} = 0.02$. Dropping perpendiculars from point H, we get the phase-H composition as: $x_{ac} = 0.23$, $x_{wat} = 0.05$, $x_{eth} = 0.72$.

(b) $\overline{FG}$ = 30¼ mm and $\overline{GH}$ = 22½ mm. The lever rule gives 30¼ n_F = 22½ n_H = 22½ (40 mol – n_F); then n_F = 17 mol and n_H = 23 mol. Phase F has 0.06(17 mol) = 1.0 mol of acetone, 0.92(17 mol) = $15._6$ mol of water, and 0.02(17 mol) = 0.3_4 mol of ether. Phase H has $5._3$ mol of acetone, $1._2$ mol of water, and $16._6$ mol of ether.

12.64 **(a)** The diagram resembles Fig. 12.33 with ether replaced by ethyl acetate.

(b) The overall composition is $x_{ac} = 0.20$, $x_{ethyl} = 0.40$, $x_{wat} = 0.40$. This point lies in the two-phase region and is just a bit below one of the tie lines graphed in (a). Drawing in a tie line through this point, we find it intersects the binodal curve at points corresponding to the following compositions. Phase α: $x^{\alpha}_{wat} = 0.91_5$, $x^{\alpha}_{ac} = 0.06_2$, $x^{\alpha}_{ethyl} = 0.02_3$. Phase β: $x^{\beta}_{wat} = 0.26_0$, $x^{\beta}_{ac} = 0.23_8$, $x^{\beta}_{ethyl} = 0.50_2$. We have $0.91_5 n^{\alpha} + 0.26_0(0.50 \text{ mol} - n^{\alpha}) = 0.20$ mol; so $n^{\alpha} = 0.107$ mol and $n^{\beta} = 0.393$ mol. (Alternatively, the lever rule could be used.) Then n^{α}_{wat} = 0.91_5(0.107 mol) = 0.098 mol, $n^{\alpha}_{ac} = 0.0066$ mol, $n^{\alpha}_{ethyl} = 0.0025$ mol; $n^{\beta}_{wat} = 0.102$ mol, $n^{\beta}_{ac} = 0.093_5$ mol, $n^{\beta}_{ethyl} = 0.197$ mol. Multiplication by the molar masses gives $m^{\alpha}_{wat} = 1.76$ g, $m^{\alpha}_{ac} = 0.38$ g, $m^{\alpha}_{ethyl} = 0.22$ g; $m^{\beta}_{wat} = 1.84$ g, $m^{\beta}_{ac} = 5.43$ g, $m^{\beta}_{ethyl} = 17.4$ g.

12.65 The overall composition of the system is given by point G, so the overall mole fraction x_A of A equals the distance $\overline{GS}$ to the side opposite vertex A. Points F and H give the compositions of the two phases F and H in equilibrium, and $x_A^F = \overline{FR}$, $x_A^H = \overline{HT}$. Equation (12.49) with α = F, β = H becomes $n^F(\overline{GS} - \overline{FR}) = n^H(\overline{HT} - \overline{GS})$ and $\boxed{n^F/n^H = \overline{HK}/\overline{GM}}$. The figure gives sin θ $= \overline{HK}/\overline{GH} = \overline{GM}/\overline{FG}$. So $\overline{HK}/\overline{GM} = \overline{GH}/\overline{FG}$ and the boxed equation becomes $n^F/n^H = \overline{GH}/\overline{FG}$ or $\overline{FG}n^F = \overline{GH}n^H$.

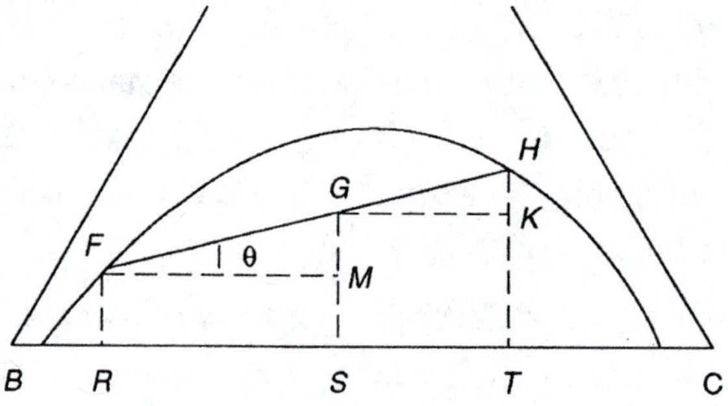

12.66 **(a)** Since the solution in beaker B has a lower vapor pressure than that in A, the evaporation rate from the A solution exceeds that from the B solution and the equilibrium state will have beaker A empty, its liquid having vaporized from A and condensed in B. The vapor pressure in the box will be that of the diluted salt solution in B.

(b) The vapor pressure of the B solution is lower than that of the A solution. Liquid will evaporate from A and condense in B until the sucrose molalities are equalized. Let a mass z of H_2O vaporize from A. Equating molalities, we have (0.01 mol)/(100 g – z) = (0.03 mol)/(100 g + z) and z = 50 g. Beaker A ends up with 50 g of water and 0.01 mol of sucrose; beaker B ends up with 150 g of water and 0.03 mol of sucrose.

12.67 The diagram has the same appearance as Fig. 12.22. With x_A going from 0 to 1 on the horizontal axis, the areas are relabeled as follows: l.s. becomes vapor, l.s. + α becomes vapor + α (where α is a dilute solution of liquid A in the solvent liquid B), l.s. + β becomes vapor + β (where β is a dilute solution of liquid B in liquid A). The horizontal line is a mixture of phase α, vapor, and phase β. The region $\alpha + \beta$ is a mixture of the two immiscible liquids α and β.

12.68 Yes. For example, an endothermic chemical reaction might occur. Another example is adding a salt to a mixture of water and ice at 0°C (see Sec. 12.10).

12.69 **(a)** $c = 2, p = 1$; $f = c - p + 2 - r - a = 2 - 1 + 2 - 0 - 0 = 3$, but since P is held fixed, there are two degrees of freedom (temperature and x_A).

(b) $c = 2, p = 2$; $f = 2 - 2 + 2 - 0 - 0 = 2$, but since P is held fixed, there is one degree of freedom. Once T is fixed, the mole-fraction composition of each phase is fixed.

(c) $c = 2, p = 3$; $f = 2 - 3 + 2 - 0 - 0 = 1$, but since P is held fixed, there are no degrees of freedom. All intensive variables (T, P, mole fractions) are fixed.

12.70 Assuming ideal vapor, we have $P_i \equiv x_{i,\text{vap}}P = \gamma_{I,i}x_i^{\ell}P_i^*$. But for an azeotrope, $x_i^{\upsilon} = x_i^l$, so $x_i^lP = \gamma_{I,i}x_i^lP_i^*$ and $\gamma_{I,i} = P/P_i^*$. $\gamma_{I,\text{ethanol}} = (760 \text{ torr})/(581 \text{ torr}) = 1.31$ and $\gamma_{I,\text{ethyl acetate}} = (760 \text{ torr})/(631 \text{ torr}) = 1.20$.

12.71 **(a)** For each substance, $\mu_i^{\alpha} = \mu_i^{\beta}$ and $\mu_i^* + RT\ln \gamma_{I,i}^{\alpha}x_i^{\alpha} = \mu_i^* + RT\ln \gamma_{I,i}^{\beta}x_i^{\beta}$, so $\gamma_{I,i}^{\alpha}x_i^{\alpha} = \gamma_{I,i}^{\beta}x_i^{\beta}$. For phase α, x_w is very close to 1 and $\gamma_{I,w}^{\alpha} \cong 1$. Hence $(1)x_w^{\alpha} = \gamma_{I,w}^{\beta}x_w^{\beta}$ and $\gamma_{I,w}^{\beta} = x_w^{\alpha}/x_w^{\beta} = 0.999595/0.00300 = 333$. For phase β, $\gamma_{I,\text{ben}}^{\beta} \cong 1$ and $\gamma_{I,\text{ben}}^{\alpha}x_{\text{ben}}^{\alpha} = (1)x_{\text{ben}}^{\beta}$, so $\gamma_{I,\text{ben}}^{\alpha} = x_{\text{ben}}^{\beta}/x_{\text{ben}}^{\alpha} = 0.99700/0.000405 = 2460$.

(b) This solution is phase β. With ideal vapor assumed, $P_i = \gamma_{I,i}x_iP_i^*$, so $P_{\text{ben}} = 1(0.997)(95.2 \text{ torr}) = 94.9$ torr and $P_w = 333(0.00300)(23.8 \text{ torr}) = 23.8$ torr. $P = P_{\text{ben}} + P_w = 118.7$ torr.

(c) This solution is phase α. $P_{\text{ben}} = 2460(0.000405)(95.2 \text{ torr}) = 94.8$ torr; $P_w = 1(0.999595)(23.8 \text{ torr}) = 23.8$ torr. $P = 118.6$ torr.

12.72 Let b stand for benzene and α and β be the two liquid phases. $\mu_b^{\alpha} = \mu_b^{\beta}$ and $\mu_b^* + RT\ln \gamma_{I,b}^{\alpha}x_b^{\alpha} = \mu_b^* + RT\ln \gamma_{I,b}^{\beta}x_b^{\beta}$, so $\gamma_{I,b}^{\alpha}x_b^{\alpha} = \gamma_{I,b}^{\beta}x_b^{\beta}$. Above solution α, $P_b^{\alpha} = \gamma_{I,b}^{\alpha}x_b^{\alpha}P_b^*$. Above solution β, $P_b^{\beta} = \gamma_{I,b}^{\beta}x_b^{\beta}P_b^*$. Use of $\gamma_{I,b}^{\alpha}x_b^{\alpha} = \gamma_{I,b}^{\beta}x_b^{\beta}$ gives $P_b^{\alpha} = P_b^{\beta}$.

12.73 Assuming an ideally dilute solution, we have $-0.64°\text{C} \approx -k_fm_B$ and $m_B \approx (0.64 \text{ K})/(1.86 \text{ K kg/mol}) = 0.34_4$ mol/kg.

(a) $\Delta T_b \approx k_b m_b = (0.513 \text{ K kg/mol})(0.34_4 \text{ mol/kg}) = 0.17_6$ K. The normal boiling point of water is 99.974°C (Sec. 1.5 and Fig. 7.1), so $T_b \approx$ 100.15°C.

(b) $P = P_A \approx x_A P_A^*$. We have $m_B = n_B/w_A = n_B/n_A M_A \approx x_B/M_A$ since the solution is dilute. Then $x_B \approx m_B M_A = (0.34_4 \text{ mol/kg})(0.01801 \text{ kg/mol}) =$ 0.0062. So $P \approx 0.9938(23.76 \text{ torr}) = 23.61$ torr.

(c) $\Pi \approx x_B RT/V^*_{m,A} = 0.0062(82.06 \text{ cm}^3 \text{ atm/mol-K})(293 \text{ K})/(18.0 \text{ cm}^3\text{/mol})$ = 8.3 atm.

12.74 **(a)** Same. **(b)** Same. **(c)** Same. **(d)** Same. **(e)** Different. **(f)** Different.

12.75 **(a)** False. See Fig. 12.20.

(b) True. See Fig. 12.2.

(c) False. See line FHK in Fig. 12.19.

(d) False. The eutectic solution in Fig. 12.19 freezes entirely at one temperature.

(e) True as x_B increases in the solution, μ_B in the solution must increase [Eq. (4.90)]. So μ_B^υ in the vapor in equilibrium with the solution must increase. Since $\mu_B^\upsilon = \mu_B^{\circ,\upsilon}(T) + RT \ln (P_B/P^\circ)$, if μ_B^υ increases at constant T, then P_B must increase.

(f) True. **(g)** False. **(h)** True. **(i)** False. **(j)** True. **(k)** False.

Reminder: Use this manual to check your work, and not to avoid working the problems.

R12.1 $I_m = \frac{1}{2}\sum_i z_i^2 m_i =$

$\frac{1}{2}[2^2(0.040) + (-1)^2(0.080) + 0.060 + 0.060]$ mol/kg $= 0.18$ mol/kg.

$$\log \gamma_\pm = -0.51(2)(1)\left[\frac{(0.18)^{1/2}}{1+(0.18)^{1/2}} - 0.30(0.18)\right] = -0.124,\ \gamma_\pm = 0.751.$$

R12.2 The standard state for species in liquid solutions was chosen to be at the pressure of the solution; consequently, ΔG° and the equilibrium constant for such reactions depend on P. Therefore the equilibrium constants for (d) and (e) depend on pressure.

R12.3 **(a)** If the system is treated as an ideal-gas reaction, there is no shift when P is changed because there is no change in total number of moles of gas. If nonideality is allowed for, the change in P will change the intermolecular interactions and will change the fugacity coefficients, and so the equilibrium composition will change with pressure.

(b) The total moles of product gases differs from the total moles of reactant gases and the equilibrium composition changes when P changes.

(c) Suppose we consider the gas as ideal and neglect the change in activity of the solids with change in P. Then Eq. (11.25) gives $K^\circ = P_{CO_2}/P^\circ$ If we isothermally increase the pressure by decreasing the volume, then, since K° depends on T only and T is constant, the equilibrium will shift to the left to reduce P_{CO_2} to its previous value. Hence the numbers of moles change when P is changed.

(d) As discussed in Prob. R12.2, the equilibrium constant depends on P and a change in P changes the equilibrium composition. This change is small unless the change in P is large.

(e) Same as (d).

R12.4 **(a)** See Eq. (6.4) with $P_i = x_i P$.

(b) See Eq. (10.97) with $f_i = \phi_i x_i P$.

(c) See Eq. (9.42).

(d) See Eqs. (9.58) and (9.59).

(e) See Eq. (10.6).

R12.5 **(a)** The negative term is eliminated when $x_R = 0.40$, and $t_{max} = 63°C$ (the critical solution temperature). At the maximum, the slope is zero and $dt/dx = 0 = -1440(x_R - 0.40)$ and $x_R = 0.40$.

(b) $25 = 63 - 720(x_R - 0.40)^2$; $x_R - 0.40 = \pm(38/720)^{1/2} = \pm 0.23$; $x_R = 0.63$ and 0.17.

(c) $n_{tot} = 5.08$ mol. $n_R = n_R^\alpha + n_R^\beta$ and
$2.30 \text{ mol} = 0.63n^\alpha + 0.17(5.08 \text{ mol} - n^\alpha)$; $n^\alpha = 3.12$ mol;
$n^\beta = 1.96$ mol. $n_R^\alpha = 0.63n^\alpha = 1.97$; $n_R^\beta = 0.17n^\beta = 0.33$;
$n_S^\alpha = 0.37n^\alpha = 1.15$; $n_S^\beta = 0.83n^\beta = 1.63$.

(d) To produce a one-phase system, we must increase the overall mole fraction of S to make it equal to $1 - 0.17 = 0.83$. If z is the number of moles of S to add, then $0.83 = (2.78 \text{ mol} + z)/(5.08 \text{ mol} + z)$ and $z = 8.45$ mol.

R12.6 (a) With activity coefficients neglected, $6.8 \times 10^{-4} = x^2/(0.15 - x)$ and $x =$ 0.0098 (mol/kg). The initial estimate of the ionic strength is 0.0098 mol/kg and the Davies equation gives $\log\gamma_\pm = -0.0444$ and $\gamma_\pm = 0.903$. Then $6.8 \times 10^{-4} = (0.903)^2x^2/(0.15 - x)$ and $x = 0.0108$. For an ionic strength of 0.0108 mol/kg, the Davies equation gives $\log\gamma_\pm = -0.0464$ and $\gamma_\pm = 0.899$. Then $6.8 \times 10^{-4} = (0.899)^2x^2/(0.15 - x)$ and $x = 0.0108$. The H^+ molality is 0.0108 mol/kg. From Prob. 11.21,
$c(H^+) = m(H^+)\rho_{H_2O} = (0.0108 \text{ mol/kg})(0.997 \text{ kg/L}) = 0.0108 \text{ mol/L}$.

(b) The initial estimate of the ionic strength is 0.20 mol/kg from the KCl. The Davies equation gives $\log\gamma_\pm = -0.127$ and $\gamma_\pm = 0.746$. Then $6.8 \times 10^{-4} = (0.746)^2x^2/(0.15 - x)$ and $x = 0.0129$. The improved estimate of the ionic strength is 0.213, and the Davies equation gives $\log\gamma_\pm = -0.128_5$; $\gamma_\pm = 0.744$. Then $6.8 \times 10^{-4} = (0.744)^2x^2/(0.15 - x)$ and $x = 0.0130$. The H^+ molality is 0.0130 mol/kg. Then
$c(H^+) = m(H^+)\rho_{H_2O} = (0.0130 \text{ mol/kg})(0.997 \text{ kg/L}) = 0.0130 \text{ mol/L}$.

R12.7 $\gamma_\pm^3 = \gamma_+\gamma_-^2$.

R12.8 If the vapor is assumed ideal, then $P_i \equiv x_i^v P = \gamma_{I,i} x_i^l P_i^*$. Then
$0.194(286 \text{ torr}) = \gamma_{I,chl}(0.291)(293 \text{ torr})$ and $\gamma_{I,chl} = 0.651$.
$(1 - 0.194)(286 \text{ torr}) = \gamma_{I,ac}(1 - 0.291)(344 \text{ torr})$ and $\gamma_{I,ac} = 0.945$.

R12.9 **(a)** $a_{m,i} = \gamma_i m_i / m^\circ$.

(b) $a_{c,i} = \gamma_i c_i / c^\circ$.

(c) $a_{\text{II},i} = \gamma_{\text{II},i} x_i$.

R12.10 See Fig. 12.19 and the accompanying discussion.

R12.11 $k_f = M_A R(T_f^*)^2/\Delta_{fus}H_{m,A} =$

(0.06207 kg/mol)(8.314 J/mol-K)(260.6 K)^2/(9960 J/mol) = 3.519 K kg/mol

(258.8 K – 260.6 K) = –(3.519 K kg/mol)m_Y and m_Y = 0.51$_2$ mol/kg.

n_Y = (0.51$_2$ mol/kg)(0.0864 kg) = 0.044$_2$ mol. M_Y = (4.25 g)/(0.044$_2$ mol) = 96 g/mol and the approximate molecular weight is 96.

R12.12 The mole fractions in the two phases in equilibrium in the miscibility gap region are $x_B^\alpha = 0.125/(1 + 0.125) = 0.111$ and $x_B^\beta = 3.00/4.00 = 0.750$. The system has $x_{B,\text{overall}} = 0.556$ and so lies in the two-phase region. The total moles of B is the sum of the B moles in each phase, so 2.50 mol = $0.111n^\alpha + 0.750(4.50\text{ mol} - n^\alpha)$ so $n^\alpha = 1.37$ mol and $n^\beta = 3.13$ mol. Then $n_B^\alpha = 0.111(1.37\text{ mol}) = 0.152$ mol and $n_B^\beta = 2.50$ mol – 0.15 mol = 2.35 mol. Also, $n_C^\alpha = (1 - 0.111)(1.37\text{ mol}) = 1.22$ mol and n_C^β = 0.250(3.13 mol) = 0.78 mol.

R12.13 **(a)** Hypothetical, since the standard state is the ideal gas at 1 bar.

(b) Real, since it is the pure solvent at the *T* and *P* of the solution.

(c) Hypothetical; see Sec. 10.4.

R12.14 For $AgCl(s) \rightleftarrows Ag^+(aq) + Cl^-(aq)$, ΔG°_{298}/(kJ/mol) =

77.09 – 131.23 + 109.79 = 55.65, so

55650 J/mol = –(8.3145 J/mol-K)(298.15 K) ln K° and

$K^\circ = 1.78 \times 10^{-10}$ at 25°C. Integration of the van't Hoff equation with ΔH° assumed constant gives $\ln(K_2^\circ/K_1^\circ) = (\Delta H^\circ/R)(T_1^{-1} - T_2^{-1})$.

ΔH°_{298}/(kJ/mol) = 105.56 – 167.16 + 127.07 = 65.47.

$\ln[K^\circ_{323}/(1.78 \times 10^{-10})] = [(65470 \text{ J/mol})/(8.3145 \text{ J/mol-K})]\text{K}^{-1} \times (298.1^{-1} - 323.1^{-1})$ and $K^\circ_{323} = 1.37 \times 10^{-9}$.

Chapter 13

13.1 **(a)** Yes; **(b)** no.

13.2 **(a)** T; **(b)** T; **(c)** T (assuming $\phi = 0$ at infinity).

13.3 $F = |Q_1Q_2|/4\pi\varepsilon_0 r^2 =$

$$\frac{2(1.60\times10^{-19}\,\text{C})(1.60\times10^{-19}\,\text{C})}{4\pi(8.854\times10^{-12}\,\text{C}^2\,\text{N}^{-1}\,\text{m}^{-2})(1.0\times10^{-10}\,\text{m})^2} = 4.6\times10^{-8}\,\text{N}$$

13.4 **(a)** $E = |Q|/4\pi\varepsilon_0 r^2 = (1.60 \times 10^{-19}\text{ C})(8.99 \times 10^9\text{ N m}^2\text{ C}^{-2})/(2.0 \times 10^{-10}\text{ m})^2$ $= 3.6 \times 10^{10}$ V/m.

(b) 0.90×10^{10} V/m.

13.5 $\phi_2 - \phi_1 = (Q/4\pi\varepsilon_0)(1/r_2 - 1/r_1) =$

$$\frac{1.60\times10^{-19}\,\text{C}}{4\pi(8.854\times10^{-12}\,\text{C}^2\,\text{N}^{-1}\,\text{m}^{-2})}\left(\frac{1}{4.0\times10^{-10}\,\text{m}} - \frac{1}{2.0\times10^{-10}\,\text{m}}\right) = -3.60\,\text{V}$$

13.6 **(a)** F; **(b)** F.

13.7 **(a)** (3.00 mol)(2 × 96485 C/mol) = 5.79×10^5 C.

(b) (0.600 mol)(–96485 C/mol) = -5.79×10^4 C.

13.8 **(a)** T.

13.9 Equation (13.23) gives $\mu_i^{\text{Li}} - \mu_i^{\text{Rb}} = -(96500\text{ C/mol})(-0.1\text{ V}) = 10^4$ J/mol.

13.10 **(a)** F; **(b)** T; **(c)** F.

13.11 $\mathcal{E} \equiv \phi_R - \phi_L = \phi_{\text{Cu}} - \phi_{\text{Cu}'} = (\phi_{\text{Cu}} - \phi_{\text{CuSO}_4(aq)}) + (\phi_{\text{CuSO}_4(aq)} - \phi_{\text{ZnSO}_4(aq)}) +$ $(\phi_{\text{ZnSO}_4(aq)} - \phi_{\text{Zn}}) + (\phi_{\text{Zn}} - \phi_{\text{Cu}'}) = 0.3\text{ V} - 0.1\text{ V} + 0 + 0.2\text{ V} = 0.4\text{ V}.$

13.12 **(a)** T; **(b)** T; **(c)** T; **(d)** F.

13.13 **(a)** 2; **(b)** 1; **(c)** 2; **(d)** 6; **(e)** 2.

13.14 ΔG°_{298}/(kJ/mol) = 0 + 4(0) + 2(–111.25) – 97.89 – 65.49 – 2(–237.129) = 88.38. $\mathcal{E}^\circ = -\Delta G^\circ/nF = -(88380 \text{ J/mol})/2(96485 \text{ C/mol}) = -0.458$ V.

13.15 **(a)** The NaCl changes the ionic strength and hence changes the activity coefficients and the activities. Therefore $\mathcal{E}$ changes.

(b) By definition, $\mathcal{E}^\circ \equiv -\Delta G^\circ/nF$. The NaCl doesn't change ΔG° and so doesn't change $\mathcal{E}^\circ$.

13.16 The half-reactions are $In \rightarrow In^{3+} + 3e^-$ and $2e^- + Hg_2SO_4(s) \rightarrow 2Hg + SO_4^{2-}$. The cell reaction is $2In(s) + 3Hg_2SO_4(s) \rightarrow 2In^{3+}(aq) + 6Hg(\ell) + 3SO_4^{2-}(aq)$. Taking the activities of the pure solids and the pure liquid as 1, we have for the activity quotient: $Q = (a_+)^2(a_-)^3 = 2^2 3^3(\gamma_\pm m_i/m^\circ)^5$, where + and – refer to $In^{+3}(aq)$ and $SO_4^{2-}(aq)$, respectively, and Eq. (13.46) was used. The Nernst equation gives $\mathcal{E} = \mathcal{E}^\circ - (RT/6F) \ln (a_+)^2(a_-)^3 =$ $\mathcal{E}^\circ - (RT/6F) \ln [(108)^{1/5}\gamma_\pm m_i/m^\circ]^5 = \mathcal{E}^\circ - (5RT/6F) \ln (2.5508\gamma_\pm m_i/m^\circ)$, where i refers to $In_2(SO_4)_3$.

13.17 **(a)** Equation (13.49) applies. We plot the left side (l.s.) of (13.49) vs. $(m/m^\circ)^{1/2}$. We have l.s. = $\mathcal{E}$ + [2(8.314 J/mol-K)(333.15 K)/(96485 C/mol)] ln (m/m°) = $\mathcal{E}$ + (0.05741 V) ln (m/m°). The data are

l.s./V	0.1985	0.1993	0.2008	0.2104
$(m/m^\circ)^{1/2}$	0.03162	0.04472	0.07071	0.31622

Plotting the three points at high dilution and drawing a straight line through them, we find the intercept to be $\mathcal{E}^\circ$ = 0.1966 V.

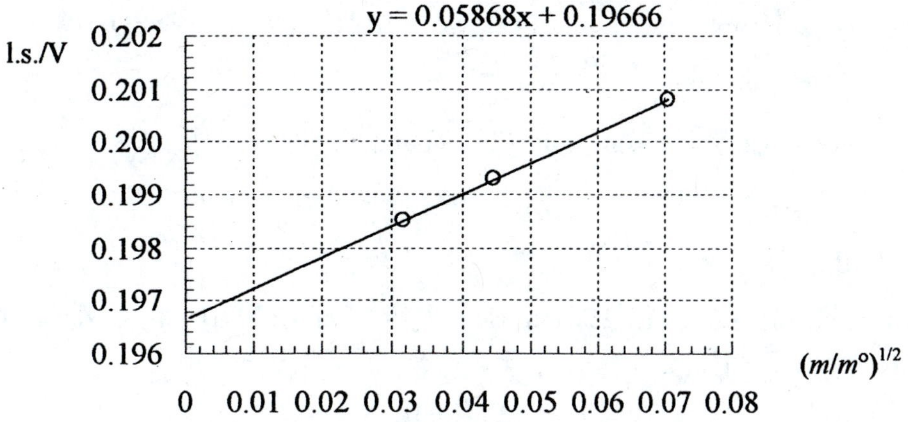

(b) From Eq. (13.49), $(2RT/F) \ln \gamma_\pm = \mathcal{E}° - \text{l.s.}$ and $\ln \gamma_\pm = (17.41_7\ V^{-1})(0.1966\ V - \text{l.s.}) = 3.424 - 17.41_7(\text{l.s./V})$. At 0.005 mol/kg, $\ln \gamma_\pm = 3.424 - 17.41_7(0.2008) = -0.073$ and $\gamma_\pm = 0.930$. At 0.1 mol/kg, $\ln \gamma_\pm = -0.240_5$ and $\gamma_\pm = 0.786$.

13.18 **(a)** $\ln \phi = B^\dagger P = BP/RT = (14.0\ \text{cm}^3/\text{mol})(750/760)\text{atm}/(82.06\ \text{cm}^3\text{-atm/mol-K})(298.1\ \text{K}) = 0.00056_5$ and $\phi = 1.00056_5$. $f = \phi P = 1.00056_5$ bar.

(b) The half-reaction is $H_2 \rightarrow 2H^+ + 2e^-$ and the error in $\mathcal{E}$ is $(RT/nF) \ln f(H_2) - (RT/nF) \ln P(H_2) = (RT/nF) \ln (f/P) = (RT/nF) \ln \phi = [(8.3145\ \text{J/mol-K})(298.15\ \text{K})/2(96485\ \text{C/mol})]0.00056_5 = 0.0000073$ V.

13.19 Equation (13.39) shows that $\mathcal{E}$ depends only on the chemical potentials μ_i of the species involved in the cell's chemical reaction, so $\mathcal{E}$ must be independent of what metals are used for the terminals.

13.20 **(a)** Doubled; **(b)** squared; **(c)** doubled, since $\ln Q^2 = 2 \ln Q$; **(d)** unchanged.

13.21 **(a)** Equation (13.21) applied to $Cu \rightleftarrows Cu^{2+}(aq) + 2e^-(Cu)$ gives $\tilde{\mu}_{Cu} = \tilde{\mu}_{Cu^{2+}(aq)} + 2\tilde{\mu}_{e^-(Cu)}$. Substitution of (13.19) for $\tilde{\mu}$ of each species gives $\mu_{Cu} = \mu_{Cu^{2+}(aq)} + 2F\phi(aq.\ CuSO_4) + 2[\mu_{e^-(Cu)} - F\phi(Cu)]$ and $\phi(Cu) - \phi(aq.\ CuSO_4) = [\mu_{Cu^{2+}(aq)} - \mu_{Cu} + 2\mu_{e^-(Cu)}]/2F$.

(b) By analogy to the result of (a), we have $\phi(\text{Zn}) - \phi(aq.\ \text{ZnSO}_4) = [\mu_{\text{Zn}^{2+}(aq)} - \mu_{\text{Zn}} + 2\mu_{e^-(\text{Zn})}]/2F$.

(c) For $e^-(\text{Cu}') \rightleftarrows e^-(\text{Zn})$, Eqs. (13.21) and (13.19) give $\tilde{\mu}_{e^-(\text{Cu}')} = \tilde{\mu}_{e^-(\text{Zn})}$, $\mu_{e^-(\text{Cu}')} - F\phi(\text{Cu}') = \mu_{e^-(\text{Zn})} - F\phi(\text{Zn})$, and $\phi(\text{Cu}') - \phi(\text{Zn}) = [\mu_{e^-(\text{Cu}')} - \mu_{e^-(\text{Zn})}]/F$.

(d) Substitution in (13.25) and use of $\mathcal{E}_J \equiv \phi(aq.\ \text{CuSO}_4) - \phi(aq.\ \text{ZnSO}_4)$ gives $\mathcal{E} = [\mu_{\text{Cu}^{2+}(aq)} - \mu_{\text{Cu}} + 2\mu_{e^-(\text{Cu})}]/2F + \mathcal{E}_J - [\mu_{\text{Zn}^{2+}(aq)} - \mu_{\text{Zn}} + 2\mu_{e^-(\text{Zn})}]/2F + [\mu_{e^-(\text{Zn})} - \mu_{e^-(\text{Cu}')}]/F = \mathcal{E}_J + (\mu_{\text{Cu}^{2+}(aq)} + \mu_{\text{Zn}} - \mu_{\text{Zn}^{2+}(aq)} - \mu_{\text{Cu}})/2F$, since $\mu_{e^-(\text{Cu})} = \mu_{e^-(\text{Cu}')}$ (Sec. 13.3). Substitution of $\mu_i = \mu_i^\circ + RT \ln a_i$ for each species gives $\mathcal{E} = \mathcal{E}_J + (\mu^\circ_{\text{Cu}^{2+}(aq)} + \mu^\circ_{\text{Zn}} - \mu^\circ_{\text{Zn}^{2+}(aq)} - \mu^\circ_{\text{Cu}})/2F + (RT/2F)(\ln a_{\text{Cu}^{2+}(aq)} + \ln a_{\text{Zn}} - \ln a_{\text{Zn}^{2+}(aq)} - \ln a_{\text{Cu}}) = \mathcal{E}_J - \Delta G^\circ/2F - (RT/2F) \ln [a_{\text{Zn}^{2+}(aq)} a_{\text{Cu}}/a_{\text{Cu}^{2+}(aq)} a_{\text{Zn}}]$, which is (13.51), since $\mathcal{E}^\circ = -\Delta G^\circ/2F$.

13.22 For electrolyte i, $\mu_i = \mu_i^\circ + RT \ln a_i$ [Eq. (10.4)] and $\mu_i = \nu_+\mu_+ + \nu_-\mu_- = \nu_+\mu_+^\circ + \nu_+ RT \ln a_+ + \nu_-\mu_-^\circ + \nu_- RT \ln a_-$ [Eqs. (10.38) and (10.4)]. Equating these two expressions for μ_i and using $\mu_i^\circ = \nu_+\mu_+^\circ + \nu_-\mu_-^\circ$ [Eq. (10.44)], we get $RT \ln a_i = \nu_+ RT \ln a_+ + \nu_- RT \ln a_- = RT \ln [(a_+)^{\nu_+}(a_-)^{\nu_-}]$, so $a_i = (a_+)^{\nu_+}(a_-)^{\nu_-}$.

13.23 **(a)** F; **(b)** T.

13.24 **(a)** $\Delta G^\circ_{298}/(\text{kJ/mol}) = 2(-4.7) + 3(0) - 3(65.49) - 2(0) = -205.9$.
$\Delta G^\circ = -2.06 \times 10^5$ J/mol.
$\mathcal{E}^\circ = -\Delta G^\circ/nF = (2.06 \times 10^5 \text{ J/mol})/6(96485 \text{ C/mol}) = 0.356$ V.

(b) The left half-reaction must be an oxidation and is $\text{Fe} \rightarrow \text{Fe}^{3+} + 3\text{e}^-$; the right half-reaction is $\text{Cu}^{2+} + 2\text{e}^- \rightarrow \text{Cu}$. Then $\mathcal{E}^\circ = \mathcal{E}^\circ_R - \mathcal{E}^\circ_L = 0.34 \text{ V} - (-0.04 \text{ V}) = 0.38$ V.

13.25 **(a)** At 25°C, $\mathcal{E}° = -1.978 \text{ V} = \mathcal{E}_R^\circ - \mathcal{E}_L^\circ = \mathcal{E}^\circ_{\text{nonesuch}} - \mathcal{E}^\circ_{\text{calomel}} = \mathcal{E}^\circ_{\text{nonesuch}} -$ 0.268 V and $\mathcal{E}^\circ_{\text{nonesuch}} = -1.710$ V.

(b) Use of $\mathcal{E}° = \mathcal{E}_R^\circ - \mathcal{E}_L^\circ$ gives at 43°C: $-0.80 \text{ V} = \mathcal{E}^\circ_{\text{nonpareil}} - \mathcal{E}^\circ_{\text{calomel}}$ and $1.70 \text{ V} = \mathcal{E}^\circ_{\text{nonpareil}} - \mathcal{E}^\circ_{\text{nonesuch}}$. Addition of these equations gives $0.90 \text{ V} = \mathcal{E}^\circ_{\text{nonpareil}} - \mathcal{E}^\circ_{\text{nonesuch}}$. For the cell nonpareil | nonesuch, $\mathcal{E}° = \mathcal{E}^\circ_{\text{nonesuch}} - \mathcal{E}^\circ_{\text{nonpareil}} = -0.90$ V

13.26 $\mathcal{E}° = \mathcal{E}_R^\circ - \mathcal{E}_L^\circ = 0.222 \text{ V} - 0 = 0.222$ V. We have $\mathcal{E} = \mathcal{E}° - (RT/nF) \ln Q$ so $\ln Q = (nF/RT)(\mathcal{E}° - \mathcal{E})$. We shall write the cell reaction using the smallest integers as coefficients, as in Eq. (13.34).

(a) $\ln Q = 2(96485 \text{ C/mol})(0.222 \text{ V} + 1.00 \text{ V})/(8.314 \text{ J/mol-K})(298.1 \text{ K}) =$ 95.1 and $Q = 2 \times 10^{41}$.

(b) With $\mathcal{E} = 1.00$ V, we get $\ln Q = -60.6$ and $Q = 5 \times 10^{-27}$.

13.27 The emf is given by Eq. (13.45) and the following paragraph as $\mathcal{E} = \mathcal{E}° + (RT/2F) \ln [P(H_2)/P°]$, where $a(H^+)a(Cl^-) = a(HCl) = 1$. We have $\ln [P(H_2)/P°] = (2F/RT)(\mathcal{E} - \mathcal{E}°) = (77.85 \text{ V}^{-1})(\mathcal{E} - 0.222 \text{ V})$.

(a) $\ln [P(H_2)/P°] = (77.85 \text{ V}^{-1})(-0.300 \text{ V} - 0.222 \text{ V}) = -40.6_4$ and $P(H_2) = 2 \times 10^{-18}$ bar.

(b) $\ln [P(H_2)/P°] = 6.07$ and $P(H_2) = 434$ bar. (This answer is only approximate, since at this high pressure we should use the fugacity, rather than the pressure, of H_2.)

13.28 **(a)** The left half-reaction is an oxidation, so the half-reactions are $Fe^{2+} \rightarrow Fe^{3+} + e^-$ and $I_2 + 2e^- \rightarrow 2I^-$. Multiplying the left half-reaction by 2 and adding it to the right one, we get as the cell reaction: $2Fe^{2+} + I_2 \rightarrow 2Fe^{3+} + 2I^-$.

(b) $\mathcal{E}° = \mathcal{E}_R^\circ - \mathcal{E}_L^\circ = 0.535 \text{ V} - 0.771 \text{ V} = -0.236$ V. $\mathcal{E} = \mathcal{E}° - (RT/nF) \ln Q$ $= -0.236 \text{ V} - [(8.314 \text{ J/mol-K})(298.1 \text{ K})/2(96485 \text{ C/mol})] \times \ln [(1.20)^2(0.100)^2/(2.00)^2 1] = -0.164$ V.

(c) $\mathcal{E} = \phi_R - \phi_L < 0$, so $\phi_R < \phi_L$ and the left-hand terminal is at the higher potential.

(d) The negative value of $\mathcal{E}$ indicates that the spontaneous cell reaction is in a direction opposite to that written in (a). Hence the half-reaction $Fe^{3+} + e^- \rightarrow Fe^{2+}$ occurs spontaneously in the left half-cell. Electrons therefore flow into the left terminal from the load.

13.29 **(a)** The left half-reaction is an oxidation, so

$$Cu \rightarrow Cu^{2+} + 2e^-$$

$$\underline{2e^- + Hg_2SO_4(c) \rightarrow 2Hg + SO_4^{2-}}$$

$$Cu + Hg_2SO_4(c) \rightarrow 2Hg + Cu^{2+} + SO_4^{2-}$$

(b) $\mathcal{E}° = \mathcal{E}_R^\circ - \mathcal{E}_L^\circ = 0.615\ \text{V} - 0.339\ \text{V} = 0.276\ \text{V}$. Equation (13.46) gives $a(Cu^{2+})a(SO_4^{2-}) = (\gamma_\pm m_i/m°)^2 = [0.043(1.00)]^2 = 0.00185$. The activities of the solids are 1 and $\mathcal{E} = 0.276\ \text{V} - (RT/2F) \ln 0.00185 = 0.357\ \text{V}$.

(c) With $\gamma_\pm = 1$, we would get $\mathcal{E} = \mathcal{E}° = 0.276$ V.

13.30 The cell reaction is $Zn + 2AgCl(c) \rightleftarrows Zn^{2+} + 2Ag + 2Cl^-$ and $\mathcal{E}° = 0.222\ \text{V} - (-0.762\ \text{V}) = 0.984$ V. Using Eq. (13.46), we have $\mathcal{E} = \mathcal{E}° - (RT/2F) \ln [a(Zn^{2+})a(Cl^-)^2] = \mathcal{E}° - (RT/2F) \ln [4(0.0100\gamma_\pm)^3] =$ $0.984\ \text{V} - [(8.314\ \text{J/mol-K})(298.15\ \text{K})/2(96485\ \text{C/mol})] \ln[4(0.0100)^3(0.708)^3]$ $= 1.157$ V.

13.31 Addition of (1) $Cr^{3+} + e^- \rightarrow Cr^{2+}$ and (2) $Cr^{2+} + 2e^- \rightarrow Cr$ gives (3) $Cr^{3+} + 3e^- \rightarrow Cr$. Hence $\Delta G_3^\circ = \Delta G_1^\circ + \Delta G_2^\circ$, which becomes $-n_3F\mathcal{E}_3^\circ =$ $-n_1F\mathcal{E}_1^\circ - n_2F\mathcal{E}_2^\circ$; so $\mathcal{E}_3^\circ = (n_1\mathcal{E}_1^\circ + n_2\mathcal{E}_2^\circ)/n_3 = [1(-0.424\ \text{V}) + 2(-0.90\ \text{V})]/3 =$ -0.74 V.

13.32 $\mathcal{E} = \mathcal{E}° - (RT/nF) \ln [a(Zn^{2+})/a(Cu^{2+})]$. The $ZnSO_4$ solution has $I_m/m° =$ $\frac{1}{2}[2^2(0.002) + 2^2(0.002)] = 0.00800$ and the Davies equation gives $\log \gamma(Zn^{2+}) = -0.163$; $\gamma(Zn^{2+}) = 0.68_8$. Also, $I_m/m° = 0.00400$ for $CuSO_4(aq)$; $\log \gamma(Cu^{2+}) = -0.119$; $\gamma(Cu^{2+}) = 0.76_0$. Use of $a_{m,i} = \gamma_\pm m_i/m°$ gives $a(Zn^{2+}) = 0.68_8(0.00200) = 0.00138$ and $a(Cu^{2+}) = 0.00076_0$.
So $\mathcal{E} = 0.339\ \text{V} - (-0.762\ \text{V}) - (RT/2F) \ln (0.00138/0.00076_0) = 1.093$ V.

13.33 **(a)** $Pt|Ag|AgCl(c)|KCl(aq)|Hg_2Cl_2(c)|Hg|Pt'$.

(b) $Pt|H_2|H_2SO_4(aq)|Hg_2SO_4(c)|Hg|Pt'$ (or we can use a $Pb|PbSO_4$ half-cell).

13.34 $Pt'|H_2|HCl(aq)|Cl_2|Pt$,
$Pt|H_2|HCl(aq)|AgCl(c)|Ag|Pt'$,
$Pt|H_2|HCl(aq)|Hg_2Cl_2(c)|Hg|Pt'$,
$Pt|Ag|AgCl(c)|HCl(aq)|Hg_2Cl_2(c)|Hg|Pt'$.

13.35 **(a)** The half-reactions are $Ag \rightarrow Ag^+(0.01m^\circ) + e^-$ and $Ag^+(0.05m^\circ) + e^- \rightarrow Ag$. The cell reaction is $Ag^+(0.05m^\circ) \rightarrow Ag^+(0.01m^\circ)$. We have $\mathcal{E}^\circ = \mathcal{E}^\circ_R - \mathcal{E}^\circ_L = 0$. Then $\mathcal{E} = 0 - (RT/F) \ln [\gamma_{+,L}(0.01)/\gamma_{+,R}(0.05)] = -(0.02569 \text{ V})[-1.6094 + \ln (\gamma_{+,L}/\gamma_{+,R})] = 0.04135 \text{ V} - (0.02569 \text{ V}) \ln (\gamma_{+,L}/\gamma_{+,R})$. We have $I_{m,L} = 0.0100$ mol/kg and $I_{m,R} = 0.0500$ mol/kg. The Davies equation for Ag^+ gives $\log \gamma_{+,L} = -0.0448$ and $\gamma_{+,L} = 0.902$; also, $\log \gamma_{+,R} = -0.0855_5$ and $\gamma_{+,R} = 0.821$. Hence $\mathcal{E} = 0.04135 \text{ V} - (0.02569 \text{ V}) \ln (0.902/0.821) = 0.0389$ V.

(b) $\mathcal{E} = \phi_R - \phi_L > 0$, so $\phi_R > \phi_L$.

(c) The electrons flow from low to high potential and hence flow into the right terminal.

13.36 $\mathcal{E} = -[(8.314 \text{ J/mol-K})(358.1 \text{ K})/2(96485 \text{ C/mol})] \ln (2521/666) = -0.0205$ V, where Eq. (13.60) was used.

13.37 **(a)** T; **(b)** F.

13.38 **(a)**

$$[Ag + Cl^- \rightarrow AgCl(s) + e^-] \times 2$$
$$Hg_2Cl_2(s) + 2e^- \rightarrow 2Hg + 2Cl^-$$
$$\overline{2Ag + Hg_2Cl_2(s) \rightarrow 2AgCl(s) + 2Hg}$$

(b) $\mathcal{E} = \mathcal{E}^\circ - (RT/2F) \ln [a(AgCl)a^2(Hg)/a^2(Ag)a(Hg_2Cl_2)] = \mathcal{E}^\circ$, since the activities of the solids can be taken as 1. So $\mathcal{E} = \mathcal{E}^\circ = \mathcal{E}^\circ_R - \mathcal{E}^\circ_L = 0.2680 \text{ V} - 0.2222 \text{ V} = 0.0458$ V.

(c) $\mathcal{E} = \mathcal{E}^\circ = 0.0458$ V.

(d) Since $\mathcal{E} = \mathcal{E}^\circ$, we have $\partial\mathcal{E}/\partial T = \partial\mathcal{E}^\circ/\partial T$. So $\Delta S^\circ = nF(\partial\mathcal{E}^\circ/\partial T)_P = 2(96485 \text{ C/mol})(0.000338 \text{ V/K}) = 65.2$ J/mol-K. $\Delta G^\circ = -nF\mathcal{E}^\circ =$

$-2(96458 \text{ C/mol})(0.0458 \text{ V}) = -8840$ J/mol. $\Delta H° = \Delta G° + T\,\Delta S° =$
-8840 J/mol + (298.1 K)(65.2 J/mol-K) = 10.6 kJ/mol.

13.39 $\mathcal{E}° = \mathcal{E}°_R - \mathcal{E}°_L = 0 - (-0.01 \text{ V}) = 0.01$ V. $\Delta G° = -nF\mathcal{E}° = -RT \ln K°$ and
$\ln K° = nF\mathcal{E}°/RT = 2(96485 \text{ C/mol})(0.01 \text{ V})/(8.314 \text{ J/mol-K})(298.1 \text{ K}) = 0.8$.
$K° = 2$.

13.40 **(a)** For the reaction $2Na^+ + H_2 \rightarrow 2Na + 2H^+$, we have $\mathcal{E}°_{298} = \mathcal{E}°_R - \mathcal{E}°_L =$
$-2.714 \text{ V} - 0 = -2.714$ V. Then $\Delta G°_{298} = -nF\mathcal{E}° = -2(96485 \text{ C/mol}) \times$
$(-2.714 \text{ V}) = 5.237 \times 10^5 \text{ J/mol} = 2(0) + 2(0) - 2\Delta_f G°_{298}(Na^+) - 2(0)$ and
$\Delta_f G°_{298}(Na^+) = -2.619 \times 10^5 \text{ J/mol} = -261.9$ kJ/mol.

(b) For the reaction $2Cl^- + 2H^+ \rightarrow Cl_2 + H_2$, we have $\mathcal{E}°_{298} = 0 - (1.360 \text{ V}) =$
-1.360 V. Then $\Delta G°_{298} = -2(96485 \text{ C/mol})(-1.360 \text{ V}) =$
$2.624 \times 10^5 \text{ J/mol} = 0 + 0 - 2\Delta_f G°_{298}(Cl^-) - 2(0)$ and $\Delta_f G°_{298}(Cl^-) =$
$-1.312 \times 10^5 \text{ J/mol} = -131.2$ kJ/mol.

(c) For the reaction $Cu^{2+} + H_2 \rightarrow Cu + 2H^+$, $\mathcal{E}°_{298} = 0.339 \text{ V} - 0 = 0.339$ V
and $\Delta G°_{298} = -2(96485 \text{ C/mol})(0.339 \text{ V}) = -6.54 \times 10^4 \text{ J/mol} =$
$0 + 2(0) - \Delta_f G°_{298}(Cu^{2+}) - 0$ and $\Delta_f G°_{298}(Cu^{2+}) = 6.54 \times 10^4 \text{ J/mol} =$
65.4 kJ/mol.

13.41 $2e^- + PbI_2(c) \rightarrow Pb + 2I^-(aq)$

$Pb \rightarrow Pb^{2+}(aq) + 2e^-$

$PbI_2(c) \rightarrow Pb^{2+}(aq) + 2I^-(aq)$

$\mathcal{E}° = -0.365 \text{ V} - (-0.126 \text{ V}) = -0.239$ V. $\Delta G° = -nF\mathcal{E}° =$
$-2(96485 \text{ C/mol})(-0.239 \text{ V}) = 4.61 \times 10^4$ J/mol. $\ln K°_{sp} = -\Delta G°/RT =$
$-(46100 \text{ J/mol})/(8.314 \text{ J/mol-K})(298.1 \text{ K}) = -18.6$ and $K°_{sp} = 8 \times 10^{-9}$.

13.42 **(a)** $\mathcal{E}° = 1.360 \text{ V} - 1.078 \text{ V} = 0.282$ V. $\Delta G° = -2(96485 \text{ C/mol})(0.282 \text{ V}) =$
-54400 J/mol. $\ln K° = -\Delta G°/RT =$
$(54400 \text{ J/mol})/(8.314 \text{ J/mol-K})(298.1 \text{ K}) = 21.9_5$ and $K° = 3 \times 10^9$.

(b) $\mathcal{E}° = 0.282$ V, $n = 1$, and $\Delta G° = -27200$ J/mol. $\ln K° = 10.9_8$ and $K° = 6 \times 10^4$.

(c) The half-reactions are $2(Ag + Cl^- \rightarrow AgCl + e^-)$ and $Cl_2 + 2e^- \rightarrow 2Cl^-$. $\mathcal{E}° = 1.360\text{ V} - 0.222\text{ V} = 1.138$ V. $\Delta G° = -2(96485\text{ C/mol})(1.138\text{ V}) = -2.196 \times 10^5$ J/mol. $\ln K° = 88.5_9$ and $K° = 3 \times 10^{38}$.

(d) This is the reverse of (c), so $\Delta G° = 2.19_6 \times 10^5$ J/mol and $K° = (3 \times 10^{38})^{-1} = 3 \times 10^{-39}$.

(e) The half-reactions are $2(Fe^{2+} \rightarrow Fe^{3+} + e^-)$ and $Fe^{2+} + 2e^- \rightarrow Fe$. $\mathcal{E}° = -0.440\text{ V} - 0.771\text{ V} = -1.21$ V. $\Delta G° = 2.33_7 \times 10^5$ J/mol, $\ln K° = -94.2_7$ and $K° = 1 \times 10^{-41}$.

13.43 $\mathcal{E}°/\text{V} = 0.23646 - 5.1144 \times 10^{-4}(t/°\text{C}) - 2.0628 \times 10^{-6}(t/°\text{C})^2 - 1.0808 \times 10^{8}(t/°\text{C})^3$.

13.44 Cell 1 is $Pt|Fe^{2+}(aq), Fe^{3+}(aq) \vdots\vdots Fe^{2+}(aq)|Fe|Pt'$.
Cell 2 is $Pt|Fe^{2+}(aq), Fe^{3+}(aq) \vdots\vdots Fe^{3+}(aq)|Fe|Pt'$.
Cell 3 is $Fe|Fe^{3+}(aq) \vdots\vdots Fe^{2+}(aq)|Fe$. All have the same cell reaction. For example, Cell 3 has the half-reactions $Fe^{2+} + 2e^- \rightarrow Fe$ and $Fe \rightarrow Fe^{3+} + 3e^-$, and multiplication of these half-reactions by 3 and 2, respectively, gives the overall reaction as $3Fe^{2+} \rightarrow Fe + 2Fe^{3+}$. For Cell 1, $\mathcal{E}° = -0.44\text{ V} - 0.77\text{ V} = -1.21$ V, $n\mathcal{E}° = 2(-1.21\text{ V}) = -2.42$ V, $\Delta G° = -nF\mathcal{E}° = 233$ kJ/mol. For Cell 2, $\mathcal{E}° = -0.04\text{ V} - 0.77\text{ V} = -0.81$ V, $n\mathcal{E}° = 3(-0.81\text{ V}) = -2.43$ V, $\Delta G° = -nF\mathcal{E}° = 234$ kJ/mol. For Cell 3, $\mathcal{E}° = -0.44\text{ V} - (-0.04\text{ V}) = -0.40$ V, $n\mathcal{E}° = 6(-0.40\text{ V}) = -2.40$ V, $\Delta G° = 232$ kJ/mol. $\Delta G°$ must be the same in all three cases, since all the cells have the same reaction. (Because of the spontaneous reaction $2Fe^{3+}(aq) + Fe(s) \rightarrow 3Fe^{2+}(aq)$, an $Fe^{3+}|Fe$ electrode is not reproducible.)

13.45 (a) The half-reactions are $Fe \rightarrow Fe^{2+} + 2e^-$ and $2(Fe^{3+} + e^- \rightarrow Fe^{2+})$. The cell reaction is $Fe + 2Fe^{3+} \rightarrow 3Fe^{2+}$.

(b) $\Delta G° = -nF\mathcal{E}° = -2(96485\text{ C/mol})[0.771\text{ V} - (-0.440\text{ V})] = -2.33_7 \times 10^5$ J/mol. $\Delta S° = nF(\partial\mathcal{E}°/\partial T)_P = 2(96485\text{ C/mol}) \times (0.00114\text{ V/K}) = 220$ J/mol-K. $\Delta H° = \Delta G° + T\,\Delta S° = -2.33_7 \times 10^5\text{ J/mol} + (298.1\text{ K})(220\text{ J/mol-K}) = -1.68 \times 10^5$ J/mol.

13.46 Equations (13.67) and (13.68) give at 10°C: $\mathcal{E}° = 0.23643$ V – $(4.8621 \times 10^{-4}$ V/K)(10 K) – $(3.4205 \times 10^{-6}$ V/K^2)(10 K)^2 + $(5.869 \times 10^{-9}$ V/K^3)(10 K)^3 = 0.23123 V. Then $\Delta G°$ = –2(96485 C/mol)(0.23123 V) = -4.4621×10^4 J/mol. Equation (13.69) gives at 10°C: $\Delta S°$ = 2(96485 C/mol)[-4.8621×10^{-4} V/K + 2(-3.4205×10^{-6} V/K^2)(10 K) +3(5.869×10^{-9} V/K^3)(10 K)^2] = –106.69 J/mol-K. Then $\Delta H° = \Delta G° + T\,\Delta S°$ = -4.4621×10^4 J/mol + (283.15 K)(–106.69 J/mol-K) = -7.4830×10^4 J/mol. From (13.66) and (13.67), $\Delta C_P^\circ = 2FT[2c + 6d(T - T_0)]$ = 2(96485 C/mol) × (283.15 K)[2(-3.4205×10^{-6} V/K^2) + 6(5.869×10^{-9} V/K^3)(10 K)] = –354.55 J/mol-K.

13.47 For the cell Ag | Ag^+:I^- | AgI(c) | Ag, we have the half-reactions $Ag \rightarrow Ag^+ + e^-$ and AgI(c) + $e^- \rightarrow$ Ag + I^-; the cell reaction is AgI(c) $\rightarrow Ag^+ + I^-$. Equation (13.63) gives $\mathcal{E}° = RT \ln K_{sp}^\circ / nF$ = (8.314 J/mol-K)(298.1 K) ln (8.2×10^{-17})/(96485 C/mol) = –0.951 V = $\mathcal{E}_{298}^\circ = \mathcal{E}_R^\circ - \mathcal{E}_L^\circ = \mathcal{E}_R^\circ - 0.799$ V. Hence $\mathcal{E}_R^\circ = -0.152$ V.

13.48 The half-reactions are $H_2 \rightarrow 2H^+ + 2e^-$ and 2(AgBr + $e^- \rightarrow$ Ag + Br^-). The cell reaction is H_2 + 2AgBr $\rightleftharpoons 2H^+ + 2Br^-$ + 2Ag. $\mathcal{E}° = 0.073$ V. Equation (13.46) gives $\mathcal{E} = \mathcal{E}° - (RT/2F) \ln \{a(H^+)^2 a(Br^-)^2/[P(H_2)/P°]\}$ = $\mathcal{E}° - (RT/F) \ln [a(H^+)a(Br^-)] = \mathcal{E}° - (RT/F) \ln (\gamma_\pm m_i/m°)^2$. Hence 0.200 V = 0.073 V – [2(8.314 J/mol-K)(298.15 K)/(96485 C/mol)] ln ($0.100\gamma_\pm$). We get ln ($0.100\gamma_\pm$) = –2.47 and $\gamma_\pm = 0.84$.

13.49 As in Prob. 13.40b, we find $\Delta_f G°(Cl^-) = -131.2$ kJ/mol. Since $\mu°$(HCl) = $\mu°$(HCl) = $\mu°(H^+) + \mu°(Cl^-)$ [see Eq. (10.44)] and $\Delta_f G°(H^+) = 0$, we have $\Delta_f G°$[HCl(aq)] = –131.2 kJ/mol.

13.50 **(a)** The half-reactions are $H_2 \rightarrow 2H^+ + 2e^-$ and 2[AgCl(c) + $e^- \rightarrow$ Ag + Cl^-]. The cell reaction is H_2 + 2AgCl(c) $\rightarrow 2H^+ + 2Cl^-$ + 2Ag. The solids' activities can be taken as 1, and $\mathcal{E} = \mathcal{E}° - (RT/2F) \ln \{[a(H^+)]^2[a(Cl^-)]^2\}$ = $\mathcal{E}° - (RT/F) \ln [a(H^+)a(Cl^-)]$. Since we are writing H^+ rather than H_3O^+, we write the water ionization as $H_2O \rightleftarrows H^+ + OH^-$. Then K_w° =

$a(H^+)a(OH^-)/a(H_2O)$ and $a(H^+) = K_w^\circ a(H_2O)/a(OH^-)$. So $a(H^+)a(Cl^-) = K_w^\circ a(H_2O)a(Cl^-)/a(OH^-) = K_w^\circ a(H_2O)\gamma(Cl^-)m(Cl^-)/\gamma(OH^-)m(OH^-)$. Substitution in the above equation for $\mathcal{E}$ gives the desired result.

(b) As $I_m \to 0$, the γ's go to 1 and $a(H_2O)$ goes to 1. Hence $\mathcal{E} \to \mathcal{E}^\circ - (RT/F)\ln[K_w^\circ m(Cl^-)/m(OH^-)]$.
So $\ln K_w^\circ = (F/RT)\{\mathcal{E}^\circ - \mathcal{E} - (RT/F)\ln[m(Cl^-)/m(OH^-)]\}^\infty =$ $[(96485\text{ C/mol})/(8.314\text{ J/mol-K})(298.15\text{ K})](-0.8279\text{ V}) = -32.22_5$ and $K_w^\circ = 1.01 \times 10^{-14}$.

13.51 (a) The half-reactions are the same as in Prob. 13.50a and the cell reaction is $H_2 + 2AgCl(c) \to 2H^+ + 2Cl^- + 2Ag$. As in Prob. 13.50a, $\mathcal{E} = \mathcal{E}^\circ - (RT/F)\ln[a(H^+)a(Cl^-)]$. For the ionization $HX \rightleftharpoons H^+ + X^-$, we have $K_a^\circ = a(H^+)a(X^-)/a(HX)$ and $a(H^+) = K_a^\circ a(HX)/a(X^-)$. Hence $a(H^+)a(Cl^-) = K_a^\circ a(HX)a(Cl^-)/a(X^-) =$ $K_a^\circ \gamma(HX)m(HX)\gamma(Cl^-)m(Cl^-)/\gamma(X^-)m(X^-)m^\circ$. Substitution in the above equation for $\mathcal{E}$ gives the desired result.

(b) As $I_m \to 0$, the γ's $\to 1$ and we get
$\ln K_a^\circ = (F/RT)\{\mathcal{E}^\circ - \mathcal{E} - (RT/F)\ln[m(Cl^-)m(HX)/m(X^-)m^\circ]\}^\infty =$ $[(96485\text{ C/mol})/(8.314\text{ J/mol-K})(298.15\text{ K})](-0.2814\text{ V}) = -10.953$ and $K_a^\circ = 1.75 \times 10^{-5}$.

13.52 $Sn + Pb^{2+} \rightleftarrows Sn^{2+} + Pb$. $\mathcal{E}^\circ = -0.126\text{ V} + 0.141\text{ V} = 0.015\text{ V}$. $\ln K^\circ = nF\mathcal{E}^\circ/RT = 2(96485\text{ C/mol})(0.015\text{ V})/(8.314\text{ J/mol-K})(298\text{ K}) = 1.17$ and $K^\circ = 3._2$. The solids' activities can be taken as 1, so $3._2 = a(Sn^{2+})/a(Pb^{2+}) \approx m(Sn^{2+})/m(Pb^{2+}) = z/(0.1 - z)$. We find $z = 0.07_6$. Hence $m(Sn^{2+}) = 0.07_6$ mol/kg and $m(Pb^{2+}) = 0.02_4$ mol/kg. The solution is reasonably dilute, so we expect the activity coefficients to be determined mainly by the ionic strength. Hence $\gamma(Pb^{2+}) \approx \gamma(Sn^{2+})$ and the activity coefficients cancel in the expressions for K°.

13.53 $\mathcal{E}_X = 612$ mV and $\mathcal{E}_S = 741$ mV. Equation (13.73) gives pH(X) = 6.86 + $[(612 - 741)10^{-3}\text{ V}](96485\text{ C/mol})/(8.314\text{ J/mol-K})(298.1\text{ K})2.3026 = 4.68$.

13.54 In Eq. (13.78), we take $\gamma^{\alpha}_{\mathrm{Na}^+} \approx \gamma^{\beta}_{\mathrm{Na}^+}$, since the ionic strengths of the two solutions are equal. Then $\phi^{\beta} - \phi^{\alpha}$ = [(8.314 J/mol-K)(298 K)/(96485 C/mol)] × ln (0.100/0.150) = –0.0104 V, where β is the $NaNO_3$–KNO_3 solution.

13.55 $\mu = \delta d$ and $\delta = \mu/d = (3.57 \times 10^{-30}\text{ C m})/(1.30 \times 10^{-10}\text{ m}) = 2.75 \times 10^{-20}$ C. $\delta/e = (2.75 \times 10^{-20}\text{ C})/(1.60 \times 10^{-19}\text{ C}) = 0.172$.

13.56 **(a)** $\Sigma_i\, Q_i x_i = (-0.5e)(-1.5\text{ Å}) + (-1.5e)(1.0\text{ Å}) = -0.75e\text{ Å} =$ $-0.75(1.6 \times 10^{19}\text{ C})(10^{10}\text{ m}) = -1.2 \times 10^{29}$ C m. The direction is in the negative x direction and the magnitude is 1.2×10^{29} C m.

(b) $\mu_x = \Sigma_i\, Q_i x_i = (-e)(1\text{ Å}) = -e$ Å. $\mu_y = \Sigma_i\, Q_i y_i = (-e)(1\text{ Å}) =$ $-e$ Å. $\mu = (\mu_x^2 + \mu_y^2)^{1/2} = 2^{1/2} e\text{ Å} = 2.3 \times 10^{-29}$ C m. The vector makes an angle of 45° with the x and y axes and points from the negative charges' region toward the positive charge.

(c) $\Sigma_i\, Q_i x_i = (-0.5e)(-2.5\text{ Å}) + (2e)(-1.0\text{ Å}) = -0.75e$ Å.

13.57 Let a, b, and c be the x, y, and z coordinates of the new origin in the original coordinate system. If x_i is the x coordinate of charge i in the original system, then μ_x in the new coordinate system equals $\Sigma_i\, Q_i(x_i - a) = \Sigma_i\, Q_i x_i -$ $a\Sigma_i\, Q_i = \Sigma_i\, Q_i x_i - a \cdot 0 = \Sigma_i\, Q_i x_i$, which is the same as in the original coordinate system. Similarly for μ_y and μ_z.

13.58 **(a)** $1/r_2 - 1/r_1 = (r_1 - r_2)/r_1 r_2 = (r_1 - r_2)(r_1 + r_2)/r_1 r_2(r_1 + r_2) =$ $(r_1^2 - r_2^2)/r_1 r_2(r_1 + r_2)$.

(b) For $r >> d$, we have $r_1 \approx r_2 \approx r$, and $1/r_2 - 1/r_1 \approx (r_1^2 - r_2^2)/2r^3$. The law of cosines for triangle PAC gives $r_2^2 = r_1^2 + d^2 - 2r_1 d \cos\theta$. (Because $r >> d$, angle PAC is very nearly equal to angle PBC.) Then $r_1^2 - r_2^2 =$ $2r_1 d\cos\theta - d^2 = d(2r_1\cos\theta - d) \approx 2rd\cos\theta$, since $d << r_1$.

(c) $\phi = (Q/r_2 - Q/r_1)/4\pi\varepsilon_0 = Q(2rd\cos\theta)/2r^3(4\pi\varepsilon_0) = (\mu\cos\theta)/4\pi\varepsilon_0 r^2$, since $\mu = Qd$.

13.59 (a) $w = \int_1^2 F\,dr = -(Q_1Q_2/4\pi\varepsilon_0)\int_1^2 r^{-2}\,dr = -(Q_1Q_2/4\pi\varepsilon_0)(1/r_1 - 1/r_2) =$

$$\frac{(1.602\times10^{-19}\text{ C})^2}{4\pi(8.854\times10^{-12}\text{ C}^2/\text{N-m}^2)}\left(\frac{1}{10^{-9}\text{m}} - \frac{1}{10^{-8}\text{m}}\right) = 2.08\times10^{-19}\text{ J}$$

(b) From Eq. (13.89), we include a factor $1/\varepsilon_r$, where (Sec. 10.7) $\varepsilon_r = 78.40$; so $w = (2.08 \times 10^{-19}\text{ J})/78.40 = 2.65 \times 10^{-21}$ J.

13.60 $\mu = 0$ for CCl_4 and Eq. (13.87) gives

$$\frac{1.24}{4.24}\,\frac{153.8\text{ g/mol}}{1.59\text{ g/cm}^3} = \frac{6.022\times10^{23}\text{ mol}^{-1}}{3(8.854\times10^{-12}\text{ C}^2/\text{N-m}^2)}\alpha$$

and $\alpha = 1.25 \times 10^{-33}$ cm^3-C^2/N-m^2 = 1.25×10^{-39} C^2-m/N. $\alpha/4\pi\varepsilon_0 =$ $(1.25 \times 10^{-39}$ C^2-m/N$)/4\pi(8.854 \times 10^{-12}$ C^2/N-m$^2) = 1.12 \times 10^{-29}$ m^3 = 11.2 Å^3.

13.61 (a) $\mu = 0$ for CH_4. Also, $PV = (m/M)RT$ and $M/\rho = RT/P$. Equation (13.87) gives

$$\frac{0.00094}{3.001}\,\frac{(82.06\times10^{-6}\text{ m}^3\text{-atm/mol-K})(273.1\text{ K})}{1.00\text{ atm}} = \frac{6.022\times10^{23}\text{ mol}^{-1}}{3(8.854\times10^{-12}\text{ C}^2/\text{N-m}^2)}\alpha$$

$\alpha = 3.10 \times 10^{-40}$ C^2 N^{-1} m. Also, $\alpha/4\pi\varepsilon_0 = (3.10 \times 10^{-40}$ C^2-m/N$)/$ $4\pi(8.854 \times 10^{-12}$ C^2/N-m$^2) = 2.78 \times 10^{-30}$ m^3 = 2.78 Å^3.

(b) $\varepsilon_r + 2 \approx 3$. $M/\rho = RT/P$. $\mu = 0$. Equation (13.87) gives $\varepsilon_r - 1 = PN_A\alpha/RT\varepsilon_0 = 4\pi PN_A(\alpha/4\pi\varepsilon_0)/RT =$ $4\pi(10.0\text{ atm})(6.022 \times 10^{23}\text{ mol}^{-1})(2.78 \times 10^{-30}\text{ m}^3)/$ $(82.06 \times 10^{-6}\text{ m}^3\text{-atm/mol-K})(373.1\text{ K}) = 0.0068_7$ and $\varepsilon_r = 1.0068_7$.

13.62 Equation (13.87) applies. We plot $M\rho^{-1}(\varepsilon_r - 1)/(\varepsilon_r + 2)$ vs. $1/T$. Noting that $M/\rho = RT/P$ for a gas, we have

$10^5 M\rho^{-1}(\varepsilon_r - 1)/(\varepsilon_r + 2)$	5.73	5.35	5.00_5	4.67	4.31
$10^3/(T/\text{K})$	2.602	2.380	2.249	2.066	1.916

where the units of the first line are m^3/mol. The plot is reasonably linear

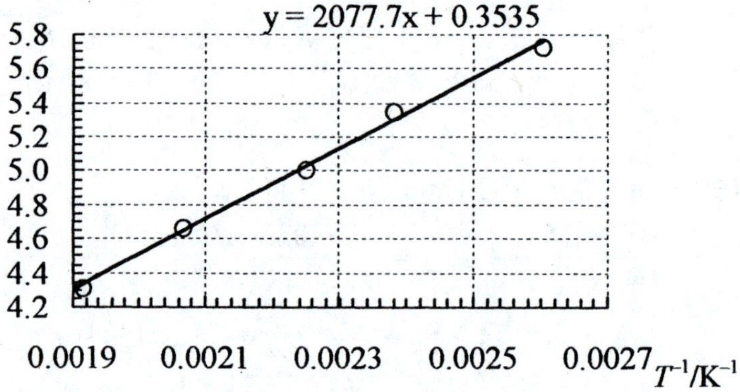

with slope 0.0208 m^3-K/mol = $N_A\mu^2/9\varepsilon_0 k$. So μ^2 = (0.0208 m^3-K/mol)9(8.854 × 10^{-12} C^2/N-m^2)(1.381 × 10^{-23} J/K)/(6.022 × 10^{23} mol^{-1}) = 3.80 × 10^{-59} C^2 m^2 and μ = 6.17 × 10^{-30} C m. The intercept is 3.5 × 10^{-6} m^3/mol = $N_A\alpha/3\varepsilon_0$, and

$$\alpha = \frac{3(8.854\times10^{-12}\ \text{C}^2/\text{N-m}^2)(3.5\times10^{-6}\ \text{m}^3/\text{mol})}{6.022\times10^{23}\ \text{mol}^{-1}}$$

α = 1.5 × 10^{-40} C^2-m/N

[Using Eqs. (19.2) and (19.3), one finds that the dipole moment is 1.85 D. Also, one find $\alpha/4\pi\varepsilon_0$ = 1.4 × 10^{-30} m^3 = 1.4 Å^3.]

13.63 **(a)** Molecular, C m; **(b)** molecular, C m^2 V^{-1}; **(c)** macroscopic, no units.

13.64 **(a)** CSe_2, because of a greater α;

(b) n-$C_{10}H_{22}$;

(c) o-dichlorobenzene, because of a greater μ.

13.65 **(a)** Let $f \equiv (\varepsilon_r - 1)/(\varepsilon_r + 2)$. $df/d\varepsilon_r = 1/(\varepsilon_r + 2) - (\varepsilon_r - 1)/(\varepsilon_r + 2)^2 = [(\varepsilon_r + 2) - (\varepsilon_r - 1)]/(\varepsilon_r + 2)^2 = 3/(\varepsilon_r + 2)^2 > 0$.

(b) We have $1 \le \varepsilon_r \le \infty$. Since $df/d\varepsilon_r > 0$ for all ε_r, the minimum value of f is at the minimum value of ε_r; at $\varepsilon_r = 1$, $f_{min} = 0$. As $\varepsilon_r \to \infty$, f approaches its maximum possible value, which is 1.

13.66 Let $P(K^+) = P_K$. Then $\phi^{int} - \phi^{ext} =$

$$0.02569 \text{ V} \ln \frac{P_K 10 + 0.04P_K 460 + 0.5P_K 40}{P_K 410 + 0.04P_K 49 + 0.5P_K 540}$$

$= (0.02569 \text{ V}) \ln (48.4/682) = -0.068 \text{ V} = -68 \text{ mV}$

13.67 The half-reactions are $Ag + Cl^-(0.0100m^\circ) \to AgCl + e^-$ and $AgCl + e^- \to Ag + Cl^-(0.100m^\circ)$. Equation (13.51) gives $\mathcal{E} = \mathcal{E}_J + 0 - (RT/F) \ln (0.100\gamma_{-,R}/0.0100\gamma_{-,L})$. $I_{m,R} = 0.100$ mol/kg and the Davies equation for Cl^- gives $\log \gamma_{-,R} = -0.107_2$ and $\gamma_{-,R} = 0.781$. Similarly, $\gamma_{-,L} = 0.902$. So $\mathcal{E} = -0.038 \text{ V} - [(8.314 \text{ J/mol-K})(298.15 \text{ K})/(96485 \text{ C/mol})] \ln 8.66 = -0.093$ V.

13.68 $I_m = 0.100$ mol/kg. The Davies equation for H^+ gives $\log \gamma_+ = -0.107_2$ and $\gamma_+ = 0.781$. Then $a(H^+) = (0.781)(0.100) = 0.0781$ and $-\log a(H^+) = 1.10_7$.

13.69 $K_c^\circ/K_m^\circ = \{[a_c(Ag^+)]^2/a_c(Cu^{2+})\}/\{[a_m(Ag^+)]^2/a_m(Cu^{2+})\} = (0.997)^2/0.997 = 0.997$. $\Delta G^\circ = -RT \ln K^\circ = -nF\mathcal{E}^\circ$ and $\mathcal{E}^\circ = (RT/nF) \ln K^\circ$, so $\mathcal{E}_m^\circ - \mathcal{E}_c^\circ = (RT/nF)(\ln K_m^\circ - \ln K_c^\circ) = -(RT/nF) \ln K_c^\circ/K_m^\circ = -(RT/2F) \ln 0.997 = 0.000039 \text{ V} = 0.039$ mV, which is insignificant.

13.70 **(a)** C; **(b)** m; **(c)** N/C = V/m; **(d)** V; **(e)** V; **(f)** C m; **(g)** dimensionless; **(h)** J/mol.

13.71 If the left and right electrodes are at different temperatures, the emf is nonzero.

Chapter 14

14.1 **(a)** T; **(b)** F.

14.2 **(a)** T; **(b)** F; **(c)** T; **(d)** T.

14.3 **(a)** $E_{tr} = \frac{3}{2}nRT = 1.5(1.00 \text{ mol})(8.314 \text{ J/mol-K})(298 \text{ K}) = 3720 \text{ J} = 889 \text{ cal}$.

(b) 3720 J.

(c) $(0.470 \text{ g})/(16.0 \text{ g/mol}) = 0.0293$ mol and $\frac{3}{2}nRT = 109$ J.

14.4 **(a)** $\langle \varepsilon_{tr} \rangle = \frac{3}{2}kT = 1.5(1.381 \times 10^{-23} \text{ J/K})(571.1 \text{ K}) = 1.18 \times 10^{-20}$ J.

(b) 1.18×10^{-20} J.

14.5 $\langle \varepsilon_{tr}(T_2) \rangle / \langle \varepsilon_{tr}(T_1) \rangle = (3kT_2/2)/(3kT_1/2) = T_2/T_1 = (373.1 \text{ K})/(273.1 \text{ K}) = 1.366$.

14.6 $v_{rms} = (3RT/M)^{1/2}$, so $v_{rms}(\text{Ne})/v_{rms}(\text{He}) = (M_{He}/M_{Ne})^{1/2} = (4.0026/20.179)^{1/2} =$ 0.4454.

14.7 $v_{rms} = (3RT_{H_2}/M_{H_2})^{1/2} = (3RT_{O_2}/M_{O_2})^{1/2}$ and $T_{H_2} = (M_{H_2}/M_{O_2})T_{O_2} =$ $(2.016/32.00)(293.1 \text{ K}) = 18.5$ K.

14.8 $E_{tr} = \frac{3}{2}nRT = \frac{3}{2}PV = 1.5(1.00 \text{ atm})(90 \times 10^6 \text{ cm}^3)(8.314 \text{ J})/(82.06 \text{ cm}^3 \text{ atm}) =$ $1.3_7 \times 10^7$ J. The answer is the same for 40°C.

14.9 **(a)** 0 to ∞; **(b)** $-\infty$ to ∞; **(c)** 0 (Fig. 14.5); **(d)** the number of molecules whose speed is in the range from v to $v + dv$.

14.10 **(a)** T; **(b)** F; **(c)** T; **(d)** F.

14.11 **(a)** $m/k = N_A m/N_A k = M/R$. The interval is small enough to be considered "infinitesimal" and Eq. (14.44) gives $dN_\upsilon =$ $4\pi N(M/2\pi RT)^{3/2} e^{-M\upsilon^2/2RT} \upsilon^2\, d\upsilon =$

$$4\pi(6.02 \times 10^{23})\left[\frac{0.0320 \text{ kg/mol}}{2\pi(8.314 \text{ J/mol-K})(300 \text{ K})}\right]^{3/2} \times$$

$$\exp\left[-\frac{(0.0320 \text{ kg/mol})(500 \text{ m/s})^2}{2(8.314 \text{ J/mol-K})(300 \text{ K})}\right](500 \text{ m/s})^2(0.001 \text{ m/s}) = 1.1 \times 10^{18}$$

(b) Considering the interval as infinitesimal and using (14.42) with x replaced by z, we have $dN_{\upsilon_z} = N(M/2\pi RT)^{1/2} \exp(-M\upsilon_z^2/2RT)\, d\upsilon_z =$ $(6.02 \times 10^{23})[0.0320 \text{ kg/mol}/2\pi(8.314 \text{ J/mol-K})(300 \text{ K})]^{1/2} \times$ $\exp[-(0.0320 \text{ kg/mol})(150 \text{ m/s})^2/2(8.314 \text{ J/mol-K})(300 \text{ K})]\,(0.001 \text{ m/s}) =$ 7.45×10^{17}.

(c) The fraction of molecules with x and z velocity components simultaneously in the ranges υ_x to $\upsilon_x + d\upsilon_x$ and υ_z to $\upsilon_z + d\upsilon_z$ is $g(\upsilon_x)g(\upsilon_z)\, d\upsilon_x\, d\upsilon_z = (dN_{\upsilon_x}/N)(dN_{\upsilon_z}/N)$ and the number of such molecules is $(dN_{\upsilon_x}/N)(dN_{\upsilon_z}/N)\,N = dN_{\upsilon_x} dN_{\upsilon_z}/N$. From part (b), dN_{υ_x} $= dN_{\upsilon_z} = 7.45 \times 10^{17}$ and $dN_{\upsilon_x} dN_{\upsilon_z}/N = (7.45 \times 10^{17})^2/(6.022 \times 10^{23}) =$ 9.22×10^{11}.

14.12 This probability equals $dN_\upsilon/N = 4\pi(M/2\pi RT)^{3/2} e^{-M\upsilon^2/2RT} \upsilon^2\, d\upsilon =$ $4\pi[(0.0160 \text{ kg/mol})/2\pi(8.314 \text{ J/mol-K})(300 \text{ K})]^{3/2} \times$ $\exp[-(0.0160 \text{ kg/mol})(400 \text{ m/s})^2/2(8.314 \text{ J/mol-K})(300 \text{ K})] \times$ $(400 \text{ m/s})^2(0.001 \text{ m/s}) = 1.24 \times 10^{-6}$.

14.13 **(a)** Fig. 14.6 gives $G(\upsilon) = 0.0018_5$ s/m at $\upsilon = 500$ m/s, so $dN_\upsilon/N =$ $G(\upsilon)\, d\upsilon = (0.0018_5 \text{ s/m})(0.0002 \text{ m/s}) = 3.7 \times 10^{-7}$ and $dN_\upsilon =$ $(6.02 \times 10^{23})(3.7 \times 10^{-7}) = 2.2 \times 10^{17}$ molecules.

(b) At 1000 K, Fig. 14.6 gives $G(\upsilon) = 0.0008$ s/m; $dN_\upsilon/N =$ $(0.0008 \text{ s/m})(0.0002 \text{ m/s}) = 1.6 \times 10^{-7}$; $dN_\upsilon = 1 \times 10^{17}$.

14.14 **(a)** $\Pr = \int_{\upsilon_1}^{\upsilon_2} G(\upsilon)\, d\upsilon \cong G(\upsilon_1) \int_{\upsilon_1}^{\upsilon_2} d\upsilon = G(\upsilon_1)(\upsilon_2 - \upsilon_1)$ if $G(\upsilon)$ is essentially constant in the interval.

(b) As in the example, $M/2RT = 3.52 \times 10^{-6}\ s^2/m^2$. At $\upsilon = 90.000$ m/s, $G(\upsilon) = \exp[-(90\ m/s)^2(3.52 \times 10^{-6}\ s^2/m^2)] \times \pi^{-3/2}(3.52 \times 10^{-6}\ s^2/m^2)^{3/2}4\pi(90\ m/s)^2 = 1.17328 \times 10^{-4}$ s/m. Replacing 90 by 90.002, we find that at 90.002 m/s, $G(\upsilon) = 1.17333 \times 10^{-4}$ s/m, and the change is 0.004%.

14.15 (a) Equation (14.44) gives $(dN_{\upsilon_2}/N)/(dN_{\upsilon_1}/N) =$ $\upsilon_2^2 e^{-m\upsilon_2^2/2kT}/\upsilon_1^2 e^{-m\upsilon_1^2/2kT} = (\upsilon_2^2/\upsilon_1^2)\exp[M(\upsilon_1^2 - \upsilon_2^2)/2RT] =$

$$\frac{500^2}{1500^2}\exp\left[\frac{(0.0320\ \text{kg/mol})(1500^2 - 500^2)\text{m}^2/\text{s}^2}{2(8.314\ \text{J/mol-K})(298.1\ \text{K})}\right] = 45000$$

(b) From (14.44), $(m/2\pi kT)^{3/2}\exp(-m\upsilon_1^2/2kT)\,4\pi\upsilon_1^2 = (m/2\pi kT)^{3/2}\exp(-m\upsilon_2^2/2kT)\,4\pi\upsilon_2^2$ and $\exp[m(\upsilon_2^2 - \upsilon_1^2)/2kT] = \upsilon_2^2/\upsilon_1^2$. Taking logs, we have $m(\upsilon_2^2 - \upsilon_1^2)/2kT = 2\ln(\upsilon_2/\upsilon_1)$ and $T = M(\upsilon_2^2 - \upsilon_1^2)/[4R\ln(\upsilon_2/\upsilon_1)] = (0.0320\ \text{kg/mol})(1500^2 - 500^2)(\text{m/s})^2/4(8.314\ \text{J/mol-K})\ln 3.00 = 1750$ K.

14.16 $\langle f_1(w) + f_2(w)\rangle = \int_{w_{min}}^{w_{max}} [f_1(w) + f_2(w)]g(w)\,dw = \int_{w_{min}}^{w_{max}} f_1(w)g(w)\,dw + \int_{w_{min}}^{w_{max}} f_2(w)g(w)\,dw = \langle f_1(w)\rangle + \langle f_2(w)\rangle$.

(b) $\langle cf\rangle = \int_{w_{min}}^{w_{max}} cf(w)\,dw = c\int_{w_{min}}^{w_{max}} f\,dw = c\langle f\rangle$.

14.17 The fraction of molecules with x and y velocity components simultaneously in the ranges υ_x to $\upsilon_x + d\upsilon_x$ and υ_y to $\upsilon_y + d\upsilon_y$ is $g(\upsilon_x)g(\upsilon_y)\,d\upsilon_x\,d\upsilon_y = (m/2\pi kT)\exp(-m\upsilon^2/2kT)\,d\upsilon_x\,d\upsilon_y$, where $\upsilon^2 = \upsilon_x^2 + \upsilon_y^2$. Molecules with speeds lying between υ and $\upsilon + d\upsilon$ have their velocity vectors lying within a thin annulus (ring) of inner radius υ and outer radius $\upsilon + d\upsilon$. The area of this annulus is $\pi(\upsilon + d\upsilon)^2 - \pi\upsilon^2 = 2\pi\upsilon\,d\upsilon$, where the $\pi(d\upsilon)^2$ term is negligible. The probability that $\mathbf{v}$ lies in this annulus is the following sum over the annulus:

$\Sigma\,(m/2\pi kT)e^{-m\upsilon^2/2kT}d\upsilon_x\,d\upsilon_y = (m/2\pi kT)e^{-m\upsilon^2/2kT}(2\pi\upsilon\,d\upsilon) =$

$(m/kT)e^{-m\upsilon^2/2kT}\upsilon\,d\upsilon$, and this is the desired probability.

14.18 **(a)** Since the integrand has the same value at $-x$ as at $+x$, the areas on each side of the origin are equal, and the integral from $-\infty$ to 0 equals the integral from 0 to ∞.

(b) For very small x, $e^{-ax^2} \approx 1$ and the graph resembles that of $y = x$. For large x, the exponential factor dominates the factor x and the function goes to 0. The graph resembles the $\upsilon = 1$ graph in Fig. 17.18. The area on the left side of the origin is negative and exactly cancels the area on the right side of the origin.

14.19 **(a)** Let $z = ax^2$. Then $dz = 2ax\,dx$ and $\int_0^\infty xe^{-ax^2}\,dx = (1/2a)\int_0^\infty e^{-z}\,dz = -(1/2a)e^{-z}\Big|_0^\infty = -(1/2a)(0-1) = 1/2a$.

(b) $(\partial/\partial a)\int_0^\infty xe^{-ax^2}\,dx = \int_0^\infty (\partial/\partial a)\,xe^{-ax^2}\,dx = -\int_0^\infty x^3e^{-ax^2}\,dx = (\partial/\partial a)(1/2a) = -1/2a^2$ and $\int_0^\infty x^3e^{-ax^2}\,dx = 1/2a^2$. Similarly, $(\partial/\partial a)\int_0^\infty x^3e^{-ax^2}\,dx = -\int_0^\infty x^5e^{-ax^2}\,dx = (\partial/\partial a)(1/2a^2) = -1/a^3$ and $\int_0^\infty x^5e^{-ax^2}\,dx = 1/a^3$.

14.20 **(a)** $\upsilon_{rms} = (3RT/M)^{1/2} = [3(8.314\text{ J/mol-K})(500\text{ K})/(0.04401\text{ kg/mol})]^{1/2} = 532\text{ m/s}$.

(b) $\langle\upsilon\rangle = (8/3\pi)^{1/2}\upsilon_{rms} = 490\text{ m/s}$.

(c) $\upsilon_{mp} = (2/3)^{1/2}\upsilon_{rms} = 435\text{ m/s}$.

14.21 $dG(\upsilon)/d\upsilon = (m/2\pi kT)^{3/2}[8\pi\upsilon e^{-m\upsilon^2/2kT} - 4\pi\upsilon^2(m\upsilon/kT)e^{-m\upsilon^2/2kT}]$. The equation $dG(\upsilon)/d\upsilon = 0$ gives $8\pi\upsilon_{mp} = 4\pi m\upsilon_{mp}^3/kT$ and $\upsilon_{mp} = (2kT/m)^{1/2} = (2RT/M)^{1/2}$. ($\upsilon = 0$ and $\upsilon = \infty$ also satisfy $G' = 0$ but these are minima.)

14.22 $\langle\upsilon^2\rangle = \int_0^\infty \upsilon^2G(\upsilon)\,d\upsilon = 4\pi(m/2\pi kT)^{3/2}\int_0^\infty \upsilon^4e^{-m\upsilon^2/2kT}\,d\upsilon = 4\pi(m/2\pi kT)^{3/2}(24\pi^{1/2}/2^5\cdot 2)(2kT/m)^{5/2} = 3kT/m = 3RT/M$, where integral 3 (with $n = 2$) of Table 14.1 was used.

14.23 $\langle\upsilon^3\rangle = \int_0^\infty \upsilon^3G(\upsilon)\,d\upsilon = 4\pi(m/2\pi kT)^{3/2}\int_0^\infty \upsilon^5e^{-m\upsilon^2/2kT}\,d\upsilon = 4\pi(m/2\pi kT)^{3/2}(2/2)(2kT/m)^3 = 2^{7/2}(kT/m)^{3/2}/\pi^{1/2} = 2^{7/2}(RT/M)^{3/2}/\pi^{1/2}$. We have $\langle\upsilon\rangle\langle\upsilon^2\rangle = (8RT/\pi M)^{1/2}(3RT/M) = 3(2^{3/2})(RT/M)^{3/2}/\pi^{1/2} \neq \langle\upsilon^3\rangle$.

14.24 **(a)** $\langle \upsilon_x \rangle = \int_{-\infty}^{\infty} \upsilon_x g(\upsilon_x)\, d\upsilon_x = (m/2\pi kT)^{1/2} \int_{-\infty}^{\infty} \upsilon_x e^{-m\upsilon_x^2/2kT} d\upsilon_x = 0$, where integral 4 of Table 14.1 was used. From Fig. 14.5, υ_x is as likely to be negative as positive, so $\langle \upsilon_x \rangle = 0$.

(b) In calculating $\langle \upsilon_x \rangle$, we average positive and negative values of υ_x and get zero. In calculating $\langle \upsilon_x^2 \rangle$, all the $\langle \upsilon_x^2 \rangle$ values are nonnegative, and we must get $\langle \upsilon_x^2 \rangle$ to be positive.

(c) $\upsilon_{x,\text{rms}} = \langle \upsilon_x^2 \rangle^{1/2} = (kT/m)^{1/2} = (RT/M)^{1/2}$, where (14.37) was used.

(d) Zero; see Fig. 14.5.

14.25 $\langle \upsilon_x^4 \rangle = \int_{-\infty}^{\infty} \upsilon_x^4 g(\upsilon_x)\, d\upsilon_x = 2(m/2\pi kT)^{1/2} \int_0^{\infty} \upsilon_x^4 e^{-m\upsilon_x^2/2kT} d\upsilon_x$. Integral 3 with $n = 2$ in Table 14.1 gives $\langle \upsilon_x^4 \rangle = 2(m/2\pi kT)^{1/2}(24\pi^{1/2}/2^5 \cdot 2)(2kT/m)^{5/2} = 3(kT/m)^2 = 3(RT/M)^2$.

14.26 Differentiation of the distribution function in Eq. (14.52) with respect to ε_{tr} gives $2\pi(\pi kT)^{-3/2}[\frac{1}{2}\varepsilon_{\text{tr}}^{-1/2} e^{-\varepsilon_{\text{tr}}/kT} - (1/kT)\varepsilon_{\text{tr}}^{1/2} e^{-\varepsilon_{\text{tr}}/kT}] = 0$ and $\varepsilon_{\text{tr,mp}} = \frac{1}{2}kT = \langle \varepsilon_{\text{tr}} \rangle/3$.

14.27 **(a)** $(2\pi)^{1/2} I(u) = \int_0^u e^{-s^2/2} ds = \int_0^u [1 - s^2/2 + (1/2!)s^4/4 - (1/3!)s^6/8 + \cdots]\, ds = u - u^3/6 + u^5/40 - u^7/336 + \cdots$.

(b) $(2\pi)^{1/2} I(0.30) = 0.30 - (0.30)^3/6 + (0.30)^5/40 - (0.30)^7/336 + \cdots = 0.2956$ and $I(0.30) = 0.118$.

14.28 **(a)** The probability is $\int_0^{\upsilon'} G(\upsilon)\, d\upsilon = (m/2\pi kT)^{3/2} \int_0^{\upsilon'} e^{-m\upsilon^2/2kT} 4\pi\upsilon^2\, d\upsilon = (4/\pi^{1/2}\upsilon_{\text{mp}}^3) \int_0^{\upsilon'} e^{-(\upsilon/\upsilon_{\text{mp}})^2} \upsilon^2\, d\upsilon = (4/\pi^{1/2}\upsilon_{\text{mp}}^3)B$, where $\upsilon_{\text{mp}} = (2kT/m)^{1/2}$ and $B = \int_0^{\upsilon'} e^{-(\upsilon/\upsilon_{\text{mp}})^2} \upsilon^2\, d\upsilon$. The integration-by-parts formula is $\int x\, dy = xy - \int y\, dx$. Let $x = \upsilon$ and $dy = e^{-(\upsilon/\upsilon_{mp})^2} \upsilon\, d\upsilon$. Then $y = -\frac{1}{2}\upsilon_{\text{mp}}^2 e^{-(\upsilon/\upsilon_{mp})^2}$ and $B = -\frac{1}{2}\upsilon_{\text{mp}}^2 \upsilon e^{-(\upsilon/\upsilon_{mp})^2}\Big|_0^{\upsilon'} + \frac{1}{2}\upsilon_{\text{mp}}^2 \int_0^{\upsilon'} e^{-(\upsilon/\upsilon_{\text{mp}})^2} d\upsilon$. Let $s = 2^{1/2}\upsilon/\upsilon_{\text{mp}}$. Then $B = -\frac{1}{2}\upsilon_{\text{mp}}^2 \upsilon' e^{-(\upsilon'/\upsilon_{\text{mp}})^2} + \frac{1}{2}\upsilon_{\text{mp}}^2 \int_0^{2^{1/2}\upsilon'/\upsilon_{\text{mp}}} e^{-s^2/2} \upsilon_{\text{mp}} 2^{-1/2}\, ds =$

$-\frac{1}{2}\upsilon_{mp}^2\upsilon' e^{-(\upsilon'/\upsilon_{mp})^2} + \upsilon_{mp}^3 2^{-3/2}(2\pi)^{1/2} I(2^{1/2}\upsilon'/\upsilon_{mp})$ and $\int_0^{\upsilon'} G(\upsilon)\, d\upsilon =$

$2I(2^{1/2}\upsilon'/\upsilon_{mp}) - 2\pi^{-1/2}(\upsilon'/\upsilon_{mp})e^{-(\upsilon'/\upsilon_{mp})^2}$.

(b) The fraction with speed in the range from 0 to $4.243\upsilon_{mp}$ is $2I(2^{1/2} \cdot 4.243) - 2\pi^{-1/2}(4.243)e^{-(4.243)^2} = 2I(6.000) - (7.27 \times 10^{-8}) =$ $2(0.4999999990) - (7.27 \times 10^{-8}) = 0.9999999253$, since Fig. 14.10 gives $0.5 - I(6) = 1.0 \times 10^{-9}$. The fraction with speeds exceeding $4.243\upsilon_{mp}$ is $1 - 0.9999999253 = 7.5 \times 10^{-8}$.

14.29 (a) The theory of relativity shows that speeds must be less than c, the speed of light in vacuum, so the correct range is 0 to c.

(b) The number of molecules whose speed according to the Maxwell distribution lies between c and infinity is essentially zero at ordinary temperatures.

14.30 From Eq. (14.58), $0.872 = (M_{O_2}/M)^{1/2} = [(32.0 \text{ g/mol})/M]^{1/2}$ and $M = 42.1$ g/mol. The only hydrocarbon with this molar mass is C_3H_6.

14.31 (a) Since $l << d$, we can take k as equal to 1 in Eq. (14.59).:We have $P_{vp} = (\Delta m/\mathscr{A}_{hole}\, \Delta t)(2\pi RT/M)^{1/2}$

$$P_{vp} = \frac{(10.5 \times 10^{-6} \text{ kg})[2\pi(8.314 \text{ J/mol-K})(1690 \text{ K})/0.04496 \text{ kg/mol}]^{1/2}}{\pi(0.001763/2)^2 \text{ m}^2 (49.5 \times 60 \text{ s})}$$

$$P_{vp} = (2.03 \text{ N/m}^2)\frac{82.06(10^{-2}\text{ m})^3 \text{ atm}}{8.314 \text{ J}}\frac{760 \text{ torr}}{1 \text{ atm}} = 0.0152 \text{ torr}$$

(b) Equation (14.67) applies. The order of magnitude of the molecular diameter d in (14.67) is a couple of angstroms. The order of magnitude of λ is

$$\frac{(82 \text{ cm}^3\text{-atm/mol-K})(1690 \text{ K})}{2^{1/2}\pi(2 \times 10^{-8} \text{ cm})^2(0.015/760) \text{ atm}\,(6 \times 10^{23}/\text{mol})} = 7 \text{ cm}$$

Since $d_{hole} = 0.2$ cm, the condition $\lambda >> d_{hole}$ is met.

14.32 $k = (1 + l/d_{hole})^{-1} = (1 + 0.13/0.66)^{-1} = 0.83_5$. $P_{vp} = (\Delta m/k\mathscr{A}_{hole}\, \Delta t)(2\pi RT/M)^{1/2}$

$$P_{vp} = \frac{(0.285 \times 10^{-6}\text{ kg})[2\pi(8.314\text{ J/mol-K})(823\text{ K})/0.7206\text{ kg/mol}]^{1/2}}{0.83_5\pi(0.00066/2)^2\text{ m}^2(500\text{ s})}$$

$$P_{vp} = (0.48_7\text{ N/m}^2)\frac{82.06\ (10^{-2}\text{m})^3\text{ atm}}{8.314\text{ J}}\frac{760\text{ torr}}{1\text{ atm}} = 0.0037\text{ torr at 823 K}$$

$$P_{vp} = \frac{(1.250 \times 10^{-6}\text{ kg})[2\pi(8.314\text{ J/mol-K})(873\text{ K})/0.7206\text{ kg/mol}]^{1/2}}{0.83_5\pi(0.00066/2)^2\text{ m}^2(500\text{ s})}$$

$$P_{vp} = (2.2_0\text{ N/m}^2)\frac{82.06\ (10^{-2}\text{m})^3\text{ atm}}{8.314\text{ J}}\frac{760\text{ torr}}{1\text{ atm}} = 0.016_5\text{ torr at 873 K}$$

(b) $d\ln P_{vp}/dT = \Delta_{vap}H_m/RT^2$ and $\ln(P_2/P_1) = (\Delta H_m/R)(T_1^{-1} - T_2^{-1})$ so

$\ln(0.016_5/0.0037) = [\Delta H_m/(8.314\text{ J/mol-K})]\text{K}^{-1}(823^{-1} - 873^{-1})$ and

$\Delta H_m = 175$ kJ/mol.

14.33 **(a)** $k = (1 + l/d_{hole})^{-1} = (1 + 2.1/3.1)^{-1} = 0.60$.

$P_{vp} = (\Delta m/k\mathcal{A}_{hole}\Delta t)(2\pi RT/M)^{1/2}$

$$P_{vp} = \frac{(4.07 \times 10^{-6}\text{ kg})[2\pi(8.314\text{ J/mol-K})(465\text{ K})/0.3913\text{ kg/mol}]^{1/2}}{0.60\pi(0.0031/2)^2\text{ m}^2(7250\text{ s})}$$

$$P_{vp} = (0.0309\text{ N/m}^2)\frac{82.06\ (10^{-2}\text{m})^3\text{ atm}}{8.314\text{ J}}\frac{760\text{ torr}}{1\text{ atm}} = 0.000232\text{ torr at 465 K}$$

$$P_{vp} = \frac{(7.32 \times 10^{-6}\text{ kg})[2\pi(8.314\text{ J/mol-K})(484.2\text{ K})/0.3913\text{ kg/mol}]^{1/2}}{0.60\pi(0.0031/2)^2\text{ m}^2(4005\text{ s})}$$

$$P_{vp} = (0.103\text{ N/m}^2)\frac{82.06\ (10^{-2}\text{m})^3\text{ atm}}{8.314\text{ J}}\frac{760\text{ torr}}{1\text{ atm}} = 0.000770\text{ torr at 484.2 K}$$

(b) $d\ln P_{vp}/dT = \Delta_{vap}H_m/RT^2$ and $\ln(P_2/P_1) = (\Delta H_m/R)(T_1^{-1} - T_2^{-1})$ so

$\ln(0.000770/0.000232) = [\Delta H_m/(8.314\text{ J/mol-K})]\text{K}^{-1}(465^{-1} - 484.2^{-1})$ and

$\Delta H_m = 117$ kJ/mol.

(c) $1 + k\mathcal{A}_{hole}/\mathcal{A}_{sample} = 1 + 0.60(3.1/10.0)^2 = 1.06$. Multiplication of the pressures by this factor gives 0.000246 torr and 0.000816 torr.

(d) $d\ln P_{vp}/dT = (d/dT)[A - B(T/\text{K})^{-1}] = \Delta_{vap}H_m/RT^2 = BT^{-2}\text{ K}^{-1}$ and

$\Delta_{vap}H_m = BR\text{ K}^{-1} = (14290)(8.3145\text{ J/mol-K})\text{K}^{-1} = 118.8$ kJ/mol.

14.34 The CO_2 partial pressure is 0.00038 atm and Eq. (14.56) gives dN_W/dt =

$$\frac{[1.0\times(10^{-2}\text{m})]^2(0.00038\text{ atm})(6.02\times10^{23}\text{mol}^{-1})}{4[82.06\,(10^{-2}\text{m})^3\text{-atm/mol-K}](298\text{ K})}\left[\frac{8(8.314\text{ J/mol-K})(298\text{ K})}{\pi(0.0440\text{ kg/mol})}\right]^{1/2}$$

$= 8.9 \times 10^{19}\ \text{s}^{-1}$.

The mass striking the leaf in 1 s is $[(8.9 \times 10^{19})/(6.02 \times 10^{23}\ \text{mol}^{-1})](44\ \text{g/mol})$ = 6.5 mg.

14.35 $P = (0.010/760)$ atm (101325 Pa/atm) = 1.33 N/m^2. As discussed in the text, we can take the evaporation rate as essentially equal to the rate at which molecules of vapor in equilibrium with the liquid strike the liquid. Equation (14.58) gives

$$\frac{dN}{dt}=\frac{(1.33\text{ N/m}^2)(6.02\times10^{23}\text{/mol})(10^{-4}\text{m}^2)}{[2\pi(0.3906\text{ kg/mol})(8.3145\text{ J/mol-K})(393\text{ K})]^{1/2}}=8.9\times10^{17}\text{s}^{-1}$$

$$(8.9\times10^{17}\text{ molecules})\frac{1\text{ mole}}{6.02\times10^{23}\text{ molecules}}\frac{390.6\text{ g}}{1\text{ mole}}=0.58\text{ mg}$$

14.36 $I \propto (dN/dt) \times \Pr(\text{i})$, where dN/dt is the effusion rate and Pr(i) is the probability of ionization. From Eq. (14.58), $dN/dt \propto P/T^{1/2}$. If x is the direction of motion of the molecular beam, the statements in the problem give $\Pr(\text{i}) \propto 1/\langle \upsilon_x \rangle \propto 1/T^{1/2}$. Hence $I \propto (P/T^{1/2}) \times (1/T^{1/2})$ and $I \propto P/T$.

14.37 **(a)** T; **(b)** F.

14.38 **(a)** $z_b(b) = (2 + 2)$ collns./s $= 4\ \text{s}^{-1}$.

(b) $Z_{bb} = [(2 + 2 + 2)\text{s}^{-1}]/(1 \times 10^{-5}\ \text{cm}^3) = 6 \times 10^5\ \text{s}^{-1}\ \text{cm}^{-3}$.
$N_b z_b(b)/V = 3(4\ \text{s}^{-1})/(1 \times 10^{-5}\ \text{cm}^3) = 12 \times 10^5\ \text{s}^{-1}\ \text{cm}^{-3} \neq Z_{bb}$.
$\frac{1}{2}N_b z_b(b)/V = \frac{1}{2}(12 \times 10^5\ \text{s}^{-1}\ \text{cm}^{-3}) = 6 \times 10^5\ \text{s}^{-1}\ \text{cm}^{-3} = Z_{bb}$.

14.39 **(a)** $\lambda = \langle \upsilon \rangle \langle t \rangle = (450\ \text{m/s})(4.0 \times 10^{-10}\ \text{s}) = 1.8 \times 10^{-7}$ m.

(b) We have $z_b(b) = 1/(5.0 \times 10^{-10}\ \text{s}) = 2 \times 10^9$ collisions per second and $z_b(c) = 1/(8.0 \times 10^{-10}\ \text{s}) = 1.25 \times 10^9$ collisions per second. $z_b(b)$ is greater than $z_b(c)$. From Eqs. (14.61) and (14.62), these rates are influenced by the molecular diameters, the molecular concentrations, and the molecular masses. Since no information is given about the concentrations (or the molecular masses), the question is defective and we cannot say which has the larger diameter. From (14.66), $\lambda_b = (385\text{ m/s})/(3.25 \times 10^9\text{ s}^{-1}) = 1.2 \times 10^{-7}\text{ m}$.

14.40 **(a)** $z_b(b) = 2^{1/2}\pi d_b^2(8RT/\pi M_b)^{1/2}P_bN_A/RT =$
$2^{1/2}\pi(3.7 \times 10^{-10}\text{ m})^2[8(8.3145\text{ J/mol-K})(298\text{ K})/\pi(0.0280\text{ kg/mol})]^{1/2} \times (1.00\text{ atm})(6.02 \times 10^{23}/\text{mol})/(82.06 \times 10^{-6}\text{ m}^3\text{-atm/mol-K})(298\text{ K}) = 7.1 \times 10^9\text{ s}^{-1}$.

(b) $Z_{bb} = \frac{1}{2}N_bz_b(b)/V = \frac{1}{2}z_b(b)(PN_A/RT) = \frac{1}{2}(7.1 \times 10^9\text{ s}^{-1})(1.00\text{ atm}) \times (6.02 \times 10^{23}/\text{mol})/(82.06\text{ cm}^3\text{-atm/mol-K})(298\text{ K}) = 8.7 \times 10^{28}\text{ s}^{-1}\text{ cm}^{-3}$.

(c) $1.0 \times 10^{-6}\text{ torr} = 1.3 \times 10^{-9}\text{ atm}$. $z_b(b)$ is proportional to P and Z_{bb} is proportional to P^2, so $z_b(b) = (7.1 \times 10^9\text{ s}^{-1})(1.3 \times 10^{-9}\text{ atm}/1\text{ atm}) = 9.3\text{ s}^{-1}$ and $Z_{bb} = (8.7 \times 10^{28}\text{ s}^{-1}\text{ cm}^{-3})(1.3 \times 10^{-9}/1)^2 = 1.5 \times 10^{11}\text{ s}^{-1}\text{ cm}^{-3}$.

14.41 Let $b = CO_2$ and $c = N_2$. $P_b = 0.97(4.7\text{ torr})(1\text{ atm}/760\text{ torr}) = 0.0060\text{ atm}$ and $P_c = 0.00019\text{ atm}$.

(a) From (14.62), $z_b(b) = 4\pi^{1/2}d_b^2(RT/M_b)^{1/2}P_bN_A/RT =$
$4\pi^{1/2}(4.6 \times 10^{-10}\text{ m})^2[(8.3145\text{ J/mol-K})(220\text{ K})/(0.044\text{ kg/mol})]^{1/2} \times (0.0060\text{ atm})(6.02 \times 10^{23}/\text{mol})/(82.06 \times 10^{-6}\text{ m}^3\text{-atm/mol-K})(220\text{ K}) = 6.1 \times 10^7\text{ s}^{-1}$.

(b) $z_c(b) = (8\pi)^{1/2}(r_b + r_c)^2[RT(M_b^{-1} + M_c^{-1})]^{1/2}P_bN_A/RT =$
$(8\pi)^{1/2}(4.1_5 \times 10^{-10}\text{ m})^2(RT)^{1/2}[(0.044\text{ kg/mol})^{-1} + (0.028\text{ kg/mol})^{-1}]^{1/2} \times (0.0060\text{ atm})N_A/RT = 5.6_5 \times 10^7\text{ s}^{-1}$.

(c) $z_c(c) = 4\pi^{1/2}(3.7 \times 10^{-10}\text{ m})^2[RT/(0.028\text{ kg/mol})]^{1/2}(0.00019\text{ atm})N_A/RT = 1.6 \times 10^6\text{ s}^{-1}$. $z_c(c) + z_c(b) = 1.6 \times 10^6\text{ s}^{-1} + 5.6_5 \times 10^7\text{ s}^{-1} = 5.8_1 \times 10^7\text{ s}^{-1}$.

(d) $Z_{bc} = N_bz_b(c)/V = N_cz_c(b)/V = z_c(b)P_cN_A/RT =$
$(5.6_5 \times 10^7\text{ s}^{-1})(0.00019\text{ atm})N_A/RT = 3.6 \times 10^{23}\text{ s}^{-1}\text{ cm}^{-3}$.

14.42 **(a)** $\lambda = RT/2^{1/2}\pi d^2PN_A = (82.06\text{ cm}^3\text{-atm/mol-K})(300\text{ K})/$
$2^{1/2}\pi(3.7 \times 10^{-8}\text{ cm})^2(750/760)\text{ atm }(6.02 \times 10^{23}/\text{mol}) = 6.8 \times 10^{-6}\text{ cm}$.

(b) $\lambda = (6.8 \times 10^{-6}\text{ cm})(750/1) = 5.1 \times 10^{-3}\text{ cm}$.

(c) $5.1 \times 10^3\text{ cm}$.

14.43 14100 ft = (14100 ft)(12 in./1 ft)(2.54 cm/1 in.)(0.01 m/1 cm) = 4298 m. $P = P_0 e^{-Mgz/RT}$ = (760 torr) exp[–(0.029 kg/mol)(9.81 m/s^2)(4298 m)/ (8.3145 J/mol-K)(290 K)] = 458 torr.

14.44 Mgz/RT = (0.029 kg/mol)(9.81 m/s^2)(30 × 12 × 0.0254 m)/ (8.314 J/mol-K)(298 K) = 0.00105. $P_{\text{bot}} - P_{\text{top}} = P_{\text{bot}}(1 - e^{-Mgz/RT})$ = (1 atm)(1 – $e^{-0.00105}$) = 0.00105 atm.

14.45 $P/P_0 = e^{-Mgz/RT} = 0.5$ and $z = -(RT/Mg) \ln 0.5$ = –[(8.3145 J/mol-K)(250 K)/(0.029 kg/mol)(9.81 m/s^2)] ln 0.5 = 5060 m.

14.46 M_n = 0.97(44) + 0.03(28) = 43.5. $P = P_0 e^{-Mgz/RT}$ = (4.7 torr)exp[–(0.0435 kg/mol)(3.7 m/s^2)(40000 m)/(8.314 J/mol-K)(180 K)] = 0.064 torr.

14.47 **(a)** With $T = T_0 - az$, (14.70) gives $\ln(P'/P_0) = -(Mg/R)\int_0^{z'} (T_0 - az)^{-1}\, dz$ = $(Mg/aR)\ln(T_0 - az)\,|_0^{z'} = (Mg/aR)\ln[(T_0 - az')/T_0] = \ln[(T_0 - az')/T_0]^{Mg/aR}$, so $P'/P_0 = [(T_0 - az')/T_0]^{Mg/aR}$.

14.48 **(a)** For $C_{V,\text{m}}$, the translation contribution is $3R/2$, the rotational contribution is $3R/2$, and the predicted vibrational contribution is $2[3(5) - 6]R/2 = 9R$, so $C_{V,\text{m}}$ is predicted to be $12R$. $C_{P,\text{m}} = C_{V,\text{m}} + R$ and $C_{P,\text{m}}$ is predicted to be $13R$.

(b) No; the actual $C_{P,\text{m}}$ is less than $13R$ at 400 K.

(c) $C_{P,\text{m}}$ will reach $13R$ at high T's.

14.49 **(a)** $\int_{-\infty}^{\infty} g(w)\, dw = 1 = \int_{-\infty}^{\infty} Ae^{-cw^2/kT}\, dw = 2A(\pi^{1/2}/2)(kT/c)^{1/2}$ and $A = (c/\pi kT)^{1/2}$, where integrals 1 and 2 in Table 14.1 were used.

(b) $\langle \varepsilon_w \rangle = \int_{-\infty}^{\infty} \varepsilon_w g(w)\, dw = A\int_{-\infty}^{\infty} cw^2 e^{-cw^2/kT}\, dw = 2Ac(2!\pi^{1/2}/2^3 1!)(kT/c)^{3/2}$ = $\frac{1}{2}kT$, where $A = (c/\pi kT)^{1/2}$ and integrals 1 and 3 of Table 14.1 were used.

14.50 $\frac{1}{2}m\upsilon^2 = \frac{3}{2}kT$ and $\upsilon_{rms} = (3kT/m)^{1/2} =$
$[3(1.38 \times 10^{-23}\text{ J/K})(298\text{ K})/(1.0 \times 10^{-13}\text{ kg})]^{1/2} = 0.00035\text{ m/s}$.

14.51 $\langle\upsilon^2\rangle = 3RT/M$ and $\langle\upsilon\rangle^2 = 8RT/\pi M$, so $\langle\upsilon^2\rangle \neq \langle\upsilon\rangle^2$.

14.52 The possible values of s are 1, 2, 3, 4, 5, 6 and the probability of each value is 1/6. Hence $\langle s\rangle = \sum_s sp(s) = 1(1/6) + 2(1/6) + 3(1/6) + 4(1/6) + 5(1/6) + 6(1/6) = 21/6 = 3.5$. Also, $\langle s^2\rangle = \sum_s s^2p(s) = 1^2(1/6) + 2^2(1/6) + 3^2(1/6) + 4^2(1/6) + 5^2(1/6) + 6^2(1/6) = 91/6 = 15.17$. $\langle s\rangle^2 = (3.5)^2 = 12.25 \neq \langle s^2\rangle$.

14.53 **(a)** $\sigma^2_{\upsilon_x} = \langle\upsilon_x^2\rangle - \langle\upsilon_x\rangle^2 = kT/m - 0$ and $\sigma_{\upsilon_x} = (kT/m)^{1/2}$.

(b) $\langle\upsilon_x\rangle = 0$ and $\sigma = (kT/m)^{1/2}$. The distribution function (14.42) is $(2\pi)^{-1/2}\sigma^{-1}e^{-\upsilon_x^2/2\sigma^2}$ and the desired fraction is $(2\pi)^{-1/2}\sigma^{-1}\int_{-\sigma}^{\sigma} e^{-\upsilon_x^2/2\sigma^2}\,d\upsilon_x = 2(2\pi)^{-1/2}\sigma^{-1}\int_0^{\sigma} e^{-\upsilon_x^2/2\sigma^2}\,d\upsilon_x =$

$2(2\pi)^{-1/2}\sigma^{-1}\int_0^1 e^{-s^2/2}\sigma\,ds = 2I(1) = 0.68$, where we made the substitution $s = \upsilon_x/\sigma$ and used Eq. (14.51) and Fig. 14.10.

14.54 $h(xy) = f(x) + g(y)$. We take $(\partial/\partial x)_y$ of this equation. Let $z = xy$. The partial derivative of the left side is $\partial h(xy)/\partial x = (dh/dz)(\partial z/\partial x) = h'(z)y$. The partial derivative of the right side is $df(x)/dx = h'(z)y$; we have $h'(z) = y^{-1}df(x)/dx$. Similarly, taking $(\partial/\partial y)_x$ of $h(xy) = f(x) + g(y)$ gives $h'(z) = x^{-1}dg(y)/dy$. Hence $x^{-1}\,dg(y)/dy = y^{-1}\,df(x)/dx$ and $y[dg(y)/dy] = x[df(x)/dx] \equiv k$. By the argument that follows Eq. (14.33), k must be a constant. Hence $df(x) = (k/x)\,dx$ and $f(x) = k\ln x + a$, where a is an integration constant.

14.55 Molecules with $\upsilon_x = b$ have the tips of their velocity vectors lying in the plane $\upsilon_x = b$. (This plane is parallel to the plane formed by the υ_y and υ_z axes and is a distance b from it.) The region corresponding to $b \le \upsilon_x \le c$ is the region between the parallel planes at $\upsilon_x = b$ and $\upsilon_x = c$.

14.56 **(a)** 1, since the probability density $G(\upsilon)$ satisfies $\int_0^\infty G(\upsilon)\,d\upsilon = 1$, Eq. (14.25).

(b) A crude approximation to the area under the 300 K curve is found by taking the area of a triangle with height 2×10^{-5} s/cm (the peak of the curve) and base 10^5 cm/s (the width of the region for which $G(\upsilon)$ has a substantial value). This triangle's area is $\frac{1}{2}(10^5 \text{ cm/s})(2 \times 10^{-5} \text{ s/cm}) = 1$.

14.57 **(a)** H_2. The lighter H_2 molecules move faster, so that their average kinetic energy equals that of O_2.

(b) $\langle \varepsilon_{tr} \rangle$ is the same for both.

(c) $\rho = PM/RT$ is greater for O_2 because of the greater mass of the O_2 molecules.

(d) From (14.67), λ decreases as the molecular diameter increases. O_2 is larger than H_2, so $\lambda_{H_2} > \lambda_{O_2}$.

(e) The H_2 molecules are moving faster and so collide more often with the wall.

14.58 Substitution of $\mu_i = \mu_i^\circ(T) + RT \ln (P_i/P^\circ)$ into $\mu_i^\alpha + M_i g z^\alpha = \mu_i^\beta + M_i g z^\beta$ gives $\mu_i^\circ(T) + RT \ln (P_i^\alpha/P^\circ) + M_i g z^\alpha = \mu_i^\circ(T) + RT \ln (P_i^\beta/P^\circ) + M_i g z^\beta$ or $RT \ln (P_i^\alpha/P_i^\beta) = M_i g(z^\beta - z^\alpha)$. $\ln(P_i^\beta/P_i^\alpha) = -M_i g(z^\beta - z^\alpha)/RT$ and $P_i^\beta = P_i^\alpha \exp[-M_i g(z^\beta - z^\alpha)/RT]$.

14.59 **(a)** See Fig. 14.7a. **(b)** See Fig. 14.6.

14.60 λ, $z_b(b)$, and Z_{bb}.

14.61 **(a)** Statistical mechanics. **(b)** Yes.

14.62 **(a)** False. **(b)** True. See Fig. 14.5. **(c)** True. **(d)** False. **(e)** False. **(f)** True. **(g)** False. **(h)** True. See Eq. (14.13). **(i)** False.

Chapter 15

15.1 (a) T; (b) T; (c) F.

15.2 (a) Equation (15.1) gives $|q| = |k@(\Delta T/\Delta x)\Delta t| =$
(0.80 J/K-cm-s)(24 cm^2)(60 s)(50 K)/(200 cm) = 288 J.

  (b) $\Delta S_{rod} = 0$ since the rod's state is not changed. The temperature of each
end of the rod differs only slightly from the temperature of the reservoir
it is in contact with, so we can use $dS = dq_{rev}/T = dq/T$ for each reservoir.
Therefore $\Delta S = \Delta S_{hot\ res.} + \Delta S_{cold\ res.} = (-288\ J)/(325\ K) + (288\ J)/(275\ K)$
= 0.161 J/K.

15.3 $C_{V,m} = 3R/2$ for this monatomic gas, and Eq. (15.12) gives

$$k = \frac{5}{16}\left(\frac{3}{2}+\frac{9}{4}\right)\frac{(8.314\ \text{J/mol-K})}{(6.02\times10^{23}/\text{mol})(2.2\times10^{-10}\ \text{m})^2}\left[\frac{(8.314\ \text{J/mol-K})(273\ \text{K})}{\pi(0.00400\ \text{kg/mol})}\right]^{1/2}$$

$k = 0.00142$ J/K-cm-s. At 100°C, $k = 0.166$ J/K-m-s.

15.4 $C_{V,m} + 9R/4 = C_{P,m} - R + 9R/4 = C_{P,m} + 5R/4 =$
(35.309 J/mol-K) + (5/4)(8.3145 J/mol-K) = 45.7 J/mol-K. Then
$k = (5/16)(45.7$ J/mol-K)$[(8.314$ J/mol-K)(298 K)/$\pi(0.0160$ kg/mol)$]^{1/2}/$
$[(6.02 \times 10^{23}$ mol$^{-1})(4.1 \times 10^{-10}$ m)$^2] = 0.031$ J/K-s-m =
31×10^{-5} J K^{-1} s^{-1} cm^{-1}, where (15.12) was used.

15.5 $\rho = M/V_m = (18.015$ g/mol)/(18.1 cm^3/mol) = 0.995 g/cm^3 = 995 kg/m^3 and
$k = 2.8(8.314$ J/mol-K)$[17.99/(17.72)(995$ kg/m$^3)(4.46 \times 10^{-10}$ m^2/N)$]^{1/2}/$
$[(6.022 \times 10^{23}$ mol$^{-1})^{1/3}(18.1$ cm^3/mol)$^{2/3}] = 6.05 \times 10^{-5}$ J K^{-1} cm^{-2} m s^{-1} =
6.05 mJ K^{-1} cm^{-1} s^{-1}.

15.6 (a) F; (b) T; (c) T; (d) T.

15.7 The maximum $\langle \upsilon_y \rangle$ for laminar flow makes $Re = 2000$, so $\langle \upsilon_y \rangle_{max} =$
$2000\eta/\rho d = 2000(0.0089$ dyn s cm$^{-2})/(1.00$ g/cm$^3)(1.00$ cm) = 18 cm/s.

15.8 **(a)** $(32/760)$ atm $(8.314\ \text{J})/(82.06 \times 10^{-6}\ \text{m}^3\ \text{atm}) = 4266\ \text{N/m}^2$. Equation (15.17) gives $\eta = (\pi r^4/8V)(|\Delta P|/|\Delta y|)t =$ $[\pi(0.00100\ \text{m})^4/8(148 \times 10^{-6}\ \text{m}^3)][(4266\ \text{N/m}^2)/(0.240\ \text{m})](120\ \text{s}) =$ $0.00566\ \text{N m}^{-2}\ \text{s} = 5.66\ \text{mPa s} = 5.66\ \text{cP}$.

(b) In time t, a volume $V = \pi r^2 \langle d \rangle$ flows through the pipe of cross-sectional area πr^2, where $\langle d \rangle$ is the average distance traveled by the fluid. We have $\langle d \rangle = \langle \upsilon_y \rangle t$ and $\langle \upsilon_y \rangle = \langle d \rangle /t = (V/\pi r^2)/t = V/\pi r^2 t =$ $(r^2/8\eta)|\Delta P|/|\Delta y|$, where Eq. (15.17) was used. So $\langle \upsilon_y \rangle =$ $[(0.00100\ \text{m})^2/8(0.00566\ \text{N m}^{-2}\ \text{s})][(4266\ \text{N/m}^2)/(0.240\ \text{m})] =$ 0.393 m/s. Then $Re = \rho\langle \upsilon_y \rangle d/\eta =$ $[(0.00135\ \text{kg})/(10^{-6}\ \text{m}^3)](0.393\ \text{m/s})(0.00200\ \text{m})/(0.00566\ \text{N m}^{-2}\ \text{s}) =$ $187 < 2000$, so the flow is laminar.

15.9 **(a)** From (15.17), $(P_2 - P_1)/(y_2 - y_1) = -(8\eta/\pi r^4)(V/t) =$ $-8[(4 \times 10^{-3}\ \text{Pa s})/\pi(0.012_5\ \text{m})^4](5000 \times 10^{-6}\ \text{m}^3)/(60\ \text{s}) = -35\ \text{Pa/m}$.

(b) $\langle \upsilon_y \rangle = V/\mathscr{A}t = (5000\ \text{cm}^3)/\pi(1.2_5\ \text{cm})^2(60\ \text{s}) = 17\ \text{cm/s}$.

(c) $Re = \rho\langle \upsilon_y \rangle d/\eta = (0.001\ \text{kg}/10^{-6}\ \text{m}^3)(0.17\ \text{m/s})(0.025\ \text{m})/(0.004\ \text{Pa s}) =$ $1100 < 2000$, so the flow is laminar and use of Eq. (15.17) is justified. For a flow rate of 30 L/min, $\langle \upsilon_y \rangle$ and Re are 6 times as large and $Re = 6400 > 2000$, so aortic flow is turbulent during vigorous activity.

15.10 $P_1^2 - P_2^2 = (1.44 - 1.00)\ \text{atm}^2 = (0.44\ \text{atm}^2)(8.314\ \text{J})^2/(82.06 \times 10^{-6}\ \text{m}^3\ \text{atm})^2 =$ $4.52 \times 10^9\ \text{N}^2/\text{m}^4$. Eq. (15.18) gives $dn/dt = \pi(0.000210\ \text{m})^4(4.52 \times 10^9\ \text{N}^2/\text{m}^4)/$ $[16(0.0000192\ \text{N s/m}^2)(8.314\ \text{J/mol-K})(273\ \text{K})(2.20\ \text{m})] = 0.000018\ \text{mol/s}$, which is 0.00058 g/s.

15.11 $\eta_{C_6H_{14}} = \eta_{H_2O}(\rho_{C_6H_{14}} t_{C_6H_{14}}/\rho_{H_2O} t_{H_2O}) = (1.002\ \text{cP})(0.659\ \text{g/cm}^3)(67.3\ \text{s})/$ $(0.998\ \text{g/cm}^3)(136.5\ \text{s}) = 0.326\ \text{cP}$.

15.12 **(a)** The pressures at the left and right ends of C exert forces $P\pi s^2$ and $-(P + dP)\pi s^2$, respectively, on C. The viscous force on C is given by (15.13) as $\eta\mathscr{A}(d\upsilon_y/ds) = \eta(2\pi s\ dy)(d\upsilon_y/ds)$. So $-(P + dP)\pi s^2 + P\pi s^2 +$ $\eta(2\pi s\ dy)(d\upsilon_y/ds) = 0$ and $d\upsilon_y/ds = (s/2\eta)(dP/dy)$. Integration gives

$\upsilon_y = (s^2/4\eta)(dP/dy) + c$, where c is a constant. Use of $\upsilon_y = 0$ at $s = r$ gives $c = -(r^2/4\eta)(dP/dy)$. Therefore $\upsilon_y = (1/4\eta)(r^2 - s^2)(-dP/dy)$.

(b) The volume of fluid in the shell that passes a given location in time dt equals the volume of a cylindrical shell with length $\upsilon_y(s)\,dt$ and inner and outer radii s and $s + ds$; this volume is $\pi(s + ds)^2 \cdot \upsilon_y(s)\,dt - \pi s^2\upsilon_y(s)\,dt = 2\pi s\upsilon_y(s)\,ds\,dt$, since $(ds)^2$ is negligible compared with ds. Integration over all shells from $s = 0$ to r gives $dV = [\int_0^r 2\pi s\upsilon_y(s)\,ds]\,dt$. Substitution of $dV = dm/\rho$ and $\upsilon_y = (1/4\eta)(r^2 - s^2)(-dP/dy)$ gives $dm/dt = (\pi\rho/2\eta)(-dP/dy)\int_0^r (r^2s - s^3)\,ds = (\pi r^4\rho/8\eta)(-dP/dy)$. Separating P and y and integrating from one end to the other, we have $-\int_{P_2}^{P_1} dP = (dm/dt)(8\eta/\pi r^4\rho)\int_{y_2}^{y_1} dy$, which becomes $dm/dt = (\pi r^4\rho/8\eta)(P_1 - P_2)/(y_2 - y_1)$. We have $dm/dt = d(\rho V)/dt = \rho\,dV/dt = \rho V/t$, since the flow rate is constant with time. Hence (15.17) follows.

(c) Substitution of $\rho = PM/RT$ into (16\5.91), separation of P and y and integration gives $(RT/M)(dm/dt)(8\eta/\pi r^4)\int_{y_1}^{y_2} dy = -\int_{P_1}^{P_2} P\,dP = \frac{1}{2}(P_1^2 - P_2^2)$; use of $(RT/M)(dm/dt) = RT\,d(m/M)/dt = RT\,dn/dt$ gives (15.18).

15.13 In time dt, the matter in the thin shell travels a distance $\upsilon_y\,dt$ in the y direction. The volume dV of matter in the thin shell that passes a fixed location in time dt equals the length $\upsilon_y\,dt$ times the shell's cross-sectional area $d\mathcal{A}$; $dV = \upsilon_y\,dt\,d\mathcal{A}$. (The shell volume is the difference between volumes of cylinders of radii $s + ds$ and s, and the volume of a cylinder equals its length times its cross-sectional area. Hence the shell volume equals the shell length times the shell cross-sectional area.) We have $\rho = dm/dV$, where dm is the thin-shell mass that passes a given location in time dt, so $dm = \rho\,dV = \rho\upsilon_y\,dt\,d\mathcal{A}$.

15.14 Equation (15.22) gives $\upsilon = 2[(7.8 - 1.0)\text{ g cm}^{-3}](980.7\text{ cm/s}^2)(0.050\text{ cm})^2/9(0.0089\text{ dyn s cm}^{-2}) = 420\text{ cm/s}$. For glycerol, $\upsilon = 2(7.8 - 1.25)\,10^{-3}\text{ kg}(10^{-2}\text{ m})^{-3}(9.807\text{ m/s}^2)(0.00050\text{ m})^2/9(0.954\text{ N s m}^{-2}) = 0.0037\text{ m/s} = 0.37\text{ cm/s}$.

15.15 From (15.25), $d^2 = 5(MRT)^{1/2}/16\pi^{1/2}N_A\eta = 0.1763(0.04401\text{ kg/mol})^{1/2}(8.314\text{ J/mol-K})^{1/2}T^{1/2}/(6.022 \times 10^{23}/\text{mol})\eta =$

1.77×10^{-25} m^2 $(T/\text{K})^{1/2}/[\eta/(\text{kg m}^{-1}\text{ s}^{-1})] = 1.77 \times 10^{-24}$ m^2 $(T/\text{K})^{1/2}/(\eta/\text{P})$, where P stands for poise. At 0°C, $d^2 = 1.77 \times 10^{-24}$ m^2 $(273)^{1/2}/(139 \times 10^{-6})$ and $d = 4.59 \times 10^{-10}$ m = 4.59 Å. At 490°C, d = 3.85 Å; at 850°C, d = 3.69 Å.

15.16 The diameters of H_2 and D_2 are the same and (15.25) shows that η is proportional to $M^{1/2}$. So $\eta_{D_2}/\eta_{H_2} = (M_{D_2}/M_{H_2})^{1/2}$ and η_{D_2} = $(4.03/2.02)^{1/2}(8.53 \times 10^{-5}\text{ P}) = 1.20 \times 10^{-4}$ P.

15.17 **(a)** $M_n = \sum_i x_i M_i = 0.5(200\text{ kg/mol}) + 0.5(600\text{ kg/mol}) = 400$ kg/mol. $M_w = (\sum_i x_i M_i^2)/(\sum_i x_i M_i) =$ $[0.5(200\text{ kg/mol})^2 + 0.5(600\text{ kg/mol})^2]/(400\text{ kg/mol}) = 500$ kg/mol.

(b) Let us take 600 kg of each species. We then have 3.0 moles of the molecular-weight 200000 species and 1.0 mole of the molecular-weight 600000 species. Hence $M_n = \sum_i x_i M_i =$ $0.75(200\text{ kg/mol}) + 0.25(600\text{ kg/mol}) = 300$ kg/mol. $M_w =$ $[0.75(200\text{ kg/mol})^2 + 0.25(600\text{ kg/mol})^2]/(300\text{ kg/mol}) = 400$ kg/mol.

15.18 We plot $(\eta_r - 1)/\rho_B$ vs. ρ_B. The points fit a straight line well, and extrapolation to $\rho_B = 0$ gives $[\eta] = \lim_{\rho_B \to 0}(\eta_r - 1)/\rho_B = 0.147$ dm^3/g = 147 cm^3/g. Then 147 cm^3/g = $(0.034\text{ cm}^3/\text{g})(M_B/M^\circ)^{0.65}$ and M_B = 390000 g/mol. $M_{r,B}$ = 390000.

15.19 **(a)** F; **(b)** F; **(c)** T; **(d)** F; **(e)** T; **(f)** F.

15.20 **(a)** $(\Delta x)_{rms} = (2Dt)^{1/2}$, so $t = (\Delta x)^2_{rms}/2D = (1\text{ cm}^2)/2(10^{-21}\text{ cm}^2/\text{s}) =$ 5×10^{20} s $= 2 \times 10^{13}$ yr.

(b) $t = (1\text{ cm}^2)/2(10^{-30}\text{ cm}^2/\text{s}) = 5 \times 10^{29}$ s $= 2 \times 10^{22}$ yr.

15.21 **(a)** $(\Delta x)_{rms} = (2Dt)^{1/2} = [2(0.52 \times 10^{-5}\text{ cm}^2/\text{s})(60\text{ s})]^{1/2} = 0.025$ cm.

(b) t = 3600 s and $(\Delta x)_{rms}$ = 0.19 cm.

(c) t = 86400 s and $(\Delta x)_{rms}$ = 0.95 cm.

15.22 Let the origin be at the $t = 0$ location of the particle and let the particle be at point (x, y, z) at time t. Then $\langle r^2 \rangle = \langle x^2 + y^2 + z^2 \rangle = \langle x^2 \rangle + \langle y^2 \rangle + \langle z^2 \rangle$. By symmetry, $\langle x^2 \rangle = \langle y^2 \rangle = \langle z^2 \rangle$, so $\langle r^2 \rangle = 3\langle x^2 \rangle = 3(2Dt) = 6Dt$ and $r_{\text{rms}} = \langle r^2 \rangle^{1/2} = (6Dt)^{1/2}$.

15.23 From (15.35), $k = 3[(\Delta x)_{\text{rms}}]^2 \pi\eta r/tT$. For $t = 30$ s, $k = 3(7.1 \times 10^{-6}\text{ m})^2\pi \times (0.0011\text{ N-s/m}^2)(2.1 \times 10^{-7}\text{ m})/(30\text{ s})(290\text{ K}) = 1.26 \times 10^{-23}$ J/K. Then $N_A = R/k = (8.314\text{ J/mol-K})/(1.26 \times 10^{-23}\text{ J/K}) = 6.6 \times 10^{23}$/mol. For $t = 60$ s, we get $k = 1.41 \times 10^{-23}$ J/K and $N_A = 5.9 \times 10^{23}\text{ mol}^{-1}$. For 90 s, $k = 1.07 \times 10^{-23}$ J/K and $N_A = 7.8 \times 10^{23}\text{ mol}^{-1}$.

15.24 $\langle \Delta x \rangle = \frac{1}{12}(-5.3 + 3.4 - 1.9 - 0.4 + 0.5 + 3.1 - 0.2 - 3.5 + 1.4 + 0.3 - 1.0 + 2.6)$ μm $= -0.08$ μm. $\langle (\Delta x)^2 \rangle = (1/12)[(-5.3)^2 + (3.4)^2 + \cdots]\ \mu\text{m}^2 = 6.28\ \mu\text{m}^2$. $(\Delta x)_{\text{rms}} = 2.5$ μm.

15.25 **(a)** Equation (15.42) and $N/V = PN_A/RT$ give $D_{jj} = (3/8\pi^{1/2})[(8.314\text{ J/mol-K})(273\text{ K})/(0.0320\text{ kg/mol})]^{1/2} \times (3.6 \times 10^{-10}\text{ m})^{-2}[82.06\,(10^{-2}\text{ m})^3\text{-atm/mol-K}](273\text{ K})/(1.00\text{ atm})(6.02 \times 10^{23}\text{ mol}^{-1}) = 0.000016\text{ m}^2/\text{s} = 0.16\text{ cm}^2/\text{s}$.

(b) Since D_{jj} is proportional to $1/P$, we have $D_{jj} = 0.016\text{ cm}^2/\text{s}$.

15.26 **(a)** The volume of each cell is $8r_j^3$, and the molar volume is $V_{m,j} = 8r_j^3 N_A$. Hence $r_j = \frac{1}{2}(V_{m,j}/N_A)^{1/3}$.

(b) $D_{jj} \approx [(6.02 \times 10^{23}/\text{mol})/(18.1 \times 10^{-6}\text{ m}^3/\text{mol})]^{1/3}(1.38 \times 10^{-23}\text{ J/K}) \times (298\text{ K})/2\pi(0.000890\text{ N-s/m}^2) = 2.4 \times 10^{-9}\text{ m}^2/\text{s} = 2.4 \times 10^{-5}\text{ cm}^2/\text{s}$.

15.27 The N_2 and H_2O molecules are not greatly different in size, so Eq. (15.38) is appropriate. We have $D_{iB}^{\infty} \approx (1.38 \times 10^{-23}\text{ J/K})(298\text{ K})/4\pi(0.00089\text{ N-s/m}^2)(1.8_5 \times 10^{-10}\text{ m}) = 2.0 \times 10^{-9}\text{ m}^2/\text{s} = 2.0 \times 10^{-5}\text{ cm}^2/\text{s}$.

15.28 **(a)** Equations (15.25) and (15.42) give $6\eta/5\rho = (6\pi/32)\langle \upsilon \rangle \lambda = (3\pi/16)\langle \upsilon \rangle \lambda = D_{jj}$.

(b) $D_{jj} = 6\eta RT/5PM = 6(0.0000297\ \text{N-s/m}^2)(82.06 \times 10^{-6}\ \text{m}^3\text{-atm/mol-K}) \times (273.1\ \text{K})/5(1.00\ \text{atm})(0.02018\ \text{kg/mol}) = 0.000040\ \text{m}^2/\text{s} = 0.40\ \text{cm}^2/\text{s}$.

15.29 Since a hemoglobin molecule is much larger than a water molecule, we use Eq. (15.37). We have $V_m/N_A = 4\pi r_i^3/3$ and $r_i = (3V_m/4N_A\pi)^{1/3} = [3(48000\ \text{cm}^3/\text{mol})/4(6.02 \times 10^{23}/\text{mol})\pi]^{1/3} = 2.67 \times 10^{-7}$ cm. Then $D_{iB}^{\infty} \approx (1.38 \times 10^{-23}\ \text{J/K})(298\ \text{K})/6\pi(0.89 \times 10^{-3}\ \text{N-s/m}^2)(2.67 \times 10^{-9}\ \text{m}) = 9.2 \times 10^{-11}\ \text{m}^2/\text{s} = 9.2 \times 10^{-7}\ \text{cm}^2/\text{s}$.

15.30 (a) Let $z = x^2$. Then $d(x^2)/dt = dz/dt = (dz/dx)(dx/dt) = 2x(dx/dt)$. Also, $d^2(x^2)/dt^2 = (d/dt)[d(x^2)/dt] = (d/dt)[2x(dx/dt)] = 2(dx/dt)^2 + 2x(d^2x/dt^2)$. Substitution of these two equations into the equation in the problem transforms it to Eq. (15.33).

(b) Averaging, we get $0 - \frac{1}{2}f\langle d(x^2)/dt\rangle = \frac{1}{2}m\langle d^2(x^2)/dt^2\rangle - \langle m(dx/dt)^2\rangle$. We have $\varepsilon_x = \frac{1}{2}m\upsilon_x^2 = \frac{1}{2}m(dx/dt)^2$, so $2\langle\varepsilon_x\rangle = \langle m(dx/dt)^2\rangle$. Also, $\varepsilon = \varepsilon_x + \varepsilon_y + \varepsilon_z$ and $\langle\varepsilon\rangle = \langle\varepsilon_x\rangle + \langle\varepsilon_y\rangle + \langle\varepsilon_z\rangle$. By symmetry, $\langle\varepsilon_x\rangle = \langle\varepsilon_y\rangle = \langle\varepsilon_z\rangle$; thus $\langle\varepsilon_x\rangle = \langle\varepsilon\rangle/3 = (3/2)kT/3 = \frac{1}{2}kT$. So $\langle m(dx/dt)^2\rangle = 2\langle\varepsilon_x\rangle = kT$.

(c) Substitution of $\langle m(dx/dt)^2\rangle = kT$ into twice the first equation in part (b) gives $-f\langle d(x^2)/dt\rangle = m\langle d^2(x^2)/dt^2\rangle - 2kT$. Let $s = d\langle x^2\rangle/dt$. Since the derivative of a sum is the sum of the derivatives, we have $s = \langle d^2(x^2)/dt^2\rangle$ and $ds/dt = \langle d^2(x^2)/dt^2\rangle$. Thus $-fs = m\, ds/dt - 2kT$ and $m\, ds/dt + fs = 2kT$.

(d) $ds/dt = (2kT - fs)/m$ and $m(2kT - fs)^{-1}\, ds = dt$. Integration gives $-(m/f)\ln(2kT - fs) = t + c$ and $2kT - fs = e^{-ft/m}e^{-fc/m}$. From part (e), $e^{-ft/m} \approx 0$, so $s \equiv d\langle x^2\rangle/dT = 2kT/f$. Integration gives $\langle x^2\rangle = 2kTt/f + b$. The integration constant b is 0 if we take $x = 0$ at $t = 0$ for each particle. Hence, $\langle x^2\rangle = (2kT/f)t$.

(e) $m = (3\ \text{g/cm}^3)[4\pi(10^{-5}\ \text{cm})^3/3] = 10^{-14}$ g and $f = 6\pi\eta r = 6\pi(0.001\ \text{N-s/m}^2)(10^{-7}\ \text{m}) = 2 \times 10^{-9}$ N-s/m. Then $e^{-ft/m} = \exp[-(2 \times 10^{-9}\ \text{N-s/m})(1\ \text{s})/(10^{-17}\ \text{kg})] = e^{-200000000} = 10^{-90000000}$.

15.31 (a) Substitution of dn_A and dn_B from (15.30) into $0 = dV = \overline{V}_A\ dn_A + \overline{V}_B\ dn_B$ gives $0 = \overline{V}_A\,[-@D_{AB}(dc_A/dx)\ dt] + \overline{V}_B\,[-@D_{BA}(dc_B/dx)\ dt]$ and $D_{AB}\,\overline{V}_A\,(dc_A/dx) + D_{BA}\,\overline{V}_B\,(dc_B/dx) = 0$.

(b) Division of (9.16) by V gives $1 = c_A\overline{V}_A + c_B\overline{V}_B$. Since there is no volume change on mixing, $\overline{V}_A$ and $\overline{V}_B$ are independent of the concentrations and hence are independent of the x coordinate. Differentiation of the last equation with respect to x gives $0 = \overline{V}_A\,(dc_A/dx) + \overline{V}_B\,(dc_B/dx)$. Substitution of $\overline{V}_A\,(dc_A/dx) = -\overline{V}_B\,(dc_B/dx)$ into the result of (a) gives $D_{AB} = D_{BA}$.

15.32 The number of moles of the diffusing species that lie between L and M equals $c_LV_L = c_L@(\Delta x)_{\text{rms}}$, where V_L is the volume between L and M. The number of moles moving left to right through plane M in time t is therefore $\frac{1}{2}c_L@(\Delta x)_{\text{rms}}$. Similarly, $\frac{1}{2}c_R@(\Delta x)_{\text{rms}}$ moles move right to left through M in time t. The net rate of flow of the diffusing species through M is thus $dn/dt = \Delta n/\Delta t = \frac{1}{2}(c_L - c_R)@(\Delta x)_{\text{rms}}/t$. Since dc/dx is constant, the average concentration c_L equals the concentration midway between planes L and M. Likewise, the average concentration c_R between M and R equals the concentration midway between M and R. The distance from the plane midway between L and M to the plane midway between M and R is $\frac{1}{2}(\Delta x)_{\text{rms}} + \frac{1}{2}(\Delta x)_{\text{rms}} = (\Delta x)_{\text{rms}}$; so the concentration gradient is $dc/dx = \Delta c/\Delta x = (c_R - c_L)/(\Delta x)_{\text{rms}}$. Substitution into (15.30) gives $\frac{1}{2}\,(c_L - c_R)@(\Delta x)_{\text{rms}}/t = -D@(c_L - c_R)/(\Delta x)_{\text{rms}}$ and $(\Delta x)_{\text{rms}} = (2Dt)^{1/2}$.

15.33 $M_i = RTs^{\infty}/D^{\infty}_{iB}(1 - \rho\bar{v}_i) =$

$$\frac{(8.314\text{ J/mol-K})(293.1\text{ K})(4.47\times10^{-13}\text{ s})}{(6.9\times10^{-11}\text{ m}^2\text{/s})[1-(0.998\text{ g/cm}^3)(0.749\text{ cm}^3\text{/g})]} = 63\text{ kg/mol} = 63000\text{ g/mol}$$

15.34 (a) The column chart shows a bell-shaped (Gaussian) curve.

(b) Row 15 contains the numbers of molecules in the various locations after 10 time intervals. The squares of the numbers in row 3 give the $(\Delta x)^2$ values. The formula =(C15*(C3)^2)/1000 is entered in cell C1020; this formula corresponds to the term n_ss^2/N in $\langle s^2\rangle = \Sigma_s\ n_ss^2/N$. Using the method given in the problem, the contents of C1020 are copied to D1020, E1020,…, GU1020. In another cell, C1020, D1020,…, GU1020 are summed to give $\langle(\Delta x)^2\rangle$. The result is 6.667 for 10 time intervals. For 100 and 1000 time intervals, we use rows 105 and 1005 instead of row 15. The results are 66.67 and 664.8. The relation of proportionality to time is well obeyed. (The deviation for the 1000 time-interval result can be ascribed to the error introduced by using a finite length for the x axis; the values in locations beyond 100 units from the origin were omitted, thereby giving a $\langle(\Delta x)^2\rangle$ value somewhat smaller than the true one.)

(d) From (15.31), $\langle(\Delta x)^2\rangle = 2Dt$. The numbers –100,…, 100 in row 3 correspond to -100×10^{-6} cm,…, 100×10^{-6} cm, so the $\langle(\Delta x)^2\rangle$ values found in (b) should be multiplied by 10^{-12} cm^2. Thus $\langle(\Delta x)^2\rangle$ = $664.8(10^{-12}\text{ cm}^2) = 2D(1000 \times 1\text{ s})$ and $D = 3.32 \times 10^{-13}$ cm^2/s.

(e) The formula in C6 becomes

```
=0.05*A5+0.25*B5+0.40*C5+0.25*D5+0.05*E5
```

This is copied to the appropriate cells and the same procedure is used as in (b) and (d).

15.35 $I = dQ/dt = Q/t$ since I is constant. $Q = It = (1.0\text{ A})(1.0\text{ s}) = 1.0$ C. The number of electrons is $(1.0\text{ C})(6.02 \times 10^{23}\text{ electrons})/(96485\text{ C}) = 6.2 \times 10^{18}$.

15.36 $R = \rho\ell/\mathscr{A} = (1.67 \times 10^{-6}\ \Omega\text{ cm})(250\text{ cm})/(0.040\text{ cm}^2) = 0.0104\ \Omega$.

15.37 $|\Delta\phi| = IR$ and $I = |\Delta\phi|/R = (25\text{ V})/(100\ \Omega) = 0.25$ A.

15.38 $E = j/\kappa = I/\mathscr{A}\kappa = (0.10\text{ A})/(10\text{ cm}^2)(0.010\ \Omega^{-1}\text{ cm}^{-1}) = 1.0$ V/cm.

15.39 **(a)** T; **(b)** T; **(c)** F; **(d)** F; **(e)** T.

15.40 $Q = It = (2.00\text{ A})(30.0 \times 60\text{ s}) = 3600$ C.

$$(3600\text{ C})\frac{1\text{ mole e}^-}{96485\text{ C}}\cdot\frac{1\text{ mole Cu}}{2\text{ mole e}^-}\frac{63.55\text{ g Cu}}{1\text{ mole Cu}} = 1.19\text{ g of Cu}$$

15.41 **(a)** $K_{\text{cell}} = \kappa R = [(0.012856/1.000495)\ \Omega^{-1}\text{ cm}^{-1}](411.82\ \Omega) = 5.2917\text{ cm}^{-1}$.

(b) $\kappa_{H_2O} = K_{cell}/R = (5.2917\text{ cm}^{-1})/(368000\ \Omega) = 0.00001438\ \Omega^{-1}\text{ cm}^{-1}$.
$\kappa = K_{cell}/R = (5.2917\text{ cm}^{-1})/(10875\ \Omega) = 0.00048659\ \Omega^{-1}\text{ cm}^{-1} = \kappa_{H_2O} + \kappa_{MX_2}$ and $\kappa_{MX_2} = (0.00048659 - 0.00001438)\ \Omega^{-1}\text{ cm}^{-1} = 0.00047221\ \Omega^{-1}\text{ cm}^{-1}$.

(c) $\Lambda_m = \kappa/c = (0.00047221\ \Omega^{-1}\text{ cm}^{-1})/(10^{-6}\text{ mol/cm}^3) = 472.21\text{ cm}^2/\Omega\text{-mol}$.

(d) $\Lambda_{eq} = \Lambda_m/\nu_+z_+ = \Lambda_m/2 = 236.10\ \Omega^{-1}\text{ cm}^2\text{ equiv}^{-1}$.

15.42 **(a)** $\Lambda_m = \kappa/c = (1.242 \times 10^{-3}\ \Omega^{-1}\text{ cm}^{-1})/[(5.000 \times 10^{-3}\text{ mol})/(1000\text{ cm}^3)] = 248.4\ \Omega^{-1}\text{ cm}^2\text{ mol}^{-1}$.

(b) $\Lambda_{eq} = \Lambda_m/\nu_+z_+ = \Lambda_m/1(2) = 124.2\ \Omega^{-1}\text{ cm}^2\text{ equiv}^{-1}$.

15.43 Equation (15.67) gives $u(Na^+) = (10.00\text{ cm})(0.002313/\Omega\text{-cm})(0.1115\text{ cm}^2)/(0.00160\text{ A})(3453\text{ s}) = 4.668 \times 10^{-4}\text{ cm}^2\text{ V}^{-1}\text{ s}^{-1}$. Equation (15.72) gives $t(Na^+) = 1(96485\text{ C/mol})[(0.02000\text{ mol})/(1000\text{ cm}^3)](0.0004668\text{ cm}^2\text{ V}^{-1}\text{ s}^{-1})/(0.002313\ \Omega^{-1}\text{ cm}^{-1}) = 0.3894$.

15.44 **(a)** Substitution of Eq. (15.67) (which applies to any ion) into (15.72) gives $t_B = |z_B|Fc_B(x\kappa\mathscr{A}/It)/\kappa = |z_B|Fc_Bx\mathscr{A}/Q$, where $Q = It$.

(b) As in Fig. 15.23, the method gives the transport number of the cation. The equation in (a) with $x\mathscr{A} = V$ gives $t(Gd^{3+}) = 3(96485\text{ C/mol}) \times [(0.03327\text{ mol})/(1000\text{ cm}^3)](1.111\text{ cm}^3)/(0.005594\text{ A})(4406\text{ s}) = 0.434$. Then $t(Cl^-) = 1 - 0.434 = 0.566$.

15.45 The coulometer reaction is $Ag^+ + e^- = Ag$. The 0.16024 g of Ag is 0.0014855 moles of Ag, so $Q = (0.0014855\text{ mol})(96485\text{ C/mol}) = 143.33$ C. Let $n = 0.0014855$ mol. Then n moles of Cl^- enter the cathode compartment due to the reduction reaction $AgCl + e^- = Ag + Cl^-$. The total number of moles of charge on the ions that cross the plane between the middle and cathode compartments is n; this charge is composed of t_-n moles of Cl^- leaving and t_+n moles of K^+ entering the cathode compartment. The net change in number of moles of Cl^- in the cathode compartment is thus $n - t_-n = (1 - t_-)n = t_+n$, which is also the change in number of moles of K^+ in this compartment. The final composition of the cathode compartment is (0.0019404)(120.99 g) = 0.23477 g of KCl and 120.99 g − 0.235 g = 120.75_5 g of H_2O. The initial composition of this compartment is 120.75_5 g of water plus x grams of KCl, where $x/(120.75_5 + x) = 0.0014941$. We get $x = 0.18069$. The change in KCl mass in the cathode compartment is 0.05408 g, which is 0.0007254 moles. Thus, $t_+n = t_+(0.0014855 \text{ mol}) = 0.0007254$ mol and $t_+ = 0.4883$. Then $t_- = 1 - t_+ = 0.5117$.

15.46 **(a)** Using Eq. (15.84), we have $\Lambda_m^\infty(\text{LiNO}_3) = \lambda_m^\infty(\text{Li}^+) + \lambda_m^\infty(\text{NO}_3^-)$ and the following three equations:

(1) $\Lambda_m^\infty(\text{LiCl}) = \lambda_m^\infty(\text{Li}^+) + \lambda_m^\infty(\text{Cl}^-)$;

(2) $\Lambda_m^\infty(\text{KNO}_3) = \lambda_m^\infty(\text{K}^+) + \lambda_m^\infty(\text{NO}_3^-)$;

(3) $\Lambda_m^\infty(\text{KCl}) = \lambda_m^\infty(\text{K}^+) + \lambda_m^\infty(\text{Cl}^-)$.

We take (1) + (2) – (3) to get $\Lambda_m^\infty(\text{LiNO}_3) = \Lambda_m^\infty(\text{LiCl}) + \Lambda_m^\infty(\text{KNO}_3) - \Lambda_m^\infty(\text{KCl}) = (90.9 + 114.5 - 105.0)\ \text{cm}^2/\Omega\text{-mol} = 100.4\ \Omega^{-1}\ \text{cm}^2\ \text{mol}^{-1}$.

(b) (4) $\Lambda_m^\infty(\text{HCl}) = \lambda_m^\infty(\text{H}^+) + \lambda_m^\infty(\text{Cl}^-)$;

(5) $\Lambda_m^\infty(\text{NaCl}) = \lambda_m^\infty(\text{Na}^+) + \lambda_m^\infty(\text{Cl}^-)$;

(6) $\Lambda_m^\infty(\text{NaC}_2\text{H}_3\text{O}_2) = \lambda_m^\infty(\text{Na}^+) + \lambda_m^\infty(\text{C}_2\text{H}_3\text{O}_2^-)$.

Taking (4) + (6) – (5), we get $\lambda_m^\infty(\text{H}^+) + \lambda_m^\infty(\text{C}_2\text{H}_3\text{O}_2^-) = \Lambda_m^\infty(\text{HCl}) + \Lambda_m^\infty(\text{NaC}_2\text{H}_3\text{O}_2) - \Lambda_m^\infty(\text{NaCl}) = (426 + 91 - 126)\ \text{cm}^2/\Omega\text{-mol} = 391\ \text{cm}^2/\Omega\text{-mol}$.

15.47 From (15.85) for NaCl, $t_+^\infty = 0.463 = (1)\lambda_{m,+}^\infty/(96.9\ \text{cm}^2\ \Omega^{-1}\ \text{mol}^{-1})$ and $\lambda_m^\infty(\text{Na}^+) = 44.9\ \text{cm}^2/\Omega\text{-mol}$. Then $\lambda_m^\infty(\text{Cl}^-) = (96.9 - 44.9)\ \text{cm}^2/\Omega\text{-mol} = 52.0\ \text{cm}^2/\Omega\text{-mol}$. We now use (15.84) for the other electrolytes. For NaNO_3, $106.4\ \text{cm}^2/\Omega\text{-mol} = 44.9\ \text{cm}^2/\Omega\text{-mol} + \lambda_m^\infty(\text{NO}_3^-)$ and $\lambda_m^\infty(\text{NO}_3^-) = 61.5\ \text{cm}^2/\Omega\text{-mol}$. For LiNO_3, $100.2\ \text{cm}^2/\Omega\text{-mol} = \lambda_m^\infty(\text{Li}^+) + 61.5\ \text{cm}^2/\Omega\text{-mol}$ and $\lambda_m^\infty(\text{Li}^+) = 38.7\ \text{cm}^2/\Omega\text{-mol}$. For NaCNS, $107.0\ \text{cm}^2/\Omega\text{-mol} = 44.9\ \text{cm}^2/\Omega\text{-mol} + \lambda_m^\infty(\text{CNS}^-)$ and $\lambda_m^\infty(\text{CNS}^-) = 62.1\ \text{cm}^2/\Omega\text{-mol}$. For HCl, $192\ \text{cm}^2/\Omega\text{-mol} = \lambda_m^\infty(\text{H}^+) + 52.0\ \text{cm}^2/\Omega\text{-mol}$ and $\lambda_m^\infty(\text{H}^+) = 140\ \text{cm}^2/\Omega\text{-mol}$. For Ca(CNS)_2, $244\ \text{cm}^2/\Omega\text{-mol} = \lambda_m^\infty(\text{Ca}^{2+}) + 2(62.1)\ \text{cm}^2/\Omega\text{-mol}$ and $\lambda_m^\infty(\text{Ca}^{2+}) = 120\ \text{cm}^2/\Omega\text{-mol}$.

15.48 **(a)** $u_B^\infty = \lambda_{m,B}^\infty / |z_B| F = (67.2\ \Omega^{-1}\ \text{cm}^2\ \text{mol}^{-1})/1(96485\ \text{C/mol}) =$
$6.96 \times 10^{-4}\ \text{cm}^2\ \text{V}^{-1}\ \text{s}^{-1}$.

(b) $\upsilon_B^\infty = u_B^\infty E = (6.96 \times 10^{-4}\ \text{cm}^2/\text{V-s})(24\ \text{V/cm}) = 0.017\ \text{cm/s}$.

(c) $r_B \approx |z_B| e / 6\pi\eta u_B^\infty = 1(1.6022 \times 10^{-19}\ \text{C})/$
$6\pi(0.008904 \times 10^{-1}\ \text{N s m}^{-2})[6.96 \times 10^{-4}\ (10^{-2}\ \text{m})^2/\text{V-s}] =$
$1.37 \times 10^{-10}\ \text{m} = 1.37\ \text{Å}$.

15.49 We use $\Lambda_m^\infty = \nu_+ \lambda_{m,+}^\infty + \nu_- \lambda_{m,-}^\infty$.

(a) $\Lambda_m^\infty = (73.5 + 71.4)\ \text{cm}^2/\Omega\text{-mol} = 144.9\ \text{cm}^2/\Omega\text{-mol}$.

(b) $\Lambda_m^\infty = [2(73.5) + 159.6]\ \text{cm}^2/\Omega\text{-mol} = 306.6\ \text{cm}^2/\Omega\text{-mol}$.

(c) $\Lambda_m^\infty = (106.1 + 159.6)\ \text{cm}^2/\Omega\text{-mol} = 265.7\ \text{cm}^2/\Omega\text{-mol}$.

(d) $\Lambda_m^\infty = [118.0 + 2(199.2)]\ \text{cm}^2/\Omega\text{-mol} = 516.4\ \text{cm}^2/\Omega\text{-mol}$.

15.50 From (15.85), $t^\infty(Mg^{2+}) = 1(106.1)/[1(106.1) + 2(71.4)] = 0.426$ and $t^\infty(NO_3^-) = 1 - 0.426 = 0.574$.

15.51 From (15.85), $t_+^\infty = \nu_+ \lambda_{m,+}^\infty / \Lambda_m^\infty = 2\lambda_{m,+}^\infty / (259.8\ \Omega^{-1}\ \text{cm}^2\ \text{mol}^{-1}) = 0.386$ and $\Lambda_m^\infty(Na^+) = 50.1\ \Omega^{-1}\ \text{cm}^2\ \text{mol}^{-1}$. From (15.84), $\Lambda_m^\infty = 259.8\ \Omega^{-1}\ \text{cm}^2\ \text{mol}^{-1} =$
$2(50.1\ \Omega^{-1}\ \text{cm}^2\ \text{mol}^{-1}) + \lambda_m^\infty(SO_4^{2-})$ and $\lambda_m^\infty(SO_4^{2-}) = 159.6\ \Omega^{-1}\ \text{cm}^2\ \text{mol}^{-1}$.

15.52 We assume the electrodes are inert, so only Ag^+ and NO_3^- carry the current in the solution. $t_- \cong t_-^\infty = \nu_- \lambda_{m,-}^\infty / (\nu_+ \lambda_{m,+}^\infty + \nu_- \lambda_{m,-}^\infty) = 71.4/(62.1 + 71.4) =$ 0.535. If 10^{-3} mol of Ag^+ is deposited according to $Ag^+ + e^- \rightarrow Ag$, then 10^{-3} Faradays of charge must have flowed through the circuit. The nitrate ions carry a fraction t_- of the current in the solution, so the nitrate ions carried $t_-(10^{-3}$ Faradays) of charge through the solution. Hence $10^{-3} t_-$ moles of NO_3^- crossed a plane in the solution, and this is $(1.00 \times 10^{-3})(0.535)$ mol = 0.535 mmol.

15.53 Because of its higher charge, Mg^{2+} is hydrated to a much greater extent than Na^+ and so its radius is much greater than that of Na^+. Therefore u in (15.70) has similar values for these two ions.

15.54 **(a)** $\lambda_m^\infty(Cl^-)$, $u^\infty(Cl^-)$;

(b) no (interionic forces differ in the two solutions).

15.55 κ.

15.56 **(a)** Use of (15.70) in (15.78) gives $\lambda_{m,B}^\infty = |z_B|Fu_B^\infty \cong |z_B|^2 eF/6\pi\eta r_B$. So $\ln \lambda_m^\infty \cong \ln(z_B^2 eF/6\pi r_B) - \ln \eta$ and $d \ln \lambda_m^\infty/dT \cong -(1/\eta)\, d\eta/dT$.

(b) $d\eta/dT \cong \Delta\eta/\Delta T = (0.8705 - 0.9111)\text{cP}/(2\text{ K}) = -0.0203$ cP/K and $(1/\eta)(d\eta/dT) \cong (1/0.8904\text{ cP})(-0.0203\text{ cP/K}) = -0.023\text{ K}^{-1}$, so $d \ln \lambda_m^\infty/dT \cong 0.023\text{ K}^{-1}$.

(c) Integration gives $\ln(\lambda_{m,2}^\infty/\lambda_{m,1}^\infty) \cong (0.023\text{ K}^{-1})(T_2 - T_1)$ and $\lambda_{m,2}^\infty = \lambda_{m,1}^\infty \exp[(0.023\text{ K}^{-1})(T_2 - T_1)] = (71.4\ \Omega^{-1}\text{ cm}^2\text{ mol}^{-1})e^{0.23} = 89.9\ \Omega^{-1}\text{ cm}^2\text{ mol}^{-1}$.

15.57 **(a)** $\Lambda_m^\infty = (73.5 + 71.4)\text{ cm}^2/\Omega\text{-mol} = 144.9\text{ cm}^2/\Omega\text{-mol}$. From (15.87), $\Lambda_m = 144.9\text{ cm}^2/\Omega\text{-mol} - [60.6 + 0.230(144.9)](\text{cm}^2/\Omega\text{-mol})(0.00200)^{1/2} = 140.7\text{ cm}^2/\Omega\text{-mol}$. $\kappa = \Lambda_m c = (140.7\text{ cm}^2/\Omega\text{-mol})(0.00200\text{ mol})/(1000\text{ cm}^3) = 0.000281\ \Omega^{-1}\text{ cm}^{-1}$.

(b) Neglecting the conductivity of the water, we have $R = \rho \ell/\mathcal{A} = \ell/\kappa\mathcal{A} = (10.0\text{ cm})/(0.000281\ \Omega^{-1}\text{ cm}^{-1})(1.00\text{ cm}^2) = 35500\ \Omega$.

15.58 $K_c = [H^+][OH^-]$, since $\gamma_\pm = 1$ at the extremely low I_m value in this solution. Equation (15.90) with $c_+ = 0$ in the denominator gives the initial estimate $c_+ \approx (5.47 \times 10^{-8}\ \Omega^{-1}\text{ cm}^{-1})/[(350 + 199)\text{ cm}^2/\Omega\text{-mol}] = 9.96 \times 10^{-11}\text{ mol/cm}^3 = 9.96 \times 10^{-8}\text{ mol/dm}^3$. We have $S = 60.6\text{ cm}^2/\Omega\text{-mol} + 0.230(549\text{ cm}^2/\Omega\text{-mol}) = 186.9\text{ cm}^2/\Omega\text{-mol}$. Using the initial estimate of c_+ in the denominator of Eq. (15.90), we have $c_+ \approx (5.47 \times 10^{-8}\ \Omega^{-1}\text{ cm}^{-1})/[549\text{ cm}^2/\Omega\text{-mol} - (186.9\text{ cm}^2/\Omega\text{-mol})(9.96 \times 10^{-8})^{1/2}] = 9.96 \times 10^{-11}\text{ mol/cm}^3$

$= 9.96 \times 10^{-8}$ mol/dm^3, as before. Hence $K_c = (9.96 \times 10^{-8} \text{ mol/dm}^3)^2 =$ $9.9_2 \times 10^{-15}$ mol^2/dm^6.

15.59 **(a)** Since $z_+ = |z_-|$, Eq. (15.90) applies. We have $\lambda^\infty_{m,+} + \lambda^\infty_{m,-} =$ (118.0 + 159.6) cm^2/Ω-mol = 277.6 cm^2/Ω-mol and $S = 8[a + b(\lambda^\infty_{m,+} + \lambda^\infty_{m,-})] = 996$ cm^2/Ω-mol. Therefore

$$c_+ = \frac{2.21\times 10^{-3}\ \Omega^{-1}\ \text{cm}^{-1}}{277.6\ \text{cm}^2/\Omega\text{-mol} - (996\ \text{cm}^2/\Omega\text{-mol})[c_+/(10^{-3}\ \text{mol/cm}^3)]^{1/2}}$$

With $c_+ = 0$ in the denominator, we get the initial estimate $c_+ \approx$ 7.96×10^{-6} mol/cm^3. With $c_+ = 7.96 \times 10^{-6}$ mol/cm^3, we get the improved estimate $c_+ \approx 1.17 \times 10^{-5}$ mol/cm^3. Further repetitions yield the successive estimates 1.30×10^{-5} mol/cm^3, 1.35×10^{-5} mol/cm^3, 1.36×10^{-5} mol/cm^3, and 1.37×10^{-5} mol/cm^3. Hence $c_+ =$ 1.37×10^{-2} mol/dm^3. The solution is dilute, so we can take the molality as 0.0137 mol/kg. The Davies equation with $I_m = 0.0548$ mol/kg then gives $\log \gamma_\pm = -0.353$ and $\gamma_\pm = 0.443$. We can neglect the difference between molality-scale and concentration-scale activity coefficients in this dilute solution, so the concentration-scale K_{sp} is $K_{sp} =$ $(0.443)^2(0.0137 \text{ mol/dm}^3)^2 = 3.7 \times 10^{-5}$ mol^2/dm^6.

(b) No. $K_{sp} = \gamma_\pm^2 [Ca^{2+}][SO_4^{2-}]$. We found the ionic concentrations and $\gamma_\pm$. The existence of the additional equilibrium $Ca^{2+} + SO_4^{2-} = CaSO_4(aq)$ does not invalidate our work.

15.60 We have $\lambda^\infty_{m,+} + \lambda^\infty_{m,-} =$ (350 + 40.8) cm^2/Ω-mol = 391 cm^2/Ω-mol and $S =$ [60.6 + 0.230(391)] cm^2/Ω-mol = 150.5 cm^2/Ω-mol. Therefore

$$c_+ = \frac{4.95\times 10^{-5}\ \Omega^{-1}\ \text{cm}^{-1}}{391\ \text{cm}^2/\Omega\text{-mol} - (150\tfrac{1}{2}\ \text{cm}^2/\Omega\text{-mol})[c_+/(10^{-3}\ \text{mol/cm}^3)]^{1/2}}$$

With $c_+ = 0$ in the denominator, we get the initial estimate $c_+ \approx$ 1.27×10^{-7} mol/cm^3. Recalculation with this c_+ value in the denominator gives $c_+ = 1.27 \times 10^{-7}$ mol/cm^3 $= 1.27 \times 10^{-4}$ mol/dm^3, which is the H_3O^+ concentration. For this dilute solution, we can take the ionic molalities as 1.27×10^{-4} mol/kg. The Davies equation then gives $\gamma_\pm = 0.987$. Neglecting the slight differences between concentration-scale and molality-scale γ's, we have

$K_c = (0.987)^2(0.000127 \text{ mol/dm}^3)^2/[(0.001028 - 0.000127) \text{ mol/dm}^3] = 1.74 \times 10^{-5} \text{ mol/dm}^3$.

15.61 We have $\lambda^{\infty}_{m,+} + \lambda^{\infty}_{m,-} = (106 + 160)$ cm²/Ω-mol = 266 cm²/Ω-mol and $S = 8[60.6 + 0.23(266)]$ cm²/Ω-mol = 974 cm²/Ω-mol. Then

$$c_+ = \frac{6.156\times10^{-5}\ \Omega^{-1}\ \text{cm}^{-1}}{266\ \text{cm}^2/\Omega\text{-mol} - (974\ \text{cm}^2/\Omega\text{-mol})[c_+/(10^{-3}\ \text{mol/cm}^3)]^{1/2}}$$

With $c_+ = 0$ in the denominator, we get the initial estimate $c_+ \approx 2.31 \times 10^{-7}$ mol/cm³. With this value of c_+, we get the improved estimate $c_+ \approx 2.45 \times 10^{-7}$ mol/cm³. Another repetition yields the final result $c_+ = 2.45_5 \times 10^{-7}$ mol/cm³ $= 2.46 \times 10^{-4}$ mol/dm³. For this dilute solution, we can take the ion molalities as 2.46×10^{-4} mol/kg, so $I_m = 9.8 \times 10^{-4}$ mol/kg. The Davies equation gives $\log \gamma_\pm = -0.061$ and $\gamma_\pm = 0.87$. We can assume the molality-scale and concentration-scale activity coefficients to be equal in this dilute solution. We have $[Mg^{2+}] = [SO_4^{2-}] = 2.46 \times 10^{-4}$ mol/dm³ and $[MgSO_4(aq)] = 2.50 \times 10^{-4}$ mol/dm³ $- 2.45_5 \times 10^{-4}$ mol/dm³ $= 4._5 \times 10^{-6}$ mol/dm³. Hence, $K_c = (4.5 \times 10^{-6} \text{ mol/dm}^3)/(0.87)^2(2.46 \times 10^{-4} \text{ mol/dm}^3)^2 = 100$ dm³/mol.

15.62 Omitting the S term and using Eq. (15.58), we have for (15.90): $c_+ = \kappa/(\lambda^{\infty}_{m,+} + \lambda^{\infty}_{m,-}) = \Lambda_m c/(\lambda^{\infty}_{m,+} + \lambda^{\infty}_{m,-})$. Then $\alpha = c_+/c = \Lambda_m/(\lambda^{\infty}_{m,+} + \lambda^{\infty}_{m,-})$.

15.63 The Onsager equation predicts a $c^{1/2}$ dependence of Λ_m in dilute solutions, so we plot Λ_m vs. $c^{1/2}$. The points give a rather good straight-line fit; the intercept at $c = 0$ is $\Lambda^{\infty}_m = 270.8$ cm²/Ω-mol.

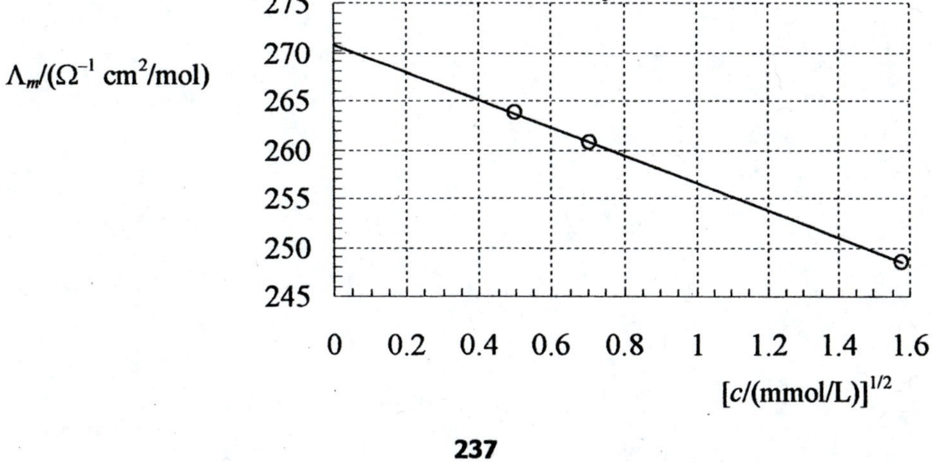

15.64 From (15.32), $(\Delta x)_{rms} = (2Dt)^{1/2}$. Substitution of the Stokes–Einstein equation (15.37) for D gives $(\Delta x)_{rms} \approx (kTt/3\pi\eta_{H_2O}r_{Mg^{2+}})^{1/2}$. Use of (15.70) for $r_{Mg^{2+}}$ gives $(\Delta x)_{rms} \approx (2kTu^{\infty}_{Mg^{2+}}t/|z_{Mg^{2+}}|e)^{1/2}$. In 1 s, $(\Delta x)_{rms} \approx$ $[2(1.38 \times 10^{-23}\text{ J/K})(298\text{ K})(55 \times 10^{-9}\text{ m}^2/\text{V-s})(1\text{ s})/2(1.60 \times 10^{-19}\text{ C})]^{1/2} =$ 3.8×10^{-5} m = 380000 Å = 0.0038 cm, where u^{∞} was taken from p. 500 of the text. In 10 s, $(\Delta x)_{rms} \approx (10/1)^{1/2}(38\ \mu\text{m}) = 120\ \mu$m.

15.65 **(a)** Increases; **(b)** increases; **(c)** increases; **(d)** increases; **(e)** increases; **(f)** increases.

15.66 Thermal conduction: q (units J), k (units J K^{-1} m^{-1} s^{-1}), T (units K), where the order is W, L, B. Viscous flow: p_y (kg m/s), η (N s/m^2), υ_y (m/s). Diffusion: n_j (mol), D_{jk} (m^2/s), c_j (mol/m^3). Electrical conduction: Q (C), κ (Ω^{-1} m^{-1}), ϕ (V).

Chapter 16

16.1 **(a)** F; **(b)** F; **(c)** T; **(d)** F; **(e)** T; **(f)** F; **(g)** T; **(h)** F. (A catalyst can appear in the rate law.)

16.2 **(a)** s^{-1}; **(b)** L mol^{-1} s^{-1}; **(c)** L^2 mol^{-2} s^{-1}.

16.3 (a).

16.4 (1/3)(–0.006 mol L^{-1} s^{-1}) = –0.002 mol L^{-1} s^{-1}.

16.5 **(a)** $n_A/V = P_A/RT$ = (0.10 atm)/(82.06 cm^3-atm/mol-K)(298 K) = $4.0_9 \times 10^{-6}$ mol/cm^3 = $4.0_9 \times 10^{-3}$ mol/dm^3. $r = k[N_2O_5]$ = $(1.73 \times 10^{-5}\ s^{-1})(4.0_9 \times 10^{-3}$ mol/dm^3) = 7.1×10^{-8} mol/dm^3-s. $J = rV = (7.1 \times 10^{-8}$ mol/dm^3-s)(12.0 dm^3) = 8.5×10^{-7} mol/s.

(b) $r = [1/(-2)]\ d[N_2O_5]/dt = 7.1 \times 10^{-8}$ mol/dm^3-s and $d[N_2O_5]/dt$ = -1.4×10^{-7} mol/dm^3-s.

(c) $(1.4 \times 10^{-7}$ mol/dm^3-s)$(6.02 \times 10^{23}$/mol)(1 s)(12 dm^3) = 1.0×10^{18}.

(d) $N_2O_5 \rightarrow 2NO_2 + \frac{1}{2}O_2$. Here, $J = -dn_{N_2O_5}/dt$ as compared to $J = -\frac{1}{2}dn_{N_2O_5}/dt$ for $2N_2O_5 \rightarrow 4NO_2 + O_2$. So $J = 2(8.5 \times 10^{-7}$ mol/s) = 17×10^{-7} mol/s. $r = J/V = 14 \times 10^{-8}$ mol/dm^3-s. $r = k[N_2O_5]$ and $k = r/[N_2O_5]$; since r is doubled, so is k, and $k = 3.46 \times 10^{-5}\ s^{-1}$. [In (c), the approximation $dn/dt \cong \Delta n/\Delta t$ was used. This approximation is justifiable, as follows: We have (0.0041 mol/L)(12 L) = 0.049 mol, which is 3×10^{22} molecules present; therefore a change of 1×10^{18} molecules will not significantly change the rate dn/dt over a period of 1 s.]

16.6 From (16.2), $J = (1/\nu_i)(dn_i/dt)$, where the stoichiometric coefficient ν_i is negative for reactants. From (4.97), $dn_i = \nu_i\ d\xi$, so $J = (1/\nu_i)\nu_i(d\xi/dt) = d\xi/dt$.

16.7 $C_B = N_B/V = (N_B/N_A)/(V/N_A) = n_B/(V/N_A) = [B]N_A$. $r = -(1/b)d[B]/dt$. $r_C = -(1/b)dC_B/dt = -(1/b)N_A d[B]/dt = N_A r$. So $r_C/r = N_A$. For simplicity, let the rate law be $r = k[\mathrm{B}]^n$. We have $r_C = k_C C_\mathrm{B}^n = k_C[\mathrm{B}]^n N_\mathrm{A}^n$. Hence $r_C/r =$ $(k_C/k)N_\mathrm{A}^n$. Also $r_C/r = N_\mathrm{A}$. Hence $(k_C/k)N_\mathrm{A}^n = N_\mathrm{A}$ and $k_C = kN_\mathrm{A}^{1-n}$.

16.8 **(a)** $P_\mathrm{A}V = n_\mathrm{A}RT$ and $P_\mathrm{A} = [\mathrm{A}]RT$. The equation $-(1/a)dP_\mathrm{A}/dt = k_P P_\mathrm{A}^n$ becomes $-(RT/a)d[\mathrm{A}]/dt = k_P[\mathrm{A}]^n(RT)^n$ and $r = -(1/a)d[\mathrm{A}]/dt =$ $(RT)^{n-1}k_P[\mathrm{A}]^n$. Comparison with $r = k[\mathrm{A}]^n$ gives $k_P = k(RT)^{1-n}$.

(b) Yes.

16.9 $r_1 = k_1[\mathrm{A}]$ and $r_2 = k_2[\mathrm{B}]$. Since [B] might be much greater than [A], it is possible for r_2 to exceed r_1 even though k_1 exceeds k_2.

16.10 **(a)** Since the first step produces one C molecule while the second step consumes two C's, the first step must occur twice for each occurrence of the second step. The second step produces one F and the third step consumes one F, so the second and third steps have the same stoichiometric number. The stoichiometric numbers are thus 2 for the first step, 1 for the second step, and 1 for the third step. Adding twice the first step to the second and third steps, we get
$2\mathrm{A} + 2\mathrm{B} + 2\mathrm{C} + \mathrm{F} + \mathrm{B} \rightarrow 2\mathrm{C} + 2\mathrm{D} + \mathrm{F} + 2\mathrm{A} + \mathrm{G}$ and the overall reaction is $3\mathrm{B} \rightarrow 2\mathrm{D} + \mathrm{G}$.

(b) Species A is consumed in the first step, regenerated in the last step, and does not appear in the overall reaction. Hence A is a catalyst. B is a reactant. C and F are reaction intermediates. B and G are products.

16.11 From Eq. (16.5), the units of k are $(\mathrm{dm^3/mol})^{n-1}\ \mathrm{s^{-1}}$. Comparison with the value of k shows that $n - 1 = 1$, $n = 2$.

16.12 $r_1 = k_1[\mathrm{A}]$. Since [A] decreases exponentially with time (Fig. 16.3), so does r_1. $r_2 = k_2[\mathrm{B}]$. [B] increases from 0 to a maximum and then decreases to zero (Fig. 16.3) and so does r_2.

16.13 **(a)** $-k_1[\mathrm{B}]$, $k_1[\mathrm{B}]$; **(b)** $-k_1[\mathrm{B}]$, $k_1[\mathrm{B}] - k_2[\mathrm{E}]$, $k_2[\mathrm{E}]$; **(c)** $-k_1[\mathrm{B}] + k_{-1}[\mathrm{E}]$, $k_1[\mathrm{B}] - k_{-1}[\mathrm{E}] - k_2[\mathrm{E}]$, $k_2[\mathrm{E}]$; **(d)** $-(k_1 + k_3)[\mathrm{B}] + k_{-1}[\mathrm{E}] + k_{-3}[\mathrm{F}]$, $k_1[\mathrm{B}] - k_{-1}[\mathrm{E}]$, $k_3[\mathrm{B}] - k_{-3}[\mathrm{F}]$.

16.14 **(a)** F; **(b)** F; **(c)** T (assuming V is constant).

16.15 **(a)** $[A]/[A]_0 = e^{-k_A t}$, so $0.65 = e^{-k_A(325\text{ s})}$ and $\ln 0.65 = -k_A(325\text{ s})$. We get $k_A = 0.0013_3\text{ s}^{-1}$. From Eq. (16.11), $k = k_A/a = k_A/2 = 0.00066\text{ s}^{-1}$.

(b) $t = -(1/k_A)\ln([A]/[A]_0) = -[1/(0.0013_3\text{ s}^{-1})]\ln(0.30\text{ or }0.10) =$ 905 s or 1731 s.

16.16 **(a)** $k_A t_{1/2} = 0.693$. From (16.11), $k_A = ak = 2(1.73 \times 10^{-5}\text{ s}^{-1}) = 3.46 \times 10^{-5}\text{ s}^{-1}$. So $t_{1/2} = 0.693/k_A = 0.693/(3.46 \times 10^{-5}\text{ s}^{-1}) = 2.00 \times 10^4\text{ s}$.

(b) $[A] = [A]_0 e^{-k_A t} = (0.010\text{ mol/dm}^3)\, e^{-(3.46\times 10^{-5}\text{ s}^{-1})(24\times 3600\text{ s})} = 5.0 \times 10^{-4}\text{ mol/dm}^3$.

16.17 $d[A]/dt = -k_A[A]^3$, $\int_1^2 [A]^{-3}\, d[A] = -\int_1^2 k_A\, dt$, $-\frac{1}{2}[A]^{-2}\Big|_1^2 = -k_A(t_2 - t_1)$, $1/[A]_2^2 - 1/[A]_1^2 = 2k_A(t_2 - t_1)$ or $1/[A]^2 - 1/[A]_0^2 = 2k_A t$, where we took $t_1 = 0$ and $t_2 = t$.

16.18 **(a)** Eqs. (16.22) and (16.19) with $A = NO_2$, $B = F_2$, $a = 2$, and $b = 1$ apply. $[A]_0 = (2.00\text{ mol})/(400\text{ dm}^3) = 0.00500\text{ mol/dm}^3$ and $[B]_0 = 0.00750\text{ mol/dm}^3$. $(a[B]_0 - b[A]_0)kt =$ $[2(0.00750) - 0.00500](\text{mol/dm}^3)(38\text{ dm}^3/\text{mol-s})(10.0\text{ s}) = 3.80$. Using $[B]/[B]_0 = 1 - ba^{-1}[A]_0/[B]_0 + ba^{-1}[A]/[B]_0$, we have for Eq. (16.22): $3.80 = \ln\{(0.667 + [66.7\text{ dm}^3/\text{mol}][A])/(200\text{ dm}^3/\text{mol})[A]\}$ and $e^{3.80} = 0.00333/(\text{dm}^3/\text{mol})[A] + 0.333$. We find $[A] = 7.5 \times 10^{-5}\text{ mol/dm}^3 = [NO_2]$. Then $n_A = (400\text{ dm}^3)(7.5 \times 10^{-5}\text{ mol/dm}^3) = 0.0300\text{ mol } NO_2$. Let $2z$ moles of NO_2 react in 10 s. Then $2 - 2z = 0.0300$ and $z = 0.985$. Then $n_{F_2} = 3 - z = 2.01_5$ moles and $n_{NO_2F} = 2z = 1.97$ mol.

(b) $r = k[NO_2][F_2] = (38\text{ dm}^3/\text{mol-s})(0.00500\text{ mol/dm}^3)(0.00750\text{ mol/dm}^3) = 0.00142\text{ mol/dm}^3\text{-s}$ initially. After 10 s, part (a) gives $[NO_2] = 7.5 \times 10^{-5}\text{ mol/dm}^3$ and $[F_2] = (2.01_5\text{ mol})/(400\text{ dm}^3) = 5.04 \times 10^{-3}\text{ mol/dm}^3$. So $r = (38\text{ dm}^3/\text{mol-s})(7.5 \times 10^{-5}\text{ mol/dm}^3)(5.04 \times 10^{-3}\text{ mol/dm}^3) = 1.44 \times 10^{-5}\text{ mol/dm}^3\text{-s}$.

16.19 **(a)** Substitution of $N = A/\lambda$ and $N_0 = A_0/\lambda$ into $N = N_0 e^{-\lambda t}$ gives $A = A_0 e^{-\lambda t}$.

(b) $N_0 = 1.00 \text{ g} \dfrac{1 \text{ mol Ra}}{226 \text{ g}} \dfrac{6.02 \times 10^{23} \text{ atoms}}{1 \text{ mol}} = 2.66 \times 10^{21}$ atoms

$\lambda = A_0/N_0 = (3.7 \times 10^{10} \text{ s}^{-1})/(2.66 \times 10^{21}) = 1.39 \times 10^{-11} \text{ s}^{-1}$

$t_{1/2} = 0.693/\lambda = 5.0 \times 10^{10} \text{ s} = 1600$ years

$A = A_0 e^{-\lambda t} = (3.7 \times 10^{10} \text{ s}^{-1})\exp[(-1.39 \times 10^{-11} \text{ s}^{-1})(3.15 \times 10^{10} \text{ s})]$

$A = 2.4 \times 10^{10}$ dis/s

(c) $N_0 = 10.0 \text{ g KCl} \dfrac{1 \text{mol KCl}}{74.55 \text{ g}} \dfrac{6.022 \times 10^{23} \text{ K atoms}}{1 \text{ mol KCl}} (0.0000117)$

$N_0 = 9.45 \times 10^{17}$ atoms of ^{40}K.

$\lambda = 0.693/t_{1/2} = 0.693/(1.28 \times 10^9 \text{ yr}) = 5.41 \times 10^{-10} \text{ yr}^{-1}$.

$A_0 = \lambda N_0 = (5.41 \times 10^{-10} \text{ yr}^{-1})(9.45 \times 10^{17}) = 5.11 \times 10^8$ dis/yr = 16.2 dis/s.

$A = A_0 e^{-\lambda t} = (16.2 \text{ s}^{-1})\exp[(-5.41 \times 10^{-10} \text{ yr}^{-1})(2.00 \times 10^8 \text{ yr})] = 14.5$ /s.

16.20 **(a)** As noted in Section 1.8, $(d/dx) \int f(x)\, dx = f(x)$. Using this identity, we have $dy/dx = e^{w(x)}(dw/dx)(\int e^{-w(x)} f(x)\, dx + c) + e^{w(x)}[e^{-w(x)} f(x)]$. Since $w(x) = \int g(x)\, dx$, we have $dw/dx = g(x)$ and $dy/dx = e^w g(\int e^{-w} f\, dx + c) + f$. The right side of the differential equation is $f + gy = f + ge^w(\int e^{-w} f\, dx + c)$, which is the same as dy/dx. Hence, y is the correct solution.

(b) In (16.39), we take $y = [\text{B}]$, $x = t$, $f(x) = k_1[\text{A}]_0 e^{-k_1 t}$, $g(x) = -k_2$. Then $w = -\int k_2\, dt = -k_2 t$ and $y = [\text{B}] = e^{-k_2 t}(\int e^{k_2 t} k_1[\text{A}]_0 e^{-k_1 t}\, dt + c) = e^{-k_2 t}\{k_1[\text{A}]_0 e^{(k_2 - k_1)t}/(k_2 - k_1) + c\} = k_1[\text{A}]_0 e^{-k_1 t}/(k_2 - k_1) + ce^{-k_2 t}$. To evaluate c, we use the fact that $[\text{B}] = 0$ at $t = 0$; so $0 = k_1[\text{A}]_0/(k_2 - k_1) + c$ and $c = -k_1[\text{A}]_0/(k_2 - k_1)$. Hence $[\text{B}] = \{k_1[\text{A}]_0/(k_2 - k_1)\}(e^{-k_1 t} - e^{-k_2 t})$.

16.21 No. A kinetically reversible reaction is one where a significant amount of back reaction occurs. A thermodynamically reversible process must go through equilibrium states only, so a kinetically reversible reaction is a thermodynamically irreversible process.

16.22 From Eq. (16.17), $r = k[\mathrm{A}]^2$ integrates to $[\mathrm{A}] = [\mathrm{A}]_0/(1 + akt[\mathrm{A}]_0)$, and $r = k[\mathrm{A}]^2$ becomes $r = k[\mathrm{A}]_0^2/(1 + akt[\mathrm{A}]_0)^2$.

16.23 $r = k[\mathrm{A}]^0 = k = -d[\mathrm{A}]/dt$, which integrates to $[\mathrm{A}] = -kt + c$. At $t = 0$, $[\mathrm{A}]_0 = 0 + c$. So $[\mathrm{A}] = [\mathrm{A}]_0 - kt$. The graph is linear.

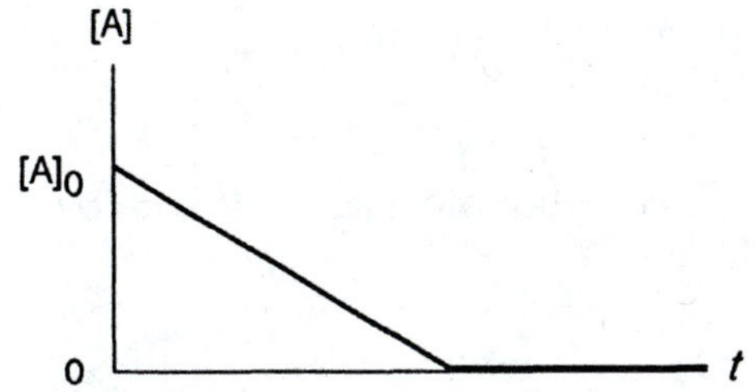

16.24 For $n = 1$, the integrated rate law is (16.30), for which [A] reaches 0 only at $t = \infty$. For $n \neq 1$, the integrated rate law is (16.28). Setting [A] = 0 in (16.28), we see there are two cases to consider. If $n < 1$, then the left side of (16.28) is zero when [A] = 0, and the time $t^\#$ required for [A] to reach 0 satisfies $0 = 1 + [\mathrm{A}]_0^{n-1}(n-1)k_{\mathrm{A}}t^\#$ and $t^\# = 1/[\mathrm{A}]_0^{n-1}(1-n)$, which is positive, since $n < 1$. Hence [A] reaches 0 in a finite time when $n < 1$. For $n > 1$, the left side of (16.28) becomes infinite when [A] equals 0 and an infinite amount of time is required for [A] to reach 0.

16.25 $a^{-1}\, d[\mathrm{A}]/dt = -k[\mathrm{A}]^2[\mathrm{B}]$. Use of (16.19) gives $\int_1^2 [\mathrm{A}]^{-2}([\mathrm{B}]_0 - ba^{-1}[\mathrm{A}]_0 + ba^{-1}[\mathrm{A}])^{-1}\, d[\mathrm{A}] = -\int_1^2 ak\, dt$. A table of integrals gives $\int [1/x^2(p + sx)]\, dx = -1/px + (s/p^2)\ln[(p + sx)/x]$. We have $x = [\mathrm{A}]$, $p = [\mathrm{B}]_0 - ba^{-1}[\mathrm{A}]_0$, and $s = ba^{-1}$, so

$$\frac{ba^{-1}}{([\mathrm{B}]_0 - ba^{-1}[\mathrm{A}]_0)^2}\ln\frac{[\mathrm{B}]_0 - ba^{-1}[\mathrm{A}]_0 + ba^{-1}[\mathrm{A}]}{[\mathrm{A}]}\Bigg|_1^2 - \frac{1}{([\mathrm{B}]_0 - ba^{-1}[\mathrm{A}]_0)[\mathrm{A}]}\Bigg|_1^2 = -ak(t_2 - t_1)$$

$$\frac{1}{a[\mathrm{B}]_0 - b[\mathrm{A}]_0}\left(\frac{1}{[\mathrm{A}]_0} - \frac{1}{[\mathrm{A}]} + \frac{b}{a[\mathrm{B}]_0 - b[\mathrm{A}]_0}\ln\frac{[\mathrm{B}]/[\mathrm{B}]_0}{[\mathrm{A}]/[\mathrm{A}]_0}\right) = -kt$$

where we used (16.19) and took $t_1 = 0$ and $t_2 = t$.

16.26 **(a)** $(1.5)^2(3) = 6.75$.

(b) $(3[\mathrm{A}]_0)^n = 27([\mathrm{A}]_0)^n$ and $n = 3$.

16.27 In equation (16.28), let $[A]/[A]_0 = \alpha$. Then $\alpha^{1-n} = 1 + [A]_0^{n-1}(n-1)k_A t_\alpha$ and $t_\alpha = (\alpha^{1-n} - 1)/[A]_0^{n-1}(n-1)k_A$. Taking the log of each side, we get the desired result for $n \neq 1$. For $n = 1$, we set $[A]/[A]_0 = \alpha$ in (16.30) to get $\alpha = e^{-k_A t_\alpha}$ and $\ln \alpha = -k_A t_\alpha$.

16.28 Check your plots against Fig. 16.6.

16.29 **(a)** We use the fractional-life method, plotting $\log t_\alpha$ vs. $\log [A]_0$. The data are

$\log t_\alpha$	2.771	2.823	2.954	3.057
$\log [A]_0$	−2.090	−2.191	−2.509	−2.726

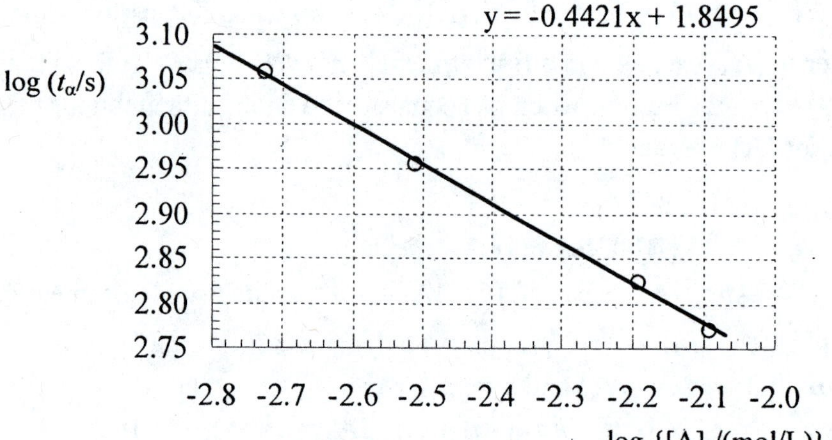

The slope of the straight line through these points is $-0.44 = 1 - n$, so $n = 1.44 \approx 1.5$. The order is 3/2.

(b) Putting $[A]/[A]_0 = \alpha = 0.69$ and $n = 1.5$ in Eq. (16.28), we get $k_A = 0.408/t_\alpha[A]_0^{1/2}$. Substitution of the four pairs of t_α and $[A]_0$ values gives $10^3 k_A/(\text{dm}^{3/2}/\text{mol}^{1/2}\text{-s})$ as 7.67, 7.64, 8.14, and 8.25. Averaging, we get $k_A = 7._9 \times 10^{-3}\ \text{dm}^{3/2}/\text{mol}^{1/2}\text{-s}$.

16.30 **(a)** Both plots give pretty good fits to a straight line. The log [A] vs. t plot (corresponding to $n = 1$) gives a better fit, but considering the inaccuracy of kinetics data, one could not absolutely rule out $n = 2$ from the plot.

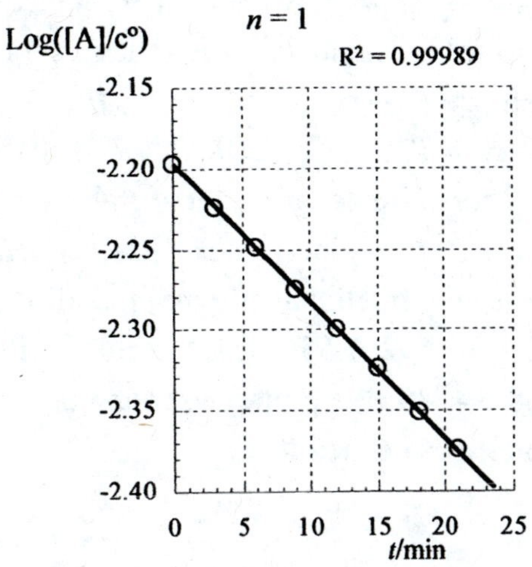

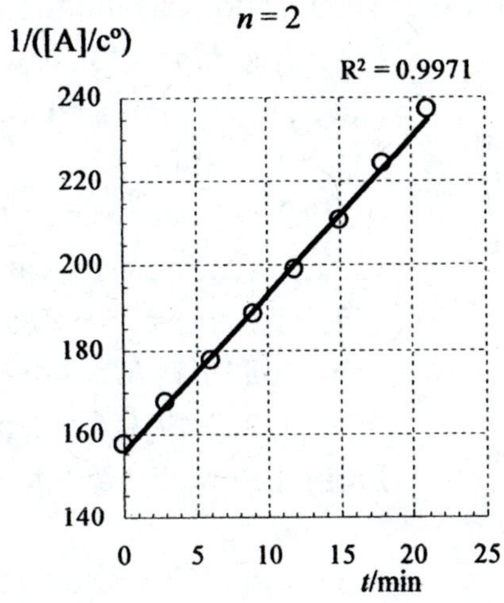

(b) A good fit is obtained for $n = 1$ and for $n = 3/2$; the $n = 2$ fit is not good. To decide between $n = 1$ and $n = 3/2$, one would need data at later times.

16.31 Equation (16.22) with $a = 1$ and $b = 1$ applies. Let $B = S_2O_3^{2-}$ and $A = C_3H_7Br$. Then $a[B]_0 - b[A]_0 = 57.1$ mmol/dm^3. We use $[A] - [A]_0 = [B] - [B]_0$ to calculate the [A] concentrations. A plot of $\ln \{([B]/[B]_0)/[A]/[A]_0)\} \equiv$ $\ln z$ vs. t is linear with slope $(a[B]_0 - b[A]_0)k$. The data are

$\ln z$	0	0.104	0.189	0.474	1.035
t/s	0	1110	2010	5052	11232

The slope is $9.2_0 \times 10^{-5}$ s^{-1}, so $k = (9.2_0 \times 10^{-5}$ s$^{-1})/(0.0571$ mol/dm$^3) =$ 1.61×10^{-3} dm^3 mol^{-1} s^{-1}.

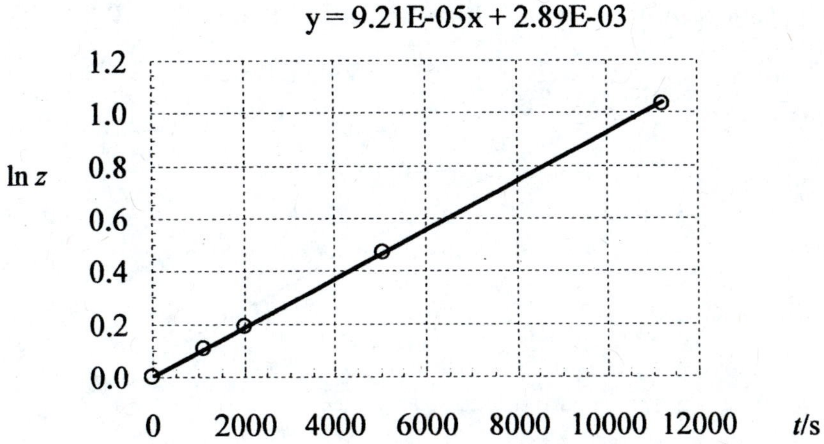

16.32 **(a)** The reaction has the form 2A → B. We have $P = P_A + P_B = (c_A + c_B)RT$. Let $2z$ mol/dm^3 of A react to form z mol/dm^3 of B. Then $c_A = c_{A,0} - 2z$ and $c_B = z = \frac{1}{2}(c_{A,0} - c_A)$. Hence $P = (c_A + \frac{1}{2}c_{A,0} - \frac{1}{2}c_A)RT = \frac{1}{2}(c_A + c_{A,0})RT$. Also, $P_0 = c_{A,0}RT$, so $c_A = (2P - P_0)/RT$ and $\alpha = c_A/c_{A,0} = 2P/P_0 - 1$, where $P_0 = 632$ torr. We calculate the α values and plot α vs. log t; comparison with the Powell-plot master curves shows the order is 2. Alternatively, we can use the fractional-life method: The calculated values of $100c_A$/(mol/dm^3) are 1.692, 1.556, 1.436, 1,345, 1.174, 1.035, 0.8929, 0.7349, 0.6252, and 0.4780. We plot c_A vs. t and take α = 0.75. From the graph, we find the times needed for

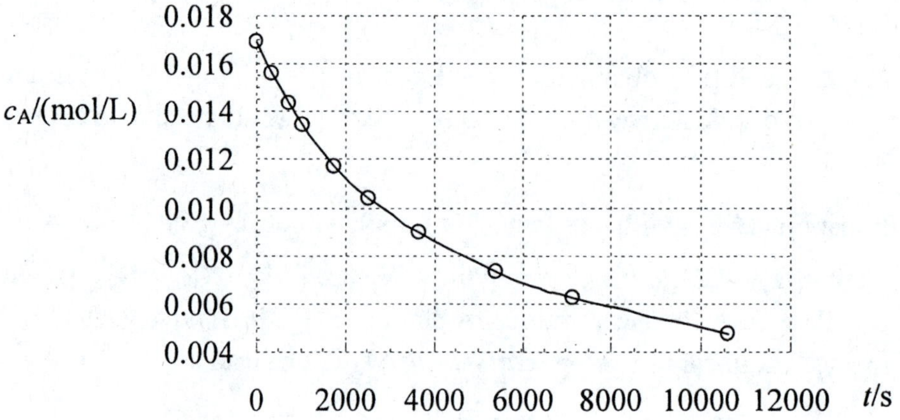

$100c_A$/(mol/dm^3) to undergo the following changes: 1.6 to 1.2, 1.4 to 1.05, 1.2 to 0.90, 1.0 to 0.75, 0.8 to 0.6. Values of log $t_{0.75}$ are

log $t_{0.75}$	3.138	3.204	3.297	3.375	3.467
log [$100c_A$/(mol/dm^3)]	0.204	0.146	0.079	0	−0.097

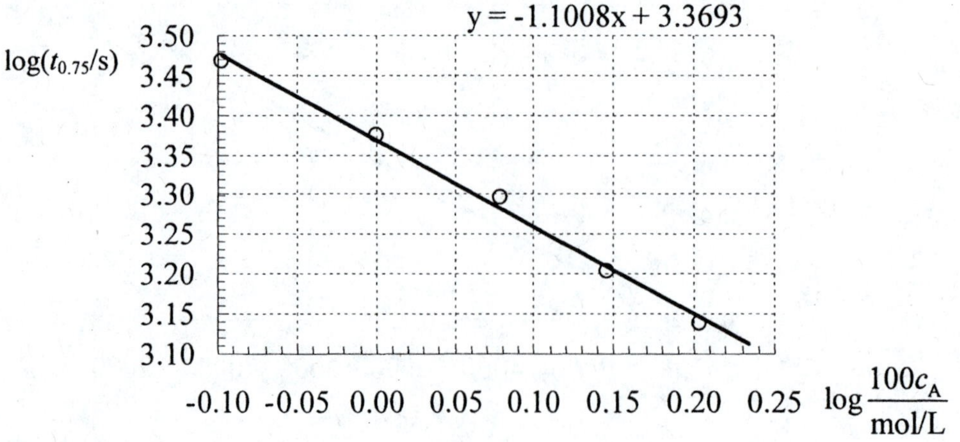

where c_A is the initial A concentration. The plot has slope $-1.1 = 1 - n$, so $n = 2$.

(b) Equation (16.16) applies. We plot $1/c_A = 1/[A]$ vs. t. The data are

$(\text{mol/dm}^3)/c_A$	59.11	64.29	69.66	74.34	85.15	
t/s	0	367	731	1038	1751	etc.

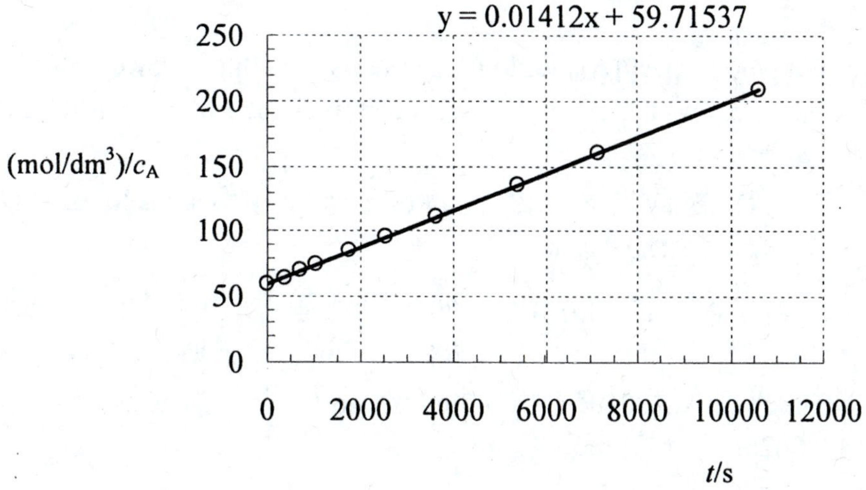

The slope is 0.0141_5 dm^3 mol^{-1} s^{-1} = $k_A = ak = 2k$, and $k = 0.0071$ dm^3 mol^{-1} s^{-1}.

16.33 (a) The first two columns of data show that tripling $[A]_0$ at constant $[B]_0$ and $[C]_0$ triples r_0, so $\alpha = 1$. The first and third columns show that tripling $[B]_0$ at constant $[A]_0$ and $[C]_0$ multiplies the rate by 9, so $\beta = 2$. The second and fourth columns show that $\gamma = 0$.

(b) $r_0 = k[A]_0[B]_0^2$, and use of the first column of data gives $k = (0.0060\ c°\ \text{s}^{-1})/(0.20\ c°)(0.30\ c°)^2 = 0.33$ dm^6 mol^{-2} s^{-1}.

(c) For the rate law (1) in Eq. (16.6), [HBr] is initially zero and initial-rate data would yield the erroneous result $r = k[H_2][Br_2]^{1/2}$.

16.34 In the first run, we have $[A]_0 >> [B]_0$, so $[A]_0$ is essentially constant. The B concentrations are 0.400, 0.200, 0.100, and 0.050 mmol/dm^3 at 0, 120, 240, and 360 s. The half-life is thus constant at 120 s; hence the order with respect to B is 1. In the second run $[B]_0$ is constant and the A concentrations are 0.400,

0.200, 0.100, and 0.050 mmol/dm^3 at 0, 69, 208, and 485 ks. The half-lives are 69, 139, and 277 ks. The half-life is doubled when the A concentration is cut in half, so $t_{1/2}$ is proportional to 1/[A] and [see Eq. (16.29)] the order with respect to A is 2. Hence $r = k[\text{A}]^2[\text{B}]$. On the first run, $r = -d[\text{B}]/dt = k[\text{A}]_0^2[\text{B}] \equiv k_\text{B}[\text{B}]$. The reaction is pseudo first order and Eq. (16.14) gives $[\text{B}] = [\text{B}]_0 e^{-k_\text{B}t}$; we have $k_\text{B} = -t^{-1} \ln ([\text{B}]/[\text{B}]_0) = -(120 \text{ s})^{-1} \ln ½ = 0.00578 \text{ s}^{-1} = k(0.400 \text{ mol/dm}^3)^2$ and $k = 0.036 \text{ dm}^6/\text{mol}^2\text{-s}$.

16.35 **(a)** We plot $\alpha = [\text{A}]/[\text{A}]_0$ vs. log t; comparison with the Powell-plot master curves shows the order is 3/2. (Alternatively, the fractional-life method can be used.)

(b) From (16.28) with $n = 1.5$, a plot of $([\text{A}]_0/[\text{A}])^{1/2}$ vs. t is linear with slope $\frac{1}{2}[\text{A}]_0^{1/2} k_\text{A}$. The data are

$([\text{A}]_0/[\text{A}])^{1/2}$	1	1.098	1.206	1.294	1.399	1.612	2.01
t/s	0	100	200	300	400	600	1000

The slope is $0.00101 \text{ s}^{-1} = 0.5(0.600 \text{ mol/dm}^3)^{1/2} k_\text{A}$ and $k_\text{A} = 0.0026_1 \text{ s}^{-1} \text{ dm}^{3/2} \text{ mol}^{-1/2} = ak = k$.

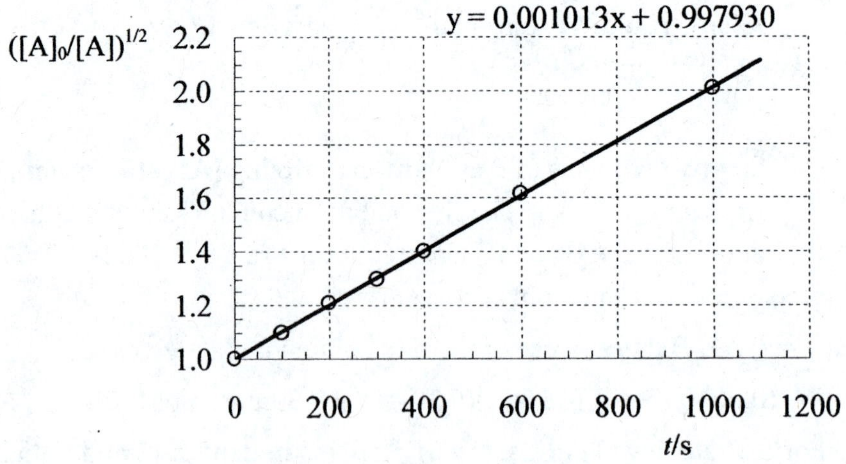

16.36 For the first run, $[\text{A}]_0 >> [\text{B}]_0$, so [A] is essentially constant. A Powell plot of $\alpha = [\text{B}]/[\text{B}]_0$ vs. log t shows the order with respect to B is 2. (Alternatively, the fractional-life method can be used.) For the second run, [A] is also essentially constant. For runs 1 and 2, we have $r_1 = -d[\text{B}]/dt = k[\text{A}]_{0,1}^{\alpha}[\text{B}]^2 \equiv j_1[\text{B}]^2$ and $r_2 = k[\text{A}]_{0,2}^{\alpha}[\text{B}]^2 \equiv j_2[\text{B}]^2$. Hence $j_1/j_2 = k[\text{A}]_{0,1}^{\alpha}/k[\text{A}]_{0,2}^{\alpha} = ([\text{A}]_{0,1}/[\text{A}]_{0,2})^{\alpha}$. The pseudo order is 2; a plot of 1/[B] vs. t is linear with slope j. For run 1, such a plot has slope 0.0119 dm^3 mol^{-1} s^{-1}; for run 2, the slope is 0.0067_1 dm^3 mol^{-1} s^{-1}. Then $(0.800/0.600)^\alpha = 0.0119/0.0067_1 = 1.77_3$ and $\alpha = (\log 1.77_3)/(\log 1.333) = 1.99 \approx 2$. Also, $j_1 = k[\text{A}]_{0,1}^2$ and $k =$ (0.0119 dm^3 mol^{-1} s^{-1})/(0.800 mol/dm^3)^2 = 0.0186 dm^9 mol^{-3} s^{-1}. (Data from the second run give 0.0186 dm^9 mol^{-3} s^{-1}.)

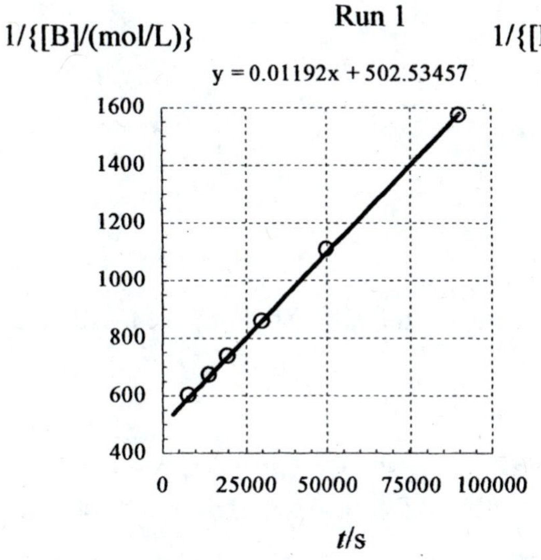

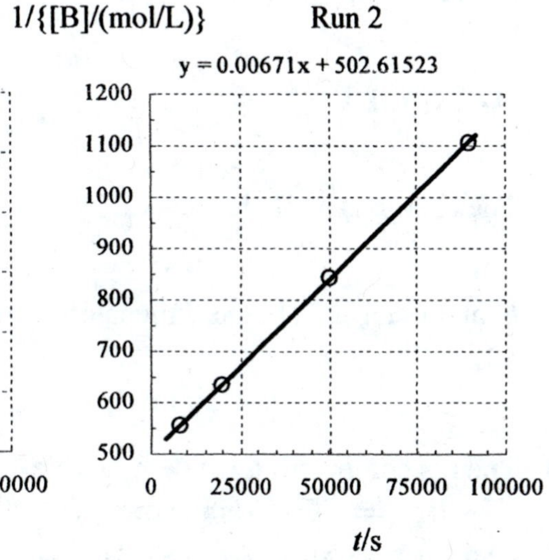

16.37 **(a)** The regression line has slope 1.289×10^{-4} L mmol^{-1} s^{-1} and $k_A = 0.128_9$ L mol^{-1} s^{-1}.

(b) The Solver gives 1.289×10^{-4} L mmol^{-1} s^{-1}.

(c) The 1/[A] vs. t regression line has slope $k_A = 1.3647 \times 10^{-4}$ L mmol^{-1} min^{-1} = $(1.3647 \times 10^{-4})(1000/60)$ L mol^{-1} s^{-1} = 2.2745×10^{-3} L mol^{-1} s^{-1}. Also (Sec. 16.3) $k = k_A/3 = 7.58 \times 10^{-4}$ L mol^{-1} s^{-1}. For a least-squares fit of the [A] vs. t data, the Solver gives $k_A = 1.2901 \times 10^{-4}$ L mmol^{-1} min^{-1} = 2.150×10^{-3} L mol^{-1} s^{-1}, and $k = k_A/3 = 7.17 \times 10^{-4}$ L mol^{-1} s^{-1}

16.38 **(a)** T; **(b)** T.

16.39 For an elementary reaction, $k_f/k_b = K_c$. We have $-RT \ln K_P^\circ = \Delta G^\circ =$ [2(51.31) – 97.89] kJ/mol = 4730 J/mol. Then $\ln K_P^\circ = -1.908$ and $K_P^\circ =$

0.148. From Eq. (6.25), $K_P^\circ = K_c^\circ RT/(\text{bar-dm}^3/\text{mol})$ and K_c° = 0.148(bar-dm^3/mol)/(0.08314 dm^3-bar/mol-K)(298.1 K) = 0.00597; so K_c = 0.00597 mol/dm^3. Then $k_b = k_f/K_c = (4.8 \times 10^4\ \text{s}^{-1})/(0.00597\ \text{mol/dm}^3) = 8.0 \times 10^6\ \text{dm}^3\ \text{mol}^{-1}\ \text{s}^{-1}$.

16.40 From (16.4), $r = -d[\text{A}]/dt = \frac{1}{2}\, d[\text{C}]/dt$, so $d[\text{A}]/dt = -r$ and $d[\text{C}]/dt = 2r$. From (1.5) for this elementary reaction, $r = k[\text{A}][\text{B}]$. Hence $d[\text{A}]/dt = -k[\text{A}][\text{B}]$ and $d[\text{C}]/dt = 2k[\text{A}][\text{B}]$.

16.41 **(a)** T; **(b)** F.

16.42 The charge is not balanced.

16.43 k_1 and k_2 have different dimensions (units) and can't be compared with each other.

16.44 The rds composition is given by rule 1 as H_2NO_2Br. One possibility is
$H^+ + Br^- \rightleftharpoons HBr$ (rapid equilib.), $HBr + HNO_2 \rightarrow ONBr + H_2O$ (slow rds),
$ONBr + C_6H_5NH_2 \rightarrow C_6H_5N_2^+ + H_2O + Br^-$.

16.45 **(a)** For runs 1 and 3, the ClO^- initial concentration is cut in half while the other initial concentrations are kept constant; since the rate is cut in half, we conclude that the order with respect to ClO^- is 1. Similarly, runs 2 and 3 tell us that the order with respect to I^- is 1. For runs 3 and 4, the OH^- initial concentration is multiplied by ¼ and the rate is quadrupled; so the order with respect to OH^- is –1. Thus $r = k[I^-][ClO^-]/[OH^-]$. For run 1, 0.00048 $c°$/s = k(0.00200 $c°$)(0.00400 $c°$)/(1.000 $c°$) and $k = 60\ \text{s}^{-1}$.

(b) We assume a mechanism with a rate-determining step (rds). From rules 1b and 2 in Sec. 16.6, the total composition of the rds reactants is $I^- + ClO^- - OH^- + xH_2O = ClIO_xH_{2x-1}^-$ and the species OH^- is a product in an equilibrium step that precedes the rds and OH^- does not appear in the rds. To keep $2x - 1$ nonnegative, x must be 1 or greater. The simplest assumption is $x = 1$. This gives the rds composition as $ClIOH^-$. A plausible mechanism that meets the preceding two requirements and that gives the proper stoichiometry is $OCl^- + H_2O \rightleftharpoons OH^- + HOCl$ (rapid equilib.); $HOCl + I^- \rightarrow HOI + Cl^-$ (rds); $HOI + OH^- \rightarrow OI^- + H_2O$ (rapid).

16.46 The rate-determining-step reactants' composition is NO_2Cl. One plausible mechanism is $NO_2Cl \rightarrow NO_2 + Cl$ (slow), $Cl + NO_2Cl \rightarrow NO_2 + Cl_2$ (rapid). Another possibility is $NO_2Cl \rightarrow NO_2 + Cl$ (slow), $Cl + Cl \rightarrow Cl_2$ (rapid), with stoichiometric numbers 2 and 1, respectively, for steps 1 and 2.

16.47 The rate-determining-steps reactants' composition is $CrTl^{5+}$. One possibility is $Cr^{2+} + Tl^{3+} \rightarrow Cr^{3+} + Tl^{2+}$ (slow), $Tl^{2+} + Cr^{2+} \rightarrow Cr^{3+} + Tl^{+}$ (rapid). Another possibility is $Cr^{2+} + Tl^{3+} \rightleftharpoons CrTl^{5+}$ (equilib.), $CrTl^{5+} \rightarrow Cr^{3+} + Tl^{2+}$ (slow), $Tl^{2+} + Cr^{2+} \rightarrow Cr^{3+} + Tl^{+}$ (rapid).

16.48 The rate-determining-step reactants' composition is NO_2F_2. A possible mechanism is $NO_2 + F_2 \rightarrow NO_2F + F$ (slow), $F + NO_2 \rightarrow NO_2F$ (rapid). Another possibility is $NO_2 + F_2 \rightleftharpoons NO_2F_2$ (equilib.), $NO_2F_2 \rightarrow NO_2F + F$ (slow), $F + NO_2 \rightarrow NO_2F$ (rapid).

16.49 $XeF_4 + NO \rightarrow XeF_3 + NOF$.

16.50 The rate-determining-step reactants have composition N_2O_5. A likely mechanism is the slow step $N_2O_5 \rightarrow NO_2 + NO_3$ followed by a series of rapid steps that yield the correct stoichiometry. One of the many possibilities for this series of rapid steps is $NO_2 + Cl_2O \rightarrow NO_2Cl + OCl$, $OCl + NO_3 \rightarrow NO_3Cl + O$, $O + O \rightarrow O_2$. (The stoichiometric number of all steps but the last is 2.)

16.51 The rate-determining-step reactants have overall composition $HgTl^{3+}$, and Hg^{2+} is a product in an equilibrium that precedes the rate-determining step. Two other mechanisms besides that given in Example 16.7 are: $Hg_2^{2+} + Tl^{3+} \rightleftharpoons HgTl^{3+} + Hg^{2+}$ (rapid equilib.), $HgTl^{3+} \rightarrow Hg^{2+} + Tl^{+}$ (slow); and $Hg_2^{2+} + Tl^{3+} \rightleftharpoons Hg^{+} + Tl^{2+} + Hg^{2+}$ (rapid equilib.), $Hg^{+} + Tl^{2+} \rightarrow Hg^{2+} + Tl^{+}$ (slow).

16.52 The reverse reaction would then proceed by the one-step mechanism $N_2 + 3H_2 \rightarrow 2NH_3$. But a tetramolecular elementary step is far too unlikely to occur.

16.53 **(a)** $d[O_2]/dt = 2k_2[O][O_3] + k_1[O_3][M] - k_{-1}[O_2][O][M]$.
$d[O_3]/dt = -k_1[O_3][M] + k_{-1}[O_2][O][M] - k_2[O][O_3]$.

(b) $d[O]/dt = 0 = k_1[O_3][M] - k_{-1}[O_2][O][M] - k_2[O][O_3]$, so $k_1[O_3][M] - k_{-1}[O_2][O][M] = k_2[O][O_3]$. Substitution into the expressions for $d[O_2]/dt$ and $d[O_3]/dt$ in (a) gives $d[O_2]/dt = 3k_2[O][O_3]$ and $d[O_3]/dt = -2k_2[O][O_3]$.

(c) From (b) we get $[O] = k_1[O_3][M]/(k_{-1}[O_2][M] + k_2[O_3])$. We have $r = -\frac{1}{2}\, d[O_3]/dt = k_2[O][O_3] = k_1k_2[O_3]^2[M]/(k_{-1}[O_2][M] + k_2[O_3]) = k_1k_2[O_3]^2/(k_{-1}[O_2] + k_2[O_3]/[M])$. Also, $r = (1/3)d[O_2]/dt = k_2[O][O_3] =$ etc.

(d) If step 1 is in equilibrium, then $k_1/k_{-1} = [O_2][O][M]/[O_3][M]$ and $[O] = k_1[O_3]/k_{-1}[O_2]$. As noted in the problem statement, $r = \frac{1}{3}\, d[O_2]/dt = k_2[O][O_3] = k_1k_2[O_3]^2/k_{-1}[O_2]$.

(e) If $k_2[O_3]/[M] << k_{-1}[O_2]$ (i.e., if $k_{-1}[O_2][M] >> k_2[O_3]$), the second term in the denominator of the steady-state expression can be neglected, thereby giving the rate-determining-step expression.

16.54 **(a)** $d[NO]/dt = 0 = k_b[NO_2][NO_3] - k_c[NO][NO_3]$ and $k_c[NO][NO_3] = k_b[NO_2][NO_3]$. $d[NO_3]/dt = 0 = k_a[N_2O_5] - k_{-a}[NO_2][NO_3] - k_b[NO_2][NO_3] - k_c[NO][NO_3] = k_a[N_2O_5] - k_{-a}[NO_2][NO_3] - k_b[NO_2][NO_3] - k_b[NO_2][NO_3]$ and $[NO_3] = k_a[N_2O_5]/(k_{-a} + 2k_b)[NO_2]$. Then $r = -\frac{1}{2}\, d[N_2O_5]/dt = -\frac{1}{2}(-k_a[N_2O_5] + k_{-a}[NO_2][NO_3]) = \frac{1}{2}k_a[N_2O_5] - \frac{1}{2}k_{-a}k_a[N_2O_5]/(k_{-a} + 2k_b) = [k_ak_b/(k_{-a} + 2k_b)][N_2O_5] = k[N_2O_5]$ and $k = k_ak_b/(k_{-a} + 2k_b)$.

(b) If step b is the rate-determining step and step a is in equilibrium, then $k_a/k_{-a} = [NO_2][NO_3]/[N_2O_5]$. The rate of the reaction equals the rate of the rate-determining step b (since the stoichiometric number of step b is 1), so $r = k_b[NO_2][NO_3] = (k_bk_a/k_{-a})[N_2O_5]$.

(c) If $k_{-a} >> 2k_b$, then the steady-state rate law of part (a) reduces to the rate-determining-step rate law of (b). (Of course, this is a necessary condition for the validity of the rate-determining-step approximation.)

(d) From this problem, we see that k of the N_2O_5 decomposition is a function of k_a, k_{-a}, and k_b. Hence, it is clear that the mechanism of the reaction in Prob. 16.50 starts off with steps a and b of the mechanism in Eq. (16.8). After step b, we need steps that give the correct stoichiometry. A possible mechanism is $N_2O_5 \rightleftharpoons NO_2 + NO_3$ (rapid equilib.), $NO_2 + NO_3 \rightarrow NO + O_2 + NO_2$ (slow, rate determining), $NO_2 + Cl_2O \rightarrow NO_2Cl + OCl$, $OCl + NO \rightarrow NO_2Cl$, $NO_2Cl + O_2 \rightarrow NO_3Cl + O$, $O + O \rightarrow O_2$. (The stoichiometric number of all steps but the last is 2.)

16.55 For (16.60), $r = \frac{1}{2}\, d[NO_2]/dt = k_2[N_2O_2][O_2]$; the initial equilibrium gives $[N_2O_2]/[NO]^2 = K_{c,1} = k_1/k_{-1}$ and $[N_2O_2] = (k_1/k_{-1})[NO]^2$, so $r = (k_2k_1/k_{-1})[NO]^2[O_2]$. For (16.61), $r = \frac{1}{2}\, d[NO_2]/dt = k_2[NO_3][NO]$ and $k_1/k_{-1} = [NO_3]/[NO][O_2]$, so $r = (k_2k_1/k_{-1})[NO]^2[O_2]$. For (16.62), $r = k[NO]^2[O_2]$.

16.56 The exact result at 1.00 s is $[A] = [A]_0e^{-kt} = (1.00 \text{ mol/L})\, e^{-(0.15 \text{ s}^{-1})(1.00 \text{ s})} =$ 0.860708 mol/L and is 0.637628 mol/L at 3.00 s. For this reaction $g([A])$ in (16.64) is $g([A]) = -k[A]^n = -k[A]$ and $[A]_1 = [A]_0 - k[A]_0\, \Delta t$. The first number in column A (the time column) is 0 and the next entry in column A is the first entry plus Δt. The first number in column B is 1 and the next entry contains the Euler formula. The results for 1 s and 3 s are 0.85873 and 0.63325 mol/L for Δt = 0.2 s and 0.85973 and 0.63546 for $\Delta t = 0.1$ s.

16.57 The exact value at 1.00 s is [Eq. (16.17)] $[A] =$ (1.00 mol/L)/[1 + (0.15 L/mol-s)(1.00 s)(1.00 mol/L)] = 0.869565 and is 0.689655 at 3.00 s. If the Prob. 16.56 spreadsheet is properly set up, one need only change n from 1 to 2 to get the Euler results for this problem, which turn out to be 0.86627 and 0.68421 mol/L for $\Delta t = 0.2$ s and 0.86795 and 0.68697 mol/L for $\Delta t = 0.1$ s.

16.58 We have $g[A] = -k[A]$ in the modified-Euler formulas. The top entries in the t and $[A]_n$ columns are 0 and 1 respectively, and the top entry in the $[A]_{n+1/2}$ column is $[A]_{1/2} = [A]_0 - k[A]_0\, \Delta t/2$. The $n = 1$ modified-Euler results at 1 and

3 s are 0.860728 and 0.637672 mol/L for $\Delta t = 0.2$ s and 0.860713 and 0.637639 mol/L for $\Delta t = 0.1$ s. Comparison with the exact values in Prob. 16.56 shows the modified-Euler results are quite good and much better than the Euler results.

16.60 **(a)** F; **(b)** F; **(c)** F.

16.61 $k = Ae^{-E_a/RT}$, $k(T_2)/k(T_1) = \exp[(E_a/R)(T_1^{-1} - T_2^{-1})]$

$$k_{720} = (0.0012\text{ dm}^3/\text{mol-s})\exp\left[\frac{177000\text{ J/mol}}{8.314\text{ J/mol-K}}\left(\frac{1}{660\text{ K}} - \frac{1}{720\text{ K}}\right)\right] = 0.018\text{ dm}^3\text{ mol}^{-1}\text{ s}^{-1}$$

16.62 $k = Ae^{-E_a/RT}$ and $\ln k = \ln A - E_a/RT$. We plot $\ln k$ vs. $1/T$. The data are

$\ln(k c° s)$	−7.52	−5.99	−4.27	−3.69	−2.75	
$10^3/T$	1.669	1.590	1.502	1.464	1.429	K^{-1}

The slope is $-19500\text{ K}^{-1} = -E_a/(8.314\text{ J/mol-K})$ and $E_a = 162$ kJ/mol. The intercept at $1/T = 0$ is $25._0 = \ln(A c° s)$; $A = 7 \times 10^{10}\text{ dm}^3\text{ mol}^{-1}\text{ s}^{-1}$. (The intercept is calculated from the slope and one point on the graph.

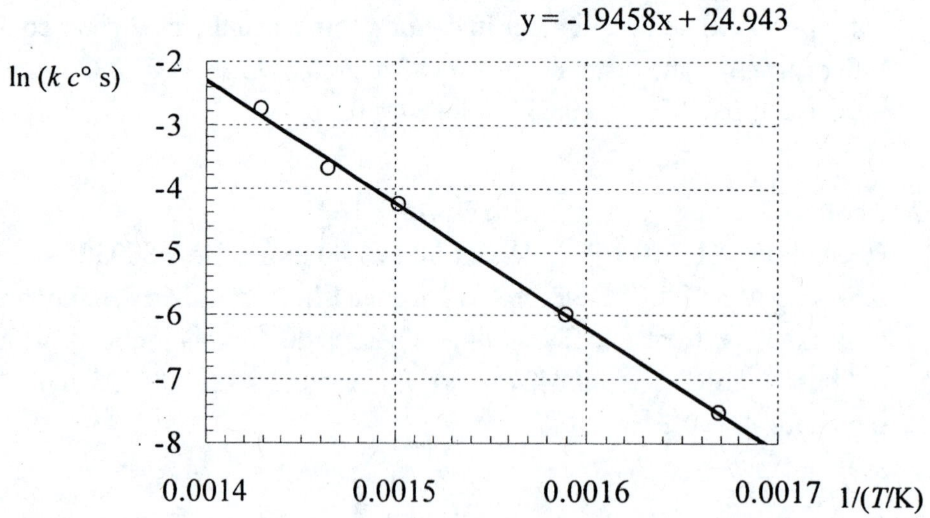

16.63 $k = Ae^{-E_a/RT}$, $k_{T_2}/k_{T_1} = \exp[(E_a/R)(1/T_1 - 1/T_2)]$, $\ln(k_{T_2}/k_{T_1}) =$ $(E_a/R)(1/T_1 - 1/T_2) = \ln(0.000030/0.0012) =$ $[E_a/(1.987 \text{ cal/mol-K})][1/(700 \text{ K}) - 1/(629 \text{ K})]$ and $E_a \approx 45._5$ kcal/mol.

$A = ke^{E_a/RT} =$

$(0.0012 \text{ dm}^3 \text{ mol}^{-1} \text{ s}^{-1}) \exp[(45500 \text{ cal/mol})/(1.987 \text{ cal/mol-K})(700 \text{ K})]$ $= 1._9 \times 10^{11} \text{ dm}^3 \text{ mol}^{-1} \text{ s}^{-1}$.

16.64 $k = Ae^{-E_a/RT}$. As $T \to \infty$, E_a/RT goes to 0 and k goes to A. As $T \to \infty$, the collision rate goes to infinity and the fraction of collisions having at least the activation energy goes to 1, so the collision-theory picture leads one to expect the rate to increase without limit as $T \to \infty$, rather than approaching an upper limit as predicted by the Arrhenius equation.

16.65 **(a)** $\ln k = \ln A - E_a/RT$. A plot of $\ln N$ vs. $1/T$ (where N is the chirping rate) has slope $-E_a/R$. We have

$\ln N$	5.18_2	4.83_6	4.60_5	
$10^3/T$	3.354	3.408	3.443	K^{-1}

The plot is linear with slope $-6.48 \times 10^3 \text{ K}^{-1} = -E_a/(1.987 \text{ cal/mol-K})$ and $E_a = 12._9$ kcal/mol.

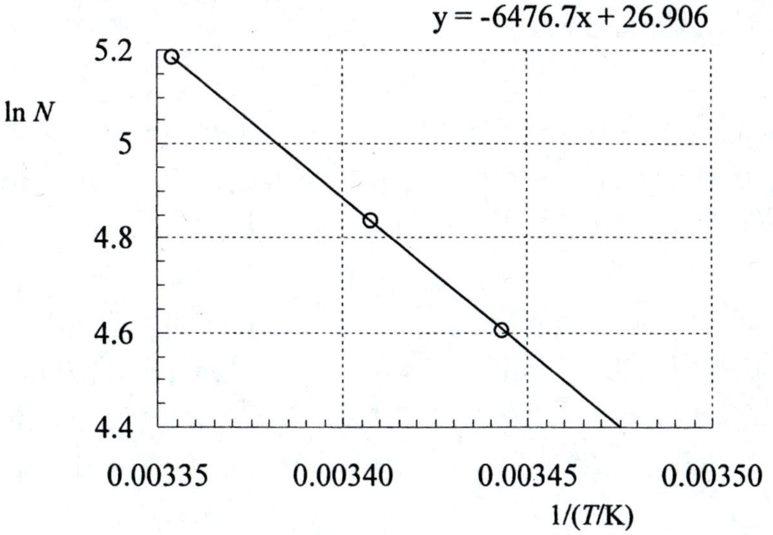

(b) At 14.0°C, $10^3/T = 3.482_5 \text{ K}^{-1}$ and the graph gives $\ln N = 4.35$. Thus $N =$ 77 per minute. The rule gives the Fahrenheit temperature as $40 + \frac{1}{4}(77) =$ 59°F. Actually, 14.0°C is 57°F, so the crickets are in error by 2°F.

16.66 **(a)** $A = 2.05 \times 10^{13}\ s^{-1}$ and $E_a = 24.65$ kcal/mol.

(b) $k(0°C) = (2.05 \times 10^{13}\ s^{-1}) \times$
exp [(–24650 cal/mol)/(1.987 cal/mol-K)(273.15 K)] = $3.87 \times 10^{-7}\ s^{-1}$.

(c) The units of k and the rate law in Eq. (16.6) show that the order is 1. Equations (16.15) and (16.11) give $k_A t_{1/2} = 0.693 = akt_{1/2} = 2kt_{1/2}$ and $t_{1/2} = 0.693/2k$. From the given $k(T)$ expression, we find $k(-50°C) = 1.47 \times 10^{-11}\ s^{-1}$ and $k(50°C) = 4.36 \times 10^{-4}\ s^{-1}$. We find $t_{1/2}(-50°C) = 2.36 \times 10^{10}$ s; $t_{1/2}(0°C) = 8.95 \times 10^5$ s; and $t_{1/2}(50°C) = 795$ s.

16.67 This fraction of collisions equals $e^{-\varepsilon/kT} = \exp(-N_A\varepsilon/RT) =$ exp[–(80000 J/mol)/(8.3145 J/mol-K)T] = exp[–9622/(T/K)].

(a) exp(–9622/300) = $e^{-32.07} = 1.2 \times 10^{-14}$;

(b) exp(–9622/310) = $e^{-31.04} = 3.3 \times 10^{-14}$;

(c) $e^{-30.07} = 8.7 \times 10^{-14}$. (Each 10°C increase nearly triples the fraction.)

16.68 From (16.43), the observed rate constant is $k = k_1 + k_2 = A_1 e^{-E_{a,1}/RT} + A_2 e^{-E_{a,2}/RT}$. From Eq. (16.68), $E_a = RT^2\, d \ln k/dT = RT^2(d/dT) \ln (k_1 + k_2) = RT^2[1/(k_1 + k_2)](dk_1/dT + dk_2/dT) = [RT^2/(k_1 + k_2)][(E_{a,1}/RT^2)A_1 e^{-E_{a,1}/RT} + (E_{a,2}/RT^2)A_2 e^{-E_{a,2}/RT}] = (E_{a,1}k_1 + E_{a,2}k_2)/(k_1 + k_2)$.

16.69 ΔH°_{298} = (–393.509 + 90.25 + 110.525 – 33.18) kJ/mol = –225.91 kJ/mol. Since $\Delta n_g = 0$, we have $\Delta H^\circ = \Delta U^\circ$, and use of $E_{a,f} - E_{a,b} = \Delta U^\circ$ gives 116 kJ/mol – $E_{a,b}$ = –226 kJ/mol. So $E_{a,b}$ = 342 kJ/mol.

16.70 The equilibrium condition for step 1 gives k_1/k_{-1} = [C][D]/[A][B]. We have $r = d[G]/dt = k_2[C]^2 = k_2(k_1[A][B]/k_{-1}[D])^2 = (k_1^2 k_2/k_{-1}^2)[A]^2[B]^2/[D]^2$, so $k = k_1^2 k_2/k_{-1}^2$ and $Ae^{-E_a/RT} = A_1^2 e^{-2E_{a,1}/RT} A_2 e^{-E_{a,2}/RT} / A_{-1}^2 e^{-2E_{a,-1}/RT}$. Then $E_a = 2E_{a,1} + E_{a,2} - 2E_{a,-1}$ = (240 + 196 – 192) kJ/mol = 244 kJ/mol.

16.71 **(a)** $k_2/k_1 = Ae^{-E_a/RT_2}/Ae^{-E_a/RT_1} = e^{(E_a/R)(1/T_1 - 1/T_2)}$. $\ln (k_2/k_1) = (E_a/R)(1/T_1 - 1/T_2)$ and ln 6.50 = [E_a/(8.314 J/mol-K)] × [(300.0 K)$^{-1}$ – (310.0 K)$^{-1}$] and E_a = 145 kJ/mol = 34.6 kcal/mol.

(b) $k_2/k_1 = \exp\,[(19000\text{ J})/(8.314\text{ J/mol-K})][(300\text{ K})^{-1} - (310\text{ K})^{-1}] = 1.28$.

16.72 (a) The linear regression slope is $-12459\text{ K}^{-1} = -E_a/R$, and $E_a = 24.76$ kcal/mol. The intercept is $30.81_6 = \ln(A/\text{s})$ and $A = 2.42 \times 10^{13}\text{ s}^{-1}$. The absolute percent errors range from 1 to 5% and the sum of the squares of the deviations is $1.9 \times 10^{-9}\text{ s}^{-2}$.

(b) Since the sum of the squares of the deviations is so small, it is helpful to minimize 10^{10} times this sum, rather than the sum itself. One finds $E_a =$ 25424 cal/mol and $A = 6.55 \times 10^{13}\text{ s}^{-1}$ as the optimum values with a sum of squares of deviations equal to 8.91×10^{-10}. The absolute percent error is 0.1% at the highest T and about 10% at the three lowest T's.

16.73 (a) For $H_2 + I_2 \rightarrow 2HI$, $\Delta G^\circ_{700} = -23.6$ kJ/mol $= -RT \ln K_P$ and $K_P = 57._7 = K_c$, since $\Delta n = 0$. Then $K_c^{1/s} = k_f/k_b = 0.064/0.0012 = 53 = (58)^{1/s}$, so $s = 1$.

(b) $K_c^{1/s} = K_c = k_f/k_b = 0.0025/0.000030 = 83$.

16.74 $k_f/k_b = K_c^{1/s}$.

(a) $k_f/k_b = K_c$. Let $r = k_bZ$ be the rate law of the reverse reaction. At equilibrium, the forward and reverse rates are equal, so $k_bZ_{eq} = k_f[BrO_3^-]_{eq}[SO_3^{2-}]_{eq}[H^+]_{eq}$ and $Z_{eq}/[BrO_3^-]_{eq}[SO_3^{2-}]_{eq}[H^+]_{eq} = k_f/k_b = K_c = [Br^-]_{eq}[SO_4^{2-}]_{eq}^3/[BrO_3^-]_{eq}[SO_3^{2-}]_{eq}^3$. Therefore $k_bZ = k_b[Br^-][SO_4^{2-}]^3[H^+]/[SO_3^{2-}]^2$.

(b) $k_f/k_b = K_c^{1/2}$. From part (a), $k_f/k_b = Z_{eq}/[BrO_3^-]_{eq}[SO_3^{2-}]_{eq}[H^+]_{eq} = K_c^{1/2} = [Br^-]_{eq}^{1/2}[SO_4^{2-}]_{eq}^{3/2}/[BrO_3^-]_{eq}^{1/2}[SO_3^{2-}]_{eq}^{3/2}$ and $k_bZ = k_b[Br^-]^{1/2}[SO_4^{2-}]^{3/2}[H^+][BrO_3^-]^{1/2}/[SO_3^{2-}]^{1/2}$.

16.75 $k_f/k_b = K_c^{1/s}$ and $\ln k_f - \ln k_b = (1/s)\ln K_c$. Differentiation of this equation with respect to T and use of Eq. (16.68) and the result of Prob. 6.21 gives $E_{a,f}/RT^2 - E_{a,b}/RT^2 = (1/s)\Delta U^\circ/RT^2$ and $E_{a,f} - E_{a,b} = \Delta U^\circ/s$.

16.76 We reverse the designations of forward and back reactions and use primes to denote the newly designated constants; i.e., $k'_f = k_b$ and $k'_b = k_f$; also, $K'_c = 1/K_c$. Since the mechanism of the reverse reaction consists of the reverse of the mechanism of the forward reaction and (as noted before Example 16.4 in the text) the reverse-reaction's rate-determining step is the reverse of that for the forward reaction, we have $s' = s$. Substitution into $k_f/k_b = K_c^{1/s}$ gives $k'_b/k'_f = (1/K'_c)^{1/s'}$ and $k'_f/k'_b = (K'_c)^{1/s'}$. Q.E.D.

16.77 Consider two chemical reactions

$a\text{A} + b\text{B} \rightleftharpoons c\text{C} + d\text{B}$ (I), $p\text{P} + q\text{Q} \rightleftharpoons r\text{R} + w\text{W}$ (II)

Let us form a third reaction that is the sum of m times the first reaction and n times the second reaction:

$ma\text{A} + mb\text{B} + np\text{P} + nq\text{Q} \rightleftharpoons mc\text{C} + md\text{B} + nr\text{R} + nw\text{W}$ (III)

The concentration-scale equilibrium constants of these reactions are

$K_\text{I} = [\text{C}]^c[\text{D}]^d/[\text{A}]^a[\text{B}]^b$, $K_\text{II} = [\text{R}]^r[\text{W}]^w/[\text{P}]^p[Q]^q$

$K_\text{III} = [\text{C}]^{cm}[\text{D}]^{dm}[\text{R}]^{rn}[\text{W}]^{wn}/[\text{A}]^{am}[\text{B}]^{bm}[\text{P}]^{pn}[\text{Q}]^{qn}$

(where all concentrations are equilibrium concentrations) and we see that $K_\text{III} = K_\text{I}^m K_\text{II}^n$. A similar result holds for a reaction formed by multiplying more than two reactions by integers and adding them. Since the elementary reactions of a mechanism multiplied by their stoichiometric numbers and then added yield the overall reaction, we have $K_c = K_1^{s_1} K_2^{s_2} \cdots K_m^{s_m} = \prod_i (K_i)^{s_i}$, where K_c is the equilibrium constant of the overall reaction and $K_1, K_2, \cdots, K_m$ are the equilibrium constants of the elementary reactions. But we know that for an elementary reaction $K_i = k_i/k_{-i}$ [Eq. (16.53)], so $K_c = \prod_i (k_i/k_{-i})^{s_i}$.

16.78 Step (b) has stoichiometric number $s = 1$, so $k_f/k_b = K_c^{1/s}$. Let $r_b = k_bZ$ be the rate law of the reverse reaction. We know that the forward reaction has rate law $r_f = k_f[N_2O_5]$. At equilibrium, $k_f[N_2O_5]_{eq} = k_bZ_{eq}$ and $Z_{eq}/[N_2O_5]_{eq} = k_f/k_b = K_c = [NO_2]^4_{eq}[O_2]_{eq}/[N_2O_5]^2_{eq}$, so $r_b = k_bZ = k_b[NO_2]^4[O_2]/[N_2O_5]$.

16.79 $1/k_{uni} = k_{-1}/k_1k_2 + 1/[M]k_1$. Since $P_0V = n_{tot}RT$, we have $[M] = P_0/RT$ and $1/k_{uni} = k_{-1}/k_1k_2 + RT/P_0k_1$. A plot of $1/k_{uni}$ vs. $1/P_0$ is linear with slope RT/k_1 and intercept k_{-1}/k_1k_2. The data are

$1/k_{uni}$	10440	9620	9260	9010	s
$10^3/P_0$	9.09	4.74	2.58	1.316	torr^{-1}

The slope is 1.83×10^5 s torr = 240 s atm = (0.08206 dm^3-atm/mol-K) × (743.1 K)/k_1 and $k_1 = 0.253\ \text{dm}^3\ \text{mol}^{-1}\ \text{s}^{-1}$. The intercept is 8770 s = $k_{-1}/k_1k_2 = 1/k_{\text{uni},P=\infty}$. So $k_{\text{uni},P=\infty} = 11.40 \times 10^{-5}\ \text{s}^{-1}$ and k_{-1}/k_2 = (8770 s)(0.253 $\text{dm}^3\ \text{mol}^{-1}\ \text{s}^{-1}$) = 2220 dm^3/mol.

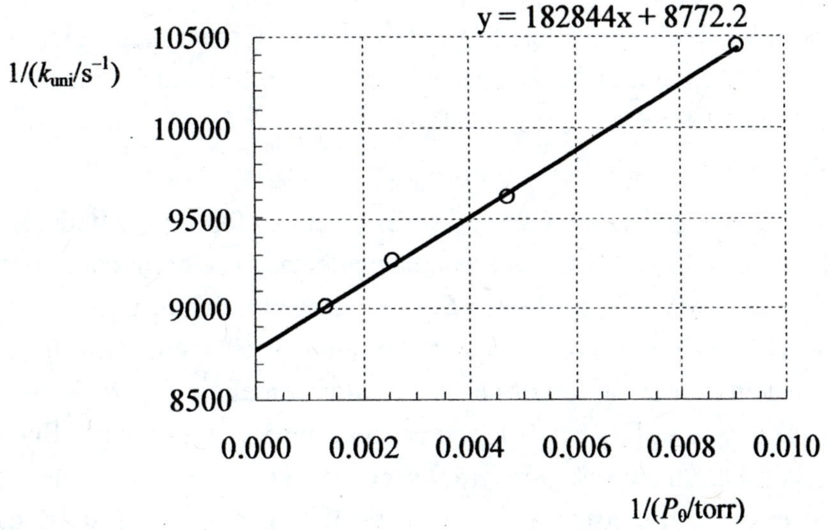

16.80 The B and C molecules are smaller than the A molecules and so undergo collisions with A less often than A molecules do.

16.81 $d[\text{Br}_2]/dt = -k_1[\text{Br}_2][\text{M}] + k_{-1}[\text{Br}]^2[\text{M}] - k_3[\text{H}][\text{Br}_2]$.
$d[\text{Br}]/dt = 2k_1[\text{Br}_2][\text{M}] - 2k_{-1}[\text{Br}]^2[\text{M}] - k_2[\text{Br}][\text{H}_2] + k_{-2}[\text{HBr}][\text{H}] + k_3[\text{H}][\text{Br}_2]$.

16.82 **(a)** Step (1) is the initiation step, (2) and (3) are the propagation steps, and (4) is the termination step.

(b) The chain-propagating steps (2) and (3) occur many times for each occurrence of step (1) or step (4). The overall reaction is therefore the sum of steps (2) and (3), namely, $CH_3CHO \rightarrow CH_4 + CO$.

(c) We use the steady-state approximation for the intermediates CH_3 and CH_3CO: $d[CH_3]/dt = 0 = k_1[CH_3CO] - k_2[CH_3][CH_3CHO] + k_3[CH_3CO] - 2k_4[CH_3]^2$. $d[CH_3CO]/dt = 0 = k_2[CH_3][CH_3CHO] -$

k_3[CH_3CO]. Addition of these two equations gives $0 = k_1[CH_3CHO] - 2k_4[CH_3]^2$ and $[CH_3] = (k_1/2k_4)^{1/2}[CH_3CHO]^{1/2}$. Then $r = d[CH_4]/dt = k_2[CH_3][CH_3CHO] = k_2(k_1/2k_4)^{1/2}[CH_3CHO]^{3/2}$. (Alternatively, we can find [CH_3CO] and use $r = d[CO]/dt$.)

16.83 The reverse of step I has a near-zero E_a, so E_a of step I equals ΔU_I°. From the Appendix, $\Delta H_I^\circ = 104$ kcal/mol and $\Delta U_I^\circ = 103\frac{1}{2}$ kcal/mol at 300 K. This $103\frac{1}{2}$ kcal/mol E_a is far higher than the 45 kcal/mol E_a of $Br_2 + M \rightarrow 2Br + M$ [see the discussion after Eq. (16.95)], so the dissociation of H_2 by M can be neglected. Step II is the reverse of step 3 in (16.88). Appendix data give $\Delta H_3^\circ = \Delta U_3^\circ = -41\frac{1}{2}$ kcal/mol $= E_{a,3} - E_{a,-3}$ and $E_{a,-3} = E_{a,\mathrm{II}} = 42\frac{1}{2}$ kcal/mol (since $E_{a,3} = 1$ kcal/mol). $E_{a,\mathrm{II}}$ is far higher than the 18 kcal/mol E_a of $Br + H_2 \rightarrow HBr + H$, so Br reacts preferentially with H_2 rather than with HBr, and reaction II can be neglected. The atom combination reaction $H + Br + M \rightarrow HBr + M$ has $E_{a,\mathrm{III}} \approx 0$. Reaction 3 in (16.88), namely, $H + Br_2 \rightarrow HBr + Br$, has $E_{a,3} \approx 1$ kcal/mol, not much different from $E_{a,\mathrm{III}}$; thus the rate constants k_3 and k_{III} are of the same order of magnitude. We have $r_3 = k_3[H][Br_2]$ and $r_{\mathrm{III}} = k_{\mathrm{III}}[H][Br][M]$. The concentration of the reactant Br_2 is high and is of the same order of magnitude as the M concentration. The very low concentration of the reactive intermediate Br makes $r_{\mathrm{III}} << r_3$, so we can neglect reaction III.

16.84 **(a)** Initiation is step 1: propagation is steps 2 and 3; termination is step −1.

(b) $[Cl]^2/[Cl_2] = k_1/k_{-1}$ and $[COCl]/[Cl][CO] = k_2/k_{-2}$. So $[Cl] = (k_1/k_{-1})^{1/2}[Cl_2]^{1/2}$ and $[COCl] = (k_2/k_{-2})(k_1/k_{-1})^{1/2}[CO][Cl_2]^{1/2}$. For the forward reaction, $r_f = d[COCl_2]/dt = k_3[COCl][Cl_2]$. (Since we want the rate law for the forward reaction, we assume negligible amount of product has formed and we do not consider the reverse of step 3.) Substitution for [COCl] from (b) gives $r_f =$ $(k_3k_2k_1^{1/2}/k_{-2}k_{-1}^{1/2})[Cl_2]^{3/2}[CO]$

(c) $r_b = -d[COCl_2]/dt = k_{-3}[COCl_2][Cl] = (k_{-3}k_1^{1/2}/k_{-1}^{1/2})[Cl_2]^{1/2}[COCl_2]$. [As a check, putting $r_f = r_b$, we get $[COCl_2]_{eq}/[CO]_{eq}[Cl_2]_{eq} = (k_2/k_{-2})(k_3/k_{-3}) = K_c$, where the result of Prob. 16.77 was used.]

16.85 **(a)** $k = k_2(k_1/k_{-1})^{1/2}$ and $Ae^{-E_a/RT} = A_2A_1^{1/2}A_{-1}^{-1/2}e^{-(E_{a,2}+\frac{1}{2}E_{a,1}-\frac{1}{2}E_{a,-1})/RT}$, so $E_a = E_{a,2} + \frac{1}{2}E_{a,1} - \frac{1}{2}E_{a,-1}$.

(b) As noted after Eq. (16.95), $E_{a,-1} \approx 0$ and $E_{a,1} \approx \Delta U_1^\circ$. Data in the Appendix give $\Delta H_1^\circ = 46.1$ kcal/mol; then $\Delta U_1^\circ = \Delta H_1^\circ - RT = 45.5$ kcal/mol at 298 K. So $E_{a,2} = E_a - \frac{1}{2}E_{a,1} = (40.6 - 22.7)$ kcal/mol = 18 kcal/mol. The ratio k_1/k_{-1} is K_c for the reaction $Br_2(g) \rightleftharpoons 2Br(g)$; data in the Appendix give $\Delta G_{298}^\circ = 38.64$ kcal/mol; (6.14) and (6.25) give $K_{P,298}^\circ = 4.7 \times 10^{-29}$ and $K_{c,298}^\circ = 1.9 \times 10^{-30}$. So $k_1/k_{-1} = 1.9 \times 10^{-30}$ mol/dm^3 at 298 K. Substitution in $Ae^{-E_a/RT} = A_2e^{-E_{a,2}/RT}(k_1/k_{-1})^{1/2}$ at 298 K gives $A_2 = (1.6 \times 10^{11}$ dm$^{3/2}$/mol$^{1/2}$-s$)(1.9 \times 10^{-30}$ mol/dm$^3)^{-1/2} \times$ exp {[(18000 – 40600) cal/mol]/(1.987 cal/mol-K)(298 K)} = 3×10^9 dm^3 mol^{-1} s^{-1}. Then $k_2 = A_2e^{-E_{a,2}/RT} =$ $(3 \times 10^9$ dm^3/mol-s$)e^{-(18000\text{ cal/mol})/RT}$.

16.86 **(a)** $[R_{tot}\cdot] = (fk_i/k_t)^{1/2}[I]^{1/2} = (0.008 \text{ mol/dm}^3)^{1/2}(0.5 \times 5 \times 10^{-5} \text{ s}^{-1})^{1/2}$ $/(2 \times 10^7 \text{ dm}^3\text{/mol-s})^{1/2} = 1.0 \times 10^{-7}$ mol/dm^3.
$\langle DP \rangle = k_p[M]/(fk_ik_t)^{1/2}[I]^{1/2} = (3 \times 10^3 \text{ dm}^3\text{/mol-s})(2 \text{ mol/dm}^3)/$ $(0.5 \times 1000 \text{ dm}^3\text{/mol-s}^2)^{1/2}(0.008 \text{ mol/dm}^3)^{1/2} = 3000$. $-d[M]/dt =$ $k_p(fk_i/k_t)^{1/2}[M][I]^{1/2} = k_p[M][R_{tot}\cdot] = (3000 \text{ dm}^3\text{/mol-s})(2 \text{ mol/dm}^3) \times$ $(1.0 \times 10^{-7} \text{ mol/dm}^3) = 0.0006$ mol/dm^3-s. $d[P_{tot}]/dt = k_t[R_{tot}\cdot]^2 =$ $(2 \times 10^7 \text{ dm}^3\text{/mol-s})(1.0 \times 10^{-7} \text{ mol/dm}^3)^2 = 2 \times 10^{-7}$ mol/dm^3-s.

(b) When termination is by disproportionation, two polymer molecules (instead of one) are formed whenever two radicals combine. This doubles $d[P_{tot}]/dt$ and hence cuts $\langle DP \rangle$ in half. Thus $d[P_{tot}]/dt = 4 \times 10^{-7}$ mol/dm^3-s and $\langle DP \rangle = 1500$. The other quantities are unchanged.

16.87 The initiation reaction $2M \rightarrow 2R\cdot$ contributes a term $2k_i[M]^2$ to $-d[M]/dt$ and (16.97) is modified to $-d[M]/dt = 2k_i[M]^2 + k_p[M][R_{tot}\cdot]$. To apply the steady-state condition $d[R_{tot}\cdot]/dt = 0$, we note that the initiation step has $(d[R_{tot}\cdot]/dt)_i =$ $2fk_i[M]^2$. So $0 = (d[R_{tot}\cdot]/dt)_i + (d[R_{tot}\cdot]/dt)_t = 2fk_i[M]^2 - 2k_t[R_{tot}\cdot]^2$ and $[R_{tot}\cdot] =$ $(fk_i/k_t)^{1/2}[M]$. Substitution in the above $-d[M]/dt$ equation gives $-d[M]/dt =$ $[2k_i + k_p(fk_i/k_t)^{1/2}][M]^2$. Use of the above $[R_{tot}\cdot]$ expression in (16.105) gives

$d[P_{tot}]/dt = k_t[R_{tot}\cdot]^2 = fk_i[M]^2$. Equation (16.104) becomes $\langle DP \rangle$ = $-(d[M]/dt)/d[P_{tot}]/dt = [2k_i + k_p(fk_i/k_t)^{1/2}]/fk_i$.

16.88 $d[A]/dt = -k_f[A] + k_b[C]^2$. Let $[A]_{eq}$ and $[C]_{eq}$ be the equilibrium concentrations under the new conditions, and let $x \equiv [A]_{eq} - [A]$. Then $dx/dt = -d[A]/dt$. Since 2 moles of C are formed when 1 mole of A reacts, we have $[C]_{eq} - [C] = -2x$. Then $-dx/dt = -k_f([A]_{eq} - x) + k_b([C]_{eq}^2 + 4x[C]_{eq} + 4x^2) =$ $-k_f[A]_{eq} + k_b[C]_{eq}^2 + xk_b(k_f/k_b + 4[C]_{eq} + 4x)$. At equilibrium, $d[A]/dt = 0$ and the first equation in this paragraph gives $-k_f[A]_{eq} + k_b[C]_{eq}^2 = 0$. Since the perturbation is small, [C] is close to $[C]_{eq}$ and we can neglect $4x$ in comparison with $4[C]_{eq}$. We then have $dx/dt = -xk_b(k_f/k_b + 4[C]_{eq}) = -\tau^{-1}x$, where $\tau =$ $(k_f + 4k_b[C]_{eq})^{-1}$. Integration gives $x = x_0e^{-t/\tau}$ or $[A] - [A]_{eq} = ([A]_0 - [A]_{eq})e^{-t/\tau}$.

16.89 **(a)** $CH_3CH_2CH_3$, CH_3CH_3, $CH_3CH_2CH_2CH_3$, and N_2.
(b) $CH_3CH_2CH_3$ and N_2 (cage effect).

16.90 Equation (16.112) gives $k_D = 2\pi(6.02 \times 10^{23}/\text{mol})(4 \times 10^{-8}\text{ cm}) \times$ $(8.4 \times 10^{-5}\text{ cm}^2/\text{s}) = 1.3 \times 10^{13}\text{ cm}^3/\text{mol-s} = 1.3 \times 10^{10}\text{ dm}^3\text{ mol}^{-1}\text{ s}^{-1}$.

16.91 **(a)** From Eqs. (16.114) and (16.115), $\ln k_D = \ln T - \ln \eta + \ln$ (const.). Then Eq. (16.68) gives $E_a = RT^2\, d \ln k_D/dT = RT^2[1/T - (1/\eta)d\eta/dT] =$ $RT - (RT^2/\eta)d\eta/dT$.
(b) $E_a = (1.987\text{ cal/mol-K})(298\text{ K}) - (1.987\text{ cal/mol-K})(298\text{ K})^2(-0.023\text{ K}^{-1})$ $= 4.7$ kcal/mol = 19 kJ/mol.

16.92 **(a)** F; **(b)** T; **(c)** F; **(d)** F.

16.93 The data give

$1/r_0$	35.7	20.8	12.5	6.45	L-s/mmol
$1/[S]_0$	0.800	0.400	0.200	0.050	L/mmol

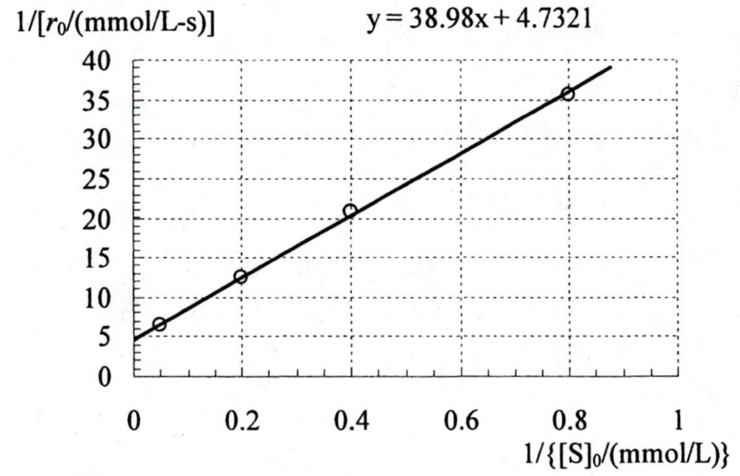

The slope is 39 s = $K_M/k_2[E]_0$ and the intercept is 4.$_7$ dm^3-s/mmol = $1/k_2[E]_0$. So $k_2 = (2.8 \times 10^{-9}$ mol/dm$^3)^{-1}(4700$ dm^3-s/mol$)^{-1} = 7.{}_6 \times 10^4$ s^{-1} and K_M = (39 s)/(4.$_7$ dm^3-s/mmol) = 8.$_3 \times 10^{-3}$ mol/dm^3.

16.94 **(a)** $k_2[E]_0 = r_0K_M/[S]_0 + r_0$ and $r_0 = -r_0K_M/[S]_0 + k_2[E]_0$.

(b)

r_0	0.028	0.048	0.080	0.155	mmol/L-s
$r_0/[S]_0$	0.0224	0.0192	0.016	0.00775	s^{-1}

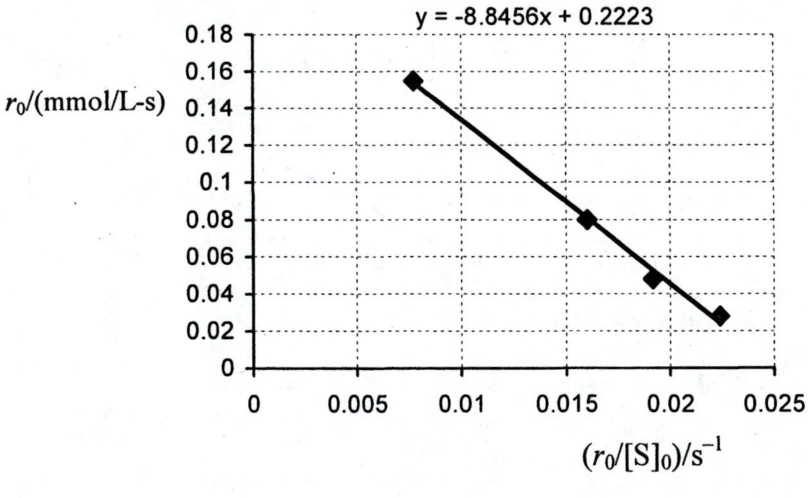

The slope equals $-K_M$, so K_M = 8.8$_5$ mmol/L. The intercept is $k_2[E]_0$ = 0.222 mmol/L-s so k_2 = (0.222 mmol/L-s)/(2.8 × 10^{-6} mmol/L) = 7.9 × 10^4 s^{-1}.

16.95 With step –2 included, Eq. (16.121) becomes

$d[\text{ES}]/dt = 0 = k_1[\text{E}][\text{S}] - k_{-1}[\text{ES}] - k_2[\text{ES}] + k_{-2}[\text{E}][\text{P}]$

Replacing [E] by $[\text{E}]_0 - [\text{ES}]$, we get

$0 = ([\text{E}]_0 - [\text{ES}])(k_1[\text{S}] + k_{-2}[\text{P}]) - (k_{-1} + k_2)[\text{ES}]$

$$[\text{ES}] = \frac{k_1[\text{S}] + k_{-2}[\text{P}]}{k_{-1} + k_2 + k_1[\text{S}] + k_{-2}[\text{P}]}[\text{E}]_0 \quad \text{(Eq. A)}$$

$$r = -d[\text{S}]/dt = k_1[\text{E}][\text{S}] - k_{-1}[\text{ES}] = k_1([\text{E}]_0 - [\text{ES}])[\text{S}] - k_{-1}[\text{ES}]$$

$r = k_1[\text{E}]_0[\text{S}] - (k_1[\text{S}] + k_{-1})[\text{ES}]$ and use of Eq. A for [ES] gives

$$r = \frac{k_1k_2[\text{S}] - k_{-1}k_{-2}[\text{P}]}{k_1[\text{S}] + k_{-2}[\text{P}] + k_{-1} + k_2}[\text{E}]_0$$

16.96 **(a)** We plot $1/\upsilon$ vs. $1/P$. The data are

$10^3\upsilon^{-1}/(\text{g/cm}^3)$	9.90	7.35	6.54	6.17	6.06	6.02
$10^2P^{-1}/\text{atm}^{-1}$	28.6	10.0	5.99	3.89	2.99	2.55

The straight-line fit is fairly good. The intercept is 0.0056_5 g/cm^3 = $1/\upsilon_{\text{mon}}$, and υ_{mon} = 177 cm^3/g. The slope is 0.0150 atm g/cm^3 = $1/\upsilon_{\text{mon}}b$, and $b = [(177\ \text{cm}^3/\text{g})(0.0150\ \text{atm g/cm}^3)]^{-1} = 0.38\ \text{atm}^{-1}$.

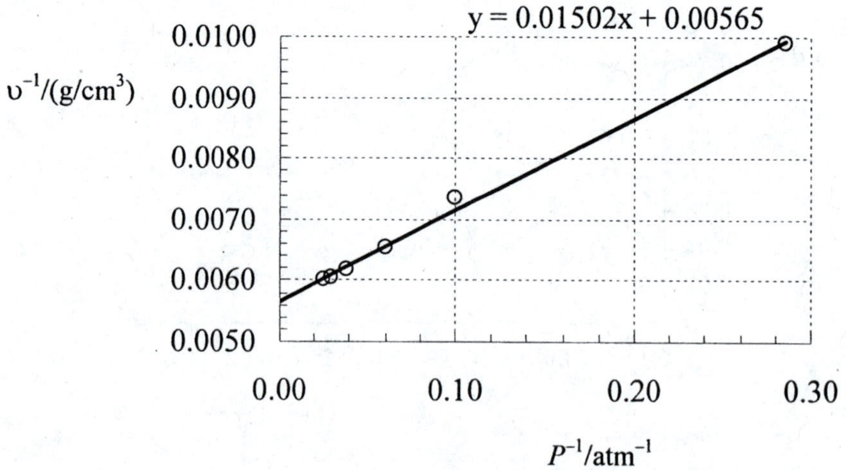

(b) To keep the units simple, we rewrite the Freundlich isotherm as $\upsilon = k(P/P^\#)^a$, where $P^\#$ = 1 atm. We plot log [υ/(cm^3/g)] vs. log ($P/P^\#$). The data are

$\log[\upsilon/(\text{cm}^3/\text{g})]$	2.004	2.134	2.185	2.210	2.217	2.220
$\log(P/P^{\#})$	0.544	1.000	1.223	1.410	1.525	1.593

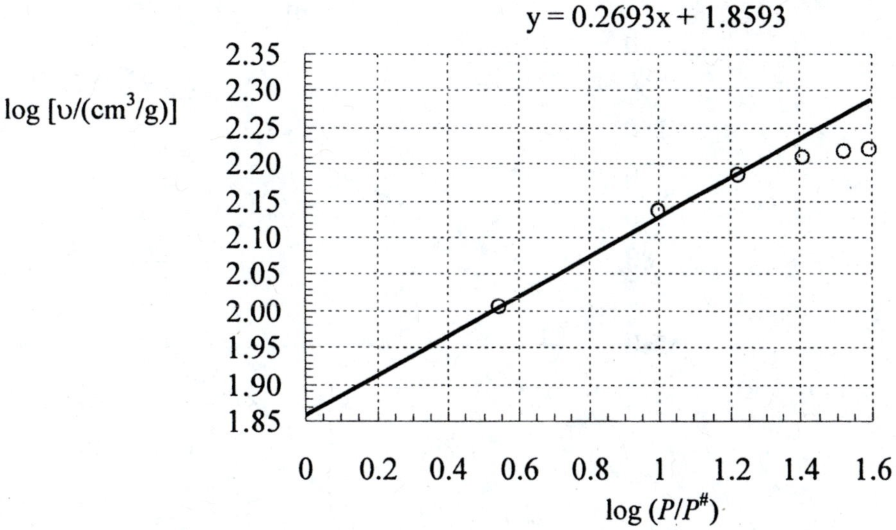

The straight-line fit is rather poor in that the points show a pronounced curvature at high P. (Recall from the text that the Freundlich isotherm doesn't work at high P.) Ignoring the three points at high P, we draw a straight line through the remaining points. The intercept is 1.86 = $\log[k/(\text{cm}^3/\text{g})]$, so $k = 72\ \text{cm}^3/\text{g}$. The slope is $0.269 = a$.

(c) For the Langmuir isotherm, $\upsilon_{\text{mon}}b = c = 67\ \text{cm}^3/(\text{g atm})$ and $\upsilon = (67\ \text{cm}^3/\text{g})(P/P^{\#})/(1 + 0.38P/P^{\#})$. With $P/P^{\#} = 7.0$, one gets $\upsilon = 128\ \text{cm}^3/\text{g}$. For the Freundlich isotherm, $\upsilon = (72\ \text{cm}^3/\text{g})(7)^{0.269} = 122\ \text{cm}^3/\text{g}$.

16.97 (a) $\upsilon = r\ln(sP^{\#}) + r\ln(P/P^{\#})$, where $P^{\#} = 1$ atm. A plot of υ vs. $\ln(P/P^{\#})$ is linear with slope r and intercept $r\ln(sP^{\#})$.

(b) The data are

$\upsilon/(\text{cm}^3/\text{g})$	101	136	153	162	165	166
$\ln(P/P^{\#})$	1.253	2.303	2.815	3.246	3.512	3.669

As with the Freundlich isotherm in Prob. 16.96b, the points at high pressures show a strong curvature, indicating that the Temkin isotherm doesn't apply at high P. (Note that it predicts $\upsilon \to \infty$ as $P \to \infty$.) We ignore the three high-pressure points and draw a straight line through the remaining points. The slope is 33.4 $cm^3/g = r$. The intercept is 59 cm^3/g = (33.4 cm^3/g) ln ($sP^{\#}$) and s = 5.85 atm^{-1}.

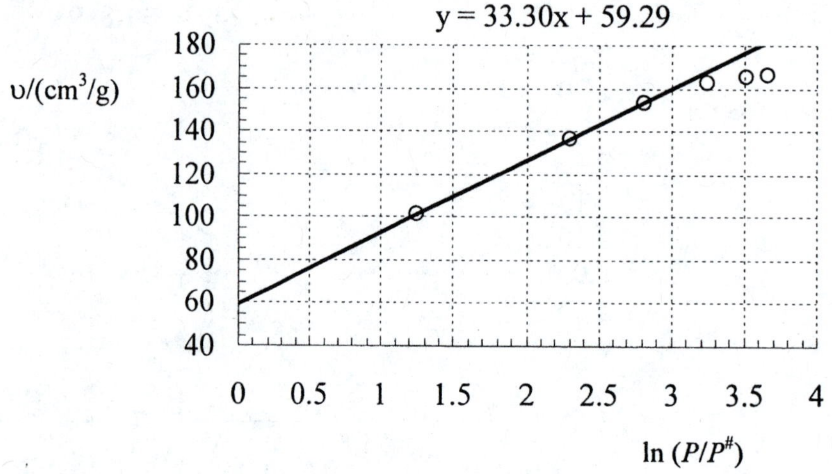

16.98 We have $1 + bP = cP/\upsilon$, so a plot of P/υ vs. P is linear.

16.99 **(a)** The curve resembles Fig. 16.24b, indicating formation of more than a monolayer. We have a type II isotherm and the BET equation is appropriate.

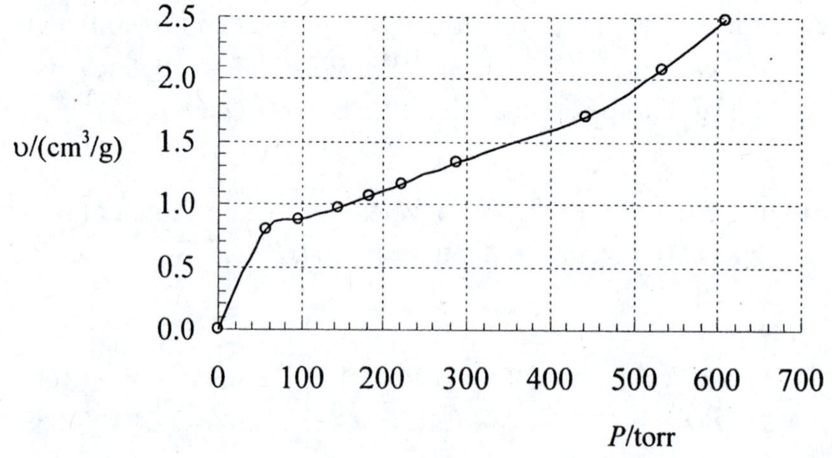

(b) We plot $P/\upsilon(P^* - P)$ vs. P/P^*, where (since 77 K is the normal boiling point) P^* = 760 torr. The data are

$10^2[P/\upsilon(P^* - P)]/(g/cm^3)$	9.97	16.4	24.1	29.9	35.8
$10^2 P/P^*$	7.37	12.5	19.1	24.1	29.3

$10^2[P/\upsilon(P^* - P)]/(\text{g/cm}^3)$	45.6	81.3	112.9	162.6
$10^2 P/P^*$	37.8	58.2	70.1	80.1

The low- and medium-pressure points fit a straight line well, but the three points at high pressure deviate greatly from this line (the BET isotherm often works poorly at high pressures) and will be ignored. We draw a straight line of slope m and intercept b through the first six points. We find $m = 1.17$ g/cm^3 and $b = 0.01_7$ g/cm^3. We have $m = 1/\upsilon_{\text{mon}} - 1/\upsilon_{\text{mon}}c$ and $b = 1/\upsilon_{\text{mon}}c$, so $\upsilon_{\text{mon}} = (m + b)^{-1} = 0.842$ cm^3/g; also, $c = (1 - m\upsilon_{\text{mon}})^{-1} = 7 \times 10^1$.

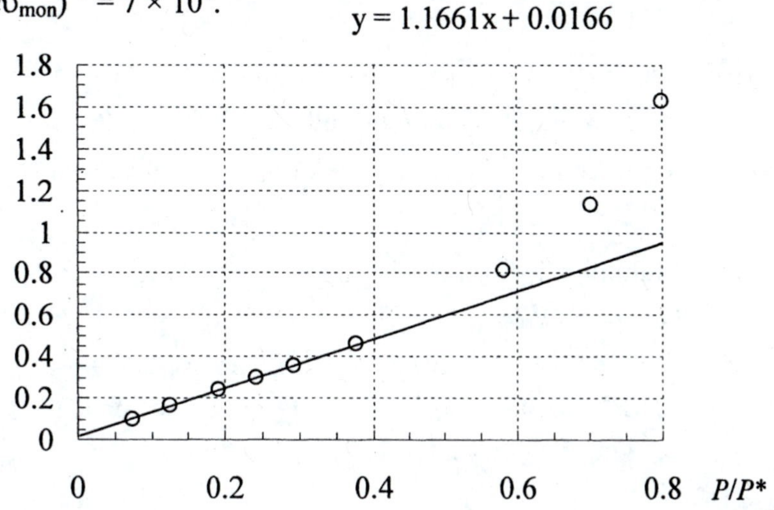

(c) $n = (1 \text{ atm})(0.842 \text{ cm}^3)/RT = 3.76 \times 10^{-5}$ moles of N_2 adsorbed per gram of sample. This is 2.26×10^{19} molecules, so the surface area of the powder is $(2.26 \times 10^{19})(16 \times 10^{-16} \text{ cm}^2) = 36000 \text{ cm}^2 = 3.6 \text{ m}^2$.

16.100 For $\theta << 1$ in Eq. (16.127), we have $1/\theta >> 1$ and $1/bP + 1 >> 1$, so that $1/bP >> 0$ and hence $1/bP >> 1$ and $bP << 1$. Therefore bP can be neglected in the denominator of (16.128) to give $\upsilon = cP$.

16.101 Let N be the number of adsorption sites. The rate of desorption of A is proportional to the number $\theta_A N$ of adsorbed A molecules and equals $k_{d,A}\theta_A N$. The A adsorption rate is proportional to the gas-A partial pressure P_A and to the number $(1 - \theta_A - \theta_B)N$ of unoccupied sites. Hence $k_{d,A}\theta_A N = k_{a,A}P_A(1 - \theta_A - \theta_B)N$ and $\theta_A = (k_{a,A}/k_{d,A})P_A(1 - \theta_A - \theta_B) = b_A P_A(1 - \theta_A - \theta_B)$, where $b_A = k_{a,A}/k_{d,A}$. Similarly, $\theta_B = b_B P_B(1 - \theta_A - \theta_B)$. Division gives $\theta_B/\theta_A = b_B P_B/b_A P_A$. Hence, $\theta_A = b_A P_A(1 - \theta_A - b_B P_B \theta_A/b_A P_A)$ and $\theta_A =$

$b_AP_A/(1 + b_AP_A + b_BP_B)$. Similarly, $\theta_B = b_BP_B/(1 + b_AP_A + b_BP_B)$. The fraction of occupied sites is $\theta_A + \theta_B$ and equals υ/υ_m, so $\upsilon/\upsilon_m = \theta_A + \theta_B = (b_AP_A + b_BP_B)/(1 + b_AP_A + b_BP_B)$.

16.102 (a) From (16.133), $d \ln P = -(\Delta\overline{H}_a/RT^2)\ dT$ at constant θ; integration gives $\ln (P_2/P_1) = (<\Delta\overline{H}_a>/R)(1/T_2 - 1/T_1)$, where $<\Delta\overline{H}_a>$ is an average $\Delta\overline{H}_a$ over the temperature range. $\ln (0.03/0.0007) = [<\Delta\overline{H}_a>/(8.314\text{ J/mol-K})][(873\text{ K})^{-1} - (773\text{ K})^{-1}]$ and $<\Delta\overline{H}_a> = -210$ kJ/mol.

(b) For $\theta = 0.10$ and 500 to 600°C, $<\Delta\overline{H}_a> = R[(873\text{ K})^{-1} - (773\text{ K})^{-1}]^{-1} \times \ln (23/8) = -59$ kJ/mol. For 600 to 700°C, $<\Delta\overline{H}_a> = R[(973\text{ K})^{-1} - (873\text{ K})^{-1}]^{-1} \ln (50/23) = -55$ kJ/mol.

16.103 (a) Multiplication of (16.132) by υc gives $Pc/(P^* - P) = (\upsilon/\upsilon_{mon}) \times [1 + (c - 1)P/P^*]$ and $\upsilon/\upsilon_{mon} = PP^*c/(P^* - P)(P^* + Pc - P)$.

(b) For $P << P^*$, we have $P^* - P \cong P^*$ and $\upsilon/\upsilon_{mon} = PP^*c/P^*(P^* + Pc) = Pc/(P^* + cP) = (P^*)^{-1}cP/[1 + c(P^*)^{-1}P]$, which is the form of the Langmuir isotherm (16.127).

16.104 This suggests that the CO is chemisorbed in two different forms (Fig. 16.22).

16.105 We plot $\log t_{1/2}$ vs. $\log P_0$. The data are

$\log t_{1/2}$	0.88_1	0.56_8	0.23_0
$\log P_0$	2.42_3	2.11_4	1.76_3

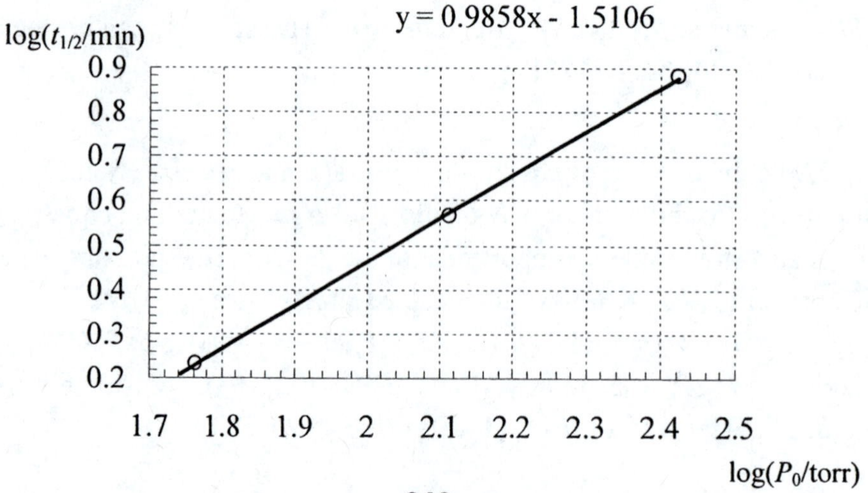

The slope is 0.99 = 1 – n and n = 0.

16.106 The overall reaction is $N_2 + 3H_2 \rightarrow 2NH_3$. To produce this overall reaction from the steps listed, we must multiply steps *a, b, c, d, e* and *f* by 1, 3, 2, 2, 2, and 2, respectively, and then add them. These are the stoichiometric numbers of steps *a* to *f*. The fact that the rate-determining step probably has stoichiometric number $s = 1$ indicates that the rate-determining step is probably step (a), $N_2 + 2* \rightarrow 2N*$.

16.107 The rate of the bimolecular desorption reaction 2A(ads) → A_2(g) is proportional to $(n_A/\mathscr{A})(n_A/\mathscr{A})$ and so is proportional to θ_A^2; $r_{des} = k_d\theta_A^2$. The rate r_{ads} of the adsorption reaction A_2(g) → 2A(ads) is proportional to the A_2–surface collision rate, which is proportional to P. Since two adjacent vacant sites are required for A_2 to be adsorbed dissociatively, the adsorption rate is also proportional to $(1 - \theta_A)(1 - \theta_A)$, the square of the fraction of vacant sites. Hence, $r_{ads} = k_a(1 - \theta_A)^2P$. So $r_{ads} = r_{des}$ and $k_a(1 - \theta_A)^2P = k_d\theta_A^2$ or $k_a^{1/2}(1 - \theta_A)P^{1/2}/k_d^{1/2} = \theta_A$. Solving for θ_A, we get $\theta_A = (k_a/k_d)^{1/2}P^{1/2}/[1 + (k_a/k_d)^{1/2}P^{1/2}]$, which is (16.130) with $b = k_a/k_d$.

16.108 (a) We assume that each CO molecule occupies one adsorption site. (This isn't always true.) The number of sites per cm^2 is $(2.3 \times 10^{-9}$ mol) × $(6.02 \times 10^{23}$/mol) = $1.3_8 \times 10^{15}$.

(b) The total number of sites is $(1.38 \times 10^{15}/\text{cm}^2)(5.00\ \text{cm}^2) = 6.9 \times 10^{15}$. The number of occupied sites is $(9.2 \times 10^{-10}\ \text{mol})(6.02 \times 10^{23}/\text{mol}) = 5.5 \times 10^{14}$. So $\theta = (5.5 \times 10^{14})/(6.9 \times 10^{15}) = 0.080$. The adsorption rate per unit area is $r_s = (5.5 \times 10^{14})/t(5.00\ \text{cm}^2) = (1.1 \times 10^{14}\ \text{cm}^{-2})t^{-1}$ and the equation in Sec. 16.19 that defines s becomes $s = (1.1 \times 10^{14}/\text{cm}^2)(2\pi MRT)^{1/2}/tPN_A$. We have $tP = (0.43 \times 10^{-6}\ \text{torr} \cdot \text{s}) \times (1\ \text{atm}/760\ \text{torr})(101325\ \text{Pa}/1\ \text{atm}) = 5.7 \times 10^{-5}\ \text{Pa} \cdot \text{s}$ and $s = (1.1 \times 10^{14}\ 10^4\ \text{m}^{-2})[2\pi(0.028\ \text{kg/mol})(8.314\ \text{J/mol-K})(300\ \text{K})]^{1/2}/(5.7 \times 10^{-5}\ \text{s N/m}^2)(6.02 \times 10^{23}) = 0.67$. Since θ is close to 0, this is approximately s_0.

16.109 (a) $t_{1/2} = 0.693/k_{des} = 0.693\, e^{E_{a,des}/RT}/A_{des} = 0.693e^{(151000\ \text{J/mol})/(8.314\ \text{J/mol-K})(300\ \text{K})}/(2.4 \times 10^{14}/\text{s}) = 5.6 \times 10^{11}$ s;

(b) 5.3×10^{-4} s.

16.110 $d = (2Dt)^{1/2} = (2D_0 e^{-E_{a,\text{mig}}/RT} t)^{1/2} =$
$[2(0.014\ \text{cm}^2/\text{s})e^{-(88000\ \text{J/mol})/(8.314\ \text{J/mol-K})(300\ \text{K})}(1\ \text{s})]^{1/2} = 3.7 \times 10^{-9}$ cm. At 100 s, d is $100^{1/2} = 10$ times d at 1 s, which is 3.7×10^{-8} cm.

16.111 (a) With $E_{a,\text{ads}} \cong 0$, Eq. (16.71) gives $E_{a,\text{des}} \cong -\Delta U^\circ_{\text{ads}} \cong -\Delta H^\circ_{\text{ads}}$.
$t_{1/2} = 0.693/k_{\text{des}} = 0.693e^{E_{a,\text{des}}/RT}/A_{\text{des}} \cong 0.693e^{|\Delta H_{\text{ads}}|/RT}/A_{\text{des}} =$
$0.693e^{(50000\ \text{J/mol})/(8.314\ \text{J/mol-K})(300\ \text{K})}/(10^{15}\ \text{s}^{-1}) = 3.5 \times 10^{-7}$ s.

(b) 180 s. **(c)** 4.6×10^{19} s.

16.112 Let the half-reaction be $M^z + ne^- = M^{z'}$ or its reverse (where z and z' are the charges on the species M). Let B in Eq. (16.134) be the species e^-. Then $|\nu_B| = n$ and $r_s = \mathscr{A}^{-1}|\nu_B|^{-1}|dn_B/dt| = (1/\mathscr{A}n)|dn_{e^-}/dt|$, where n_{e^-} (not to be confused with n, the number of electrons in the half-reaction) is the number of moles of electrons. Since the Faraday constant is the absolute value of the charge per mol of electrons, we have
$|Q| = Fn_{e^-}$, so $|dn_{e^-}/dt| = F^{-1}|dQ/dt| = I/F$ and $r_s = (1/\mathscr{A}nF)I = j/nF$.

16.113 (a) Let the steps be called I, II, III, IV. Step III requires two ^{3}He nuclei, but only one ^{3}He is produced in step II. Hence the stoichiometric number of step II is 2; this requires that the stoichiometric number of step I be 2. The stoichiometric number of step IV is also 2, so as to get rid of the two positrons formed when step I is multiplied by 2. Multiplication of steps I, II, III, and IV by 2, 2, 1, and 2 gives as the overall reaction
$4\,{}^1_1\text{H} + 2\,{}^{\ 0}_{-1}\text{e}^- \rightarrow {}^4_2\text{He} + 2\nu + 6\gamma$.

(b) ΔU_m is -2.6×10^{12} J/mol. There are $(3.9 \times 10^{26}\ \text{J})/(2.6 \times 10^{12}\ \text{J/mol}) = 1.5 \times 10^{14}$ moles of ^{4}He produced each second.

(c) The number of neutrinos produced each second is $2(1.5 \times 10^{14}\ \text{mol})(6.02 \times 10^{23}/\text{mol}) = 1.8 \times 10^{38}$. The area of a sphere of radius 1.5×10^8 km is $4\pi(1.5 \times 10^{11}\ \text{m})^2 = 2.8 \times 10^{23}\ \text{m}^2$. One $\text{cm}^2 = 10^{-4}\ \text{m}^2$ and the number of neutrinos hitting a square centimeter of the earth in 1 second is $(1.8 \times 10^{38})[(10^{-4}\ \text{m}^2)/(2.8 \times 10^{23}\ \text{m}^2)] = 6 \times 10^{10}$ (which is a lot of neutrinos).

16.114 (a) For this elementary reaction, $r = k_{max}[B][C] = -d[B]/dt = -(d/dt)(n_B/V) = -(d/dt)(N_B/N_AV) = -(1/V)(dN_B/dt)/N_A = Z_{BC}/N_A$ and $k_{max} = Z_{BC}/N_A[B][C]$. At each collision, one B molecule disappears, and Z_{BC} is the collision rate per unit volume, so $Z_{BC} = -(1/V)(dN_B/dt)$.

(b) We have $[B] = N_B/VN$ and $[C] = N_C/VN_A$, so (14.63) gives $k_{max} = Z_{BC}/N_A[B][C] = N_A(r_B + r_C)^2(8RT\pi)^{1/2}(M_B^{-1} + M_C^{-1})^{1/2} = (6.02 \times 10^{23}/\text{mol})(4 \times 10^{-10}\ \text{m})^2[8(8.314\ \text{J/mol-K})(300\ \text{K})\pi]^{1/2} \times [(0.030\ \text{kg/mol})^{-1} + (0.050\ \text{kg/mol})^{-1}]^{1/2} = 1.8 \times 10^8\ \text{m}^3\ \text{s}^{-1}\ \text{mol}^{-1} = 1.8 \times 10^{11}\ \text{dm}^3\ \text{mol}^{-1}\ \text{s}^{-1}$.

16.116 (a) T. **(b)** T. **(c)** T. **(d)** T. **(e)** T. From (16.2) and (4.97), $J = (1/\nu_i)\, dn_i/dt = d\xi/dt$. **(f)** F. See Eq. (16.77). **(g)** T. **(h)** F; e.g., see (16.60). **(i)** T. **(j)** T. **(k)** F. **(l)** F (r increases with t in an explosion). **(m)** F. **(n)** F (See Sec. 16.12.).

R16.1 $\langle \varepsilon_{tr} \rangle = \frac{3}{2}kT = 1.5(1.381 \times 10^{-23}\ \text{J/K})(298\ \text{K}) = 6.17 \times 10^{-21}\ \text{J}$.

$\langle \varepsilon_{tr} \rangle = \frac{3}{2}kT = \langle \frac{1}{2}m\upsilon^2 \rangle$ and $\langle \upsilon^2 \rangle^{1/2} = (3kT/m)^{1/2} = (3RT/M)^{1/2} = [3(8.314\ \text{J/mol-K})(298.1\ \text{K})/0.0320\ \text{kg/mol})]^{1/2} = 482\ \text{m/s}$.

R16.2 $dN_B/dt = \lambda_1 N_X$ and $dN_C/dt = \lambda_2 N_X$. Then $dN_X/dt = -dN_B/dt - dN_C/dt = -\lambda_1 N_X - \lambda_2 N_X = -(\lambda_1 + \lambda_2)N_X = -\lambda_X N_X$, so $\lambda_X = \lambda_1 + \lambda_2$. Then $\lambda_X t_{1/2,X} = 0.693$ and $t_{1/2,X} = 0.693/\lambda_X = 0.693/(\lambda_1 + \lambda_2)$.

R16.3 What was intended was to find $\langle \upsilon^4 \rangle$, but in the first printing of the text, this appears as $\langle \upsilon \rangle^4$. For $\langle \upsilon \rangle^4$, we have $\langle \upsilon \rangle = \int_0^\infty \upsilon G(\upsilon)\, d\upsilon = 4\pi(m/2\pi kT)^{3/2} \times \int_0^\infty \upsilon^3 e^{-m\upsilon^2/2kT}\, d\upsilon = 4\pi(m/2\pi kT)^{3/2}[2(m/2kT)^2]^{-1} = (8RT/\pi M)^{1/2}$ and $\langle \upsilon \rangle^4 = 64R^2T^2/\pi^2M^2$. For $\langle \upsilon^4 \rangle$, $\langle \upsilon^4 \rangle = \int_0^\infty \upsilon^4 G(\upsilon)\, d\upsilon = 4\pi(m/2\pi kT)^{3/2} \int_0^\infty \upsilon^6 e^{-m\upsilon^2/2kT}\, d\upsilon = 4\pi(m/2\pi kT)^{3/2}[6!\pi^{1/2}/2^7 3!(m/2kT)^{7/2}] = 15(kT/m)^2$, where Integral 3 in Table 14.1 was used.

R16.4 $k_2/k_1 = Ae^{-E_a/RT_2}/Ae^{-E_a/RT_1} = e^{(E_a/R)(1/T_1 - 1/T_2)}$ so $\ln(k_2/k_1) = (E_a/R)(1/T_1 - 1/T_2)$ and $\ln(k_3/k_1) = (E_a/R)(1/T_1 - 1/T_3)$. Then $\dfrac{\ln(k_3/k_1)}{\ln(k_2/k_1)} = \dfrac{(1/T_1 - 1/T_3)}{(1/T_1 - 1/T_2)}$ and

$$\frac{\ln(k_{320}/k_{302})}{\ln(k_{315}/k_{302})} = \frac{302^{-1} - 320^{-1}}{302^{-1} - 315^{-1}} = 1.363\text{; } \ln(k_{320}/k_{302}) = [\ln(3.40)](1.363) =$$

1.668 and $k_{320}/k_{302} = 5.30$. Alternatively, we can solve for E_a/R to get E_a/R = 8955 K and use this to solve the problem.

R16.5 Like the rms speed, the average speed is proportional to $T^{1/2}$, so doubling T, multiplies the average speed by $2^{1/2} = 1.414$.

R16.6 **(a)** V; **(b)** C m; **(c)** Pa s = N m^{-2} s; **(d)** m; **(e)** (m/s)/(N/C) = (m/s)/(V/m) = $m^2\ s^{-1}\ V^{-1}$; **(f)** $(mol/m^3)^{-2}\ s^{-1}$; **(g)** V = J/C; **(h)** (C m)/(N/C) = (C m)/(V/m) = C m^2/V = $C^2\ m^2$/J.

R16.7 $\eta = (5\pi/32)\rho\langle v\rangle\lambda$ and $k = (5\pi/32)\rho\langle v\rangle\lambda(C_{V,m} + \frac{9}{4}R)/M$, so $k/\eta = (C_{V,m} + \frac{9}{4}R)/M$. Since T and P do not differ greatly from room T and P, we can use the 25°C, 1 bar value for $C_{V,m}$. The Appendix gives $C_{V,m}$ = $C_{P,m} - R$ = (35.309 – 8.314) J/mol-K = 27.0 J/mol-K and
k = (27.0 J/mol-K)(10.3 × 10^{-6} Pa s)/(0.0160 kg/mol) = 0.0174 J $s^{-1}\ K^{-1}\ m^{-1}$.

R16.8 The half reactions are Ag(s) + Cl^-(aq) → AgCl(s) + e- and Cd^{2+}(aq) + $2e^-$ → Cd(s). The cell reaction is 2Ag(s) + $2Cl^-$(aq) + Cd^{2+}(aq) → AgCl(s) + Cd(s). Table 13.1 gives $\mathcal{E}°$ = –0.402 V – 0.222 V = –0.624 V. We have Q^{-1} = $a_+(a_-)^2 = (\gamma_+ m_+/m°)(\gamma_- m_-/m°)^2 = (\gamma_\pm)^3 m_+ m_-^2/(m°)^3$, where + and – refer to Cd^{2+} and Cl^-, and where the activities of the solids have been taken as equal to 1,. The ionic strength is $\frac{1}{2}[2^2(0.010) + (0.020)]$ mol/kg = 0.030 mol/kg, and the Davies equation gives $\log\gamma_\pm = -0.141$ and $\gamma_\pm = 0.722$. So $Q^{-1} = (0.722)^3 0.01(0.02)^2 = 1.51\times10^{-6}$ and $Q = 6.6_4 \times 10^6$. Then
$\mathcal{E}$ = –0.624 V –[(8.314 J/mol-K)(298.1 K)/2(96485 C/mol)] $\ln 6.6_4 \times 10^6$ = –0.826 V.

R16.9 From (15.25), $d^2 = 5(MRT)^{1/2}/16\pi^{1/2}\eta N_A$ =

$$\frac{5[(0.03007\text{ kg/mol})(8.314\text{ J/mol-K})(273.1\text{ K})]^{1/2}}{16\pi^{1/2}(85.5\times10^{-7}\text{ Pa s})(6.022\times10^{23}\text{ mol}^{-1})} = 2.83\times10^{-19}\text{ m}^2$$

and $d = 5.3 \times 10^{-10}$ m.

R16.10 From (15.70), $r_B \approx |z_B|e/(6\pi\eta u_B^\infty) =$
$2(1.60 \times 10^{-19}\text{ C})/[6\pi(0.00089\text{ Pa s})(55 \times 10^{-9}\text{ m}^2/\text{V-s})] = 3.47 \times 10^{-10}$ m.

R16.11 One possibility is

$$A + C \rightarrow I + R$$
$$I + B \rightarrow J$$
$$J \rightarrow S + C$$

R16.12 $r = -\frac{1}{3}(d[H_2]/dt) = -\frac{1}{3}(-0.00056\text{ mol/L-s}) = 0.00019\text{ mol/L-s}.$

R16.13 $\mathscr{E}° = -0.549\text{ V} - (-0.762\text{ V}) = 0.213\text{ V}.$
$\Delta G° = -nF\mathscr{E}° = -6(96485\text{ C/mol})(0.213\text{ J/C}) = -123._3\text{ kJ/mol} =$
$-RT\ln K°$, so $\ln K° = [(123300\text{ J/mol})/(8.3145\text{ J/mol-K})(298.1\text{ K})] = 49.75$
and $K° = 4.0 \times 10^{21}$.

R16.14 If there is a rate-determining step, it must have the total composition $O_2 + 2NO = N_2O_4$. One possibility is (1) $O_2 + 2NO \rightarrow 2NO_2$ (slow rds), (2) $NO_2 + SO_2 \rightarrow NO + SO_3$ (rapid), where the stoichiometric number of the second step is 2. Another possibility is (1) $O_2 + NO \rightleftharpoons NO_3$ (rapid eq), (2) $NO_3 + NO \rightarrow 2NO_2$ (slow rds), (3) $NO_2 + SO_2 \rightarrow NO + SO_3$ (rapid), where the stoichiometric number of the third step is 2. Another possibility is (1) $O_2 + NO \rightleftharpoons NO_3$ (rapid eq), (2) $NO_3 + NO \rightarrow N_2O_4$ (slow rds), (3) $N_2O_4 + SO_2 \rightarrow SO_3 + N_2O_3$ (rapid), (4) $N_2O_3 + SO_2 \rightarrow SO_3 + N_2O_2$ (rapid), (5) $N_2O_2 \rightarrow 2NO$ (rapid).

R16.15 $[A]^{-n}\,d[A] = -k\,dt$; integration gives $([A]^{-n+1} - [A]_0^{-n+1})/(-n+1) = -kt$, so $[A]^{-n+1} = [A]_0^{-n+1} + (n-1)kt$ and $[A] = \{[A]_0^{-n+1} + (n-1)kt\}^{1/(1-n)}$.

R16.16 (a) $r = k_2[D][E]$. $d[D]/dt = 0 = k_1[A][B] - k_{-1}[C][D] - k_2[D][E]$ and $[D] = k_1[A][B]/\{k_{-1}[C] + k_2[E]\}$ and $r = \dfrac{k_2k_1[A][B][E]}{k_{-1}[C] + k_2[E]}$

(b) $r = k_2[\text{D}][\text{E}]$. $r_1 = r_{-1}$ and $k_1[\text{A}][\text{B}] = k_{-1}[\text{C}][\text{D}]$ so $[\text{D}] = k_1[\text{A}][\text{B}]/k_{-1}[\text{C}]$ and $r = k_2k_1[\text{A}][\text{B}][\text{E}]/k_{-1}[\text{C}]$. Note that if $k_{-1}[\text{C}] >> k_2[\text{E}]$, then the rate in (a) becomes equal to the rate in (b).

R16.17 **(a)** F. The emf is the open-circuit potential difference between the terminals (which have the same composition).

(b) F. See Sec. 16.12.

(c) F. See Fig. 15.22.

(d) T.

(e) T.

R16.18 **(a)** No; **(b)** Yes. **(c)** No. **(d)** No. **(e)** Yes. **(f)** No. **(g)** Yes.

R16.19 $\Delta G° = -nF\mathcal{E}°$ and $(\partial\,\Delta G°/\partial T)_P = -nF(\partial\mathcal{E}°/\partial T)_P$. From $dG = -S\,dT + V\,dP$, we have $(\partial G/\partial T)_P = -S$, so $(\partial\,\Delta G°/\partial T)_P = -\Delta S°$, and $\Delta S° = nF(d\mathcal{E}°/dT)$.

Chapter 17

17.1 **(a)** $dR/d\nu = 0 = \dfrac{2\pi h}{c^2}\left[\dfrac{3\nu^2}{e^{h\nu/kT}-1} - \dfrac{\nu^3(h/kT)e^{h\nu/kT}}{(e^{h\nu/kT}-1)^2}\right]$

So $3 = (h\nu_{max}/kT)\,e^{h\nu_{max}/kT}/(e^{h\nu_{max}/kT} - 1) = xe^x/(e^x - 1)$, where $x = h\nu_{max}/kT$. Then $3e^x - 3 = xe^x$ and multiplication by e^{-x} gives $x + 3e^{-x} = 3$.

(b) For $x = 0, 1, 2, 3$, the function $x + 3e^{-x}$ equals 3, 2.104, 2.406, and 3.149. So the nonzero root lies between 2 and 3. We have $(3 - 2.406)/(3.149 - 2.406) = 0.80$, so interpolation gives $x \approx 2.80$. For $x = 2.80, 2.81, 2.82, 2.83$, we find $x + 3e^{-x} = 2.98243, 2.99061, 2.99882, 3.00704$. The root lies between 2.82 and 2.83, and interpolation gives $x = 2.821_4$. The Excel Solver gives 2.82143937… . (Use Options to change the Precision to 10^{-15}.)

(c) At 300 K, $\nu_{max} = kTx/h = (1.3807 \times 10^{-23}\text{ J/K})(300\text{ K})(2.821)/(6.626 \times 10^{-34}\text{ J s}) = 1.76 \times 10^{13}\text{ s}^{-1}$. At 3000 K, ν_{max} is 10 times as large, namely, $1.76 \times 10^{14}\text{ s}^{-1}$. From Fig. 20.2, these frequencies lie in the infrared.

(d) $T = h\nu_{max}/xk = (6.626 \times 10^{-34}\text{ J s})(3.5 \times 10^{14}\text{ s}^{-1})/2.821(1.38 \times 10^{-23}\text{ J/K}) = 6000\text{ K}$.

(e) $\nu_{max} = kTx/h = (1.38 \times 10^{-23}\text{ J/K})(306\text{ K})2.821/(6.626 \times 10^{-34}\text{ J s}) = 1.80 \times 10^{13}\text{ s}^{-1}$. Infrared.

17.2 **(a)** The total emission per unit time and per unit area is $\int_0^\infty R(\nu)\,d\nu = (2\pi h/c^2)\int_0^\infty [\nu^3/(e^{h\nu/kT} - 1)]\,d\nu$. Let $z = h\nu/kT$; then $dz = (h/kT)\,d\nu$. We have $\int_0^\infty R(\nu)\,d\nu = (2\pi h/c^2)(kT/h)^4 \int_0^\infty [z^3/(e^z - 1)]\,dz = (2\pi h/c^2)(kT/h)^4\pi^4/15 = 2\pi^5k^4T^4/15c^2h^3$.

(b) The emission rate is $(2\pi^5k^4T^4/15c^2h^3)(4\pi r^2) =$

$$\frac{8\pi^6(1.381\times 10^{-23}\text{ J/K})^4(5800\text{ K})^4(0.7\times 10^9\text{ m})^2}{15(2.998\times 10^8\text{ m/s})^2(6.626\times 10^{-34}\text{ J s})^3} = 3.9_6 \times 10^{26}\text{ J/s}$$

(similar to the value given in Prob. 16.113).

(c) In 1 year, $\Delta E = (3.9_6 \times 10^{26}\text{ J/s})(365.25 \times 24 \times 60 \times 60\text{ s}) = 1.2_5 \times 10^{34}\text{ J}$. So $\Delta m = \Delta E/c^2 = (1.2_5 \times 10^{34}\text{ J})/(2.998 \times 10^8\text{ m/s})^2 = 1.4 \times 10^{17}\text{ kg}$.

17.3 **(a)** $h\nu = \Phi + \frac{1}{2}m\upsilon^2$ and $h\nu_{thr} = \Phi$. For K we have $\nu_{thr} = \Phi/h =$ $(2.2\text{ eV})(1.60 \times 10^{-19}\text{ J/eV})/(6.626 \times 10^{-34}\text{ J s}) = 5.3 \times 10^{14}\text{ s}^{-1}$ and $\lambda_{thr} = c/\nu_{thr} = (3.0 \times 10^{10}\text{ cm/s})/(5.3 \times 10^{14}\text{ s}^{-1}) = 5.7 \times 10^{-5}$ cm. For Ni we find $\nu_{thr} = 1.2 \times 10^{15}\text{ s}^{-1}$ and $\lambda_{thr} = 2.5 \times 10^{-5}$ cm.

(b) K-yes; Ni-no.

(c) $\frac{1}{2}m\upsilon^2 = h\nu - \Phi = (6.63 \times 10^{-34}\text{ J s})(3.00 \times 10^8\text{ m/s})/(4.00 \times 10^{-7}\text{ m}) -$ $(2.2\text{ eV})(1.60 \times 10^{-19}\text{ J/eV}) = 1.4 \times 10^{-19}\text{ J} = 0.9$ eV.

17.4 $E = h\nu = hc/\lambda = (6.626 \times 10^{-34}\text{ J s})(2.998 \times 10^8\text{ m/s})/(700 \times 10^{-9}\text{ m}) =$ 2.8×10^{-19} J.

17.5 $E_{photon} = h\nu = hc/\lambda = (6.626 \times 10^{-34}\text{ J s})(2.998 \times 10^8\text{ m/s})/(590 \times 10^{-9}\text{ m}) =$ 3.37×10^{-19} J. Then $100\text{ J/s} = N(3.37 \times 10^{-19}\text{ J})$ and $N = 2.97 \times 10^{20}$ photons/s.

17.6 $h\nu = \Phi + \frac{1}{2}m\upsilon^2 = \Phi + K_{max}$ and $K_{max} = h\nu - \Phi$.

$10^{12}K_{max}$/ergs	3.41	2.56	1.95	0.75
$10^{-14}\nu/\text{s}^{-1}$	9.593	8.213	7.408	5.490

where we used $\nu = c/\lambda$. The slope is $6.5_3 \times 10^{-27}$ erg s $= 6.5_3 \times 10^{-34}$ J s $= h$. Φ can be found from the graph as the value of $h\nu$ at $K_{max} = 0$ or as the negative of the intercept at $\nu = 0$. We find $\Phi = 2.8_5 \times 10^{-12}$ erg $= 1.8$ eV.

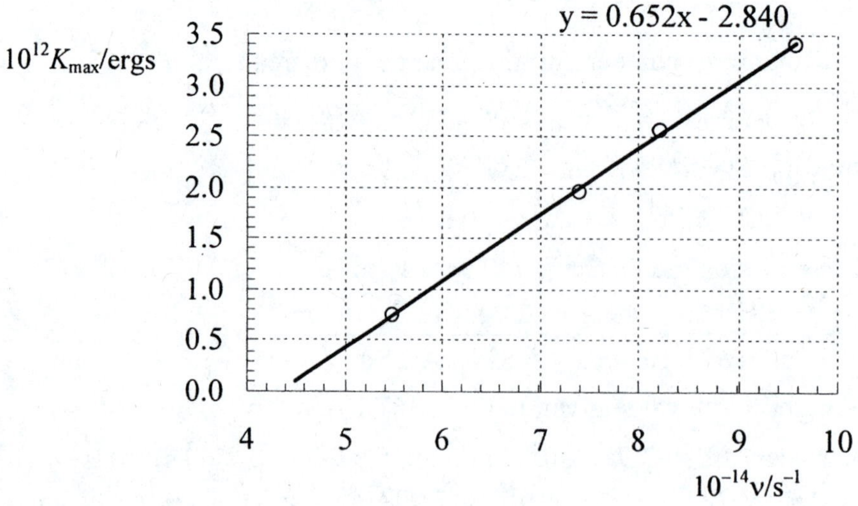

17.7 **(a)** $\lambda = h/m\upsilon = (6.626 \times 10^{-34}\text{ J s})/(1.67 \times 10^{-27}\text{ kg})(6.0 \times 10^{4}\text{ m/s}) = 6.6 \times 10^{-12}\text{ m}$.

(b) $\lambda = (6.626 \times 10^{-34}\text{ J s})/(0.050\text{ kg})(1.20\text{ m/s}) = 1.1 \times 10^{-32}\text{ m}$.

17.8 **(a)** $\sin\theta = \lambda/w$ and $\lambda = h/m\upsilon$, so $\sin\theta = h/m\upsilon w =$

$$\frac{6.26\times10^{-34}\text{ J}\cdot\text{s}}{(9.11\times10^{-31}\text{ kg})(6.0\times10^{6}\text{ m/s})(2400\times10^{-10}\text{ m})} =$$

5.05×10^{-4} and $\theta = 0.0289° = 5.05 \times 10^{-4}$ rad.

(b) Let z be the width of the central maximum. Figure 17.4 gives $\tan\theta = \overline{PE}/\overline{DE} = \frac{1}{2}z/(40\text{ cm})$ and $z = 2(40\text{ cm})\tan 0.0289° = 0.040$ cm.

(c) $\Delta p_x = 2h/w = 2(6.63 \times 10^{-34}\text{ J s})/(2400 \times 10^{-10}\text{ m}) = 5.5 \times 10^{-27}\text{ kg m/s}$.

17.9 $\Delta x\,\Delta p_x \gtrsim h$ and $\Delta p_x \gtrsim h/\Delta x = (6.6 \times 10^{-34}\text{ J s})/(1 \times 10^{-10}\text{ m}) = 6._6 \times 10^{-24}\text{ kg m s}^{-1}$. We have $\Delta p_x = \Delta(m\upsilon_x) = m\,\Delta\upsilon_x$ and $\Delta\upsilon_x = \Delta p_x/m \gtrsim (6._6 \times 10^{-24}\text{ kg m s}^{-1})/(9.1 \times 10^{-31}\text{ kg}) = 7 \times 10^{6}$ m/s, which is very large.

17.10 The time t and the 9 spatial coordinates $x_1, y_1, z_1, x_2, y_2, z_2, x_3, y_3, z_3$ of the three particles.

17.11 **(a)** F; **(b)** F; **(c)** T; **(d)** T; **(e)** T; **(f)** T.

17.12 $|z|^2 = zz^*$ and $|z| = (zz^*)^{1/2}$.

(a) $|-2| = 2$;

(b) $|3 - 2i| = [(3 - 2i)(3 + 2i)]^{1/2} = (9 + 4)^{1/2} = 13^{1/2}$;

(c) $|\cos\theta + i\sin\theta| = [(\cos\theta + i\sin\theta)(\cos\theta - i\sin\theta)]^{1/2} = (\cos^2\theta + \sin^2\theta)^{1/2} = 1^{1/2} = 1$;

(d) $|-3e^{-i\pi/5}| = [(-3e^{-i\pi/5})(-3e^{i\pi/5})]^{1/2} = (9e^{0})^{1/2} = 3$.

17.13 Let $\text{ls}_{17.10}$ and $\text{rs}_{17.10}$ denote the left side and right side of (17.10), respectively. We have $\text{ls}_{17.10} = \text{rs}_{17.10}$. To see if $c\Psi$ is a solution of (17.10), we replace Ψ in (17.10) by $c\Psi$ and see if (17.10) is satisfied. With $c\Psi$ as the proposed solution, the left side of (17.10) becomes $(-\hbar/i)(\partial/\partial t)(c\Psi) = c(-\hbar/i)(\partial\Psi/\partial t) = c\,\text{ls}_{17.10}$.

The right side becomes
$-(\hbar^2/2m_1)[\partial^2(c\Psi)/\partial x_1^2 + \cdots] - \cdots -(\hbar^2/2m_n)[\partial^2(c\Psi)/\partial x_n^2 + \cdots] + Vc\Psi =$
$c\ \mathrm{rs}_{17.10}$. Since Ψ is a solution of (17.10), we have $\mathrm{ls}_{17.10} = \mathrm{rs}_{17.10}$. Then $c\ \mathrm{ls}_{17.10} = c\ \mathrm{rs}_{17.10}$ and so $c\Psi$ satisfies (17.10).

17.14 $\int_a^b \int_c^d [\int_s^t f(r)g(\theta)h(\phi)\, dr]\, d\theta\, d\phi = \int_a^b \int_c^d g(\theta)h(\phi)[\int_s^t f(r)\, dr]\, d\theta\, d\phi =$
$\int_s^t f(r)\, dr \int_a^b [\int_c^d g(\theta)h(\phi)\, d\theta]\, d\phi = \int_s^t f(r)\, dr \int_c^d g(\theta)\, d\theta \int_a^b h(\phi)\, d\phi$

17.15 The result of Prob. 17.14 with r, θ, ϕ replaced by x, y, z, gives $\int |\Psi|^2\, d\tau =$
$\int_{-\infty}^{\infty} (2/\pi c^2)^{1/2} e^{-2x^2/c^2} (2/\pi c^2)^{1/2} e^{-2y^2/c^2} (2/\pi c^2)^{1/2} e^{-2z^2/c^2}\, dx\, dy\, dz =$
$(2/\pi c^2)^{1/2} \int_{-\infty}^{\infty} e^{-2x^2/c^2}\, dx \cdot (2/\pi c^2)^{1/2} \int_{-\infty}^{\infty} e^{-2y^2/c^2}\, dy \cdot (2/\pi c^2)^{1/2} \int_{-\infty}^{\infty} e^{-2z^2/c^2}\, dz$
Use of integrals 1 (with $n = 0$) and 2 in Table 14.1 gives
$(2/\pi c^2)^{1/2} \int_{-\infty}^{\infty} e^{-2x^2/c^2}\, dx = 2(2/\pi c^2)^{1/2} \int_0^{\infty} e^{-2x^2/c^2}\, dx =$
$2(2/\pi c^2)^{1/2} \pi^{1/2}(1/2)(c^2/2)^{1/2} = 1$. By symmetry the y and z integrals also equal 1 and Ψ is normalized.

17.16 The 9 spatial coordinates $x_1, y_1, z_1, x_2, y_2, z_2, x_3, y_3, z_3$ of the three particles.

17.17 **(a)** T; **(b)** F; **(c)** T; **(d)** T.

17.18 The time-dependent Schrödinger equation is more general, since the time-independent equation applies only to stationary states.

17.19 Let $f = f_1 + if_2$ and $g = g_1 + ig_2$, where f_1 is the real part of f, and f_2 is the coefficient of the imaginary part of f. Then $(fg)^* = [(f_1 + if_2)(g_1 + ig_2)]^* =$
$[f_1g_1 - f_2g_2 + i(f_2g_1 + f_1g_2)]^* = f_1g_1 - f_2g_2 - i(f_2g_1 + f_1g_2)$. Also, $f^*g^* =$
$(f_1 + if_2)^*(g_1 + ig_2)^* = (f_1 - if_2)(g_1 - ig_2) = f_1g_1 - f_2g_2 - i(f_2g_1 + f_1g_2) = (fg)^*$.

17.20 Let $\mathrm{ls}_{17.24}$ and $\mathrm{rs}_{17.24}$ denote the left and right side of (17.24). To see if $k\psi$ is a solution of (17.24) we replace ψ in (17.24) by $k\psi$ and see if (17.24) is satisfied. With $k\psi$ as the proposed solution, the left side of (17.24) becomes
$(-\hbar^2/2m_1)[\partial^2(k\psi)/\partial x_1^2 + \cdots] - \cdots + V(k\psi) =$

$k[(-\hbar^2/2m_1)(\partial^2\psi/\partial x_1^2 + \cdots) - \cdots + V\psi] = k\ \text{ls}_{17.24}$. The right side becomes $Ek\psi = k(E\psi) = k\ \text{rs}_{17.24}$. Equation (17.24) is $\text{ls}_{17.24} = \text{rs}_{17.24}$ and multiplication by k gives $k\ \text{ls}_{17.24} = k\ \text{rs}_{17.24}$, so $k\psi$ is a solution of (17.24).

17.21 **(a)** Yes. The integral $\int_{-\infty}^{\infty} e^{-2ax^2}\,dx$ with $a > 0$ is finite, as shown by integral 1 with $n = 0$ and integral 2 in Table 14.1.

(b) No. The integral of the square of this function is infinite.

(c) No. $\int_{-\infty}^{\infty} x^{-2}dx = \int_{-\infty}^{0} x^{-2}dx + \int_{0}^{\infty} x^{-2}dx = -x^{-1}\,|_{-\infty}^{0} - x^{-1}\,|_{0}^{\infty} = \infty - 0 - (0 - \infty) = \infty$.

(d) No. For $x < 0$, $|x| = -x$, and the integral from $-\infty$ to ∞ of the square of this function is $\int_{-\infty}^{0} [1/(-x)^{1/2}]\,dx + \int_{0}^{\infty} (1/x^{1/2})\,dx =$ $-2(-x)^{1/2}\Big|_{-\infty}^{0} + 2x^{1/2}\Big|_{0}^{\infty} = 0 - (-2)(\infty) + 2(\infty) - 0 = \infty$.

(e) False. Consider the function $f \equiv e^{-ax^2}/|x|^{1/4}$, which is the product of the functions in (a) and (d). The integrand is infinite at $x = 0$. Despite this, the integral $\int_{-\infty}^{\infty} |f|^2\,dx$ is not infinite. For x between -1 and 1, the integrand is less than $|x|^{-1/2}$ (except at $x = 0$), so $\int_{-1}^{1} e^{-2ax^2}|x|^{-1/2}\,dx <$ $\int_{-1}^{1} |x|^{-1/2}\,dx = \int_{-1}^{0} |-x|^{-1/2}\,dx + \int_{0}^{1} |x|^{-1/2}\,dx = -2(-x)^{1/2}\Big|_{-1}^{0} + 2x^{1/2}\Big|_{0}^{1} =$ $2 + 2 = 4$. So $0 < \int_{-1}^{1} |f|^2\,dx < 4$. Over the rest of the integration range we deal with $\int_{-\infty}^{-1} e^{-2ax^2}|x|^{-1/2}\,dx$ and $\int_{1}^{\infty} e^{-2ax^2}|x|^{-1/2}\,dx$. Here, the $1/|x|^{1/2}$ factor makes the integrand less than e^{-2ax^2} and, since integration of e^{-2ax^2} gives a finite area under the curve (see part a), the integrals $\int_{-\infty}^{-1} e^{-2ax^2}|x|^{-1/2}\,dx$ and $\int_{1}^{\infty} e^{-2ax^2}|x|^{-1/2}\,dx$ are finite. Hence, $\int_{-\infty}^{\infty} |f|^2\,dx$ is finite and f is quadratically integrable even though it is infinite at the origin.

17.22 $E = n^2h^2/8ma^2$; $|\Delta E| = h\nu = hc/\lambda$; $\lambda = hc/|\Delta E| = hc[8ma^2/h^2|n_2^2 - n_1^2|] =$ $8ma^2c/h|n_2^2 - n_1^2| = \dfrac{8(1.0\times10^{-30}\text{ kg})(6.0\times10^{-10}\text{ m})^2(3.0\times10^{8}\text{ m/s})}{(6.63\times10^{-34}\text{ J}\cdot\text{s})(25-16)} =$ $1.4_5 \times 10^{-7}$ m.

17.23 **(a)** $\int_0^{a/4} |\psi|^2\, dx = (2/a) \int_0^{a/4} \sin^2 (n\pi x/a)\, dx =$

$(2/a)[\frac{1}{2}x - \frac{1}{4}(a/n\pi) \sin (2n\pi x/a)]\Big|_0^{a/4} = \frac{1}{4} - (1/2n\pi) \sin \frac{1}{2}n\pi$ where we used $\int \sin^2 cx\, dx = \frac{1}{2}x - \frac{1}{4}c^{-1} \sin 2cx$.

(b) For $n = 1, 2, 3$, we get 0.091, 0.250, and 0.303, respectively.

17.24 The interval 0.0001 Å is much, much smaller than the box length, so we can consider this to be an "infinitesimal" interval. The probability is then $|\psi|^2\, dx = (2/a) \sin^2 (n\pi x/a)\, dx$.

(a) For $n = 1$, we get $[2/(2\text{ Å})] \sin^2 [\pi(1.6\text{ Å})/(2.0\text{ Å})](0.0001\text{ Å}) = 3.45 \times 10^{-5}$.

(b) For $n = 2$, we get 9.05×10^{-5}.

17.25

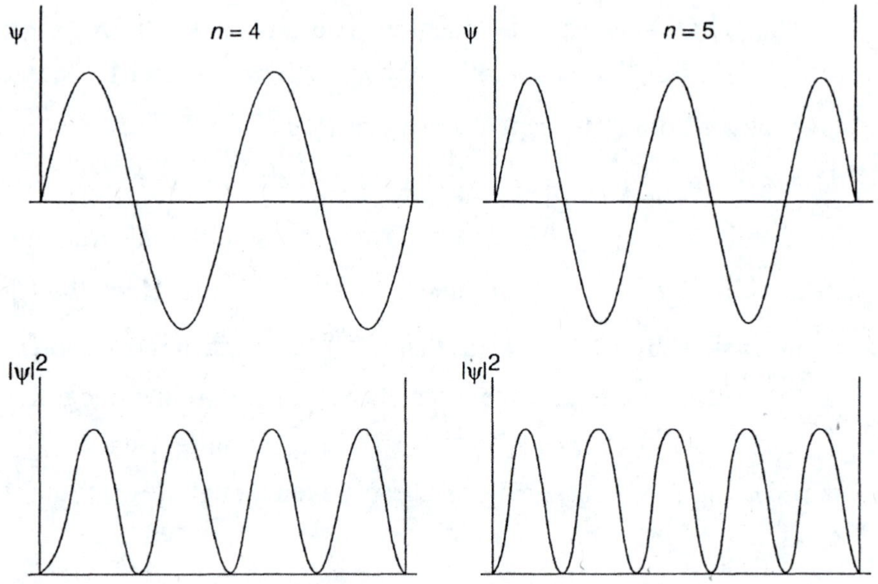

Note that ψ^2 has zero slope at the nodes.

17.26 $d^2\psi/dx^2 = 0$. Integration with respect to x gives $d\psi/dx = c$ and a second integration gives $\psi = cx + d$, where c and d are integration constants. The continuity condition at $x = 0$ requires that $\psi = 0$ at $x = 0$, so $0 = c(0) + d$ and

$d = 0$. Then $\psi = cx$. The continuity condition at $x = a$ requires that $\psi = 0$ at $x = a$, and $0 = ca$, so $c = 0$. Then $\psi = 0$.

17.27 The lowest frequency transition corresponds to $n = 1 \to 2$. Use of (17.7) gives $2^2h^2/8ma^2 - 1^2h^2/8ma^2 = h\nu$ and $a = (3h/8m\nu)^{1/2} =$ $[3(6.63 \times 10^{-34}\text{ J s})]^{1/2}/[8(9.11 \times 10^{-31}\text{ kg})(2.0 \times 10^{14}\text{ s}^{-1})]^{1/2} = 1.2 \times 10^{-9}$ m.

17.28 $h\nu = E_{\text{upper}} - E_{\text{lower}} = (n_u^2 - n_\ell^2)h^2/8ma^2$, where u and ℓ stand for upper and lower. $\nu = (n_u^2 - n_\ell^2)h/8ma^2$. $\nu_{3\to4} = (4^2 - 3^2)h/8ma^2 = 7h/8ma^2$. $\nu_{6\to9} = (9^2 - 6^2)h/8ma^2 = 45h/8ma^2$. So $\nu_{6\to9}/\nu_{3\to4} = 45/7$ and $\nu_{6\to9} = (45/7)(4.00 \times 10^{13}\text{ s}^{-1}) = 2.57 \times 10^{14}\text{ s}^{-1}$.

17.29 $\int_0^a \psi_i^* \psi_j\, dx = 2a^{-1} \int_0^a \sin(n_i\pi x/a)\sin(n_j\pi x/a)\, dx$, $n_i \neq n_j$. A table of integrals gives $\int \sin cx \sin bx\, dx = [1/2(c-b)] \sin[(c-b)x] - [1/2(c+b)] \sin[(c+b)x]$, provided $c^2 \neq b^2$. So $\int_0^a \psi_i^* \psi_j\, dx =$

$$\frac{2}{a}\left[\frac{\sin[(n_i - n_j)\pi x/a]}{2(n_i - n_j)\pi/a} - \frac{\sin[(n_i + n_j)\pi x/a]}{2(n_i + n_j)\pi/a}\right]\Bigg|_0^a = 0$$

since $\sin[(n_i - n_j)\pi] = 0$, $\sin[(n_i + n_j)\pi] = 0$, and $\sin 0 = 0$.

17.30 The left side of (17.28) becomes $d^2\psi/dx^2 = (2/a)^{1/2}(-1)(n^2\pi^2/a^2)\sin(n\pi x/a)$. With use of (17.34), the right side of (17.28) becomes $-(2m/\hbar^2)(n^2h^2/8ma^2) \times (2/a)^{1/2}\sin(n\pi x/a) = -(\pi^2n^2/a^2)(2/a)^{1/2}\sin(n\pi x/a)$, which equals the left side.

17.31 For (a), $|\psi|$ is a maximum at $(\frac{1}{4}a, \frac{1}{2}a)$ and at $(\frac{3}{4}a, \frac{1}{2}a)$, where a is the box length. For (b), $|\psi|$ is a maximum at $(\frac{1}{4}a, \frac{1}{4}a)$, $(\frac{1}{4}a, \frac{3}{4}a)$, $(\frac{3}{4}a, \frac{1}{4}a)$, $(\frac{3}{4}a, \frac{3}{4}a)$.

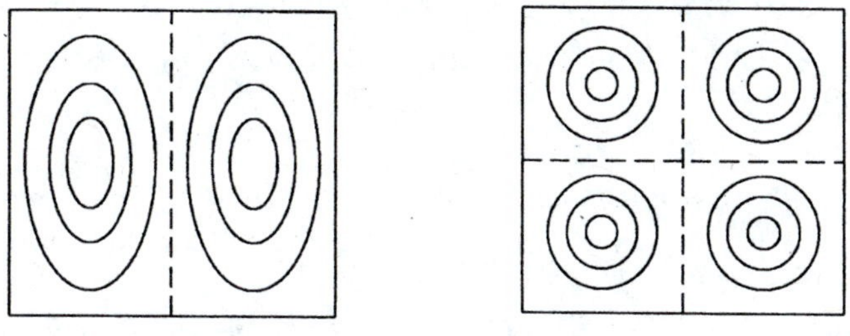

(a) (b)

where the dashed lines are nodal lines and the x axis is horizontal.

17.32 **(a)** $n_x^2 + n_y^2 + n_z^2 = 21$ is satisfied by $n_x n_y n_z$ = 124, 142, 214, 241, 412, 421 and no others, so the degeneracy is 6.

(b) $n_x^2 + n_y^2 + n_z^2 = 24$ is satisfied only by $n_x n_y n_z$ = 422, 242, 224, and the degeneracy is 3.

17.33 **(a)** $E = (h^2/8ma^2)(n_x^2 + n_y^2 + n_z^2)$. There are 17 states with $n_x^2 + n_y^2 + n_z^2 \le 16$, namely: $n_x n_y n_z$ = 111, 211, 121, 112, 122, 212, 221, 311, 131, 113, 222, 123, 132, 213, 231, 312, 321.

(b) These states give a total of 6 different values for $n_x^2 + n_y^2 + n_z^2$, namely, $n_x^2 + n_y^2 + n_z^2$ = 3, 6, 9, 11, 12, 14, so there are 6 energy levels in the given range.

17.34 **(a)** T; **(b)** F; **(c)** F; **(d)** F; **(e)** F; **(f)** F; **(g)** T.

17.35 **(a)** Yes; **(b)** no; **(c)** yes; **(d)** no.

17.36 Since the wave function is an eigenfunction of the Hamiltonian (energy) operator, we must get the eigenvalue $25h^2/8ma^2$.

17.37 **(a)** $\hat{A}\hat{B}f(x) - \hat{B}\hat{A}f(x) = (d^2/dx^2)[xf(x)] - x[(d^2/dx^2)f(x)] =$ $(d/dx)[xf'(x) + f(x)] - xf''(x) = xf''(x) + f'(x) + f'(x) - xf''(x) = 2f'(x)$.

(b) $(\hat{A} + \hat{B})(e^{x^2} + \cos 2x) = (d^2/dx^2 + x)[e^{x^2} + \cos 2x] =$ $(d^2/dx^2)(e^{x^2} + \cos 2x) + x(e^{x^2} + \cos 2x) = d^2(e^{x^2})/dx^2 + d^2(\cos 2x)/dx^2 +$ $xe^{x^2} + x\cos 2x = 2e^{x^2} + 4x^2e^{x^2} - 4\cos 2x + xe^{x^2} + x\cos 2x$.

17.38 **(a)** $(\quad)^2$ and $(\quad)^*$ are nonlinear; the others are linear.

(b) $\hat{H}(f + g) = [(-\hbar^2/2m_1)\nabla_1^2 - \cdots + V](f + g) =$ $[(-\hbar^2/2m_1)\nabla_1^2 - \cdots](f + g) + V(f + g) = (-\hbar^2/2m_1)\nabla_1^2 f - (\hbar^2/2m_1)\nabla_1^2 g$ $- \cdots + Vf + Vg = [-(\hbar^2/2m_1)\nabla_1^2 - \cdots + V]f + [-(\hbar^2/2m_1)\nabla_1^2 - \cdots + V]g$

$= \hat{H}f + \hat{H}g$, where the definition of the sum of operators and the fact that $(\partial^2/\partial x_1^2)(f+g) = \partial^2 f/\partial x_1^2 + \partial^2 g/\partial x_1^2$ were used. Similarly, one finds $\hat{H}(cf) = c\hat{H}f$. So $\hat{H}$ is linear.

17.39 **(a)** When $\hat{B}$ operates on $g(x)$, it turns g into another function, which we shall call $f(x)$. When $\hat{A}$ operates on $f(x)$, we get another function, so $\hat{A}\hat{B}g(x)$ is a function. **(b)** Operator. **(c)** Function. **(d)** Operator. **(e)** Function.

17.40 **(a)** $\hat{p}_x^3 = [(\hbar/i)(\partial/\partial x)]^3 = -(\hbar^3/i)(\partial^3/\partial x^3)$.

(b) $\hat{p}_z^4 = [(\hbar/i)(\partial/\partial z)]^4 = \hbar^4(\partial^4/\partial z^4)$.

17.41 $(d^2/dx^2)(\sin 3x) = (d/dx)(3\cos 3x) = -9(\sin 3x)$, so $\sin 3x$ is an eigenfunction of d^2/dx^2 with eigenvalue -9. $(d^2/dx^2)(6\cos 4x) = -96\cos 4x = -16(6\cos 4x)$, so $6\cos 4x$ is an eigenfunction of d^2/dx^2 with eigenvalue -16. $(d^2/dx^2)(5x^3) = 30x$, which does not equal a constant times $5x^3$; so $5x^3$ is not an eigenfunction of d^2/dx^2. $(d^2/dx^2)x^{-1} = 2x^{-3} \neq (\text{const.})x^{-1}$. $(d^2/dx^2)(3e^{-5x}) = 75e^{-5x} = 25(3e^{-5x})$, and the eigenvalue is 25. $(d^2/dx^2)\ln 2x = -1/x^2 \neq (\text{const.})\ln 2x$.

17.42 **(a)**

$\langle p_x \rangle = \int_{-\infty}^{\infty} \psi^* \hat{p}_x \psi\, dx = \int_{-\infty}^{\infty} \psi^*(\hbar/i)(\partial/\partial x)\psi\, dx =$
$(\hbar/i)\int_{-\infty}^{0} \psi^*(\partial\psi/\partial x)\, dx + (\hbar/i)\int_0^a \psi^*(\partial\psi/\partial x)\, dx +$
$(\hbar/i)\int_a^{\infty} \psi^*(\partial\psi/\partial x)\, dx =$
$(\hbar/i)\int_0^a \psi^*(\partial\psi/\partial x)\, dx = (\hbar/i)(2/a)(n\pi/a)\int_0^a \sin(n\pi x/a)\cos(n\pi x/a)\, dx =$
$(2n\pi\hbar/ia^2)(a/n\pi)\tfrac{1}{2}\sin^2(n\pi x/a)\big|_0^a = 0$, since $\sin n\pi = 0$ for $n =$ 1, 2, 3, ... (We used the fact that $\psi^* = 0$ outside the box.)

(b) $\langle x \rangle = \int_0^a \psi^* x \psi\, dx = \int_0^a (2/a)^{1/2}\sin(n\pi x/a)\, x\, (2/a)^{1/2}\sin(n\pi x/a)\, dx =$
$(2/a)\int_0^a x\sin^2(n\pi x/a)\, dx$. A table of integrals gives
$\int x\sin^2 cx\, dx = \tfrac{1}{4}x^2 - (x/4c)\sin 2cx - (1/8c^2)\cos 2cx$, so $\langle x \rangle =$
$(2/a)[\tfrac{1}{4}x^2 - (ax/4n\pi)\sin(2n\pi x/a) - (a^2/8n^2\pi^2)\cos(2n\pi x/a)]\big|_0^a =$
$(2/a)(\tfrac{1}{4}a^2 - a^2/8n^2\pi^2 + a^2/8n^2\pi^2) = a/2$, since $\sin 2n\pi = 0$ and $\cos 2n\pi = 1$ for $n = 1, 2, \ldots$.

(c) $\langle x^2 \rangle = \int_0^a \psi^* x^2 \psi \, dx = (2/a) \int_0^a x^2 \sin^2 (n\pi x/a) \, dx$. A table of integrals gives $\int x^2 \sin^2 cx \, dx = x^3/6 - (x^2/4c - 1/8c^3) \sin 2cx - (x/4c^2) \cos 2cx$, so $\langle x^2 \rangle = (2/a)[x^3/6 - (ax^2/4n\pi - a^3/8n^3\pi^3) \sin (2n\pi x/a) - (a^2x/4n^2\pi^2) \cos (2n\pi x/a)] \big|_0^a = (2/a)(a^3/6 - a^3/4n^2\pi^2) = a^2/3 - a^2/2n^2\pi^2$.

17.43 The time-independent Schrödinger equation $\hat{H}\psi = E\psi$ for (17.64) is $(\hat{H}_1 + \hat{H}_2 + \cdots + \hat{H}_r)\psi = E\psi$ and $(\hat{H}_1\psi + \hat{H}_2\psi + \cdots + \hat{H}_r\psi) = E\psi$ (1). Taking $\psi = f_1(q_1)f_2(q_2) \cdots f_r(q_r)$, we have $\hat{H}_1\psi = \hat{H}_1[f_1(q_1)f_2(q_2) \cdots f_r(q_r)] = (f_2 \cdots f_r)\hat{H}_1 f_1$, since $\hat{H}_1$ involves only q_1. Equation (1) becomes $(f_2 \cdots f_r)\hat{H}_1 f_1 + (f_1 f_3 \cdots f_r)\hat{H}_2 f_2 + \cdots + (f_1 \cdots f_{r-1})\hat{H}_r f_r = E f_1 f_2 \cdots f_r$. Division by $f_1 f_2 \cdots f_r$ gives $(1/f_1)\hat{H}_1 f_1 + (1/f_2)\hat{H}_2 f_2 + \cdots + (1/f_r)\hat{H}_r f_r = E$ (2). By the same kind of argument used after Eq. (17.39), each term on the left side of equation (2) must be a constant. Calling these constants $E_1, E_2, \ldots, E_r$, we have $(1/f_1)\hat{H}_1 f_1 = E_1$ or $\hat{H}_1 f_1 = E_1 f_1$, etc., and equation (2) gives $E_1 + E_2 + \cdots + E_r = E$.

17.44 $\psi = (2/a) \sin (n_1\pi x_1/a) \sin (n_2\pi x_2/a)$. $E = n_1^2 h^2/8m_1a^2 + n_2^2 h^2/8m_2a^2$.

17.45 **(a)** nu, frequency; **(b)** vee, quantum number.

17.46 $\nu_{\text{light}} = (E_{\text{upper}} - E_{\text{lower}})/h = [(\upsilon_{\text{upper}} + \frac{1}{2})h\nu_{\text{osc}} - (\upsilon_{\text{lower}} + \frac{1}{2})h\nu_{\text{osc}}]/h = (\upsilon_{\text{upper}} - \upsilon_{\text{lower}})\nu_{\text{osc}} = (8 - 7)\nu_{\text{osc}} = \nu_{\text{osc}} = 6.0 \times 10^{13} \text{ s}^{-1}$.

17.47 Squaring the curves in Fig. 17.18, we get the following curves (note the unequal peak heights):

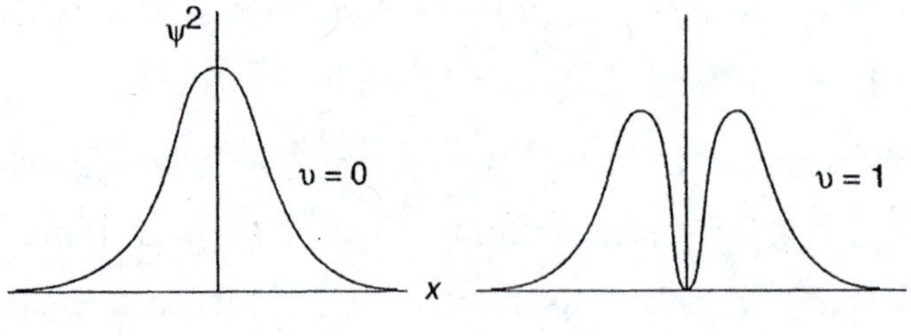

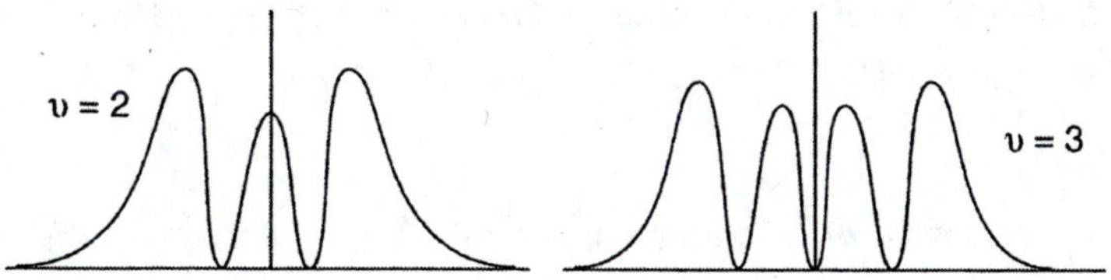

17.48 **(a)** From Fig. 17.18, ψ_0 is a maximum at $x = 0$; likewise, ψ_0^2 is a maximum at $x = 0$ and this is the most probable value of x.

(b) $d\psi_1^2/dx = 0 = (4\alpha^3/\pi)^{1/2}(2xe^{-\alpha x^2} - 2\alpha x^3 e^{-\alpha x^2})$, so $x = \pm 1/\alpha^{1/2} = \pm(\hbar/2\pi\nu m)^{1/2}$. ($x = 0$ is a minimum.)

17.49 $d^2\psi_0/dx^2 = (\alpha/\pi)^{1/4}(d^2/dx^2)e^{-\alpha x^2/2} = (\alpha/\pi)^{1/4}(d/dx)(-\alpha x e^{-\alpha x^2/2}) = (\alpha/\pi)^{1/4}(-\alpha e^{-\alpha x^2/2} + \alpha^2 x^2 e^{-\alpha x^2/2}) = (\alpha^2 x^2 - \alpha)\psi_0 = (16\pi^4\nu^2 m^2 x^2/h^2 - 4\pi^2\nu m/h)\psi_0$. Equation (17.73) gives $k = 4\pi^2\nu^2 m$, so $\frac{1}{2}kx^2\psi_0 = 2\pi^2\nu^2 m x^2\psi_0$. Then $-(\hbar^2/2m)(d^2\psi_0/dx^2) + \frac{1}{2}kx^2\psi_0 = (-2\pi^2\nu^2 m x^2 + \frac{1}{2}h\nu)\psi_0 + 2\pi^2\nu^2 m x^2\psi_0 = \frac{1}{2}h\nu\psi_0 = E_0\psi_0$.

17.50 $\int_{-\infty}^{\infty}\psi_1^*\psi_1\,dx = (4\alpha^3/\pi)^{1/2}\int_{-\infty}^{\infty}x^2 e^{-\alpha x^2}\,dx = 2(4\alpha^3/\pi)^{1/2}\int_0^{\infty}x^2 e^{-\alpha x^2}\,dx = 2(4\alpha^3/\pi)^{1/2}(2\pi^{1/2}/2^3\alpha^{3/2}) = 1$, See integrals 1 and 3 (with $n = 1$) in Table 14.1.

17.51 **(a)** $\langle x\rangle = \int_{-\infty}^{\infty}\psi^* x\psi\,dx = (\alpha/\pi)^{1/2}\int_{-\infty}^{\infty}xe^{-\alpha x^2}\,dx = 0$, where integral 4 (with $n = 0$) in Table 14.1 was used. This result is obvious from Fig. 17.18.

(b) $\langle x^2\rangle = \int_{-\infty}^{\infty}\psi^* x^2\psi\,dx = (\alpha/\pi)^{1/2}\int_{-\infty}^{\infty}x^2 e^{-\alpha x^2}\,dx = 2(\alpha/\pi)^{1/2}\int_0^{\infty}x^2 e^{-\alpha x^2}\,dx = 2(\alpha/\pi)^{1/2}(2\pi^{1/2}/2^3\alpha^{3/2}) = 1/2\alpha = h/8\pi^2\nu m$, where we used integrals 1 and 3 in Table 14.1.

(c) $\langle p_x\rangle = \int_{-\infty}^{\infty}\psi^*\hat{p}_x\psi\,dx = (\alpha/\pi)^{1/2}\int_{-\infty}^{\infty}e^{-\alpha x^2/2}(\hbar/i)(\partial/\partial x)e^{-\alpha x^2/2}\,dx = (\alpha/\pi)^{1/2}(\hbar/i)(-\alpha)\int_{-\infty}^{\infty}xe^{-\alpha x^2}\,dx = 0$ (from integral 4).

17.52 **(a)** $E = K + V = \frac{1}{2}m(dx/dt)^2 + \frac{1}{2}kx^2 = \frac{1}{2}m\{(k/m)^{1/2}A\cos[(k/m)^{1/2}t + b]\}^2 + \frac{1}{2}kA^2\sin^2[(k/m)^{1/2}t + b] = \frac{1}{2}kA^2\{\cos^2[(k/m)^{1/2}t + b] + \sin^2[(k/m)^{1/2}t + b]\} = \frac{1}{2}kA^2$.

(b) $m\, d^2x/dt^2 = m(d^2/dt^2)\{A \sin [(k/m)^{1/2}t + b]\} = -mA(k/m) \sin [(k/m)^{1/2}t + b]$
$= -k\{A \sin [(k/m)^{1/2}t + b]\} = -kx$

17.53 (a) $\nu = (1/2\pi)(k/m)^{1/2}$ and $k = 4\pi^2\nu^2 m = 4\pi^2(2.4\ \text{s}^{-1})^2(0.045\ \text{kg}) = 10.2$ N/m.

(b) $E = \frac{1}{2}kA^2 = 0.5(10.2\ \text{N/m})(0.04\ \text{m})^2 = 0.0082\ \text{J} = (\upsilon + \frac{1}{2})h\nu$, so $\upsilon + \frac{1}{2} =$
$(0.0082\ \text{J})/(6.626 \times 10^{-34}\ \text{J} \cdot \text{s})(2.4\ \text{s}^{-1}) = 5.2 \times 10^{30} = \upsilon$.

17.54 (a) The Hamiltonian is the sum of three one-dimensional harmonic-oscillator Hamiltonians, one for each coordinate; the separation-of-variables theorem [Eqs. (17.65) and (17.66)] gives the energy as the sum of three one-dimensional-harmonic-oscillator energies:
$E = E_x + E_y + E_z = (\upsilon_x + ½)h\nu_x + (\upsilon_y + ½)h\nu_y + (\upsilon_z + ½)h\nu_z$, where
$\upsilon_x = 0, 1, 2, \ldots, \upsilon_y = 0, 1, 2, \ldots, \upsilon_z = 0, 1, 2, \ldots$, and
$\nu_x = (1/2\pi)(k_x/m)^{1/2},\ \nu_y = (1/2\pi)(k_y/m)^{1/2},\ \nu_z = (1/2\pi)(k_z/m)^{1/2}$, where m is the particle's mass.

(b) The lowest level has $\upsilon_x = \upsilon_y = \upsilon_z = 0$ and $E = \frac{1}{2}h(\nu_x + \nu_y + \nu_z)$.

17.55 $H = (1/2\mu)(\mu^2\upsilon_x^2 + \mu^2\upsilon_y^2 + \mu^2\upsilon_z^2) + V + (1/2M)(M^2\upsilon_X^2 + M^2\upsilon_Y^2 + M^2\upsilon_Z^2) =$
$V + \frac{1}{2}\mu(\upsilon_x^2 + \upsilon_y^2 + \upsilon_z^2) + \frac{1}{2}M(\upsilon_X^2 + \upsilon_Y^2 + \upsilon_Z^2)$. Using Eq. (17.77), we have
$\upsilon_x = dx/dt = dx_2/dt - dx_1/dt = \upsilon_{x,2} - \upsilon_{x,1}$. Similarly, $\upsilon_y = \upsilon_{y,2} - \upsilon_{y,1}$ and
$\upsilon_z = \upsilon_{z,2} - \upsilon_{z,1}$. Since $X = (m_1x_1 + m_2x_2)/M$, we have $\upsilon_X = dX/dt =$
$[m_1(dx_1/dt) + m_2(dx_2/dt)]/M = (m_1\upsilon_{x,1} + m_2\upsilon_{x,2})/(m_1 + m_2)$; similar equations hold for υ_Y and υ_Z. So $H = V + \frac{1}{2}[m_1m_2/(m_1 + m_2)] \times$
$(\upsilon_{x,1}^2 - 2\upsilon_{x,1}\upsilon_{x,2} + \upsilon_{x,2}^2 + \cdots) + \frac{1}{2}(m_1 + m_2)(m_1 + m_2)^{-2} \times$
$(m_1^2\upsilon_{x,1}^2 + 2m_1m_2\upsilon_{x,1}\upsilon_{x,2} + m_2^2\upsilon_{x,2}^2 + \cdots) =$
$V + \frac{1}{2}(m_1 + m_2)^{-1}[(m_1 + m_2)m_1\upsilon_{x,1}^2 + (m_1 + m_2)m_2\upsilon_{x,2}^2 + \cdots] =$
$V + \frac{1}{2}(m_1\upsilon_{x,1}^2 + m_2\upsilon_{x,2}^2 + m_1\upsilon_{y,1}^2 + m_2\upsilon_{y,2}^2 + m_1\upsilon_{z,1}^2 + m_2\upsilon_{z,2}^2) =$
$V + \frac{1}{2}m_1\upsilon_1^2 + \frac{1}{2}m_2\upsilon_2^2 = V + m_1^2\upsilon_1^2/2m_1 + m_2^2\upsilon_2^2/2m_2 = p_1^2/2m_1 + p_2^2/2m_2 + V$.
(The dots indicate similar terms in y and z.)

17.56 (a) $\mu = m_1m_2/(m_1 + m_2) = [(12.0\ \text{g/mol})/N_A][(16.0\ \text{g/mol})/N_A]/$
$[(28.0\ \text{g/mol})/N_A] = (6.86\ \text{g/mol})/N_A = 1.14 \times 10^{-23}$ g.

(b) $I = \mu d^2 = (1.14 \times 10^{-26}\ \text{kg})(1.13 \times 10^{-10}\ \text{m})^2 = 1.45 \times 10^{-46}\ \text{kg m}^2$.

(c) $E_{rot} = J(J+1)\hbar^2/2I$. $\hbar^2/2I = (6.626 \times 10^{-34}$ J s$)^2/8\pi^2(1.45 \times 10^{-46}$ kg m$^2)$ $= 3.83 \times 10^{-23}$ J. For $J = 0, 1, 2, 3$, we have $E_{rot} = 0$, 7.66×10^{-23} J, 23.0×10^{-23} J, 46.0×10^{-23} J, respectively. The levels are $(2J+1)$-fold degenerate, so the degeneracies are 1, 3, 5, 7.

(d) For $J = 0$ to 1, $\Delta E = 7.66 \times 10^{-23}$ J $- 0 = 7.66 \times 10^{-23}$ J $= h\nu =$ $(6.626 \times 10^{-34}$ J s$)\nu$ and $\nu = 1.16 \times 10^{11}$ s^{-1}. For $J = 1$ to 2, $\Delta E =$ $(23.0 - 7.66)10^{-23}$ J $= h\nu$ and $\nu = 2.32 \times 10^{11}$ s^{-1}.

17.57 **(a)** Let $Nx(a-x)$ be the normalized function. So $\int_0^a [Nx(a-x)]^* Nx(a-x)\,dx$ $= 1$ and $|N| = 1/[\int_0^a x^2(a-x)^2\,dx]^{1/2}$. From Example 17.8, $\int_0^a x^2(a-x)^2\,dx = a^5/30$, so $|N| = (30/a^5)^{1/2}$.

(b) $\langle x^2 \rangle \cong (30/a^5) \int_0^a x(a-x)x^2x(a-x)\,dx = (30/a^5) \int_0^a (a^2x^4 - 2ax^5 + x^6)\,dx =$ $(30/a^5)(a^7/5 - a^7/3 + a^7/7) = 30a^2/105 = 2a^2/7 = 0.2857a^2$. The true value is found by setting $n = 1$ in Prob. 17.42c to give $\langle x^2 \rangle = a^2(1/3 - 1/2\pi^2) =$ $0.2827a^2$. The error is 1.1%.

17.58 The value of k that minimizes the variational integral W satisfies $\partial W/\partial k = 0$. We have $\partial W/\partial k = 0 = (\hbar^2/ma^2)[(8k+1)/(2k-1) - 2(4k^2+k)/(2k-1)^2] =$ $(\hbar^2/ma^2)(8k^2 - 8k - 1)/(2k-1)^2$ and $8k^2 - 8k - 1 = 0$. The solutions are $k =$ 1.112372 and -0.112372. The negative value of k makes $\phi = \infty$ at $x = 0$ and so is rejected. For $k = 1.112372$, $W = (\hbar^2/ma^2)(4k^2+k)/(2k-1) = 4.94949\hbar^2/ma^2$ $= 4.94949h^2/4\pi^2ma^2 = 0.125372h^2/ma^2$ compared with the true value $h^2/8ma^2 =$ $0.125h^2/ma^2$. The percent error is only 0.30%.

17.59 **(a)** $\int_0^a \phi^*\phi\,dx = \int_0^a x^2(a-x)^2x^2(a-x)^2\,dx =$ $\int_0^a (a^4x^4 - 4a^3x^5 + 6a^2x^6 - 4ax^7 + x^8)\,dx = a^9/5 - 2a^9/3 + 6a^9/7 - a^9/2 +$ $a^9/9 = a^9/630$. We have $\hat{H}\phi = (\hbar^2/2m)(d^2/dx^2)(x^2a^2 - 2ax^3 + x^4) =$ $-(\hbar^2/2m)(2a^2 - 12ax + 12x^2)$. So $\int_0^a \phi^* \hat{H}\phi\,dx =$
$-(\hbar^2/m) \int_0^a x^2(a-x)^2(a^2 - 6ax + 6x^2)\,dx =$
$-(\hbar^2/m) \int_0^a (a^4x^2 - 8a^3x^3 + 19a^2x^4 - 18ax^5 + 6x^6)\,dx =$
$-(\hbar^2/m)(a^7/3 - 2a^7 + 19a^7/5 - 3a^7 + 6a^7/7) = \hbar^2a^7/105m = h^2a^7/420\pi^2m$.
Then $\int \phi^* \hat{H}\phi\,dx / \int \phi^*\phi\,dx = (h^2a^7/420\pi^2m)(630/a^9) = (3/2\pi^2)(h^2/ma^2) =$

$0.152h^2/ma^2 \approx E_{gs}$. The true E_{gs} is $h^2/8ma^2 = 0.125h^2/ma^2$. The error is 22%.

(b) It is discontinuous at $x = a$, since $\phi = 0$ outside the box.

17.60 $\hat{H} = \hat{H}^0 + \hat{H}'$, where $\hat{H}^0$ is the particle-in-a-box Hamiltonian operator and $\hat{H}' = kx$ for $0 \le x \le a$. We have $E_n^{(1)} = \int \psi_n^* \hat{H}'\psi_n \, d\tau =$
$(2/a) \int_0^a kx \sin^2(n\pi x/a)\, dx =$
$(2k/a)[\frac{1}{4}x^2 - \frac{1}{4}(ax/n\pi)\sin(2n\pi x/a) - (a^2/8n^2\pi^2)\cos(2n\pi x/a)]_0^a = \frac{1}{2}ak$, where $\sin 2n\pi = 0$ and $\cos 2n\pi = 1$ were used. $E^{(0)} = n^2h^2/8ma^2$, so $E^{(0)} + E^{(1)} = n^2h^2/8ma^2 + \frac{1}{2}ak$.

17.61 **(a)** T; **(b)** F; **(c)** T; **(d)** F; **(e)** This question is defective. What was intended was to ask whether $\sum_m b_m c_m \delta_{mn} = b_n c_n$ (Eq. A) is true. Because the δ_{mn} factor makes all terms zero except the one with $m = n$, Eq. A is true.

17.62 We are given that $\hat{B}$ and $\hat{C}$ are Hermitian operators, so from (17.92) we have $\int f^* \hat{B}g \, d\tau = \int g(\hat{B}f)^* \, d\tau$ (1) and $\int f^* \hat{C}g \, d\tau = \int g(\hat{C}f)^* \, d\tau$ (2). To prove that $\hat{B} + \hat{C}$ is Hermitian, we shall prove that it satisfies (17.92):
$\int f^*(\hat{B}+\hat{C})g \, d\tau = \int g[(\hat{B}+\hat{C})f]^* \, d\tau$ (A). The left side of equation (A) is
$\int f^*(\hat{B}+\hat{C})g \, d\tau = \int f^*(\hat{B}g + \hat{C}g)\, d\tau = \int f^* \hat{B}g \, d\tau + \int f^* \hat{C}g \, d\tau =$
$\int g(\hat{B}f)^* \, d\tau + \int g(\hat{C}f)^* \, d\tau = \int g[(\hat{B}f)^* + (\hat{C}f)^*]\, d\tau = \int g[(\hat{B}f) + (\hat{C}f)]^* \, d\tau =$
$\int g[(\hat{B}+\hat{C})f]^* \, d\tau$, which completes the proof of Eq. (A). In the proof, we used the definition (17.51) of the sum of operators, the integral identity (1.53), the given equations (1) and (2), and the identity $(z_1 + z_2)^* = z_1^* + z_2^*$, which is easily proved by writing z_1 and z_2 as $a_1 + ib_1$ and $a_2 + ib_2$, respectively, where the a's and b's are real.

17.63 With $\Psi = f + cg$, Eq. (17.91) becomes $\int (f + cg)^* \hat{M}(f + cg)\, d\tau =$
$\int (f + cg)[\hat{M}(f + cg)]^* \, d\tau$. Using the identity $(z_1 + z_2)^* = z_1^* + z_2^*$ (which is easily proved by writing z_1 and z_2 as $a_1 + ib_1$ and $a_2 + ib_2$, respectively, where the a's and b's are real) and using the result of Prob. 17.19 and the fact that $\hat{M}$

is a linear operator, we get

$\int f^* \hat{M}f\, d\tau + c^* \int g^* \hat{M}f\, d\tau + c\int f^* \hat{M}g\, d\tau + c^* c\int g^* \hat{M}g\, d\tau =$

$$\int f(\hat{M}f)^*\, d\tau + c\int g(\hat{M}f)^*\, d\tau + c^* \int f(\hat{M}g)^*\, d\tau + cc^* \int g(\hat{M}g)^*\, d\tau$$

Equation (17.91) with Ψ replaced by either f or by g shows that the first integral on the left side of this equation equals the first integral on the right side, and that the last integral on the left side equals the last integral on the right side. Therefore we are left with

$c^* \int g^* \hat{M}f\, d\tau + c\int f^* \hat{M}g\, d\tau = c\int g(\hat{M}f)^*\, d\tau + c^* \int f(\hat{M}g)^*\, d\tau$

Putting $c = 1$ we get

$\int g^* \hat{M}f\, d\tau + \int f^* \hat{M}g\, d\tau = \int g(\hat{M}f)^*\, d\tau + \int f(\hat{M}g)^*\, d\tau$

Putting $c = i$ and then dividing by i, we get (since $i^* = -i$)

$-\int g^* \hat{M}f\, d\tau + \int f^* \hat{M}g\, d\tau = \int g(\hat{M}f)^*\, d\tau - \int f(\hat{M}g)^*\, d\tau$

Adding the last two equations and dividing by 2, we get $\int f^* \hat{M}g\, d\tau =$ $\int g(\hat{M}f)^*\, d\tau$.

17.64 **(a)** $\int f^* \hat{x}g\, d\tau = \int_{-\infty}^{\infty} f^* xg\, dx = \int_{-\infty}^{\infty} gx^* f^*\, dx = \int_{-\infty}^{\infty} g(xf)^*\, dx$, where $(ab)^* = a^*b^*$ [the equation after (17.19)] and the fact that x is real ($x = x^*$) were used.

(b) $\int f^* \hat{p}_x g\, d\tau = \int_{-\infty}^{\infty} f^*(\hbar/i)(\partial g/\partial x)\, dx$. Let $u = f^*$ and $d\upsilon = (\partial g/\partial x)\, dx$. Then $\upsilon = g$ and the integration-by-parts formula $\int u\, d\upsilon = u\upsilon - \int \upsilon\, du$ gives $\int f^* \hat{p}_x g\, d\tau = (\hbar/i)f^* g\,|_{-\infty}^{\infty} - (\hbar/i)\int_{-\infty}^{\infty} g(\partial f^*/\partial x)\, dx$. Equation (17.92) requires that f and g be well-behaved, which includes the requirement of quadratic integrability. In order to be quadratically integrable, the functions f and g must go to zero as x goes to $\pm\infty$. So $\int f^* \hat{p}_x g\, d\tau =$ $\int_{-\infty}^{\infty} g[(\hbar/i)\partial f/\partial x]^*\, dx = \int_{-\infty}^{\infty} g(\hat{p}_x f)^*\, dx$, which completes the proof.

17.65 We must prove that $\hat{M}(c_1 f_1 + c_2 f_2) = b(c_1 f_1 + c_2 f_2)$. Using the linearity equations given after Example 17.5, we have $\hat{M}(c_1 f_1 + c_2 f_2) = \hat{M}(c_1 f_1) +$ $\hat{M}(c_2 f_2) = c_1\hat{M}f_1 + c_2\hat{M}f_2 = c_1 bf_1 + c_2 bf_2 = b(c_1 f_1 + c_2 f_2)$, where the given eigenvalue equations for $\hat{M}$ were used.

17.66 $\int g_1^* g_2\,d\tau = \int f_1^*(f_2 + kf_1)\,d\tau = \int f_1^* f_2\,d\tau + k\int f_1^* f_1\,d\tau =$
$\int f_1^* f_2\,d\tau - [\int f_1^* f_2\,d\tau / \int f_1^* f_1\,d\tau]\int f_1^* f_1\,d\tau = 0.$

17.67 $F \equiv x^2(1-x)$, $G \equiv \sum_{n=1}^{m} c_n\psi_n$, where $\psi_n = 2^{1/2}\sin(n\pi x)$. From Eq. (17.98), $c_n = \int \psi_n^* F\,d\tau = 2^{1/2}\int_0^1 (x^2 - x^3)\sin(n\pi x)\,dx$. A table of integrals (or use of the website integrals.wolfram.com or a calculator that can do symbolic integration) gives $\int x^2 \sin kx\,dx = k^{-3}(2 - k^2x^2)\cos kx + 2k^{-2}x\sin kx$ and $\int x^3 \sin kx\,dx = k^{-3}(6x - k^2x^3)\cos kx + k^{-4}(3k^2x^2 - 6)\sin kx$. We get $c_n = -2^{1/2}(n\pi)^{-3}[4(-1)^n + 2]$, where $\sin n\pi = 0$ and $\cos n\pi = (-1)^n$ were used. So $G = \sum_{n=1}^{m}(-2)(n\pi)^{-3}[4(-1)^n + 2]\sin n\pi x$. We set up a spreadsheet with x values going from 0 to 1 in steps of 0.02 in column A, the values of F at these points in column B, and the values of the first, second,…, fifth terms in the series G in columns C, D, E, F, and G. In column H we sum the first three terms of the series and in column I we sum the first 5 terms of the series. The data in columns B, H, and I are graphed versus x on the same plot. The five-term sum gives a more accurate representation of F than the three-term function. For example, some values are

x	0.1	0.2	0.4	0.6	0.8	0.9
F	0.0090	0.0320	0.0960	0.144	0.128	0.081
3 terms	0.0153	0.0344	0.0914	0.148	0.126	0.072
5 terms	0.0106	0.0308	0.0972	0.143	0.130	0.079

17.68 **(a)** The series is a sum of terms and the complex conjugate of a sum is the sum of the complex conjugates. This if $z_1 = x_1 + iy_1$ and $z_2 = x_2 + iy_2$, where x_1, y_1, x_2, and y_2 are real, then $(z_1 + z_2)^* = (x_1 + iy_1 + x_2 + iy_2)^* = [x_1 + x_2 + i(y_1 + y_2)]^* = x_1 + x_2 - i(y_1 + y_2) = x_1 - iy_1 + x_2 - iy_2 = z_1^* + z_2^*$. So $\phi^* = \sum_k (c_k\psi_k)^*$. Then the use of the result of Prob. 17.19 gives $\phi^* = \sum_k c_k^*\psi_k^* = \sum_j c_j^*\psi_j^*$, where the last equality holds because the summation index is a dummy variable (Sec. 1.8).

(b) $1 = \int \phi^*\phi\,d\tau = \int \sum_j c_j^*\psi_j^* \sum_k c_k\psi_k\,d\tau = \sum_j\sum_k \int c_j^*\psi_j^* c_k\psi_k\,d\tau =$
$\sum_j\sum_k c_j^* c_k \int \psi_j^*\psi_k\,d\tau = \sum_j\sum_k c_j^* c_k\delta_{jk}$, since the integral of a sum is

the sum of the integrals, and we used the orthonormality of the wave functions. When the sum over k is performed, the δ_{jk} factor makes all terms equal to zero except for the one term where $k = j$. Thus the sum over k equals $c_j{}^*c_j = |c_j|^2$ and $1 = \sum_j |c_j|^2 = \sum_k |c_k|^2$.

(c) $W = \int \phi^* \hat{H} \phi \, d\tau = \int \sum_j c_j^* \psi_j^* \hat{H} \sum_k c_k \psi_k \, d\tau = \int \sum_j c_j^* \psi_j^* \sum_k c_k \hat{H} \psi_k \, d\tau$

where we used the linearity of the operator $\hat{H}$. Now $\hat{H}\psi_k = E_k\psi_k$, so $W = \sum_j \sum_k c_j^* c_k E_k \int \psi_j^* \psi_k \, d\tau = \sum_j \sum_k c_j^* c_k E_k \delta_{jk} = \sum_j |c_j|^2 E_j = \sum_k |c_k|^2 E_k$.

(d) Since $|c_k|^2 E_k \geq |c_k|^2 E_{gs}$, each term in the sum $W = \sum_k |c_k|^2 E_k$ is equal to or greater than the corresponding term in the sum $\sum_k |c_k|^2 E_{gs}$, the first sum must be greater than or equal to the second sum, and so $W \geq \sum_k |c_k|^2 E_{gs}$. But part (b) gives $\sum_k |c_k|^2 = 1$, so $W \geq E_{gs}$.

17.69 **(a)** Since $|\psi|^2\,dx$ is a probability and probabilities are dimensionless, $|\psi|^2$ has units of length^{-1} and ψ has units of length$^{-1/2}$. The SI units of ψ are $\text{m}^{-1/2}$ for a one-particle one-dimensional system.

(b) $|\psi|^2\,dx\,dy\,dz$ is dimensionless and ψ has units of length$^{-3/2}$.

(c) $|\psi|^2\,dx_1\,dy_1\,dz_1\,dx_2\,dy_2\,dz_2$ is dimensionless and ψ has units of length^{-3}.

17.70 The blackbody function (17.2) depends on the combinations of constants h/c^2 and h/k. In 1900, c was known reasonably accurately, so by fitting the observed blackbody curves Planck obtained values for both h and k. Use of $R = N_A k$ then gave N_A. Use of $F = N_A e$ then gave e.

17.71 **(a)** $E = n^2h^2/8ma^2$ and doubling a multiplies E by ¼.

(b) $E = J(J+1)\hbar^2/2I = J(J+1)\hbar^2/2\mu d^2$ and doubling d multiplies E by ¼.

(c) $E = h\nu = h(1/2\pi)(k/m)^{1/2}$ and doubling m multiplies E by $1/\sqrt{2}$.

17.72 **(a)** T. **(b)** T. The future state is predicted by integrating the time-dependent Schrödinger equation. **(c)** T. **(d)** T. **(e)** F. "Sum" must be replaced by "product" to make the statement true. **(f)** T. **(g)** F. **(h)** F. **(i)** F.

Chapter 18

18.1 **(a)** F; **(b)** T; **(c)** T.

18.2 The equation shows that $4\pi\varepsilon_0$ has the same units as Q_1Q_2/rV, which are $C^2/(m\ J) = C^2\ N^{-1}\ m^{-2}$, since 1 J = 1 N m.

18.3 $c^2 = 1/\mu_0\varepsilon_0$, so $1/4\pi\varepsilon_0 = \mu_0 c^2/4\pi = (4\pi \times 10^{-7}\ N\ s^2/C^2)c^2/4\pi = 10^{-7}c^2\ N\ s^2\ C^{-2}$.

18.4 $V = Q_1Q_2/4\pi\varepsilon_0 r$.

(a) $V = (1.602 \times 10^{-19}\ C)^2/4\pi(8.854 \times 10^{-12}\ C^2/N\text{-}m^2)(3.0 \times 10^{-10}\ m) = 7.7 \times 10^{-19}\ J = (7.7 \times 10^{-19}\ J)\ (1\ eV/1.602 \times 10^{-19}\ J) = 4.8\ eV$.

(b) Let the electrons be numbered 1 and 2, let the proton be p and let e denote the proton charge. Then $V = V_{12} + V_{1p} + V_{2p} = (1/4\pi\varepsilon_0) \times [(-e)^2/(3.0 \times 10^{-10}\ m) - e^2/(4.0 \times 10^{-10}\ m) - e^2/(5.0 \times 10^{-10}\ m)] = [(1.602 \times 10^{-19}\ C)^2/4\pi(8.854 \times 10^{-12}\ C^2/N\text{-}m^2)](-1.167 \times 10^9\ m^{-1}) = -2.7 \times 10^{-19}\ J = (-2.7 \times 10^{-19}\ J)(1\ eV/1.602 \times 10^{-19}\ J) = -1.7\ eV$.

18.5 $V = 4\pi r^3/3$, so $V_{nuc}/V_{atom} = r^3_{nuc}/r^3_{atom} = (10^{-12}\ cm)^3/(10^{-8}\ cm)^3 = 1 \times 10^{-12}$.

18.6 **(a)** T; **(b)** T; **(c)** F; $\psi_{1s} \neq 0$ at $r = 0$. **(d)** T; **(e)** F; **(f)** F; **(g)** T; **(h)** F; **(i)** F.

18.7 **(a)** θ; **(b)** r; **(c)** ϕ.

18.8 **(a)** T; **(b)** T; **(c)** F.

18.9 **(a)** 0, 1, 2, 3, 4.

(b) −5, −4, −3, −2, −1, 0, 1, 2, 3, 4, 5.

18.10 **(a)** The only $n = 1$ state is $1s_0$, so the degeneracy is 1 (i.e., nondegenerate), if spin is not considered.

(b) The states $2s$, $2p_1$, $2p_0$, $2p_{-1}$ have the same energy, so the degeneracy is 4.

(c) The states $3s$, $3p_1$, $3p_0$, $3p_{-1}$, $3d_2$, $3d_1$, $3d_0$, $3d_{-1}$, $3d_{-2}$ have the same energy and the degeneracy is 9. (The general formula is n^2.)

18.11 These are hydrogenlike species, so $E = -(Z^2/n^2)(13.60 \text{ eV})$. The ionization energy IE is $-E$ for the ground state, $n = 1$.

(a) IE = $2^2(13.60 \text{ eV}) = 54.4$ eV and the ionization potential IP is 54.4 V.

(b) IE = $3^2(13.60 \text{ eV}) = 122.4$ eV and IP = 122.4 V.

18.12 $|\Delta E| = -(13.60 \text{ eV})(1.602 \times 10^{-19} \text{ J/1 eV})(1/3^2 - 1/2^2) = 3.026 \times 10^{-19} \text{ J} = h\nu$ and $\nu = (3.026 \times 10^{-19} \text{ J})/(6.626 \times 10^{-34} \text{ J s}) = 4.567 \times 10^{14}\text{/s}$.
$\lambda = c/\nu = (2.9979 \times 10^8 \text{ m/s})/(4.567 \times 10^{14}\text{/s}) = 6.564 \times 10^{-7} \text{ m} = 656.4 \text{ nm}$.

18.13 For a hydrogen atom, $\mu = m_e m_p/(m_e + m_p) =$

$$\frac{(9.1095 \times 10^{-31} \text{ kg})(1.67265 \times 10^{-27} \text{ kg})}{(9.1095 \times 10^{-31} \text{ kg} + 1.67265 \times 10^{-27} \text{ kg})} = 9.1045 \times 10^{-31} \text{ kg}$$

We have $a = 4\pi\varepsilon_0 \hbar^2/\mu e^2 = 4\pi(8.854 \times 10^{-12} \text{ C}^2/\text{N-m}^2)(6.6262 \times 10^{-34} \text{ J} \cdot \text{s})^2/$ $4\pi^2(9.1045 \times 10^{-31} \text{ kg})(1.6022 \times 10^{-19} \text{ C})^2 = 5.295 \times 10^{-11} \text{ m} = 0.5295$ Å.

18.14 This is a hydrogenlike species, so its energy levels are given by Eq. (18.14) as $E = -(Z^2/n^2)[e^2/(4\pi\varepsilon_0)2a] = -(Z^2/n^2)[\mu e^4/2(4\pi\varepsilon_0)^2\hbar^2]$. Let m_e be the electron mass; the positron has mass m_e. So $\mu_{\text{positronium}} = m_e^2/(m_e + m_e) = m_e/2$, as compared with $\mu \approx m_e$ for an H atom. Since E is proportional to μ, each positronium energy level is half the corresponding H-atom energy. The positronium ionization potential is thus $\frac{1}{2}(13.6 \text{ V}) = 6.8$ V.

18.15 Using Table 18.1 and Eq. (18.27), $\langle r \rangle = \int \psi^* r \psi \, d\tau = \pi^{-1}(Z/a)^3 \times$ $\int_0^{2\pi} \int_0^{\pi} \int_0^{\infty} e^{-Zr/a} r e^{-Zr/a} r^2 \sin\theta \, dr \, d\theta \, d\phi = (Z^3/\pi a^3) \int_0^{\infty} r^3 e^{-2Zr/a} \, dr \int_0^{\pi} \sin\theta \, d\theta \int_0^{2\pi} d\phi$. A table of integrals gives $\int z^3 e^{bz} \, dz = e^{bz}(z^3/b - 3z^2/b^2 + 6z/b^3 - 6/b^4)$, so $\int_0^{\infty} r^3 e^{-2Zr/a} \, dr = 6/(-2Z/a)^4 = 3a^4/8Z^4$, since $e^{-2Zr/a}$ vanishes at $r = \infty$. (This result also follows from the definite integral $\int_0^{\infty} z^n e^{-bz} \, dz = n!/b^{n+1}$ found in most tables.) Then $\langle r \rangle = (Z^3/\pi a^3)(3a^4/8Z^4)(2)(2\pi) = 3a/2Z$.

18.16 $e^{i\phi} = 1 + i\phi + (i\phi)^2/2! + (i\phi)^3/3! + (i\phi)^4/4! + (i\phi)^5/5! + \cdots =$
$1 + i\phi - \phi^2/2! - i\phi^3/3! + \phi^4/4! + i\phi^5/5! + \cdots$.
Also, $\cos\phi + i\sin\phi = (1 - \phi^2/2! + \phi^4/4! - \cdots) + i(\phi - \phi^3/3! + \phi^5/5! - \cdots) =$
$1 + i\phi - \phi^2/2! - i\phi^3/3! + \phi^4/4! + i\phi^5/5! + \cdots = e^{i\phi}$.

18.17 $2p_x = 2^{-1/2}(2p_1 + 2p_{-1}) = 2^{-1/2}(1/8\pi^{1/2})(Z/a)^{5/2}e^{-Zr/2a}r\sin\theta\,(e^{i\phi} + e^{-i\phi})$. We have $e^{i\phi} + e^{-i\phi} = \cos\phi + i\sin\phi + \cos(-\phi) + i\sin(-\phi) = \cos\phi + i\sin\phi + \cos\phi - i\sin\phi = 2\cos\phi$, so $2p_x = (2^{1/2}/2^3\pi^{1/2})(Z/a)^{5/2}e^{-Zr/2a}r\sin\theta\cos\phi = \pi^{-1/2}(Z/2a)^{5/2}xe^{-Zr/2a}$, where Eq. (18.7) was used. Also, $2p_y = (2p_1 - 2p_{-1})/i\sqrt{2} = (2^{-1/2}/i)(1/8\pi^{1/2})(Z/a)^{5/2}e^{-Zr/2a}\,r\sin\theta\,(e^{i\phi} - e^{-i\phi})$. We have $e^{i\phi} - e^{-i\phi} = \cos\phi + i\sin\phi - (\cos\phi - i\sin\phi) = 2i\sin\phi$, so $2p_y = (2^{1/2}/2^3\pi^{1/2})(Z/a)^{5/2}e^{-Zr/2a}\,r\sin\theta\sin\phi = \pi^{-1/2}(Z/2a)^{5/2}ye^{-Zr/2a}$.

18.18 **(a)** $a_1 + ib_1 = a_2 + ib_2$. We must have $a_1 = a_2$ and $b_1 = b_2$.

(b) $(2\pi)^{-1/2}e^{im\phi} = (2\pi)^{-1/2}e^{im(\phi+2\pi)} = (2\pi)^{-1/2}e^{im\phi}e^{2\pi mi}$, so $1 = e^{2\pi mi} = \cos(2\pi m) + i\sin(2\pi m)$, where (18.24) was used. Equating the real parts and the imaginary parts of this last equation [as shown in part (a)], we have $\cos(2\pi m) = 1$ and $\sin(2\pi m) = 0$. The cosine function equals 1 only for angles of 0, $\pm2\pi$, $\pm4\pi$, $\pm6\pi, \ldots$ and the sine vanishes at these angles. Therefore $2\pi m = 0, \pm2\pi, \pm4\pi, \ldots$ and $m = 0, \pm1, \pm2, \ldots$.

18.19 ψ_{2p_z} has the form bze^{-dr}; b and d are constants. Along the z axis, $x = 0 = y$ and $r = (x^2 + y^2 + z^2)^{1/2} = (z^2)^{1/2} = |z|$. Thus $\psi_{2p_z} = bze^{-d|z|}$ along the z axis. Near $z = 0$, $e^{-d|z|} \approx 1$ and $\psi_{2p_z} \approx bz$ (a straight line through the origin). For large values of $|z|$, the exponential causes ψ to fall to zero. Also, $|\psi_{2p_z}|^2 = |b|^2z^2e^{-2d|z|}$ along the z axis. $|\psi|^2$ is parabolic near the origin and is positive for negative values of z. The graphs are:

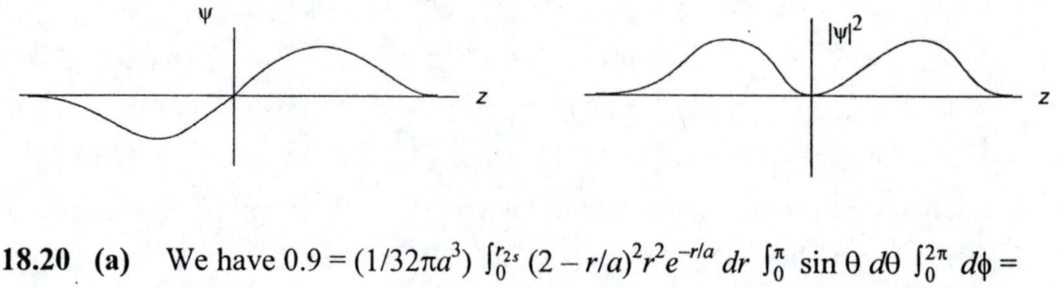

18.20 **(a)** We have $0.9 = (1/32\pi a^3)\int_0^{r_{2s}}(2 - r/a)^2r^2e^{-r/a}\,dr\int_0^{\pi}\sin\theta\,d\theta\int_0^{2\pi}d\phi =$
$e^{-r_{2s}/a}[-r_{2s}^2/2a^2 - r_{2s}/a - 1 - r_{2s}^4/8a^4] + 1$, where a table of integrals was

used. Let $\upsilon \equiv r_{2s}/a$. We must solve $e^{-\upsilon}(\upsilon^2/2 + \upsilon + 1 + \upsilon^4/8) = 0.1$. Trial and error (or a spreadsheet Solver) gives $\upsilon = 9.125$ and $r_{2s} = 9.125a =$ 4.83 Å.

(b) $0.95 = e^{-r_{2s}/a}[-r_{2s}^2/2a^2 - r_{2s}/a - 1 - r_{2s}^4/8a^4] + 1$ and $e^{-\upsilon}(\upsilon^2/2 + \upsilon + 1 + \upsilon^4/8) = 0.05$. We find $\upsilon = 10.283$ and $r_{2s} = 10.283a =$ 5.44 Å.

18.21 $\partial e^{-Zr/a}/\partial x = (\partial e^{-Zr/a}/\partial r)(\partial r/\partial x)$. We have $r = (x^2 + y^2 + z^2)^{1/2}$ and $\partial r/\partial x = \frac{1}{2}(x^2 + y^2 + z^2)^{-1/2}(2x) = x/r$. So $\partial e^{-Zr/a}/\partial x = -(Zx/ra)e^{-Zr/a}$. $\partial^2 e^{-Zr/a}/\partial x^2 = (\partial/\partial x)(-Zxe^{-Zr/a}/ra) = -Ze^{-Zr/a}/ra - x[(\partial/\partial r)(Ze^{-Zr/a}/ra)](\partial r/\partial x) = -Ze^{-Zr/a}/ra + (x^2Z^2/r^2a^2)e^{-Zr/a} + (Zx^2/ar^3)e^{-Zr/a}$. Similar equations hold for $\partial^2/\partial y^2$ and $\partial^2/\partial z^2$ of $e^{-Zr/a}$. Then $(\partial^2/\partial x^2 + \partial^2/\partial y^2 + \partial^2/\partial z^2)e^{-Zr/a} = -3Ze^{-Zr/a}/ra + [(x^2 + y^2 + z^2)Z^2/r^2a^2]e^{-Zr/a} + [Z(x^2 + y^2 + z^2)/ar^3]e^{-Zr/a} = -3Ze^{-Zr/a}/ra + (Z^2/a^2)e^{-Zr/a} + (Z/ra)e^{-Zr/a} = -2Ze^{-Zr/a}/ra + (Z^2/a^2)e^{-Zr/a} = -2\mu e^2Ze^{-Zr/a}/4\pi\varepsilon_0 r\hbar^2 + (Z^2/a^2)e^{-Zr/a}$, where (18.14) was used. The Hamiltonian operator is (18.5) and $\hat{H}\psi_{1s} = -(\hbar^2/2\mu)(\partial^2/\partial x^2 + \partial^2/\partial y^2 + \partial^2/\partial z^2)[\pi^{-1/2}(Z/a)^{3/2}e^{-Zr/a}] - (Ze^2/4\pi\varepsilon_0 r)[\pi^{-1/2}(Z/a)^{3/2}e^{-Zr/a}] = -(Ze^2/4\pi\varepsilon_0 r)\pi^{-1/2}(Z/a)^{3/2}e^{-Zr/a} - \pi^{-1/2}(Z/a)^{3/2}(\hbar^2/2\mu)[-2\mu e^2Ze^{-Zr/a}/4\pi\varepsilon_0 r\hbar^2 + (Z^2/a^2)e^{-Zr/a}] = -(\hbar^2Z^2/2\mu a^2)\pi^{-1/2}(Z/a)^{3/2}e^{-Zr/a} = -[Z^2e^2/2(4\pi\varepsilon_0)a][\pi^{-1/2}(Z/a)^{3/2}e^{-Zr/a}] = E_{1s}\psi_{1s}$.

18.22 $\langle V\rangle = \langle -Ze^2/4\pi\varepsilon_0 r\rangle = -(Ze^2/4\pi\varepsilon_0)\int \psi_{1s}^* r^{-1}\psi_{1s}\, d\tau = -(Z^4e^2/4\pi\varepsilon_0 a^3\pi)\int_0^{2\pi}\int_0^{\pi}\int_0^{\infty} e^{-Zr/a}r^{-1}e^{-Zr/a}\, r^2 \sin\theta\, dr\, d\theta\, d\phi = -(Z^4e^2/4\pi\varepsilon_0 a^3\pi)\int_0^{\infty} re^{-2Zr/a}\, dr \int_0^{\pi}\sin\theta\, d\theta \int_0^{2\pi} d\phi$. Using either the definite integral $\int_0^{\infty} r^n e^{-br}\, dr = n!/b^{n+1}$ or the indefinite integral $\int re^{-br}\, dr = -e^{-br}(r/b + 1/b^2)$, we get $\int_0^{\infty} re^{-2Zr/a}\, dr = a^2/4Z^2$. Then $\langle V\rangle = -(Z^4e^2/4\pi\varepsilon_0 a^3\pi)(a^2/4Z^2)2(2\pi) = -Z^2e^2/4\pi\varepsilon_0 a$ for the ground state.

18.23 **(a)** We have: $\langle r\rangle = \int \psi_{2p_z}^* r\psi_{2p_z}\, d\tau = (Z^5/32\pi a^5)\int_0^{2\pi}\int_0^{\pi}\int_0^{\infty} re^{-Zr/2a}\cos\theta\, r\, re^{-Zr/2a}\cos\theta\, r^2\sin\theta\, dr\, d\theta\, d\phi = (Z^5/32\pi a^5)\int_0^{\infty} r^5e^{-Zr/a}\, dr \int_0^{2\pi} d\phi \int_0^{\pi}\cos^2\theta\sin\theta\, d\theta$. A table of definite integrals gives $\int_0^{\infty} r^n e^{-br}\, dr = n!/b^{n+1}$ for $b > 0$ and n a positive integer. So $\int_0^{\infty} r^5e^{-Zr/a}\, dr = 5!a^6/Z^6$. Let $t = \cos\theta$; then $dt = -\sin\theta\, d\theta$ and

$\int_0^\pi \cos^2\theta \sin\theta \, d\theta = -\int_1^{-1} t^2 \, dt = 2/3$. So
$\langle r \rangle = (Z^5/32\pi a^5)(120a^6/Z^6)(2/3)(2\pi) = 5a/Z$.

(b) The $2p_z$ and $2p_x$ orbitals have the same shape and the same size and differ only in spatial orientation. Since r does not depend on spatial orientation, $\langle r \rangle$ must be the same for the $2p_x$ and $2p_z$ states.

(c) $\langle r \rangle_{2p_x} = \int \psi_{2p_x}^* r \psi_{2p_x} \, d\tau = (Z^5/32\pi a^5) \int_0^\infty r^5 e^{-Zr/a} \, dr \int_0^{2\pi} \cos^2\phi \, d\phi \times$
$\int_0^\pi \sin^3\theta \, d\theta$. The r integral was found in (a). A table of integrals gives $\int \cos^2\phi \, d\phi = \frac{1}{2}\phi + \frac{1}{4} \sin 2\phi$ and $\int \sin^3\theta \, d\theta = -\frac{2}{3} \cos\theta - \frac{1}{3} \cos\theta \sin^2\theta$ and we find $\langle r \rangle_{2p_x} = (Z^5/32\pi a^5)(120a^6/Z^6)(\pi)(4/3) = 5a/Z$.

18.24 Equations (18.28) and (18.19) give the ground-state H-atom radial distribution function as $R_{1s}^2 r^2 = 4(Z/a)^3 r^2 e^{-2Zr/a}$. The maximum is found by setting the derivative equal to zero: $0 = 4(Z/a)^3[2re^{-2Zr/a} - (2Zr^2/a)e^{-2Zr/a}]$ and $r = a/Z$. (The root $r = 0$ is a minimum.)

18.25 Let $c = 2.00$ Å. To find the desired probability, we integrate $\psi^*\psi \, d\tau$ over the volume of a sphere of radius c. The angles go over their full ranges and r goes from 0 to c. Table 18.1 and Eq. (18.27) give the probability as $(1/\pi a^3) \times$ $\int_0^c \int_0^\pi \int_0^{2\pi} e^{-2r/a} r^2 \sin\theta \, dr \, d\theta \, d\phi = (1/\pi a^3) \int_0^c r^2 e^{-2r/a} \, dr \int_0^\pi \sin\theta \, d\theta \int_0^{2\pi} d\phi$. The radial integral has the same form as the radial integral in Prob. 18.20 except that r_{2s} is replaced by c. So $\int_0^c r^2 e^{-2r/a} \, dr = -e^{-2c/a}(\frac{1}{2}ac^2 + \frac{1}{2}a^2c + \frac{1}{4}a^3) + \frac{1}{4}a^3$. The θ and ϕ integrals are given in Prob. 18.20, and the desired probability is $-e^{-2c/a}(2c^2/a^2 + 2c/a + 1) + 1$. We have $c/a = (2.00 \text{ Å})/(0.5295 \text{ Å}) = 3.77_7$, and the probability is $1 - e^{-2(3.777)}[2(3.777)^2 + 2(3.777) + 1] = 0.981$.

18.26 $\int_0^{2\pi} |\Phi|^2 \, d\phi = \int_0^{2\pi} \Phi^*\Phi \, d\phi = (1/2\pi) \int_0^{2\pi} e^{-im\phi} e^{im\phi} \, d\phi = (1/2\pi) \int_0^{2\pi} d\phi = 1$.

18.27 **(a)** T; **(b)** T; **(c)** T.

18.28 Let θ be the angle between the positive z axis and an angular-momentum vector. For $m = +1$ in Fig. 18.10, $\cos\theta = L_z/|\mathbf{L}| = \hbar/\sqrt{2}\hbar = 1/\sqrt{2} = 0.7071$ and $\theta = 45°$. For $m = 0$, $\theta = 90°$. For $m = -1$, $\theta = 180° - 45° = 135°$.

18.29 **(a)** $L = rp \sin \beta$, where β is the angle between **r** and **p**. For circular motion, the velocity vector **v** is perpendicular to the radius, and so is **p** $= m$**v**; thus $\beta = 90°$ and $\sin \beta = 1$. Since $p = m\upsilon$, we get $L = m\upsilon r$.

(b) The **L** vector is perpendicular to both **r** and **p** and **r** and **p** lie in the plane of the circular motion. Hence **L** is perpendicular to the plane of the circular orbit.

18.30 **(a)** From Sec. 18.4, $|\mathbf{L}| = [l(l+1)]^{1/2}\hbar = 0$, since $l = 0$ for the 1s state.

(b) From Sec. 17.3, Bohr had $|\mathbf{L}| = m\upsilon r = nh/2\pi = h/2\pi$ for the ground state. The Bohr theory had the wrong value of $|\mathbf{L}|$.

18.31 $|\mathbf{L}| = [l(l+1)]^{1/2}\hbar = [1(2)]^{1/2}\hbar = \sqrt{2}\hbar = \sqrt{2}(6.626 \times 10^{-34}\text{ J s})/2\pi = 1.491 \times 10^{-34}$ J s.

18.32 $|\mathbf{S}| = [s(s+1)]^{1/2}\hbar = [0.5(1.5)]^{1/2}(6.626 \times 10^{-34}\text{ J s})/2\pi = 9.13 \times 10^{-35}$ J s.

18.33 Let θ be the angle between the z axis and a spin vector. For $m_s = +\frac{1}{2}$, we have $\cos \theta = \frac{1}{2}\hbar/\sqrt{3/4}\hbar = 1/\sqrt{3} = 0.57735$ and $\theta = 54.7°$. For $m_s = -\frac{1}{2}$, $\theta = 180° - 54.7° = 125.3°$.

18.34 **(a)** Electronic orbital angular momentum; $|\mathbf{L}| = [l(l+1)]^{1/2}\hbar$.

(b) z component of electronic orbital angular momentum; $L_z = m\hbar$.

(c) Electronic spin angular momentum; $|\mathbf{S}| = [s(s+1)]^{1/2}\hbar$.

(d) z component of electronic spin angular momentum; $S_z = m_s\hbar$.

18.35 **(a)** For $s = 3/2$, Eq. (18.32) gives $m_s = 3/2, 1/2, -1/2$, and $-3/2$. The possible z components of the spin are $m_s\hbar$. The length of the spin vector is $\sqrt{s(s+1)}\hbar = \frac{1}{2}\sqrt{15}\hbar$. The possible orientations are

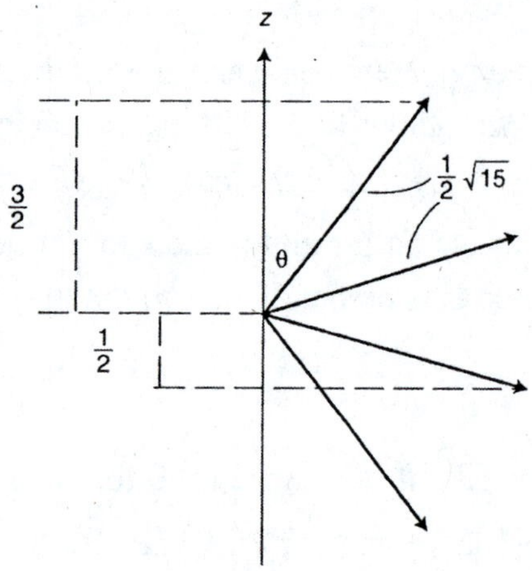

(b) $\cos\theta = 1.5\hbar/\frac{1}{2}\sqrt{15}\hbar =$ 0.7746 and $\theta = 39.2°$.

18.36 **(a)** F; **(b)** F; **(c)** F; **(d)** T.

18.37 **(a)** Neither, since $f(2)g(1) \neq \pm f(1)g(2)$.

(b) Symmetric, since $g(2)g(1) = g(1)g(2)$.

(c) Antisymmetric, since $f(2)g(1) - g(2)f(1) = -[f(1)g(2) - g(1)f(2)]$.

(d) Symmetric. **(e)** Antisymmetric.

18.38 **(a)** Antisymmetric. **(b)** Symmetric. **(c)** Antisymmetric, since interchange of 1 and 2 multiplies h by –1 and multiplies k by – 1. **(d)** Neither.

18.39 The true ground-state energy of He is –79.0 eV (Sec. 18.6). The variational value –86.7 eV is less than the true E_{gs}; this violates the variation theorem (17.86), so there must be an error in the calculation.

18.40 There is one electron, so $S = s = \frac{1}{2}$ and $2s + 1 = 2$; also, $L = l$.

(a) ^2S. **(b)** ^2P. **(c)** ^2D.

18.41 $2S + 1 = 4$, so $S = 3/2$. The code letter F means $L = 3$.

18.42 **(a)** Total electronic orbital angular momentum; $|\mathbf{L}| = [L(L+1)]^{1/2}\hbar$.

(b) Total electronic spin angular momentum; $|\mathbf{S}| = [S(S+1)]^{1/2}\hbar$.

(c) z component of total electronic spin angular momentum; $S_z = M_S\hbar$.

18.43 **(a)** *D* means $L = 2$, so $|\mathbf{L}| = [L(L+1)]^{1/2}\hbar = 6^{1/2}\hbar$.

(b) $2S + 1 = 3$ and $S = 1$, so $|\mathbf{S}| = [S(S+1)]^{1/2}\hbar = 2^{1/2}\hbar$.

18.44 **(a)** Electrons in filled subshells can be ignored. The 3*d* electron has $s = ½$ and $l = 2$. With only one electron outside filled subshells, there is only one term, namely ^2D, with $L = 2$ and $S = ½$.

(b) ^2P.

18.45 Let θ be the angle between the *z* axis and a spin vector. For spin function α(1) and vector $\mathbf{S}_1$, we have $\cos\theta = \frac{1}{2}\hbar/(3/4)^{1/2}\hbar = 1/3^{1/2} = 0.57735$ and $\theta = 54.7°$ (as in Prob. 18.35). Likewise, $\theta = 54.7°$ for $\mathbf{S}_2$. For α(1)α(2), the total spin vector **S** has magnitude $2^{1/2}\hbar$ and *z* component $\hbar$, so $\cos\theta = \hbar/2^{1/2}\hbar = 1/2^{1/2} = 0.7071$ and $\theta = 45°$ for **S**. **S** lies closer to the *z* axis than do $\mathbf{S}_1$ and $\mathbf{S}_2$. $\mathbf{S}_1$ and $\mathbf{S}_2$ lie on the surface of a cone making angle 54.7° with the *z* axis and **S** lies within this cone:

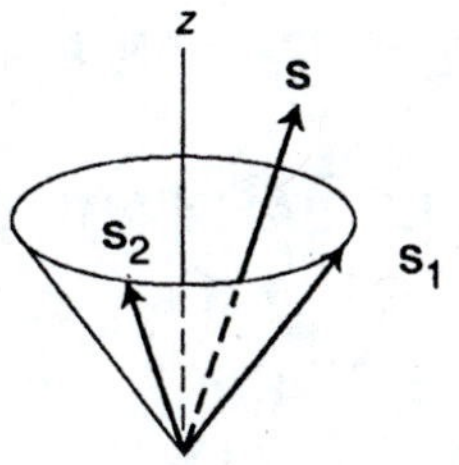

18.46 **(a)** $l = 1$ for a 3*p* electron and $l = 2$ for a 3*d* electron. The maximum and minimum *L* values are 2 + 1 and $|2 - 1|$ and the possible *L* values are 3, 2, and 1. A 3*p* electron has $s = \frac{1}{2}$ and so does a 3*d* electron, so the maximum and minimum *S* values are $\frac{1}{2} + \frac{1}{2}$ and $|\frac{1}{2} - \frac{1}{2}|$; the possible *S* values are 1 and 0. (Electrons in filled subshells were ignored.)

(b) Combining each *S* value with each *L* value, we have as the terms: ^3F, ^3D, ^3P, ^1F, ^1D, and ^1P, where $2S + 1 = 3$ and 1 for $S = 1$ and 0, and *P*, *D*, *F* denote $L = 1, 2, 3$.

18.47 **(a)** $L = 0$ and $S = \frac{1}{2}$, so $J = \frac{1}{2}$ and the only level is $^2S_{1/2}$.

(b) $L = 1$ and $S = 3/2$. So $J = 5/2, 3/2, 1/2$, and the levels are $^4P_{5/2}$, $^4P_{3/2}$, $^4P_{1/2}$.

(c) $L = 3$ and $S = 2$, so the levels are $^5F_5$, $^5F_4$, $^5F_3$, $^5F_2$, $^5F_1$.

(d) $L = 2$ and $S = 1$; the levels are ${}^3D_3$, ${}^3D_2$, ${}^3D_1$.

18.48 Let the electrons be numbered 1, 2, and 3. The nuclear charge is $3e$. As was done with He in Eq. (18.35), we use the electron mass m_e in the Hamiltonian. Then $\hat{H} = -(\hbar^2/2m_e)\nabla_1^2 - (\hbar^2/2m_e)\nabla_2^2 - (\hbar^2/2m_e)\nabla_3^2 - 3e^2/4\pi\varepsilon_0 r_1 - 3e^2/4\pi\varepsilon_0 r_2 - 3e^2/4\pi\varepsilon_0 r_3 + e^2/4\pi\varepsilon_0 r_{12} + e^2/4\pi\varepsilon_0 r_{13} + e^2/4\pi\varepsilon_0 r_{23}$, where r_1 is the distance between electron 1 and the nucleus.

18.49 Let f and g denote the $n = 1$ and $n = 2$ spatial functions, i.e., $f = (2/a)^{1/2} \sin(\pi x/a)$ and $g = (2/a)^{1/2} \sin(2\pi x/a)$. With interelectronic repulsion ignored, the spatial wave function is a product of one-electron spatial functions. Analogous to Eqs. (18.45) and (18.46), we form the linear combinations $2^{-1/2}[f(1)g(2) + f(2)g(1)]$ and $2^{-1/2}[f(1)g(2) - f(2)g(1)]$ that don't distinguish between the electrons. To satisfy the Pauli principle, the symmetric spatial function must be combined with the antisymmetric two-electron spin function (18.41) and the antisymmetric spatial function must be combined with one of the symmetric spin functions. The approximate wave functions are therefore

$2^{-1/2}[f(1)g(2) + f(2)g(1)]2^{-1/2}[\alpha(1)\beta(2) - \beta(1)\alpha(2)]$

$2^{-1/2}[f(1)g(2) - f(2)g(1)]\alpha(1)\alpha(2)$

$2^{-1/2}[f(1)g(2) - f(2)g(1)]\beta(1)\beta(2)$

$2^{-1/2}[f(1)g(2) - f(2)g(1)]2^{-1/2}[\alpha(1)\beta(2) + \beta(1)\alpha(2)]$

The first wave function has $S = 0$. The second, third and fourth have $S = 1$ and have the same energy as one another (since they have the same spatial factor). According to Hund's rule, the $S = 1$ functions lie lower.

18.50 The ground-state configuration is $1s^2 2s^2$. To make the approximate wave function antisymmetric, we use a Slater determinant. Analogous to Eq. (18.54), we have

$$\psi_{gs} \approx N \begin{vmatrix} 1s(1)\alpha(1) & 1s(1)\beta(1) & 2s(1)\alpha(1) & 2s(1)\beta(1) \\ 1s(2)\alpha(2) & 1s(2)\beta(2) & 2s(2)\alpha(2) & 2s(2)\beta(2) \\ 1s(3)\alpha(3) & 1s(3)\beta(3) & 2s(3)\alpha(3) & 2s(3)\beta(3) \\ 1s(4)\alpha(4) & 1s(4)\beta(4) & 2s(4)\alpha(4) & 2s(4)\beta(4) \end{vmatrix}$$

where N is a normalization constant (equal to $1/\sqrt{24}$).

18.51 H (1s), Li (1$s^2$2s), B (1$s^2$2$s^2$2p), C (1$s^2$2$s^2$2p^2), N (1$s^2$2$s^2$2p^3), O (1$s^2$2$s^2$2p^4), and F (1$s^2$2$s^2$2p^5) all have one or more unpaired electrons and so have $S \neq 0$ and have paramagnetic ground states. He (1s^2), Be (1$s^2$2s^2), and Ne (1$s^2$2$s^2$2p^6) have all electrons paired, have $S = 0$ and $L = 0$ and do not have paramagnetic ground states. (Ne has two 2p electrons with $m = +1$, two 2p electrons with $m = 0$, and two 2p electrons with $m = -1$, and so has total orbital angular-momentum quantum number $L = 0$.)

18.52 We want the energy needed for ${}_{18}Ar^{17+} \rightarrow Ar^{18+}$. The ion Ar^{17+} has one electron and so is a hydrogenlike species. From Eq. (18.18) with $n = 1$, the ionization potential is $(18)^2(13.6\text{ V}) = 4406$ V.

18.53 $\varepsilon = -(Z_{\text{eff}}^2/n^2)(13.6\text{ eV})$.

(a) In Li, the first ionization potential is for removal of a 2s electron, so $5.4\text{ eV} = (Z_{\text{eff}}^2/2^2)(13.6\text{ eV})$ and $Z_{\text{eff}} = 1.26$.

(b) $9.3\text{ eV} = (Z_{\text{eff}}^2/2^2)(13.66\text{ eV})$ and $Z_{\text{eff}} = 1.65$. (The increase over Li is due to the poor screening of one 2s electron by the other.)

18.54 **(a)** If $s = 3/2$, the m_s values are 3/2, 1/2, –1/2, –3/2. For $s = 3/2$, the electrons are still fermions and the Pauli exclusion principle still holds. The four values of m_s mean that 4 electrons (instead of 2) can go in each orbital. The 1s, 2s, and 2p subshells would therefore hold 4, 4, and 12 electrons (double their capacities for s = ½). The ground-state configurations are 1s^3, 1$s^4$2$s^4$2p, and 1$s^4$2$s^4$2p^9.

(b) For $s = 1$, the electrons would be bosons and there would be no restriction on the number of electrons in a spin-orbital. The ground-state configurations would therefore be 1s^3, 1s^9, and 1s^{17}.

18.55 **(a)** The outer electron in K is further from the nucleus, so Na has the higher ionization potential.

(b) The ineffective screening of one 4s electron by the other makes Z_{eff} greater in Ca than in K, so Ca has the higher ionization potential.

(c) Cl.

(d) Kr.

18.56 For $Z = 10$, the figure gives $\sqrt{\varepsilon/\varepsilon_H} = 8$, $2._2$, and $1._6$. Since $\varepsilon_H = -13.6$ eV, we get $\varepsilon_{1s} \approx -870$ eV, $\varepsilon_{2s} \approx -66$ eV, and $\varepsilon_{2p} \approx -35$ eV. Substitution in $\varepsilon = -(Z_{\text{eff}}^2/n^2)(13.6 \text{ eV})$ gives $Z_{\text{eff},1s} \approx 8$, $Z_{\text{eff},2s} \approx 4._4$, $Z_{\text{eff},2p} \approx 3._2$.

18.57 **(a)** T; **(b)** T.

18.58 Nitrogen, with 3 unpaired electrons.

18.59 Ionization energy data in Sec. 18.8 show that $\Delta E = 5.1$ eV for $\text{Na} \rightarrow \text{Na}^+ + \text{e}^-$, so $E(\text{Na}^+ + \text{e}^-) > E(\text{Na})$ and $E(\text{Na}^+ + 2\text{e}^-) > E(\text{Na} + \text{e}^-)$. Electron affinity data give $\Delta E = -0.5$ eV for $\text{Na} + \text{e}^- \rightarrow \text{Na}^-$, so $E(\text{Na}^-) < E(\text{Na} + \text{e}^-)$. The lowest-energy (most stable) system is Na^-; the highest-energy system is $\text{Na}^+ + 2\text{e}^-$.

18.60 **(a)** Sr; **(b)** F; **(c)** K; **(d)** C; **(e)** Cl^-. Cl^- and Ar are isoelectronic and the higher Z in Ar means a smaller size.

18.61 $\phi = \sum_{k=1}^{5} b_k g_k$, where $g_k = N_k \exp(-\zeta_k r/a_0)$. Replacement of Z by ζ_k in the $1s$ orbital in Table 18.1 gives $N_k = \pi^{-1/2}(\zeta_k/a_0)^{3/2}$. So $\phi = b_1 g_1 + b_2 g_2 + \cdots + b_5 g_5 =$ $0.768\pi^{-1/2}(1.417/a_0)^{3/2}\exp(-1.417r/a_0) + 0.233\pi^{-1/2}(2.377/a_0)^{3/2}\exp(-2.377r/a_0)$ $+ 0.041\pi^{-1/2}(4.396/a_0)^{3/2}\exp(-4.396r/a_0) -$ $0.010\pi^{-1/2}(6.527/a_0)^{3/2}\exp(-6.527r/a_0) + 0.002\pi^{-1/2}(7.943/a_0)^{3/2}\exp(-7.943r/a_0)$.

18.62 $$D_1 = 2^{-1/2}\begin{vmatrix} 1s(1)\alpha(1) & 2s(1)\alpha(1) \\ 1s(2)\alpha(2) & 2s(2)\alpha(2) \end{vmatrix} =$$

$2^{-1/2}[1s(1)\alpha(1)2s(2)\alpha(2) - 1s(2)\alpha(2)2s(1)\alpha(1)] =$ $2^{-1/2}[1s(1)2s(2) - 1s(2)2s(1)]\alpha(1)\alpha(2)$, which is the $S = 1$, $M_S = 1$ function in Fig. 18.13. Replacement of α by β in the preceding equations shows that D_4 equals the $S = 1$, $M_S = -1$ function.

$$D_2 = 2^{-1/2}\begin{vmatrix} 1s(1)\alpha(1) & 2s(1)\beta(1) \\ 1s(2)\alpha(2) & 2s(2)\beta(2) \end{vmatrix} =$$

$2^{-1/2}[1s(1)2s(2)\alpha(1)\beta(2) - 2s(1)1s(2)\beta(1)\alpha(2)]$. Interchange of α and β in D_2 gives $D_3 = 2^{-1/2}[1s(1)2s(2)\beta(1)\alpha(2) - 2s(1)1s(2)\alpha(1)\beta(2)]$. We have $2^{-1/2}(D_2 + D_3) = 2^{-1/2}\{1s(1)2s(2)2^{-1/2}[\alpha(1)\beta(2) + \beta(1)\alpha(2)] - 2s(1)1s(2)2^{-1/2} \times$

$[\beta(1)\alpha(2) + \alpha(1)\beta(2)]\}$, which is the $S = 1$, $M_s = 0$ function. Similarly, $2^{-1/2}(D_2 - D_3)$ is found to be the $S = 0$, $M_S = 0$ function.

18.63 $\int_0^{2\pi} \int_0^{\pi} \int_0^{a} r^2 \sin\theta \, dr \, d\theta \, d\phi = \int_0^{2\pi} d\phi \int_0^{\pi} \sin\theta \, d\theta \int_0^{a} r^2 \, dr = 2\pi(2)(a^3/3) = \frac{4}{3}\pi a^3$.

18.64 **(a)** E is proportional to $-Z^2$, so $E_{\text{H}} > E_{\text{He}^+}$.

(b) The ionization energy of K^+.

(c) The energy-level spacing for these one-electron species is proportional to Z^2, so ν is proportional to Z^2. Thus $\nu_{\text{He}^+} > \nu_{\text{H}}$ and $\lambda_{\text{H}} > \lambda_{\text{He}^+}$.

(d) These quantities are equal.

18.65 **(a)** Particle in a box; rigid rotor.

(b) Harmonic oscillator.

(c) Hydrogenlike atom.

18.66 **(a)** $d\tau = \frac{4}{3}\pi r^3 = \frac{4}{3}\pi(0.0010 \text{ Å})^3 = 4.19 \times 10^{-9} \text{ Å}^3$. $|\psi|^2 = (1/\pi a^3)e^{-2r/a} = [\pi(0.5295 \text{ Å})^3]^{-1} = 2.14 \text{ Å}^{-3}$. $|\psi|^2 \, d\tau = 9.0 \times 10^{-9}$.

(b) $|\psi|^2 = (1/\pi a^3)e^{-2(0.50 \text{ Å}/a)} = 0.324$. $d\tau = 4.19 \times 10^{-9} \text{ Å}^3$. $|\psi|^2 \, d\tau = 1.4 \times 10^{-9}$.

(c) $|\psi|^2 = 1.35 \times 10^{-8}$ and $|\psi|^2 \, d\tau = 5.7 \times 10^{-17}$.

18.67 From (18.28) and (18.19), this probability is $\Pr = |R_{1s}(r)|^2 r^2 \, dr = (4/a^3)e^{-2r/a} r^2 \, dr$.

(a) $\Pr = 4(0.5295 \text{ Å})^{-3} e^{-2(0.100/0.5295)}(0.100 \text{ Å})^2(0.001 \text{ Å}) = 0.00018_5$.

(b) $4(0.5295 \text{ Å})^{-3}\exp[-2(0.500/0.5295)](0.500 \text{ Å})^2(0.001 \text{ Å}) = 0.00102$.

(c) 0.00062.

(d) 4.2×10^{-9}. (See also Fig. 18.8.)

18.68 **(a)** $d\tau = dx$ and $-\infty \le x \le \infty$.

(b) $d\tau = dx\,dy\,dz$. x, y, and z each range from $-\infty$ to ∞, but since $|\psi|^2 = 0$ outside the box, we need integrate over only the region $0 \le x \le a,\ 0 \le y \le b,\ 0 \le z \le c$.

(c) $d\tau = r^2 \sin\theta\, dr\, d\theta\, d\phi$. $0 \le r \le \infty,\ 0 \le \theta \le \pi,\ 0 \le \phi \le 2\pi$.

18.69 Yes. The gravitational force is far smaller than the electrostatic force and so can be neglected. $|F_{grav}|/|F_{el}| = (Gm_em_p/r^2)/(e^2/4\pi\varepsilon_0 r^2) = 4\pi\varepsilon_0 Gm_em_p/e^2 = 4\pi(8.85 \times 10^{-12}\ C^2/N\text{-}m^2)(6.67 \times 10^{-11}\ m^3/s^2\text{-}kg)(9.1 \times 10^{-31}\ kg) \times (1.67 \times 10^{-27}\ kg)/(1.6 \times 10^{-19}\ C)^2 = 4 \times 10^{-40}$.

18.70 **(a)** $\psi_{2p_z} = \frac{1}{4}(2\pi)^{-1/2}a^{-5/2}re^{-r/2a}\cos\theta$. The maximum value of $\cos\theta$ occurs at $\theta = 0$, where $\cos\theta = 1$. Setting $\cos\theta = 1$ in ψ and then taking $\partial\psi/\partial r = 0$, we get $\psi_{2p_z}/\partial r = 0 = \frac{1}{4}(2\pi)^{-1/2}a^{-5/2}[e^{-r/2a} - (1/2a)re^{-r/2a}]$. Solving for r, we get $r = 2a$. Setting $r = 2a$ and $\cos\theta = 1$ in ψ_{2p_z}, we get $\psi_{max} = \frac{1}{4}(2\pi)^{-1/2}a^{-5/2}(2a)e^{-1} = \pi^{-1/2}(2a)^{-3/2}e^{-1}$.

(b) This problem is incompletely stated, in that it is intended that the calculations be done for points in the yz plane, where $x = 0$. In the yz plane, $|\psi_{2p_z}/\psi_{max}| = \frac{1}{2}(|z|/a)e\exp\{-\frac{1}{2}[(y/a)^2 + (z/a)^2]^{1/2}\} = k$.
A BASIC program is

```
15 INPUT "PSI/PSIMAX";K
25 PRINT "PSI/PSIMAX=";K
35 FOR Z = 0.01 TO 10 STEP 0.01
45 W=LOG(Z/2)+1–LOG(K)
55 IF W<Z/2 THEN 85
65 Y=SQR(4*W*W*–Z*Z)
75 PRINT "Z/A=";Z;" Y1/A=";
   Y;" Y2/A=";–Y
85 NEXT Z
95 STOP
```

18.71 As a hint, elements were named for the physicists in (a) and (d).

18.72 **(a)** F; **(b)** F. **(c)** F; this is true only for *identical* fermions. **(d)** F. **(e)** F. **(f)** F; **(g)** F; only well-behaved solutions are possible stationary states. **(h)** T. **(i)** F; this is true only if the states have the same energy.

Chapter 19

19.1 The table of bond radii in Sec. 19.1 gives the following estimates.

(a) 0.30 Å + 0.77 Å = 1.07 Å for CH, 0.77 Å + 0.66 Å = 1.43 Å for CO, and 0.66 Å + 0.30 Å = 0.96 Å for OH.

(b) 0.30 Å + 0.77 Å = 1.07 Å for HC and 0.60 Å + 0.55 Å = 1.15 Å for CN.

19.2 Each BF bond has some double-bond character, as shown by the Lewis structure 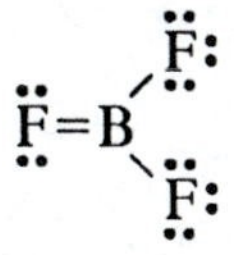 and two others.

19.3 **(a)** The $TeBr_2$ Lewis dot formula has four electron pairs around Te and two lone pairs on Te. The geometry is bent with bond angle somewhat less than 109½°.

(b) Hg has electron configuration $\cdots 5d^{10}6s^2$ and has 2 valence electrons. So Hg has two valence pairs in the Lewis structure, and $HgCl_2$ is linear.

(c) With 3 pairs on Sn, $SnCl_2$ is bent with angle a bit less than 120°.

(d) With 5 pairs on Xe, XeF_2 is linear.

(e) The dot formula has four pairs on Cl, so the ion is bent with angle somewhat less than 109½°.

19.4 **(a)** With 5 pairs on Br, BrF_3 is T-shaped (Fig. 19.2b).

(b) Three pairs on Ga. Trigonal planar.

(c) 4 pairs on O. Trigonal pyramidal with angles a bit less than 109½°.

(d) Four pairs on P. Trigonal pyramidal with angles a bit less than 109½°.

19.5 **(a)** 4 pairs on Sn. Tetrahedral.

(b) 5 pairs on Se. Seesaw shape.

(c) 6 pairs on Xe. Square planar.

(d) 4 pairs on B. Tetrahedral.

(e) 6 pairs on Br. Square planar.

19.6 **(a)** 5 pairs on As. Trigonal bipyramidal.

(b) 6 pairs on Br. Square-based pyramid.

(c) 6 pairs on Sn. Octahedral.

19.7 Each multiple bond is counted as one pair.

(a) "3" pairs on O. Bent, with angle close to 120°.

(b) "3" pairs on N. Trigonal planar.

(c) "3" pairs on S. Trigonal planar.

(d) "3" pairs on S. Bent. Angle close to 120°.

(e) "4" pairs on S. Approximately tetrahedral.

(f) "4" pairs on S. Pyramidal with angles close to 109½°.

(g) "4" pairs on I. Pyramidal with angles close to 109½°.

(h) "5" pairs on S. Trigonal bipyramidal.

(i) "4" pairs on Xe. Trigonal pyramidal with angles close to 109½°.

(j) "6" pairs on Xe. Square-based pyramid.

19.8 **(a)** There are four electron pairs around the methyl carbon, so the HCH and HCC bond angles will be close to 109½°. (Because the four groups attached to the methyl carbon are not all the same, we cannot expect the angles at this C to be exactly 109½°.) There are "2" pairs around the CN carbon, so the CCN angle will be 180°.

(b) There are "3" pairs around the CH_2 carbon and the bond angles at this carbon will be close to 120°. Because of the greater repulsions exerted by the double-bond's electrons, the HCH angle will be a bit less than 120° and the HCC angles a bit more than 120°. The bond angles at the CH carbon will be close to 120°. The angles at the CH_3 carbon will be close to 109½°.

(c) The HCH and HCN angles at the methyl carbon are close to 109½°; There are 4 pairs around the N and the HNC angles are a bit less than 109½°.

(d) Near 109½° for the HCH and HCO angles. A bit less than 109½° for the HOC angle.

(e) A bit less than 109½° for each FOO angle.

19.9 **(a)** By rule 2, the OH hydrogen will eclipse the CO double bond, which makes *D*(HOCH) equal to 180°. The molecule is planar. A Newman projection with the O behind the C is

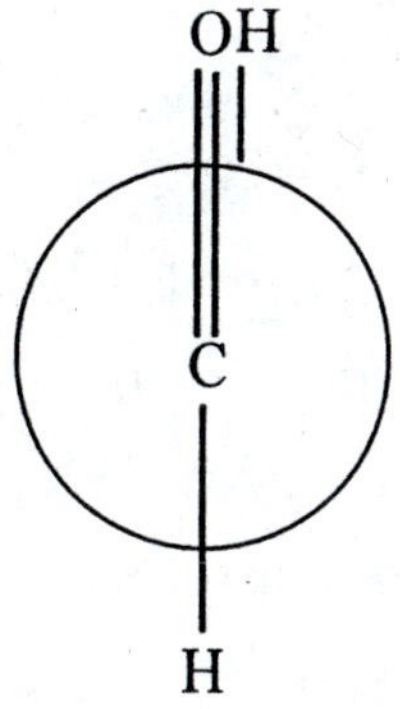

(b) By rule 1, the bonds are staggered. The two possible conformations are

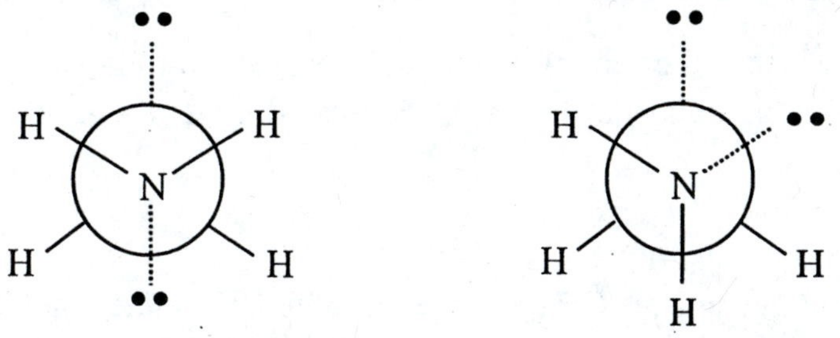

I **II**

The rules don't allow us to decide whether I or II is more stable. The smallest *D*(HNNH) angle is 60° in both I and II. Experiment shows that in fact II is more stable than I.

(c) Rule 2 says the OH hydrogen eclipses the double bond in this planar molecule. *D*(HOCC) = 0. A Newman projection with O behind C is

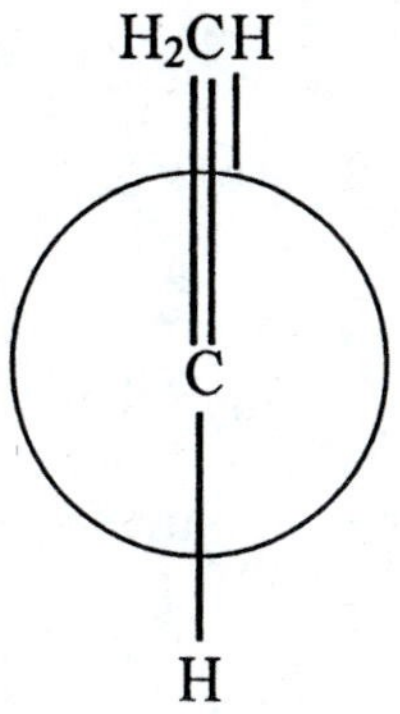

19.10 The Lewis dot structure is $:\ddot{\underset{..}{O}}-\ddot{O}=\ddot{O}: \longleftrightarrow :\ddot{O}=\ddot{O}-\ddot{\underset{..}{O}}:$. The bond angle is close to 120°. The lone pair on the central atom makes the dipole moment nonzero.

19.11 **(a)** $\Delta_{at}H^\circ_{298} \approx \Delta_{at}H^\circ_{298,re} - \Delta_{at}H^\circ_{298,pr} =$

$[2(415) + 812 + 2(436)]$ kJ/mol – $[344 + 6(415)]$ kJ/mol = –320 kJ/mol.
The true value is $[-84.68 - 226.73 - 2(0)]$ kJ/mol = –311.4 kJ/mol.

(b) $\Delta H^\circ_{298} \approx [946 + 3(436) - 2(3)391]$ kJ/mol = –92 kJ/mol.
In truth, $\Delta H^\circ_{298} = 2(-46.1)$ kJ/mol = –92.2 kJ/mol.

19.12 We assume tetrahedral angles. As noted in Sec. 19.1, the vector sum of three CH moments in a CH_3 group equals the moment of one CH bond. The H_3CF dipole moment is thus the sum of the moments $\overset{+}{H}-\overset{-}{C}$ and $\overset{+}{C}-\overset{-}{F}$. We have 1.85 D = 0.4 D + μ_{CF} and μ_{CF} = 1.4$_5$ D. Similarly, 1.62 D = 0.4 D + μ_{CI} and μ_{CI} = 1.2 D.

19.13 **(a)** The net moment of the CH_3 group equals the CH moment and the dipole moment is the sum of the moments $\overset{+}{H}-\overset{-}{C}$ and $\overset{+}{C}-\overset{-}{Cl}$.
So $\mu \approx$ 0.4 D + 1.5 D = 1.9 D.

(b) $\mu \approx$ 0.4 D + 1.5 D = 1.9 D.

(c) $\mu \approx$ 0.4 D + 1.5 D = 1.9 D. (Here, agreement with experiment is poor.)

(d) 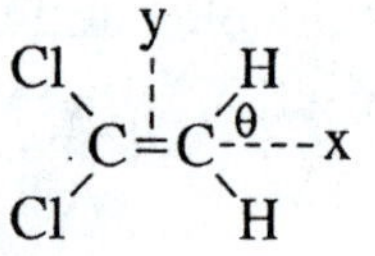

$\mu_x \approx 2\mu_{ClC} \cos\theta + 2\mu_{CH} \cos\theta =$

2(1.5 D) cos 60° + 2(0.4 D) cos 60° = 1.9 D. $\mu_y = 0$. Hence $\mu \approx$ 1.9 D.

19.14 The moments listed in Sec. 19.1 give the H_3CCN moment as the sum of the $\overset{+}{H}-\overset{-}{C}$ and $\overset{+}{C}\equiv\overset{-}{N}$ bond moments, namely, as μ = 0.4 D + 3.5 D = 3.9 D. If we now assume the polarity $\overset{-}{H}-\overset{+}{C}$, then the H_3C moment is oppositely directed from the CN moment, and we would have 3.9 D = –0.4 D + $\mu_{C\equiv N}$ and $\mu_{C\equiv N}$ = 4.3 D (instead of 3.5 D).

19.15 **(a)** $\Delta_{CH}/(\text{kJ/mol}) = 415 - \frac{1}{2}(344 + 436) = 25$. $|x_C - x_H| = 0.102(25)^{1/2} = 0.5$ (compared with 0.3 in Table 19.2).

(b) $\Delta_{CO}/(\text{kJ/mol}) = 350 - \frac{1}{2}(344 + 143) = 106\frac{1}{2}$. $|x_C - x_O| = 0.102(106)^{1/2}$ $= 1.0_5$ (compared with 0.9 in Table 19.2).

(c) $\Delta_{CCl}/(\text{kJ/mol}) = 328 - \frac{1}{2}(344 + 243) = 34\frac{1}{2}$. $|x_C - x_{Cl}| = 0.102(34\frac{1}{2})^{1/2}$ $= 0.6$ (compared with 0.7 in Table 19.2).

19.16 **(a)** For H, there is only one valence electron and $\langle E_{i,\text{val}}\rangle = 13.6$ eV (Sec. 18.3), so $x_H = 0.169(13.6) = 2.30$.

(b) Li has one valence electron whose ionization energy is given by the table in Sec. 19.8 as 5.4 eV, so $x_{Li} = 0.169(5.4) = 0.91$.

(c) Be has electron configuration $1s^2 2s^2$ and each $2s$ electron has ionization energy 9.3 eV (Sec. 18.8), so $x_{Be} = 0.169(9.3) = 1.5_7$.

(d) $x_{Na} = 0.169(5.1) = 0.86$.

19.17 $x_H = 1.66(1/0.667)^{1/3} + 0.37 = 2.27$. $x_{Li} = 1.66(1/24.3)^{1/3} + 0.37 = 0.94$. $x_{Be} = 1.66(2/5.60)^{1/3} + 0.37 = 1.55$. $x_B = 1.66(3/3.0_3)^{1/3} + 0.37 = 2.0_2$. $x_C = 1.66(4/1.7_6)^{1/3} + 0.37 = 2.5_5$. N: 3.1_2. O: 3.6_2. F: 4.2_3.

19.18 **(a)** $|x_A - x_B| + |x_B - x_C| = x_A - x_B + x_B - x_C = x_A - x_C = |x_A - x_C|$. Substitution of Eq. (19.3) into this equation gives $0.102[\Delta_{AB}/(\text{kJ/mol})]^{1/2} + 0.102[\Delta_{BC}/(\text{kJ/mol})]^{1/2} = 0.102[\Delta_{AC}/(\text{kJ/mol})]^{1/2}$, so $\Delta_{AB}^{1/2} + \Delta_{BC}^{1/2} = \Delta_{AC}^{1/2}$.

(b) Table 19.1 gives $\Delta_{ON}/(\text{kJ/mol}) = 175 - \frac{1}{2}(143 + 159) = 24$; $\Delta_{NC}/(\text{kJ/mol}) = 292 - \frac{1}{2}(344 + 159) = 40.5$; $\Delta_{OC}/(\text{kJ/mol}) = 350 - \frac{1}{2}(143 + 344) = 106.5$. $\Delta_{ON}^{1/2} + \Delta_{NC}^{1/2} = 11._3\ (\text{kJ/mol})^{1/2}$; $\Delta_{OC}^{1/2} = 10._3\ (\text{kJ/mol})^{1/2}$. The relation in (a) is obeyed fairly well.

19.19 **(a)** 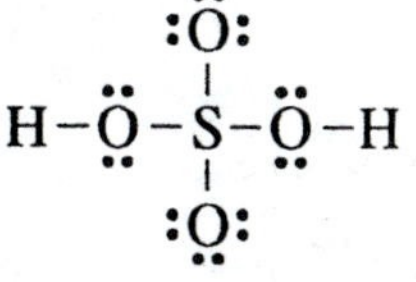

(b) In the dot formula of (a), the S has $\frac{1}{2}(8) = 4$ valence electrons, as compared with 6 valence electrons in a free S atom. The formal charge on S is +2 for this dot formula. This is an unlikely value for a nonmetallic element.

(c)

$$:\!\ddot{F}\!:\ :\!\ddot{F}\!:$$
$$:\!\ddot{F}\!-\!S\!-\!\ddot{F}\!:$$
$$:\!\ddot{F}\!:\ :\!\ddot{F}\!:$$

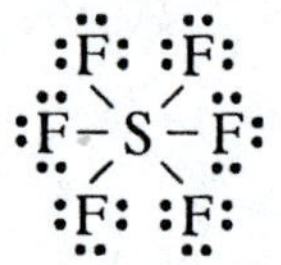

(d) The SF_6 dot formula shows that S can share as many as 12 valence electrons. (This is due to the presence of $3d$ orbitals on S.) A dot

formula for H_2SO_4 that gives S a zero formal charge is $H-\ddot{O}-\overset{\displaystyle :O:}{\underset{\displaystyle :O:}{S}}-\ddot{O}-H$.

Here S has $\frac{1}{2}(12) = 6$ valence electrons, as in a free S atom.

(e) A dot formula for SO_4^{2-} that gives S a zero formal charge is

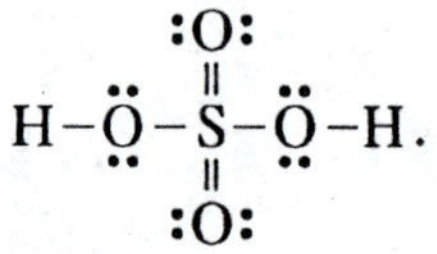

$$\left[\ :\ddot{O}-\overset{\displaystyle :O:}{\underset{\displaystyle :O:}{S}}-\ddot{O}:\ \right]^{2-}$$

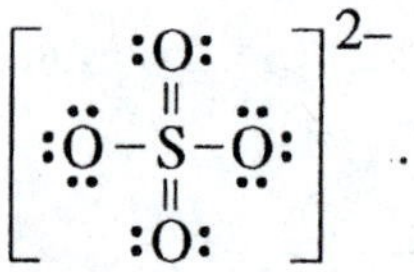

. In addition, there are other resonance structures in which the double bonds and single bonds are permuted. Each sulfur–oxygen bond is intermediate between a single bond and a double bond.

19.20 $:C\!\equiv\!O:$ The carbon has $2 + \frac{1}{2}(6) = 5$ valence electrons, as compared with 4 in the free C atom. The formal charge on C is -1. (This formal charge opposes the greater electronegativity of O and produces a dipole moment with the polarity $\overset{-\ +}{CO}$.)

19.21 (a) Because of the electronegativity difference between H and Cl, we expect the $H-Cl$ bond energy to be larger than the average of the $H-H$ and $Cl-Cl$ bond energies, so $\Delta H°$ is negative.

(b) $\Delta H° < 0$, for reasons similar to those in (a).

19.22 (a) T **(b)** F; **(c)** F; **(d)** T; **(e)** T.

19.23 $\hat{K}_e = -(\hbar^2/2m_e)\nabla_1^2 - (\hbar^2/2m_e)\nabla_2^2$; $\hat{K}_N = -(\hbar^2/2m_A)\nabla_A^2 - (\hbar^2/2m_B)\nabla_B^2$; $\hat{V}_{NN} = Z_A Z_B e^2/4\pi\varepsilon_0 R_{AB}$; $\hat{V}_{Ne} = -Z_A e^2/4\pi\varepsilon_0 r_{1A} - Z_B e^2/4\pi\varepsilon_0 r_{1B} - Z_A e^2/4\pi\varepsilon_0 r_{2A} - Z_B e^2/4\pi\varepsilon_0 r_{2B}$; $\hat{V}_{ee} = e^2/4\pi\varepsilon_0 r_{12}$.

19.24 **(a)** $KF \xrightarrow{1} K^+ + F^- \xrightarrow{2} K + F$. According to the model, the energy needed to dissociate KF to $K^+ + F^-$ is $\Delta E_1 \approx e^2/4\pi\varepsilon_0 R_e = (1.602 \times 10^{-19}\ \text{C})^2/4\pi(8.854 \times 10^{-12}\ \text{C}^2/\text{N-m}^2)(2.17 \times 10^{-10}\ \text{m}) = 1.06_3 \times 10^{-18}\ \text{J} = 6.63\ \text{eV}$. The energy change for step 2 is $\Delta E_2 = -4.34$ eV + 3.40 eV = –0.94 eV. The net ΔE is 6.63 eV – 0.94 eV = 5.69 eV.

(b) According to the model, $\mu \approx eR_e = (1.602 \times 10^{-19}\ \text{C})(2.17 \times 10^{-10}\ \text{m}) = 3.48 \times 10^{-29}$ C m = 10.4 D, where (19.2) was used.

(c) Both compounds are essentially completely ionic with charges of +1 and –1 on the cation and anion. The larger size of Cl^- as compared with F^- makes R_e greater in KCl and gives KCl the greater dipole moment (which is approximately eR_e).

19.25 **(a)** At R_e, $\partial E_e/\partial R = 0 = -12B/R^{13} + e^2/4\pi\varepsilon_0 R^2$ and $B = e^2 R_e^{11}/12(4\pi\varepsilon_0)$.

(b) At equilibrium, the electronic energy is $E_{e,\text{eq}} = B/R_e^{12} - e^2/4\pi\varepsilon_0 R_e = e^2 R_e^{11}/12(4\pi\varepsilon_0)R_e^{12} - e^2/4\pi\varepsilon_0 R_e = -11e^2/12(4\pi\varepsilon_0)R_e = -11(1.602 \times 10^{-19}\ \text{C})^2/12(4\pi)(8.854 \times 10^{-12}\ \text{C}^2/\text{N-m}^2)(2.36 \times 10^{-10}\ \text{m}) = -8.96 \times 10^{-19}\ \text{J} = -5.59$ eV. According to the model, it requires 5.59 eV to dissociate the NaCl molecule to $Na^+ + Cl^-$. ΔE for $Na^+ + Cl^- \rightarrow Na + Cl$ is –5.14 eV + 3.61 eV = –1.53 eV (where data was taken from Example 19.1 in the text). The model gives D_e of NaCl as 5.59 eV – 1.53 eV = 4.06 eV.

(c) The Pauli repulsion decreases D_e. Since 4.06 eV is less than the true D_e, the function B/R^{12} overestimates the Pauli repulsion. For $E_e = A/R^m - e^2/4\pi\varepsilon_0 R$, we have at R_e, $\partial E_e/\partial R = 0 = -mA/R^{m+1} + e^2/4\pi\varepsilon_0 R^2$ and $A = e^2 R_e^{m-1}/4\pi\varepsilon_0 m$. So $E_{e,\text{eq}} = A/R_e^m - e^2/4\pi\varepsilon_0 R_e = e^2 R_e^{m-1}/4\pi\varepsilon_0 m R_e^m - e^2/4\pi\varepsilon_0 R_e = -(1 - 1/m)e^2/4\pi\varepsilon_0 R_e$. Then 4.25 eV = $(1 - 1/m)e^2/4\pi\varepsilon_0 R_e$ – 5.14 eV + 3.61 eV. We have $e^2/4\pi\varepsilon_0 R_e = (1.602 \times 10^{-19}\ \text{C})^2/4\pi(8.854 \times 10^{-12}\ \text{C}^2/\text{N-m}^2)(2.36 \times 10^{-10}\ \text{m}) = 9.77 \times 10^{-19}\ \text{J} = 6.10$ eV, so $1 - 1/m = 0.948$ and $m = 19$.

19.26 **(a)** T; **(b)** F; **(c)** T; **(d)** T.

19.27 **(a)** Even; **(b)** neither; **(c)** odd; **(d)** even; **(e)** odd.

19.28 The box size is small enough to be considered "infinitesimal." The probability is $|\psi|^2\, dV \approx (2 + 2S)^{-1}(1s_A + 1s_B)^2(10^{-6}\text{ Å}^3)$, where $S = e^{-R/a_0}(1 + R/a_0 + R^2/3a_0^2)$. At the equilibrium separation of $R = 1.06$ Å = $2.00a_0$, we find $S = e^{-2}(1 + 2 + 4/3) = 0.586$ and $(2 + 2S)^{-1} = 0.315$.

(a) At nucleus A, $r_A = 0$ and $r_B = R = 2.00a_0$, so $1s_A = \pi^{-1/2}(1/a_0)^{3/2}e^0 = 1.466\text{ Å}^{-3/2}$ and $1s_B = \pi^{-1/2}(1/a_0)^{3/2}e^{-2.00} = 0.198\text{ Å}^{-3/2}$. Then $|\psi|^2\, dV \approx 0.315(1.466 + 0.198)^2\text{ Å}^{-3}(10^{-6}\text{ Å}^3) = 8.7 \times 10^{-7}$.

(b) At the midpoint of the internuclear axis, $r_A = r_B = R/2 = 1.00a_0$, $1s_A = 1s_B = \pi^{-1/2}(1/a_0)^{3/2}e^{-1.00} = 0.539\text{ Å}^{-3/2}$ and $|\psi|^2\, dV \approx 0.315(0.539 + 0.539)^2(10^{-6}) = 3.7 \times 10^{-7}$.

(c) $r_A = R/3 = 2a_0/3$, $r_B = 2R/3 = 4a_0/3$, $1s_A = 0.753\text{ Å}^{-3/2}$, $1s_B = 0.386\text{ Å}^{-3/2}$, $|\psi|^2\, dV \approx 4.1 \times 10^{-7}$.

19.29 The MO electron configuration is $(\sigma_g 1s)^2(\sigma_u^* 1s)^2$. To achieve antisymmetry, we use a Slater determinant. Analogous to Eqs. (18.54) and (19.20) and the Be wave function in Prob. 18.50, we write

$$N\begin{vmatrix} \sigma_g 1s(1)\alpha(1) & \sigma_g 1s(1)\beta(1) & \sigma_u^* 1s(1)\alpha(1) & \sigma_u^* 1s(1)\beta(1) \\ \sigma_g 1s(2)\alpha(2) & \sigma_g 1s(2)\beta(2) & \sigma_u^* 1s(2)\alpha(2) & \sigma_u^* 1s(2)\beta(2) \\ \sigma_g 1s(3)\alpha(3) & \sigma_g 1s(3)\beta(3) & \sigma_u^* 1s(3)\alpha(3) & \sigma_u^* 1s(3)\beta(3) \\ \sigma_g 1s(4)\alpha(4) & \sigma_g 1s(4)\beta(4) & \sigma_u^* 1s(4)\alpha(4) & \sigma_u^* 1s(4)\beta(4) \end{vmatrix}$$

where N is a normalization constant.

19.30 We use the homonuclear diatomic MOs in Fig. 19.15.

(a) $(\sigma_g 1s)^2(\sigma_u^* 1s)$.

(b) $(\sigma_g 1s)^2(\sigma_u^* 1s)^2(\sigma_g 2s)^2$.

(c) $(\sigma_g 1s)^2(\sigma_u^* 1s)^2(\sigma_g 2s)^2(\sigma_u^* 2s)^2$.

(d) $(\sigma_g 1s)^2(\sigma_u^* 1s)^2(\sigma_g 2s)^2(\sigma_u^* 2s)^2(\pi_u 2p)^4$.

(e) $(\sigma_g 1s)^2(\sigma_u^* 1s)^2(\sigma_g 2s)^2(\sigma_u^* 2s)^2(\pi_u 2p)^4(\sigma_g 2p)^2$. He_2^+ has an unpaired electron and so is paramagnetic. All the others have filled shells and are not paramagnetic.

(f) (e) with $(\pi_g^* 2p)^4$ added.

19.31 **(a)** $(2 - 1)/2 = ½$. **(b)** $(4 - 2)/2 = 1$. **(c)** $(4 - 4)/2 = 0$. **(d)** $(8 - 4)/2 = 2$.

(e) $(10 - 4)/2 = 3$ (in agreement with the dot formula :N≡N:). **(f)** 1.

19.32 **(a)** The N_2 MO configuration is given in Prob. 19.30e. The highest occupied N_2 MO is a bonding MO, so N_2^+ has one less bonding electron than N_2. Therefore N_2 has the higher D_e.

(b) The O_2 MO configuration is shown in Fig. 19.18. The highest-occupied shell, $\pi_g^* 2p$, is half-filled and is antibonding. Therefore O_2^+ has one fewer antibonding electron than O_2 and has two fewer antibonding electrons than O_2^-, so O_2^+ has the highest D_e.

19.33 $(\hat{H} + c)\psi = \hat{H}\psi + c\psi = E\psi + c\psi = (E + c)\psi$, where the definition of the sum of operators was used.

19.34 **(a)** We feed the valence electrons into the MOs of Fig. 19.22, where $n = 2$ and $n' = 2$. NCl has $5 + 7 = 12$ valence electrons and has the valence-electron configuration $(\sigma s)^2(\sigma^* s)^2(\pi p)^4(\sigma p)^2(\pi^* p)^2$. NCl^+ and NCl^- have 11 and 13 electrons, respectively, and have the configurations $(\sigma s)^2(\sigma^* s)^2(\pi p)^4(\sigma p)^2(\pi^* p)$ and $(\sigma s)^2(\sigma^* s)^2(\pi p)^4(\sigma p)^2(\pi^* p)^3$, respectively.

(b) $(8 - 4)/2 = 2$ for NCl (which is similar to O_2); $(8 - 3)/2 = 2.5$ for NCl^+; $(8 - 5)/2 = 1.5$ for NCl^-.

(c) Each of these species has a partly filled π^* shell, so each has $S \neq 0$ and each is paramagnetic.

19.35 **(a)** C1s, C2s, C2p_x, C2p_y, C2p_z, O1s, O2s, O2p_x, O2p_y, O2p_z.

(b) C1s, C2s, C2p_z, O1s, O2s, O2p_z contribute to σ MOs. C2p_x, C2p_y, O2p_x, O2p_y contribute to π MOs.

19.36

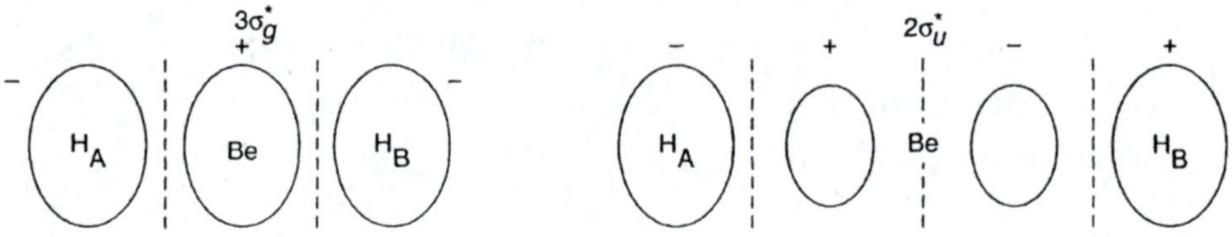

19.37 **(a)** g, σ. **(b)** g, δ. **(c)** g, δ. **(d)** g, π.

19.38 **(a)** σ, since it has no nodal planes that contain the internuclear (z) axis. **(b)** π, since it has one nodal plane containing the internuclear axis. **(c)** π. **(d)** σ. **(e)** σ (see Fig. 18.6.) **(f)** δ, since it has two nodal planes containing the internuclear axis. **(g)** δ. **(h)** π. **(i)** π.

19.39 **(a)** H_A1s, H_B1s, C1s, C2s, C2p_x, C2p_y, C2p_z, O1s, O2s, O2p_x, O2p_y, O2p_z.

(b) The dot formula 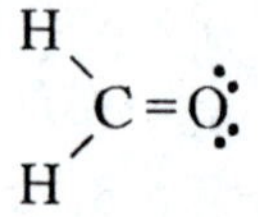 suggests the following description of the occupied localized MOs. We use sp^2 hybrid AOs on C to form the CH bonds and the σ bond of the double bond. These sp^2 hybrids are formed from C2s and the in-plane p orbitals C2p_y and C2p_z. The bonding σ MO between H_A and C is a linear combination of H_A1s, C2s, C2p_y, and C2p_z. The bonding σ MO between H_B and C is a linear combination of H_B1s, C2s, C2p_y, and C2p_z. As we did with carbon, we form in-plane sp^2 hybrids on oxygen, using O2s, O2p_y, and O2p_z; these hybrids go to form the σ bond of the double bond and the lone-pair AOs on oxygen. Overlap between C2p_x and O2p_x forms the π bond of the CO double bond. The σ bond of the CO double bond is formed by overlap of those sp^2 hybrids on C and O that point along the z (CO) axis; the C2p_y and O2p_y AOs do not contribute to these sp^2 hybrids—the C2p_y and O2p_y AOs each have one nodal plane containing the z axis and cannot contribute to the CO σ MO. Therefore the CO localized σ bonding MO is a linear combination of C2s, C2p_z, O2s, and O2p_z. The lone-pair localized MOs on O are formed from the O2s, O2p_y, and O2p_z AOs. There is an inner-shell localized MO on C that is essentially identical to the C1s AO and an inner-shell localized MO on O that is identical to O1s.

19.40 **(a)** $3H_2(g) + 6C(\text{graphite}) \xrightarrow{a} 6H(g) + 6C(g) \xrightarrow{b} C_6H_6(g)$. Appendix data give $\Delta H^\circ_{a,298} = [6(217.96) + 6(716.68) - 3(0) - 6(0)]$ kJ/mol = 5607.8 kJ/mol. Viewing C_6H_6 as containing three CC single bonds and three CC double bonds, we use the bond energies in Table 19.1 to get $\Delta H^\circ_{b,298} \approx -[6(415) + 3(344) + 3(615)]$ kJ/mol = –5367 kJ/mol. Then $\Delta_f H^\circ_{298} = \Delta H^\circ_{a,298} + \Delta H^\circ_{b,298} \approx 241$ kJ/mol. The Appendix gives the experimental value as 83 kJ/mol, so benzene is far more stable than it would be if it were composed of isolated single and double bonds.

(b) $5H_2(g) + 6C(\text{graphite}) \xrightarrow{a} 10H(g) + 6C(g) \xrightarrow{b} C_6H_{10}(g)$. Appendix data give $\Delta H^\circ_{a,298} = [10(217.96) + 6(716.68) - 5(0) - 6(0)]$ kJ/mol = 6479.7 kJ/mol. The bond-energy table gives $\Delta H^\circ_{b,298} \approx -[10(415) + 5(344) + 615]$ kJ/mol = –6485 kJ/mol. Then $\Delta_f H^\circ_{298} = \Delta H^\circ_{a,298} + \Delta H^\circ_{b,298} \approx -5._3$ kJ/mol. The Appendix gives the experimental value as –5.4 kJ/mol.

19.41 **(a)** The following sketches show a view from above the molecular plane. The dashed lines denote nodal planes perpendicular to the molecular plane. The molecular plane is a nodal plane for each MO. The lobes below the molecular plane have signs opposite those of corresponding lobes above the plane.

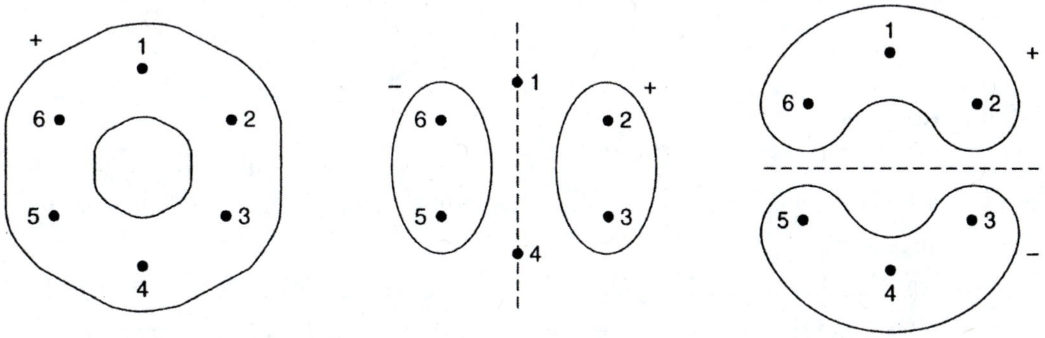

(b) The MO $p_1 + p_2 + p_3 + p_4 + p_5 + p_6$ has the fewest nodes and builds up the most electron probability density between the nuclei; this MO is lowest in energy.

19.42 Let $f \equiv (2s + 2p_z)/\sqrt{2}$. Along the z axis we have $x = 0$, $y = 0$, $r = (x^2 + y^2 + z^2)^{1/2} = (z^2)^{1/2} = |z|$ and $f = 2^{-1/2}(1/4)(2\pi)^{-1/2} \times [a^{-3/2}(|z|/a - 2)e^{-|z|/2a} + a^{-5/2}ze^{-|z|/2a}] = (1/8\pi^{1/2}a^{3/2})e^{-|w|/2}(w + |w| - 2)$, where $w \equiv z/a$. We have $1/8\pi^{1/2}a^{3/2} = 0.183$ Å$^{-3/2}$ and we find

f	−0.050	−0.082	−0.135	−0.222	−0.366	−0.143	0
z/a	−4	−3	−2	−1	0	0.5	1

f	0.086	0.135	0.163	0.149	0.120	0.091	0.066
z/a	1.5	2	3	4	5	6	7

The graph shows a sharp negative peak at $z/a = 0$ and a broad positive region for $z/a > 1$.

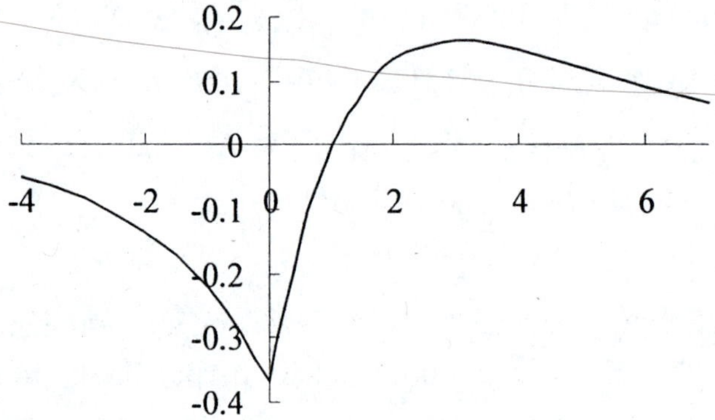

19.43 **(a)** T; **(b)** T; **(c)** F.

19.44 **(a)** Yes.

(b) No, since the HOOH dihedral angle must also be specified. Changing this dihedral angle without changing any bond distances or angles changes the structure.

(c) Yes. Imagine adding the H atoms one a time. When H_1 is added to N, the NH_1 bond distance specifies the structure. When H_2 is now added, the structure is fully specified by the two bond distances and the HNH angle (as is true for H_2O). When the third H is added, the values of the N-H_3 bond distance, the H_1NH_3 bond angle and the H_2NH_3 bond angle give three conditions that determine the three spatial coordinates of atom H_3.

(d) No. The ClCCCl dihedral angle must be specified.

(e) No. Twisting one CH_2 group with respect to the other generates different structures but does not change any bond distances or angles.

19.45 Let $f_{n_1}(x_1) \equiv (2/a)^{1/2} \sin(n_1\pi x_1/a)$ (for $0 \le x_1 \le a$) denote a particle-in-a-box wave function, where n_1 is the quantum number and x_1 is the coordinate. For noninteracting particles, the wave function is the product of wave functions for each particle, and one's first impulse might be to take ψ as $f_{n_1}(x_1)f_{n_2}(x_2)$. However, this function is not symmetric when $n_1 \ne n_2$, and spin-0 particles are bosons and require symmetric wave functions. Therefore, the normalized wave function [analogous to (18.45)] is $\psi = 2^{-1/2}[f_{n_1}(x_1)f_{n_2}(x_2) + f_{n_1}(x_2)f_{n_2}(x_1)]$. To get ρ, we integrate $|\psi|^2$ over the coordinates of all particles but one and multiply by the number of particles: $\rho =$

$2(2^{-1})[\int_0^a [f_{n_1}^2(x_1)f_{n_2}^2(x_2) + 2f_{n_1}(x_1)f_{n_2}(x_2)f_{n_1}(x_2)f_{n_2}(x_1) + f_{n_1}^2(x_2)f_{n_2}^2(x_1)]\,dx_2$.

$= f_{n_1}^2(x_1)\int_0^a f_{n_2}^2(x_2)\,dx_2 + 2f_{n_1}(x_1)f_{n_2}(x_1)\int_0^a f_{n_1}(x_2)f_{n_2}(x_2)\,dx_2 +$

$f_{n_2}^2(x_1)\int_0^a f_{n_1}^2(x_2)\,dx_2$

Use of the normalization condition for f and the orthogonality property (17.36) gives $\rho(x_1) = f_{n_1}^2(x_1) + 0 + f_{n_2}^2(x_1)$. Dropping the unnecessary subscript 1, we have $\rho(x) = f_{n_1}^2(x) + f_{n_2}^2(x) = (2/a)\sin^2(n_1\pi x/a) + (2/a)\sin^2(n_2\pi x/a)$ for $0 \le x \le a$, and $\rho = 0$ elsewhere. Also, $\int_{-\infty}^{\infty} \rho(x)\,dx = \int_0^a \rho(x)\,dx = 1 + 1 = 2$, because of normalization.

19.46 **(a)** Yes; **(b)** no; **(c)** yes; **(d)** yes.

19.47 **(a)** E_e in (19.8) depends in the coordinates of the nuclei; for fixed nuclear positions, E_e is a constant.

(b) The probability density ρ is a function of the spatial coordinates x, y, z. (In a molecule, it also depends parametrically on the nuclear locations.)

(c) E_{xc} in (19.44) is the sum of definite integrals (which are numbers) and E_{xc} is a constant for a fixed nuclear configuration. (Of course, it changes when the nuclear positions change.)

(d) υ_{xc} in (19.49) is a function of x, y, z (and also depends parametrically on the nuclear coordinates).

19.48 Multiplication of (19.7) by ψ_e^* and integration over all space gives $\int \psi_e^* \hat{H}_e \psi_e \, d\tau = E_e \int \psi_e^* \psi_e \, d\tau$. Because of normalization, the integral on the right side equals 1 and $\int \psi_e^* \hat{H}_e \psi_e \, d\tau = E_e$. Use of (19.6) gives $E_e =$ $\int \psi_e^* (\hat{K}_e + \hat{V}_{Ne} + \hat{V}_{ee} + \hat{V}_{NN}) \psi_e \, d\tau = \int \psi_e^* \hat{K}_e \psi_e \, d\tau + \int \psi_e^* \hat{V}_{Ne} \psi_e \, d\tau +$ $\int \psi_e^* \hat{V}_{ee} \psi_e \, d\tau + \int \psi_e^* \hat{V}_{NN} \psi_e \, d\tau = \langle K_e \rangle + \langle V_{Ne} \rangle + \langle V_{ee} \rangle + \langle V_{NN} \rangle$.

19.49 **(a)** T; **(b)** T; **(c)** F; **(d)** T; **(e)** T.

19.50 The potential energy of interaction between the infinitesimal charge elements dQ_1 and dQ_2 of the continuous distribution is given by Eq. (18.1) as $dQ_1 \, dQ_2/4\pi\varepsilon_0 r_{12}$, where r_{12} is the distance between dQ_1 and dQ_2. If ρ is the electron probability density, then $-e\rho$ is the charge density (the charge per unit volume) and multiplication by the infinitesimal volume gives $dQ_1 =$ $-e\rho(x_1, y_1, z_1) \, dx_1 \, dy_1 \, dz_1$ and $dQ_2 = -e\rho(x_2, y_2, z_2) \, dx_2 \, dy_2 \, dz_2$. Integration of $dQ_1 \, dQ_2/4\pi\varepsilon_0 r_{12}$ over the coordinates x_2, y_2, z_2 of dQ_2 gives the energy of interaction of dQ_1 with the entire continuous charge distribution. If we then integrate over the coordinates x_1, y_1, z_1 of dQ_1, we get the total energy of interaction of all the infinitesimal elements of charge with one another, except that we must divide by 2 to avoid counting each interaction twice. Thus the interaction energy is given by Eq. (19.48).

19.51 We have $g = -(3/4)(3/\pi)^{1/3}\rho^{4/3}$ so $\partial g/\partial \rho = -(3/4)(3/\pi)^{1/3}(4/3)\rho^{1/3}$ $-(3/\pi)^{1/3}\rho^{1/3}$ and $\partial g/\partial \rho_x = 0$. Then, since $F = E_x^{\text{LDA}}$, we have $\delta E_x^{\text{LDA}}/\delta\rho = -(3/\pi)^{1/3}\rho^{1/3}$.

19.52 There are 2 pi electrons in each double bond, plus the lone pair on N, for a total of $2k + 4$ pi electrons. These fill the lowest $k + 2$ pi MO's and the transition is from $n = k + 2$ to $n = k + 3$. There are $2k + 2$ conjugated single and double bonds and addition of an extra bond length at each end gives a box length of

$(2k + 4)(1.40\ \text{Å}) = a$. Then $|\Delta E| = (h^2/8ma^2)[(k+3)^2 - (k+2)^2] = (2k+5)(h^2/8ma^2)$. $|\Delta E| = h\nu = hc/\lambda$ and $\lambda = hc/|\Delta E| = 8ma^2c/(2k+5)h = 8m(2k+4)^2(1.40\ \text{Å})^2c/(2k+5)h = 8(9.11 \times 10^{-31}\ \text{kg})(2k+4)^2(1.40)^2 \times (10^{-10}\ \text{m})^2(3.00 \times 10^8\ \text{m/s})/(2k+5)(6.63 \times 10^{-34}\ \text{J s}) = (64.6\ \text{nm})(2k+4)^2/(2k+5)$. For $k = 1$, $\lambda = (64.6\ \text{nm})(2+4)^2/(2+5) = 332\ \text{nm}$.

19.53 For $CH_4(g)$, $\Delta_f H^\circ_{298} \cong N_A(E_e - \sum_a n_a\alpha_a) = (6.022 \times 10^{23}/\text{mol})[-40.19517 - 1(-37.88449) - 4(-0.57077)]$ hartrees $\times$ $(4.35974 \times 10^{-18}$ J/hartree$) = -72.5$ kJ/mol. For $H_2O_2(g)$, $E_e - \sum_a n_a\alpha_a = [-150.76479 - 2(-0.57077) - 2(-74.78852)]$ hartrees $= -0.04621$ hartrees and $\Delta_f H^\circ_{298} \cong -121.3$ kJ/mol.

19.54 The molecule is CH_3CHO, with H4, H5, and H6 bonded to C1 and H3 and O bonded to C2. The atoms C1, C2, H3, and O lie in the same plane. Atom H4 eclipses the O atom and H3 is staggered between H5 and H6.

19.55 **(a)** The OH hydrogen lies on the same side of the C—O bond as the carbonyl oxygen in one conformer and on the opposite side in the other conformer. For the first-mentioned conformer, a possible Z-matrix is

```
C1
O2  1  1.43
O3  1  1.23  2  120.0
H4  2  0.96  1  108.0  3    0.0
H5  1  1.07  2  120.0  4  180.0
```

where bond lengths were estimated from the bond radii in Sec. 19.1. The Z-matrix of the second conformer is the same except that the dihedral angles in row four and row five are changed from 0.0 and 180.0 to 180.0 and 0.0, respectively.

(b) One possible answer is

```
C1
X2  1  1.0
O3  1  1.23  2  90.0
O4  1  1.23  2  90.0  3  180.0
```

19.56 For the Windows version of *Gaussian*, the input in Fig. 19.38, appropriately modified, can be used. For part (a), only the basis set needs to be changed. For part (b), HF/3-21G in the Route Section should be changed to B3LYP/6-31G*. In using *Gaussian*, you are asked to specify a filename for the output. Note that filenames with an asterisk are forbidden. Results are (atoms as in Fig. 19.39):

	HF/3-21G	HF/6-31G*	HF/6-31G**	B3LYP/6-31G*
μ/D	2.12	1.87	1.83	1.69
R(CO)/Å	1.441	1.399	1.398	1.419
R(OH)/Å	0.966	0.946	0.942	0.969
R(CH4)/Å	1.085	1.088	1.088	1.101
R(CH5)/Å	1.079	1.081	1.082	1.093
∠COH/°	110.3	109.4	109.6	107.7
∠OCH4/°	112.2	112.0	112.1	112.7
∠OCH5/°	106.3	107.2	107.3	106.7
D(H6COH3)	61.4°	61.2°	61.2°	61.5°
D(H5COH3)	180°	180°	180°	180°

19.57 **(a)** The Z-matrices of Prob. 19.55a are used. One finds –188.7623096 hartrees for the conformer with the OH hydrogen near the carbonyl O and –188.7525454 hartrees for the other conformer. This is an energy difference of 0.0097642 hartrees. From Eq. (19.1) and the equation after (18.18), one hartree corresponds to 27.211(23.061 kcal/mol) = 627.51 kcal/mol, so the conformer with the OH hydrogen close to the CO oxygen is predicted to be more stable by 6.1 kcal/mol.

(b) The calculated energies are –189.7554562 hartrees for the conformer with the OH hydrogen near the carbonyl O and –189.7471663 hartrees for the other conformer. This is an energy difference of 0.0082899 hartrees or 5.2 kcal/mol.

19.58 MM2 in CS Chem 3D Ultra 9.0 and MMFF94s in Spartan (values in parentheses) give *R*(CO) = 1.419 Å (1.416 Å), *R*(OH) = 0.960 Å (0.972 Å), all CH distances = 1.112 Å (1.098 Å), ∠COH = 109.2° (107.1°), ∠HCO = 108.6° and 108.4° (109.5°, 109.0°), *D*(HCOH) = 180° and ±59.8° (180°, ±60.4°).

19.59 MM2 in CS Chem 3D Ultra 9.0 gives a steric energy of 2.173 kcal/mol for the geometry-optimized trans (anti) conformer and 3.034 kcal/mol for the optimized gauche conformer. The predicted energy difference is 0.86 kcal/mol, with the trans being of lower energy. MMFF94s in Spartan gives a steric energy of –21.238 kJ/mol for the anti form and –17.965 kJ/mol for gauche; the energy difference is 0.78 kcal/mol.

19.60 MM2 in CS Chem 3D Ultra 9.0 gives a steric energy of 1.371 kcal/mol for the cis isomer and 2.524 kcal/mol for the trans isomer. The cis isomer is predicted to be more stable than the trans by 1.2 kcal/mol, which is not what one would expect from chemical intuition. [For references on this "cis effect," see N. C. Craig et al., *J. Phys. Chem. A*, **102**, 6745 (1998).] MMFF94s in Spartan erroneously predicts the trans isomer to be more stable.

19.61 **(a)** There are seven bonds and hence 7 terms in V_{str}.

(b) There are 6 bond angles at each carbon atom and therefore a total of 12 bond angles and 12 terms in V_{bend}. [Each carbon is bonded to four atoms and the bond angle at a carbon is described by specifying the two atoms at the ends of the angle; the number of ways of selecting two objects from four objects is ½(4)(3) = 6.]

(c) Each Cl at one carbon has a 1,4 relation with three Cl atoms on the second carbon. There are thus a total of 3(3) = 9 terms in V_{tors}.

(d) There are no 1,5 or higher interactions and as in (c), there are 9 pairs of atoms that have a 1,4 relation, so there are 9 terms in V_{vdW}.

(e) 9, as in (d).

19.62 **(a)** $\approx$ 400 kJ/mol; **(b)** $=(3/2)RT \approx$ 4 kJ/mol; **(c)** $\approx$ 12 kJ/mol; **(d)** $\approx$ 600 kJ/mol; **(e)** = 1300 kJ/mol, which corresponds to 13.6 eV/molecule. So (b) < (c) < (a) < (d) < (e).

19.63 **(a)** T. **(b)** F. **(c)** F. **(d)** F. **(e)** F. **(f)** F. **(g)** T. **(h)** T. **(i)** T.

R19.1 $n = 5$, $l = 2$, m can be –2, –1, 0, 1, or 2.

R19.2 **(a)** T. **(b)** F (As shown by the discussion of measuring the position, only probabilities can be predicted, unless the state function happens to be an eigenfunction of $\hat{B}$.) **(c)** T.

R19.3 **(a)** Yes. **(b)** No.

R19.4 **(a)** d/dx; **(b)** $(\quad)^2$.

R19.5 **(a)** $[2(3)]^{1/2}\hbar = 6^{1/2}\hbar = 6^{1/2}(6.626 \times 10^{-34}\text{ J s})/2\pi = 2.58 \times 10^{-34}$ J s.

(b) $[\frac{1}{2}(\frac{3}{2})]^{1/2}\hbar = 3^{1/2}\hbar/2 = 3^{1/2}(6.626 \times 10^{-34}\text{ J s})/4\pi = 9.13 \times 10^{-35}$ J s.

R19.6 See Eqs. (18.40), (18.41), the paragraph that follows them, and Fig. 18.13.

R19.7 **(a)** See Eq. (18.35). **(b)** $e^2/4\pi\varepsilon_0 r_{12}$. **(c)** See Eq. 18.42.

R19.8 **(a)** The angles at C are close to 109½°; the angle at O is a bit less than 109½°.

(b) 90°.

(c) A bit less than 90° and a bit less than 120°.

(d) A bit less than 109½° at O; a bit more than 120° for ∠OCO and for ∠HCO(carbonyl O), a bit less than 120° for ∠HCO(hydroxyl O).

R19.9 Well predicted: geometries, dipole moments, barriers to internal rotation, electron probability densities; poorly predicted: dissociation energies.

R19.10 $\hat{A}\hat{B}(x^2y^2z^2) = (\partial^2/\partial x^2)(x^5y^2z^2) = 20x^3y^2z^2$

$\hat{B}\hat{A}(x^2y^2z^2) = x^3(\partial^2/\partial x^2)(x^2y^2z^2) = x^3 2y^2z^2 = 2x^3y^2z^2$.

$\hat{A}\hat{B}(x^2y^2z^2) - \hat{B}\hat{A}(x^2y^2z^2) = 18x^3y^2z^2$

R19.11 Only (b).

R19.12 $L = 1$ and $|\mathbf{L}| = (1 \cdot 2)^{1/2}\hbar = 2^{1/2}\hbar$. $S = 1$ and $|\mathbf{S}| = (1 \cdot 2)^{1/2}\hbar = 2^{1/2}\hbar$.

R19.13 $(\hbar/i)(x\,\partial/\partial y - y\,\partial/\partial x)$.

R19.14 If ϕ is any normalized, well-behaved function of the coordinates of the system, then $\int \phi^* \hat{H}\phi\, d\tau \ge E_{gs}$, where E_{gs} is the ground-state energy of the system with Hamiltonian operator $\hat{H}$.

R19.15 (This is the same problem as 17.59a.)

$\int \phi^*\phi\, d\tau = \int_0^l x^4(l-x)^4\, dx = \int_0^l (l^4x^4 - 4l^3x^5 + 6l^2x^6 - 4lx^7 + x^8)\, dx =$

$l^9/5 - 4l^9/6 + 6l^9/7 - 4l^9/8 + l^9/9 = l^9/630$

$\int \phi^* \hat{H}\phi\, d\tau = -(\hbar^2/2m)\int_0^l x^2(l-x)^2(d^2/dx^2)(l^2x^2 - 2lx^3 + x^4)\, dx =$

$-(\hbar^2/2m)\int_0^l x^2(l-x)^2(2l^2 - 12lx + 12x^2)\, dx =$

$-(\hbar^2/2m)\int_0^l (2l^4x^2 - 16l^3x^3 + 38l^2x^4 - 36lx^5 + 12x^6)\, dx =$

$-(\hbar^2/2m)(2l^7/3 - 16l^7/4 + 38l^7/5 - 36l^7/6 + 12l^7/7) = (\hbar^2 l^7/m)\frac{1}{105}$

$\int \phi^* \hat{H}\phi\, d\tau \Big/ \int \phi^*\phi\, d\tau = (\hbar^2/ml^2)(630/105) = 3h^2/2\pi^2ml^2 = 0.152h^2/ml^2$. The exact ground-state energy is $(1/8)\,h^2/ml^2 = 0.125\,h^2/ml^2$, so the percent error is 22%. The tedious algebra is best done using a computer algebra program such as Mathematica, MathCad, or Maple or using a graphing calculator with such capabilities.

R19.17 **(a)** See Fig. 17.7. **(b)** See Fig. 17.16. **(c)** See the lowest curve in Fig. 19.5.

R19.18 **(a)** When more than one state has the same energy, that energy level is said to be degenerate, and the degree of degeneracy of the level is the number of states that have that energy.

(b) A linear operator $\hat{B}$ is one that obeys the relations $\hat{B}(cf) = c\hat{B}f$ and $\hat{B}(f+g) = \hat{B}f + \hat{B}g$ for all functions f and g and all constants c.

R19.19 **(a)** molecular mechanics; **(b)** ab initio; **(c)** ab initio; **(d)** semiempirical; **(e)** density functional.

Chapter 20

20.1 **(a)** F; **(b)** F; **(c)** T; **(d)** F; **(e)** T.

20.2 $E = (1.00 \text{ eV})(1.602 \times 10^{-19} \text{ J/1 eV}) = 1.602 \times 10^{-19} \text{ J} = h\nu$, so
$\nu = (1.602 \times 10^{-19} \text{ J})/(6.626 \times 10^{-34} \text{ J s}) = 2.42 \times 10^{14} \text{ s}^{-1}$.
$\lambda = c/\nu = (2.998 \times 10^{10} \text{ cm/s})/(2.42 \times 10^{14}/\text{s}) = 1.24 \times 10^{-4} \text{ cm}$.
$\sigma = 1/\lambda = (1.24 \times 10^{-4} \text{ cm})^{-1} = 8.06 \times 10^{3} \text{ cm}^{-1}$.

20.3 $c_{H_2O} = c/n_{H_2O} = (2.998 \times 10^{10} \text{ cm/s})/1.33 = 2.25 \times 10^{10}$ cm/s. The frequency is unchanged in water and $\lambda_{H_2O} = c_{H_2O}/\nu = c_{H_2O}\lambda_{\text{vac}}/c = \lambda_{\text{vac}}/n_{H_2O} =$ (589 nm)/1.33 = 443 nm. $\nu = \nu_{\text{vac}} = c/\lambda_{\text{vac}} = (2.9979 \times 10^{8} \text{ m/s})/(589 \times 10^{-9} \text{ m})$ $= 5.09 \times 10^{14}$ Hz.

20.4 **(a)** F; **(b)** F; **(c)** T; **(d)** T; **(e)** T.

20.5 **(a)** s^{-1} or Hz; **(b)** m^{-1}; **(c)** m/s; **(d)** no units; **(e)** m^2/mol.

20.6 Using identities before Eq. (20.6), we get
$\mu_{mn} = (2Q/a)\frac{1}{2}\int_0^a \{x \cos [(m-n)\pi x/a] - x \cos [(m+n)\pi x/a]\}\, dx =$

$$\frac{Q}{a}\left[\frac{a^2}{(m-n)^2\pi^2}\cos\frac{(m-n)\pi x}{a} + \frac{xa}{(m-n)\pi}\sin\frac{(m-n)\pi x}{a}\right.$$

$$\left. - \frac{a^2}{(m+n)^2\pi^2}\cos\frac{(m+n)\pi x}{a} - \frac{xa}{(m+n)\pi}\sin\frac{(m+n)\pi x}{a}\right]\Bigg|_0^a$$

which reduces to Eq. (20.6), since $\sin j\pi = 0$ for j an integer and $\cos 0 = 1$.

20.7 $\nu_{\text{light}} = |\Delta E|/h = [(\upsilon_2 + \frac{1}{2})h\nu_{\text{vib}} - (\upsilon_1 + \frac{1}{2})h\nu_{\text{vib}}]/h = (\upsilon_2 - \upsilon_1)\nu_{\text{vib}} = \nu_{\text{vib}}$.

20.8 **(a)** $\hat{\mu} = Qx$, since this is a one-particle, one-dimensional system.
$Q \int \psi_0^* x \psi_1\, dx = Q(\alpha/\pi)^{1/4}(4\alpha^3/\pi)^{1/4}\int_{-\infty}^{\infty} x^2 e^{-\alpha x^2}\, dx =$

$Q\alpha(2/\pi)^{1/2}2(2!)\pi^{1/2}/2^3 1!\alpha^{3/2} = Q/(2\alpha)^{1/2}$, where integrals 1 and 3 of Table 14.1 were used.

(b) $Q(\alpha/\pi)^{1/4}(\alpha/4\pi)^{1/4} \int_{-\infty}^{\infty} (2\alpha x^3 - x)e^{-\alpha x^2}\, dx = 0$, where integral 4 was used.

(c) $Q(\alpha/\pi)^{1/4}(\alpha/9\pi)^{1/4} \int_{-\infty}^{\infty} (2\alpha^{3/2}x^4 - 3\alpha^{1/2}x^2)e^{-\alpha x^2}\, dx =$ $Q(\alpha/3\pi)^{1/2}[2\alpha^{3/2}2(4!)\pi^{1/2}/2^5 2!\alpha^{5/2} - 3\alpha^{1/2}2(2!)\pi^{1/2}/2^3 1!\alpha^{3/2}] = 0$. The results (a)–(c) are consistent with $\Delta\upsilon = \pm 1$.

20.9 $E_C - E_A = hc/\lambda_{AC}$. $E_C - E_B = hc/\lambda_{BC}$. $E_B - E_A = hc/\lambda_{AB} =$ $(E_C - E_A) - (E_C - E_B) = hc/\lambda_{AC} - hc/\lambda_{BC}$, so $1/\lambda_{AB} = 1/\lambda_{AC} - 1/\lambda_{BC} =$ $(485\text{ nm})^{-1} - (884\text{ nm})^{-1} = 0.000931\text{ nm}^{-1}$ and $\lambda_{AB} = 1075$ nm.

20.10 Since E increases as $n^2 + n$, the spacing between levels increases as n increases. Therefore the lowest-frequency absorption is due to the transition from $n = 1$ to $n = 3$. We have $\nu_{lowest} = 80\text{ GHz} = |\Delta E|/h = [b(3)(5) - b(1)(3)]/h$ $= 12b/h$ and $b = (80\text{ GHz})h/12$. The next-lowest absorption frequency is that from $n = 2$ to $n = 4$ and its frequency is $\nu = |\Delta E|/h = [4(6)b - 2(4)b]/h =$ $16b/h = 16(80\text{ GHz})h(12)^{-1}/h = (16/12)(80\text{ GHz}) = 107$ GHz.

20.11 **(a)** $\nu = |\Delta E|/h = (h^2/8ma^2)(n_2^2 - n_1^2)/h = (h/8ma^2)(n_2^2 - n_1^2) =$ $(6.626 \times 10^{-34}\text{ J s})(n_2^2 - n_1^2)/8(9.109 \times 10^{-31}\text{ kg})(2.00 \times 10^{-10}\text{ m})^2 =$ $(2.273 \times 10^{15}/\text{s})(n_2^2 - n_1^2)$. The selection rule is that Δn is odd, so the lowest frequencies result from $n = 1 \rightarrow 2$, $n = 1 \rightarrow 4$, and $n = 1 \rightarrow 6$. The frequencies are $(2.273 \times 10^{15}/\text{s})(4 - 1) = 6.82 \times 10^{15}$ Hz, 3.41×10^{16} Hz, and 7.96×10^{16} Hz.

(b) The smallest values of $n_2^2 - n_1^2$ with $n_2 - n_1$ odd are for $n = 1 \rightarrow 2$, $n = 2 \rightarrow 3$, and $n = 3 \rightarrow 4$. We get 6.82×10^{15} Hz, 1.14×10^{16} Hz, and 1.59×10^{16} Hz.

20.12 $\nu = (E_{upper} - E_{lower})/h = [an_{upper}(n_{upper} + 4) - an_{lower}(n_{lower} + 4)]/h =$ $a[(n_{lower} + 3)(n_{lower} + 3 + 4) - n_{lower}(n_{lower} + 4)]/h = (6n_{lower} + 21)a/h$, where $n_{lower} = 0, 1, 2, \ldots$

20.13 Using u and ℓ for upper and lower, we have
$\nu = h^{-1}A[K_u(K_u + 3) - K_\ell(K_\ell + 3)] = Ah^{-1}[(K_\ell + 1)(K_\ell + 4) - K_\ell(K_\ell + 3)] =$

$2Ah^{-1}(K_\ell + 2)$, where $K_\ell = 1, 2, 3, \ldots$. Then 60 GHz = $2Ah^{-1}(4)$; $A = 7.5h$ GHz; ν = (15 GHz)(K_ℓ + 2). 135 GHz = (15 GHz)(K_ℓ + 2) and $K_\ell = 7$.

20.14 $T = 10^{-A}$. For $A = 0.1$, $T = 10^{-0.1} = 0.79$ and 21% is absorbed. For $A = 1$, $T = 0.10$ and 90% is absorbed. For $A = 10$, $T = 10^{-10}$ and 99.99999999% is absorbed.

20.15 $c = n/V = P/RT$ and $T = I_\lambda/I_{\lambda,0} = 10^{-\varepsilon_\lambda c_B \ell} = 10^{-\varepsilon_\lambda P_B \ell / RT}$.

(a) $\varepsilon_\lambda P_B \ell / RT$ = (10^4 dm^3/mol-cm)[(10/760) atm](1.0 cm)/ (0.08206 dm^3-atm/mol-K)(298 K) = 5.3_8. $T = 10^{-5.38} = 4._2 \times 10^{-6}$.

(b) $T = 10^{-53.8} = 1._6 \times 10^{-54}$.

20.16 $T = I_\lambda/I_{\lambda,0} = 10^{-\varepsilon_\lambda c_B \ell}$.

(a) $\varepsilon_\lambda c_B \ell$ = (150 dm^3/mol-cm)(10^{-3} mol/dm^3)(1.0 cm) = 0.15 and $T = 10^{-0.15} = 0.71$.

(b) $T = 10^{-1.5} = 0.03_2$.

20.17 $A = \log(I_0/I) = \log(T^{-1}) = \log(1/0.083) = 1.08_1$.
c = (0.080 g/cm^3)(1 mol/14600 g)(10^3 cm^3/1 dm^3) = 0.0055 mol/dm^3. $T = I/I_0 = 10^{-\varepsilon cl}$ and $\varepsilon = -(1/cl) \log T$ = −(0.0055 mol/dm^3)$^{-1}$(0.010 cm)$^{-1}$ log 0.083 = 2.0 × 10^4 dm^3 mol^{-1} cm^{-1}.

20.18 $A_2/A_1 = [\log(I_{\lambda,0}/I_\lambda)_2]/[\log(I_{\lambda,0}/I_\lambda)_1] = \varepsilon_\lambda c_{B,2} l_2/\varepsilon_\lambda c_{B,1} l_1 = c_{B,2}/c_{B,1} = 2$.
So $\log(I_{\lambda,0}/I_\lambda)_2 = 2 \log(1/0.60) = 0.444$ and $I_{\lambda,0}/I_\lambda = 2.78$. We have $T = 1/2.78 = 0.36$ and 36% of the light is transmitted.

20.19 Let the subscripts 3 and 4 denote $Fe(CN)_6^{3-}$ and $Fe(CN)_6^{4-}$, respectively. Use of $A = (\varepsilon_3 c_3 + \varepsilon_4 c_4)l$ and $c_3 + c_4 = 1.00 \times 10^{-3}$ mol/dm^3 gives 0.701 = [(800 dm^3/mol-cm)c_3 + (320 dm^3/mol-cm)(0.00100 mol/dm^3 − c_3)](1.00 cm). We find $c_3 = 7.94 \times 10^{-4}$ mol/dm^3. Then $c_3/c_{3,0}$ = 0.000794/0.00100 = 0.794. The % reacted is 20.6%.

20.20 **(a)** T; **(b)** T; **(c)** F; **(d)** F; **(e)** T; **(f)** T; **(g)** F; **(h)** T.

20.21 **(a)** F; **(b)** T; **(c)** F.

20.22 Division of Eq. (20.25) by hc gives $D_0/hc = D_e/hc - \frac{1}{2}\tilde{\nu}_e + \frac{1}{4}\tilde{\nu}_e x_e$.

(a) $D_0/hc = [79890 - \frac{1}{2}(2359) + \frac{1}{4}(14)]\ \text{cm}^{-1} = 78714\ \text{cm}^{-1}$;
$D_0 = (78714\ \text{cm}^{-1})(100\ \text{cm/m})(6.6261 \times 10^{-34}\ \text{J s})(2.9979 \times 10^8\ \text{m/s}) = 1.5636 \times 10^{-18}\ \text{J} = 9.759\ \text{eV}$.

(b) $D_0/hc = [90544 - \frac{1}{2}(2170) + \frac{1}{4}(13)]\ \text{cm}^{-1} = 89462\ \text{cm}^{-1}$;
$D_0 = 1.7771 \times 10^{-18}\ \text{J} = 11.092\ \text{eV}$.

20.23 **(a)** D_e is the depth of the electronic energy curve E_e and k_e equals $E_e''(R_e)$. In the Born–Oppenheimer approximation, $E_e(R)$ is found by solving the electronic Schrödinger equation (19.7) in which the nuclei are fixed; the nuclear masses do not occur in (19.7) and (19.6). Hence, $E_e(R)$ is the same for $^2H^{35}Cl$ and $^1H^{35}Cl$, which have the same nuclear charges. From Eq. (20.25), D_0 differs from D_e by the zero-point vibrational energy. The vibrational frequency ν_e equals $(1/2\pi)(k/\mu)^{1/2}$. The reduced mass μ differs substantially for $^2H^{35}Cl$ and $^1H^{35}Cl$, so their ν_e's differ and their D_0's differ.

(b) For $^1H^{35}Cl$, $D_0/hc = D_e/hc - \frac{1}{2}\tilde{\nu}_e + \frac{1}{4}\tilde{\nu}_e x_e =$
$[37240 - \frac{1}{2}(2990.9) + \frac{1}{4}(52.8)]\ \text{cm}^{-1} = 35758\ \text{cm}^{-1}$;
$D_0 = 7.103_1 \times 10^{-19}\ \text{J} = 4.433_4\ \text{eV}$. From Eqs. (20.23) and (17.79): $\nu_{e,\text{DCl}}/\nu_{e,\text{HCl}} = (\mu_{\text{HCl}}/\mu_{\text{DCl}})^{1/2}$; $\mu_{\text{HCl}} = 1.00782(34.969)(\text{g/mol})/35.977N_A = (0.97959\ \text{g/mol})/N_A$; $\mu_{\text{DCl}} = (1.9044\ \text{g/mol})/N_A$. So $\tilde{\nu}_{e,\text{DCl}} =$
$(2990.9\ \text{cm}^{-1})(0.97959/1.9044)^{1/2} = 2145.1\ \text{cm}^{-1}$. Also, $D_{e,\text{DCl}} = D_{e,\text{HCl}}$.
For $^2H^{35}Cl$ we then have $D_0/hc = [37240 - \frac{1}{2}(2145) + \frac{1}{4}(53)]\ \text{cm}^{-1} =$
$36181\ \text{cm}^{-1}$ and $D_0 = 7.187 \times 10^{-19}\ \text{J} = 4.486\ \text{eV}$ (where we neglected the change in $\nu_e x_e$).

20.24 $\frac{1}{2}h\nu_e = \frac{1}{2}kA^2$. But $\nu_e = (1/2\pi)(k/\mu)^{1/2}$, so $k = 4\pi^2\mu\nu_e^2$; $\frac{1}{2}h\nu_e = 2\pi^2\mu\nu_e^2A^2$ and $A = (h/4\pi^2\mu\nu_e)^{1/2}$. For $H^{35}Cl$, $\mu = (1.0)(35.0)\text{g}/(36.0)(6.02 \times 10^{23}) = 1.6 \times 10^{-24}\ \text{g}$ and $\nu_e = (2.998 \times 10^{10}\ \text{cm/s})(2991\ \text{cm}^{-1}) = 9.0 \times 10^{13}\ \text{s}^{-1}$, so $A = [(6.63 \times 10^{-34}\ \text{J s})/4\pi^2(1.6 \times 10^{-27}\ \text{kg})(9.0 \times 10^{13}/\text{s})]^{1/2} = 1.1 \times 10^{-11}\ \text{m} =$

0.11 Å. For $^{14}N_2$, $\mu = 14(14)\text{g}/28N_A = 1.16 \times 10^{-23}$ g, $\nu_e = 7.1 \times 10^{13}\ \text{s}^{-1}$ and $A = 0.045$ Å.

20.25 Use of (20.31) gives $\lambda = c/\nu = c/2(J+1)B$, so 2.00 cm $= c/2(3)B$ and $B = c/(12.0\ \text{cm})$. We then have $\lambda = c/2(J+1)c(12.0\ \text{cm})^{-1} = (6.0\ \text{cm})/(J+1)$. $\lambda_{6\to 7} = (6.0\ \text{cm})/7 = 0.86$ cm.

20.26 From Eqs. (20.31), (20.33), and (20.15): $\nu_{J\to J+1} = 2(J+1)B_0 = 2B_0 = 2h/8\pi^2 I_0 = h/4\pi^2\mu R_0^2$ and $R_0 = (h/\mu\nu)^{1/2}/2\pi$.

(a) Use of the table of atomic masses gives $\mu = m_1m_2/(m_1+m_2) = (1.0078250)(78.91834)\text{g}/(79.92616_5)(6.02214 \times 10^{23}) = 1.652431 \times 10^{-24}$ g for $^1H^{79}Br$ and $\mu = 1.652945 \times 10^{-24}$ g for $^1H^{81}Br$. For $^1H^{79}Br$, $R_0 = [(6.62607 \times 10^{-34}\ \text{J s})/(1.652431 \times 10^{-27}\ \text{kg})(500.7216 \times 10^9/\text{s})]^{1/2}/2\pi = 1.424257$ Å. For $^1H^{81}Br$, we get $R_0 = 1.424257$ Å.

(b) For $J = 1$ to 2, $\nu = 2(J+1)B_0 = 4B_0 = 2\nu_{J=0\to 1} = 2(500.7216\ \text{GHz}) = 1001.4432$ GHz with centrifugal distortion neglected.

(c) For the $J = 0$ to 1 transition, $\nu_{DBr}/\nu_{HBr} = 2B_{0,DBr}/2B_{0,HBr} = \mu_{HBR}/\mu_{DBr}$, since R_0 is essentially unchanged on isotopic substitution. $\mu_{DBr} = (2.014102)(78.91834)\text{g}/(80.93244_2)(6.02214 \times 10^{23}) = 3.261264 \times 10^{-24}$ g. $\mu_{HBr}/\mu_{DBr} = 1.652431/3.261264 = 0.506684$ and $\nu_{DBr} = 0.506684(500.7216\ \text{GHz}) = 253.708$ GHz.

20.27 From Eq. (20.31), $\nu_{J\to J+1} = 2B_0(J+1)$, so $B_0(^{39}K^{37}Cl) = (22410\ \text{MHz})/2(3) = 3735$ MHz.

(a) $\nu = 2(3735\ \text{MHz})1 = 7470$ MHz.

(b) The reasoning in Prob. 20.23a shows that R_e for the two isotopic species is the same; further, R_0 should differ only very slightly for the two species (see, for example, Prob. 20.26). Equations (20.33) and (20.15) then give $B_0(^{39}K^{35}Cl)/B_0(^{39}K^{37}Cl) = I_0(^{39}K^{37}Cl)/I_0(^{39}K^{35}Cl) = \mu(^{39}K^{37}Cl)/\mu(^{39}K^{35}Cl) = 18.9693/18.4292 = 1.02931$, where the reduced masses were found from $m_Am_B/(m_A+m_B)$. Then $B_0(^{39}K^{35}Cl) = 1.02931(3735\ \text{MHz}) = 3844\tfrac{1}{2}$ MHz and $\nu_{J=0\to 1} = 2(3844\tfrac{1}{2}\ \text{MHz}) = 7689$ MHz for $^{39}K^{35}Cl$.

20.28 The equation preceding (20.37) is $\tilde{\nu} = \tilde{\nu}_{\text{origin}} + \tilde{B}_e J'(J'+1) - \tilde{B}_e J''(J''+1)$. For P branch lines, $J' = J'' - 1$ and $J'(J'+1) - J''(J''+1) = (J''-1)J'' - J''(J''+1) = -2J''$, so $\tilde{\nu}_P = \tilde{\nu}_{\text{origin}} - 2\tilde{B}_e J''$.

20.29 Eq. (20.39) multiplied out gives $\tilde{\nu}_R = \tilde{\nu}_{\text{origin}} + 2\tilde{B}_e(J''+1) - \tilde{\alpha}_e \upsilon'(J''^2 + 3J'' + 2) + \tilde{\alpha}_e \upsilon''(J''^2 + J'') - \tilde{\alpha}_e(J''+1)$. With centrifugal distortion neglected, Eq. (20.26) gives $\tilde{\nu}_R = (E_{J''+1,\upsilon'} - E_{J'',\upsilon''})/hc =$ $\tilde{\nu}_e(\upsilon' + \frac{1}{2}) - \tilde{\nu}_e x_e(\upsilon' + \frac{1}{2})^2 + \tilde{B}_e(J''+1)(J''+2) -$ $\tilde{\alpha}_e(\upsilon' + \frac{1}{2})(J''+1)(J''+2) - \tilde{\nu}_e(\upsilon'' + \frac{1}{2}) + \tilde{\nu}_e x_e(\upsilon'' + \frac{1}{2})^2 - \tilde{B}_e J''(J''+1) +$ $\tilde{\alpha}_e(\upsilon'' + \frac{1}{2})J''(J''+1) = \tilde{\nu}_e(\upsilon' - \upsilon'') - \tilde{\nu}_e x_e(\upsilon'^2 + \upsilon' - \upsilon''^2 - \upsilon'') +$ $\tilde{B}_e(2J''+2) + \tilde{\alpha}_e \upsilon''(J''^2 + J'') - \tilde{\alpha}_e \upsilon'(J''^2 + 3J'' + 2) - \tilde{\alpha}_e(J''+1)$, which Eq. (20.34) shows to be the same as the above multiplied-out form of Eq. (20.39), verifying (20.39). When multiplied out (20.40) is $\tilde{\nu}_P = \tilde{\nu}_{\text{origin}} - 2B_e J'' + \tilde{\alpha}_e \upsilon'(J'' - J''^2) + \tilde{\alpha}_e \upsilon''(J'' + J''^2) + \tilde{\alpha}_e J''$. Then $\tilde{\nu}_P = (E_{J''-1,\upsilon'} - E_{J'',\upsilon''})/hc$ and use of (20.26) leads to the multiplied-out form of (20.40).

20.30 $\nu_e = (1/2\pi)(k_e/\mu)^{1/2}$ and $k_e = 4\pi^2\nu_e^2\mu = 4\pi^2c^2\tilde{\nu}_e^2\mu = 4\pi^2(2.9979 \times 10^8 \text{ m/s})^2 \times (158000 \text{ m}^{-1})^2(0.01599491 \text{ kg})/2(6.0221 \times 10^{23}) = 1176$ N/m, where we used $\mu = m_A m_A/(m_A + m_A) = m_A/2$.

20.31 **(a)** From Eq. (20.34), $\tilde{\nu}_{\text{origin}}(0 \to 1) = \tilde{\nu}_e - 2\tilde{\nu}_e x_e$ and $\tilde{\nu}_{\text{origin}}(0 \to 2) = 2\tilde{\nu}_e - 6\tilde{\nu}_e x_e$. Hence $3\tilde{\nu}_{\text{origin}}(0 \to 1) - \tilde{\nu}_{\text{origin}}(0 \to 2) = \tilde{\nu}_e =$ $3(2886.0 \text{ cm}^{-1}) - 5668.0 \text{ cm}^{-1} = 2990.0 \text{ cm}^{-1}$. Then $\tilde{\nu}_e x_e = \frac{1}{2}[\tilde{\nu}_e - \tilde{\nu}_{\text{origin}}(0 \to 1)] = \frac{1}{2}(2990.0 - 2886.0)\text{cm}^{-1} = 52.0 \text{ cm}^{-1}$.

(b) From (20.34), $\tilde{\nu}_{\text{origin}}(0 \to 6) = 6(2990.0 \text{ cm}^{-1}) - 42(52.0 \text{ cm}^{-1}) =$ 15756 cm^{-1}.

20.32 **(a)** The distance of a line from the band origin is the change in rotational energy for the line's transition. The selection rule is $\Delta J = \pm 1$ and the spacings between rotational levels increase as J increases. Hence the two lines closest to the band origin involve the $J = 0$ and 1 levels. The 2139

cm^{-1} line is lower in frequency and energy than the band origin and so must be the $J = 1 \to 0$ line, where the rotational energy is decreasing. The 2147 cm^{-1} line has $J = 0 \to 1$. The 2151 cm^{-1} line has $J = 1 \to 2$. The 2135.5 cm^{-1} line has $J = 2 \to 1$.

(b) Let a, b, c, d be the four given wavenumbers in order of increasing wavenumber. From (20.40) and (20.39), we have

$a = \tilde{\nu}_{\text{origin}} - (2\tilde{B}_e - 2\tilde{\alpha}_e)2 - 4\tilde{\alpha}_e = \tilde{\nu}_{\text{origin}} - 4\tilde{B}_e$

$b = \tilde{\nu}_{\text{origin}} - (2\tilde{B}_e - 2\tilde{\alpha}_e)1 - \tilde{\alpha}_e = \tilde{\nu}_{\text{origin}} - 2\tilde{B}_e + \tilde{\alpha}_e$

$c = \tilde{\nu}_{\text{origin}} + (2\tilde{B}_e - 2\tilde{\alpha}_e)1 - \tilde{\alpha}_e = \tilde{\nu}_{\text{origin}} + 2\tilde{B}_e - 3\tilde{\alpha}_e$

$d = \tilde{\nu}_{\text{origin}} + (2\tilde{B}_e - 2\tilde{\alpha}_e)2 - 4\tilde{\alpha}_e = \tilde{\nu}_{\text{origin}} + 4\tilde{B}_e - 8\tilde{\alpha}_e$

Three spreadsheet cells are designated for $\tilde{\nu}_{\text{origin}}$, $\tilde{B}_e$, and $\tilde{\alpha}_e$. The initial guess for $\tilde{\nu}_{\text{origin}}$ can be taken as the average of b and c, namely 2143.25 cm^{-1}. The initial guesses for $\tilde{B}_e$ and $\tilde{\alpha}_e$ can be taken as zero. The four formulas for a, b, c, and d are entered into four cells and the squares of the deviations of the a, b, c, and d formula values from the observed values are calculated, and summed. The Solver is set up to minimize the sum of the squares of the deviations by varying $\tilde{\nu}_{\text{origin}}$, $\tilde{B}_e$, and $\tilde{\alpha}_e$ subject to the constraints that $\tilde{B}_e$ and $\tilde{\alpha}_e$ be positive. An excellent fit to the observed lines is found with $\tilde{\nu}_{\text{origin}} = 2143.2695\ cm^{-1}$, $\tilde{B}_e = 1.93023$ cm^{-1}, and $\tilde{\alpha}_e = 0.016411$.

(c) From (20.16) and (20.15), $B_e = h/8\pi^2\mu R_e^2$ and $R_e = (h/8\pi^2\mu B_e)^{1/2}$.
$\mu = [(12)(15.994915)/27.994915]\text{g}/(6.02214 \times 10^{23}) = 1.138500 \times 10^{-23}$ g.
$B_e = \tilde{B}_e c = (1.93023\ cm^{-1})(2.997925 \times 10^{10}\ cm/s) = 5.78668 \times 10^{10}\ s^{-1}$.
$R_e = [(6.62607 \times 10^{-34}\ J\ s)/8\pi^2(1.138500 \times 10^{-26}\ kg)(5.78668 \times 10^{10}\ s^{-1})]^{1/2}$
$= 1.12863 \times 10^{-10}$ m = 1.12863 Å, which is smaller than R_0 in Example 20.3.

20.33 With centrifugal distortion neglected, the $\upsilon = 0$ vibration-rotation levels are given by Eq. (20.26) as $E_{\text{vib-rot}} = \frac{1}{2}h\nu_e - \frac{1}{4}h\nu_e x_e + hB_e J(J+1) - \frac{1}{2}h\alpha_e J(J+1)$. For $J = 0$, $E_{\text{vib-rot}}(0) = \frac{1}{2}h\nu_e - \frac{1}{4}h\nu_e x_e$. We have $E_{\text{vib-rot}}(J) - E_{\text{vib-rot}}(0) =$ $h(B_e - \frac{1}{2}\alpha_e)J(J+1) = (\tilde{B}_e - \frac{1}{2}\tilde{\alpha}_e)hcJ(J+1)$. $(\tilde{B}_e - \frac{1}{2}\tilde{\alpha}_e)hc/kT =$ $[10.594 - ½(0.31)]\ cm^{-1}(100\ cm/m)(6.6261 \times 10^{-34}\ J\ s)(2.9979 \times 10^8\ m/s)/$

$(1.3807 \times 10^{-23}$ J/K)(300 K) = 0.050064. The degeneracy of each level is $2J + 1$, so the Boltzmann distribution law gives the relative populations as $N_J/N_0 = (2J+1)e^{-\Delta E/kT} = (2J+1)\exp[-(\tilde{B}_e - \frac{1}{2}\tilde{\alpha}_e)hcJ(J+1)/kT] = (2J+1)e^{-0.050064J(J+1)}$. We find

J	1	2	3	4	5	6
N_J/N_0	2.714	3.703	3.839	3.307	2.450	1.588

20.34 **(a)** $E_b - E_a = h\nu_{a\to b} = (6.626 \times 10^{-34}\text{ J s})(1.54 \times 10^{12}\text{ s}^{-1}) = 9.64 \times 10^{-18}\text{ J}$. $N_b/N_a = (g_b/g_a)\exp[-(E_b - E_a)/kT] = (g_b/g_a)\exp[-(9.64 \times 10^{-18}\text{ J})/(1.3806 \times 10^{-23}\text{ J/K})T] = (g_b/g_a)\exp[-(6.98 \times 10^5\text{ K})/T]$. $N_b/N_a = \exp[-(6.98 \times 10^5\text{ K})/(300\text{ K})] = \exp(-2326.7)$. $\ln(N_b/N_a) = -2326.7 = 2.3026\log(N_b/N_a)$ and $N_b/N_a \approx 10^{-1010.5} = 0.3 \times 10^{-1010}$, where (1.71) was used, and where more significant figures than are justified were used. Since N_b/N_a is far less than Avogadro's number, we can say that $N_b/N_a = 0$. (See also the discussion after Example 6.6.)

(b) $g_b/g_a = 5/3$, so $N_b/N_a \approx (5/3)(0.3 \times 10^{-1010}) = 0.5 \times 10^{-1010}$. As in (a), $N_b/N_a = 0$.

(c) $N_b/N_a = \exp[-(6.98 \times 10^5\text{ K})/(1000\text{ K})] = \exp(-698)$. $\ln(N_b/N_a) = -698 = 2.3026\log(N_b/N_a)$ and $N_b/N_a \approx 10^{-303.1} = 0.8 \times 10^{-303}$. $N_b/N_a \approx (5/3)(0.8 \times 10^{-303}) = 1.3 \times 10^{-303}$. As in (a), $N_b/N_a = 0$.

20.35 B_e—rotational constant; α_e—vibration–rotation interaction; D—centrifugal distortion; $\nu_e x_e$—anharmonicity.

20.36 **(a)** From Fig. 19.18, O_2 has 4 more bonding electrons than antibonding electrons, O_2^+ has 5 net bonding electrons, and O_2^- has 3 net bonding electrons. Therefore O_2^+ has the strongest bond and the largest k_e and O_2^- has the smallest k_e.

(b) Use of Fig. 19.15 shows that N_2 has 6 net bonding electrons and N_2^+ has 5 net bonding electrons, so N_2 has the stronger bond, the larger k_e and the larger ν_e since $\nu_e = (1/2\pi)(k_e/\mu)^{1/2}$.

(c) N_2 has a triple bond and O_2 a double bond. The N_2 bond is stronger and N_2 has the larger k_e.

(d) $E_{rot} = J(J+1)\hbar^2/2I$, where $I = \mu R_e^2$. An Na atom is heavier and larger than an Li atom, so Na_2 has the larger μ and the larger R_e. So $E_{rot,J=1}$ is greater for Li_2.

20.37 **(a)** T; **(b)** T; **(c)** T; **(d)** F.

20.38 **(a)** A C_2 axis and two symmetry planes.

(b) A C_3 axis (through the CCl bond) and three symmetry planes (each one containing C, Cl, and one F).

(c) The molecule is square planar. A C_4 axis perpendicular to the molecular plane; an S_4 axis and a C_2 axis, each coincident with the C_4 axis; a symmetry plane coincident with the molecular plane; four symmetry planes perpendicular to the molecular plane; four C_2 axes in the molecular plane (two pass through pairs of opposite F's and two lie between the F's); a center of symmetry.

(d) The structure is trigonal bipyramidal. A C_3 axis; an S_3 axis coincident with the C_3 axis; a (horizontal) plane of symmetry containing the equatorial Cl's; three planes of symmetry that each contain the two axial Cl's; three C_2 axes, each lying in the horizontal symmetry plane.

(e) The VSEPR method shows the structure is a square-based pyramid. A C_4 axis and four symmetry planes.

(f) A C_2 axis perpendicular to the molecular plane; a center of symmetry; two C_2 axes in the molecular plane; three symmetry planes—one coincident with the molecular plane and two perpendicular to it. See Prob. 20.40c.

(g) A C_∞ axis through the nuclei and an infinite number of symmetry planes that each contain the C_∞ axis.

(h) A C_∞ axis (which is also an S_∞ axis), an infinite number of symmetry planes through this axis, a symmetry plane perpendicular to this axis, a center of symmetry, an infinite number of C_2 axes perpendicular to the molecular axis.

20.39 **(a)** The symmetry elements (Prob. 20.38a) are a C_2 axis and two symmetry planes, which we call σ_a and σ_b. The symmetry operations are $\hat{E}$, $\hat{C}_2$, $\hat{\sigma}_a$, $\hat{\sigma}_b$.

(b) $\hat{E}$, $\hat{C}_3$, $\hat{C}_3^2$, $\hat{\sigma}_a$, $\hat{\sigma}_b$, $\hat{\sigma}_c$.

20.40 **(a)** Moves a nucleus at x, y, z to $-x, -y, z$.

(b) From x, y, z, to $x, y, -z$.

(c) The $\hat{S}_2$ rotation about the z axis consists of a $\hat{C}_2$ rotation about z followed by reflection in the xy plane. From the answers to (a) and (b) this moves a point at x, y, z to $-x, -y, -z$. We see that $\hat{S}_2 = \hat{i}$.

20.41 **(a)** The three principal axes intersect at the B nucleus (which is the center of mass). One principal axis is perpendicular to the molecular plane (and coincides with the C_3 axis). The other two principal axes lie in the molecular plane; one of these can be taken to coincide with a BF bond, and the other is perpendicular to this one. (As in XeF_4, the orientation of the principal axes is not unique.)

(b) The three principal axes intersect at the center of mass, which lies on the C_2 axis. One principal axis coincides with the C_2 axis. The second lies in the molecular plane and is perpendicular to the C_2 axis. The third is perpendicular to the molecular plane.

(c) The three principal axes intersect at the C nucleus. One principal axis coincides with the molecular axis (which is a C_∞ axis); the other two can be taken as any two axes through the C that are perpendicular to the molecular axis and perpendicular to each other.

29.42 **(a)** SF_6 has more than one noncoincident C_4 axis and is a spherical top.

(b) IF_5 (which is a square-based pyramid) has one C_4 axis and is a symmetric top.

(c) One C_2 axis. Asymmetric top.

(d) One C_3 axis. Symmetric top.

(e) One C_6 axis. Symmetric top.

(f) One C_∞ axis. Symmetric top.

(g) One C_3 axis. Symmetric top.

(h) Symmetric.

20.43 The principal axes intersect at the center of mass, which is the B nucleus. One principal axis is the C_3 axis. For this axis, $I_c = \sum_i m_i r_{i,c}^2 =$ $3(18.998 \text{ amu})(1.313 \text{ Å})^2 = 98.3 \text{ amu Å}^2$. The other two principal axes lie in the molecular plane and we can take one of them to coincide with a B–F bond. Hence, $I_a = \sum_i m_i r_{i,a}^2 = 2(18.998 \text{ amu})[(1.313 \text{ Å})(\sin 60°)]^2 = 49.1 \text{ amu Å}^2$. With one C_3 axis, BF_3 is a symmetric top, so $I_b = I_a = 49.1 \text{ amu Å}^2$. I_b can also be calculated by taking distances from the F atoms to an in-plane line through B and perpendicular to a B–F bond: $I_b =$ $(19.0 \text{ amu})(1.313 \text{ Å})^2 + 2[(\cos 60°)(1.313 \text{ Å})]^2(19.0 \text{ amu}) = 49.1 \text{ amu Å}^2$.

20.44 One principal axis of this symmetric top is the C_3 axis through the axial bonds. For this axis, $I = 3(34.97 \text{ amu})(2.02 \text{ Å})^2 = 428 \text{ amu Å}^2$. Another principal axis can be taken to coincide with an equatorial P–Cl bond, and for this axis $I =$ $2(34.97 \text{ amu})(2.12 \text{ Å})^2 + 2(34.97 \text{ amu})[(2.02 \text{ Å})(\sin 60°)]^2 = 528 \text{ amu Å}^2$. Since this is a symmetric top, the moments of inertia about the principal axes that are perpendicular to the C_3 axis are equal, and the third principal moment is 528 amu Å^2.

20.45 The molecule is a symmetric top with $I_a \neq I_b = I_c$. From Eq. (20.45), $E_{rot}/h =$ $BJ(J+1) + (A - B)K^2 = [\tilde{B}J(J+1) + (\tilde{A} - \tilde{B})K^2]c$.

(a) For $J = 0$, $K = 0$ and $E_{rot}/h = 0$. For $J = 1$, $K = -1, 0, 1$. For $J = 1$ and $K = 0$, $E_{rot}/h = 2\tilde{B}c = 2(0.05081 \text{ cm}^{-1})(2.9979 \times 10^{10} \text{ cm/s}) =$ $3.046 \times 10^9 \text{ s}^{-1}$. For $J = 1$ and $|K| = 1$, $E_{rot}/h = [2\tilde{B} + (\tilde{A} - \tilde{B})]c =$ $(\tilde{B} + \tilde{A})c = (0.2418 \text{ cm}^{-1})c = 7.249 \times 10^9 \text{ s}^{-1}$.

(b) From Eq. (20.47), $\nu = 2B(J+1) = 2\tilde{B}c\,(J+1) = 2\tilde{B}c,\ 4\tilde{B}c, \ldots =$ $3.046 \times 10^9 \text{ s}^{-1}, 6.093 \times 10^9 \text{ s}^{-1}, \ldots = 3046 \text{ MHz}, 6093 \text{ MHz}$.

20.46 **(a)** Let the molecule lie on the positive half of the z axis with the origin at the oxygen nucleus. Then $z_{com} = [12(1.160 \text{ Å}) + 31.972071(2.720 \text{ Å})]/$ $59.966986 = 1.682 \text{ Å}$.

(b) $I_0 = \Sigma_i m_i r_i^2 = [(15.994915)(1.682)^2 + 12(1.682 - 1.160)^2 + 31.972071(2.720 - 1.682)^2](\text{g Å})^2/(6.02214 \times 10^{23}) = 1.377_7 \times 10^{-22}\ \text{g Å}^2 = 1.377_7 \times 10^{-45}\ \text{kg m}^2$.

(c) $\nu_{J \to J+1} = 2(J+1)B_0$. $B_0 = h/8\pi^2 I_0 = (6.62608 \times 10^{-34}\ \text{J s})/8\pi^2(1.377_7 \times 10^{-45}\ \text{kg m}^2) = 6.091_2 \times 10^9\ \text{s}^{-1}$. $\nu_{0\to 1} = 2B_0 = 12.18$ GHz; $\nu_{1\to 2} = 4B_0 = 24.36$ GHz; $\nu_{2\to 3} = 6B_0 = 36.55$ GHz.

20.47 $B_0 = \tilde{B}_0 c = (0.39021\ \text{cm}^{-1})(2.99792 \times 10^{10}\ \text{cm/s}) = 1.16982 \times 10^{10}\ \text{s}^{-1}$. $I_0 = h/8\pi^2 B_0 = (6.62607 \times 10^{-34}\ \text{J s})/8\pi^2(1.16982 \times 10^{10}\ \text{s}^{-1}) = 7.1738 \times 10^{-46}\ \text{kg m}^2$. The center of mass is at the carbon atom and the principal axes pass through this atom. If d is the CO bond length, then $I_0 = \Sigma_i m_i r_i^2 = m_O d^2 + m_O d^2 = 2m_O d^2$, so $d = (I_0/2m_O)^{1/2} =$ $[(7.1738 \times 10^{-46}\ \text{kg m}^2)(6.02214 \times 10^{23})/2(15.994915 \times 10^{-3})\text{kg}]^{1/2} = 1.162 \times 10^{-10}\ \text{m} = 1.62$ Å.

20.48 **(a)** From the VSEPR method, SO_2 is nonlinear; $3\mathcal{N} - 6 = 9 - 6 = 3$.

(b) Linear. $3\mathcal{N} - 5 = 7$. **(c)** $3\mathcal{N} - 6 = 9$.

20.49 We look for sets of integers i, j, k such that $3657i + 1595j + 3756k$ is slightly greater than 7252. Systematic trial and error (best done by first setting $j = 0$ and looking for i and k values that fit, then setting $j = 1$ and looking for i and k, then setting $j = 2$, etc.) gives the $\upsilon_1'\upsilon_2'\upsilon_3'$ possibilities for the 7252 cm^{-1} band as (calculated frequencies in parentheses) 200 (7314), 101 (7413), 002 (7512).

20.50 **(a)** $\frac{1}{2}\Sigma_i h\nu_i = \frac{1}{2}hc\,\Sigma_i \tilde{\nu}_i = \frac{1}{2}hc(1340 + 667 + 667 + 2349)\ \text{cm}^{-1} = 4.99 \times 10^{-20}\ \text{J} = 0.311$ eV.

(b) $\frac{1}{2}hc(3657 + 1595 + 3756)\ \text{cm}^{-1} = 8.95 \times 10^{-20}\ \text{J} = 0.558$ eV.

20.51 **(a)** Inactive, since the dipole moment remains unchanged.

(b) Active, since the dipole moment changes.

20.52 $\nu = (1/2\pi)\sqrt{k/\mu}$.

(a) The C≡C bond is stronger and has the greater k and the greater ν.

(b) C—H has the smaller μ and the greater ν.

(c) Bending vibrations are generally lower-frequency than stretching, so C—H stretching has the greater ν.

20.53 $\tilde{\nu} = \nu/c = (1/2\pi)(k/\mu)^{1/2}/c$. Isotopic substitution does not affect the electrons and hence doesn't affect k. We have $\mu = m_1m_2/(m_1 + m_2) \approx m_1m_2/m_2 = m_1$, where $m_1 = m_H$ (or m_D) and m_2 is the mass of the rest of the molecule, and we used $m_2 >> m_1$. Therefore $\mu_{CD} \approx 2\mu_{CH}$ and $\tilde{\nu}_{CD} \approx \tilde{\nu}_{CH}/2^{1/2} = (2900 \text{ cm}^{-1})/2^{1/2} = 2050 \text{ cm}^{-1}$.

20.54 $\nu = (1/2\pi)(k/\mu)^{1/2}$ and $k = 4\pi^2\nu^2\mu = 4\pi^2\tilde{\nu}^2c^2\mu$. $\mu_{CH} = 12(1) \text{ g}/13(6.02 \times 10^{23}) = 1.53 \times 10^{-24}$ g, and $\mu_{CO} = 12(16) \text{ g}/28(6.02 \times 10^{23}) = 1.14 \times 10^{-23}$ g. So $k_{CH} = 4\pi^2(3000 \text{ cm}^{-1})^2(100 \text{ cm/m})^2(3.00 \times 10^8 \text{ m/s})^2(1.53 \times 10^{-27} \text{ kg}) = 489$ N/m. Also, $k_{CO} = 4\pi^2(1750 \text{ cm}^{-1})^2(100 \text{ cm/m})^2(3.00 \times 10^8 \text{ m/s})^2 \times (1.14 \times 10^{-26} \text{ kg}) = 1240$ N/m.

20.55 **(a)** T; **(b)** T; **(c)** F; **(d)** T (provided it is not a spherical top); **(e)** T. **(f)** T; **(g)** F.

20.56 **(a)** The rotational levels of a linear molecule are $E_{rot} = BhJ(J+1)$ [Eq. (20.45) with $K = 0$] and the pure rotational Raman selection rule is $\Delta J = \pm 2$. So $\nu_0 - \nu_{scat} = \Delta E/h = \pm(Bh/h)[(J+2)(J+3) - J(J+1)] = \pm(4J+6)B$, where $J = 0, 1, 2, \ldots$. The spacing between adjacent lines is $[4(J+1)+6]B - (4J+6)B = 4B$.

(b) $4\tilde{B} = 7.99 \text{ cm}^{-1}$ and $\tilde{B} = 1.998 \text{ cm}^{-1} = 199.8 \text{ m}^{-1}$. We have $\tilde{B} = B/c = h/8\pi^2Ic = h/8\pi^2c\mu R^2$ and $R = (h/8\pi^2c\mu\tilde{B})^{1/2}$. Also, $\mu = m_1m_2/(m_1+m_2) = m_1^2/2m_1 = m_1/2$ and $R =$

$$\left[\frac{2(6.626 \times 10^{-34} \text{ J s})(6.022 \times 10^{23} \text{ mol}^{-1})}{8\pi^2(2.998 \times 10^8 \text{ m/s})(0.01401 \text{ kg/mol})(199.8 \text{ m}^{-1})}\right]^{1/2} =$$

1.098×10^{-10} m = 1.098 Å (as in Table 20.1).

(c) The lowest J is 0 and $\nu_0 - \nu_{scat} = [4(0)+6]B = 6B$.

(d) $\nu_0 = c/\lambda_0 = (2.9979 \times 10^8 \text{ m/s})/(540.8 \times 10^{-9} \text{ m}) = 5.543_5 \times 10^{14} \text{ s}^{-1}$. $\nu_0 - \nu_{scat} = \pm 6B = \pm 6\tilde{B}c = \pm 6(199.8 \text{ m}^{-1})(2.998 \times 10^8 \text{ m/s}) = \pm 0.00359 \times 10^{14} \text{ s}^{-1}$. $\nu_{scat} = 5.543_5 \times 10^{14} \text{ s}^{-1} \pm 0.00359 \times 10^{14} \text{ s}^{-1} =$

$5.547_1 \times 10^{14}$ s^{-1} and $5.539_9 \times 10^{14}$ s^{-1}. Then $\lambda_{scat} = c/\nu_{scat} = 540.4_5$ nm and 541.1_5 nm.

20.57 **(a)** F; **(b)** T; **(c)** F.

20.58 For the Balmer series, $n_b = 2$ in Eq. (20.50); for the series limit, $n_a = \infty$ and $1/\lambda = R/4 = (109678 \text{ cm}^{-1})/4$. Then $\lambda = 3.647 \times 10^{-5}$ cm = 364.7 nm.

20.59 For the Paschen series, $n_b = 3$ in Eq. (20.50); the first three lines have n_a = 4, 5, and 6. So $1/\lambda = R(1/9 - 1/16)$, $R(1/9 - 1/25)$, $R(1/9 - 1/36)$. We get λ = 1.8756×10^{-4} cm, 1.2822×10^{-4} cm, 1.0941×10^{-4} cm.

20.60 Li^{2+} is a hydrogenlike atom and Eq. (18.14) gives the energy levels. The Li nucleus is substantially heavier than the H nucleus, so we can take μ equal to the electron mass. So $E = -9[2\pi^2 me^4/(4\pi\varepsilon_0)^2 n^2 h^2]$ and $1/\lambda = \nu/c = |\Delta E|/hc = 9[2\pi^2 me^4/(4\pi\varepsilon_0)^2 ch^3](1/n_b^2 - 1/n_a^2) = 9(109736 \text{ cm}^{-1})(1/1 - 1/4) = 7.4072 \times 10^5 \text{ cm}^{-1}$. Then $\lambda = 1.3500 \times 10^{-6}$ cm.

20.61 Using the notation of Fig. 20.38 and Eq. (20.52), we have D_0 of the B state as $D_0' = h\nu_{cont} - h\nu_{00} = (6.6261 \times 10^{-34} \text{ J s})(2.998 \times 10^8 \text{ m/s}) \times [(1750.5 \times 10^{-10} \text{ m})^{-1} - (2026.0 \times 10^{-10} \text{ m})^{-1}] = 1.543 \times 10^{-19} \text{ J} = 0.963$ eV. D_0 of the ground state is given by (20.52) as $D_0'' = (6.6261 \times 10^{-34} \text{ J s})(2.9979 \times 10^8 \text{ m/s})(1750.5 \times 10^{-10} \text{ m})^{-1} \times (1 \text{ eV})/(1.60218 \times 10^{-19} \text{ J}) - 1.970 \text{ eV} = 5.11$ eV.

20.62 $N_{J=1}/N_{J=0} = 0.181 = (3/1)e^{-\Delta\varepsilon/kT}$ and $\Delta\varepsilon/kT = -\ln(0.181/3) = 2.81$. So $T = 0.356\Delta\varepsilon/k$. From (20.15), (20.33), and (20.32), $\Delta\varepsilon = 1(2)\hbar^2/2I - 0 = h^2/4\pi^2 I = 2\tilde{B}_0 hc = 2(\tilde{B}_e - \frac{1}{2}\tilde{\alpha}_e)hc = 2(1.931 - 0.009) \text{ cm}^{-1} (1 \text{ cm}/0.01 \text{ m}) \times (6.626 \times 10^{-34} \text{ J s})(2.9979 \times 10^8 \text{ m/s}) = 7.63_6 \times 10^{-23}$ J. $T = 0.356\,\Delta\varepsilon/k = 0.356(7.63_6 \times 10^{-23} \text{ J})/(1.3807 \times 10^{-23} \text{ J/K}) = 1.97$ K.

20.63 **(a)** T; **(b)** T; **(c)** F; **(d)** T; **(e)** F; **(f)** F; **(g)** T; **(h)** T; **(i)** F; **(j)** T; **(k)** F; **(l)** T.

20.64 $F = BQ\upsilon \sin\theta = (1.5 \text{ T})(1.60 \times 10^{-19} \text{ C})(3.0 \times 10^6 \text{ m/s}) \sin\theta =$ $(7.2 \times 10^{-13} \text{ N}) \sin\theta$.

(a) $F = 0$. **(b)** $F = (7.2 \times 10^{-13} \text{ N}) \sin 45° = 5.1 \times 10^{-13}$ N.

(c) 7.2×10^{-13} N. **(d)** 0.

20.65 From the paragraph before (20.61), $|\mathbf{m}| = Q\upsilon r/2 =$ $(2.0 \times 10^{-16} \text{ C})(2.0 \times 10^5 \text{ m/s})(25 \times 10^{-10} \text{ m})/2 = 5.0 \times 10^{-20}$ J/T.

20.66 (a) $e/2m_p = (1.602176 \times 10^{-19} \text{ C})/2(1.672622 \times 10^{-27} \text{ kg}) = 4.78942 \times 10^7$ Hz/T, where (20.54) was used.

(b) $\gamma/2\pi = (4.78942 \times 10^7 \text{ Hz/T})5.58569/2\pi = 42.5775 \times 10^6$ Hz/T.

20.67 (a) Equations (20.65) and (20.63) give $E = -\gamma\hbar BM_I =$ $-1.792(4.7894 \times 10^7 \text{ s}^{-1}/\text{T})1.792(6.626 \times 10^{-34} \text{ J s})(2\pi)^{-1}(1.50 \text{ T})M_I =$ $-(1.35_8 \times 10^{-26} \text{ J})M_I$. Since $I = 3/2$, $M_I = 3/2, 1/2, -1/2$, and $-3/2$. The levels are -2.04×10^{-26} J, -0.679×10^{-26} J, 0.679×10^{-26} J, 2.04×10^{-26} J.

(b) B is doubled, so the four energies in (a) are each doubled..

20.68 (a) From (20.67) and (20.63), $\nu = |\gamma|B/2\pi =$ $(4.7894 \times 10^7 \text{ s}^{-1}/\text{T})1.792(1.50 \text{ T})/2\pi = 2.049 \times 10^7/\text{s} = 20.49$ MHz.

(b) 27.32 MHz.

20.69 From (20.67), $\nu_A/\nu_B = |\gamma_A|/|\gamma_B|$ and $\nu(^{13}\text{C}) = (10.708/42.577)(600 \text{ MHz}) =$ 151 MHz.

20.70 (a) Equations (20.67) and (20.63) give $B = 2\pi\nu/|\gamma| =$ $2\pi(60 \times 10^6 \text{ s}^{-1})/[5.5857(4.7894 \times 10^7 \text{ s}^{-1}/\text{T})] = 1.41$ T.

(b) $(300/60)(1.41 \text{ T}) = 7.05$ T.

20.71 (a) $M_I = +½$ and $-½$. The energy separation is given by (20.65) as $\Delta E =$ $|\gamma|\hbar B = (4.7894 \times 10^7 \text{ s}^{-1}/\text{T})5.5857(6.626 \times 10^{-34} \text{ J s})(2\pi)^{-1}(1.41 \text{ T}) =$ 3.98×10^{-26} J. The levels are nondegenerate and the population ratio is

$e^{-\Delta E/kT}$ = exp [–(3.98 × 10^{-26} J)/(1.381 × 10^{-23} J/K)(298 K)] = exp (–0.00000967) = 0.9999903.

(b) An increase in B increases the separation between energy levels, thereby producing a greater population difference between the initial and final states. Hence the absorption intensity increases.

20.72 From Eq. (20.72), $\nu_i - \nu_j = 10^{-6}(300 \times 10^6 \text{ Hz})(1.0) = 300$ Hz.

20.73 From the tables in the text, δ is 2 to 3 for the CH_3 protons and is 9 to 11 for the CHO proton; J is 1 to 3 Hz. The CH_3 doublet has three times the total intensity of the CHO quartet. Thus (all splittings are about 2 Hz)

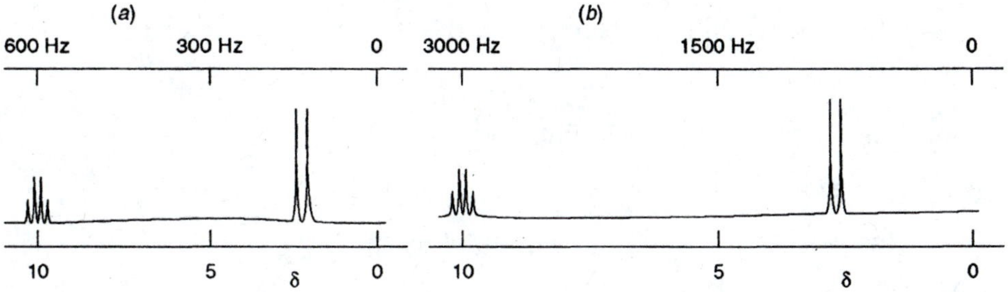

20.74 **(a)** From (20.70) and (20.68), increasing δ_i means decreasing σ_i and increasing ν_i, so ν_i increases to the left.

(b) To the left.

(c) To the right.

(d) From (20.70) and (20.69), increasing δ_i means decreasing σ_i and decreasing $B_{0,i}$, so $B_{0,i}$ increases to the right.

20.75 **(a)** One singlet peak.

(b) One proton NMR peak that is split into a doublet by the F.

(c) One singlet.

(d) The four methylene protons give a quartet of relative intensity 2; the six methyl protons give a triplet of relative intensity 3.

(e) The $(CH_3)_2$ protons give a doublet of relative intensity 6; the CH proton gives a septet of relative intensity 1.

(f) Two equal-intensity singlet peaks. The CH_3 groups are not equivalent and don't split each other.

(g) Three quartet peaks of equal intensity. In each of the three quartets, the 4 lines have equal intensity but only two of the 3 spacings are equal.

(h) The CH_3 protons give a triplet of relative intensity 3; the CHO proton gives a triplet of relative intensity 1; the CH_2 protons give an octet of relative intensity 2.

20.76 (a) This molecule has two different proton–^{19}F coupling constants: one for the ^{1}H and ^{19}F nuclei that are cis to each other, and a different coupling constant for ^{1}H and ^{19}F nuclei that are trans to each other.

(b) Here, the F atoms lie in a plane that is perpendicular to the plane of the CH_2 group, and there is only one ^{1}H–^{19}F coupling constant.

20.77 All peaks except in (b) are singlets.

(a) One peak.

(b) One peak that is split into a doublet by the F.

(c) One peak.

(d) Two peaks of equal intensity.

(e) Two peaks with 2:1 intensity ratio.

(f) Three equal-intensity peaks.

(g) Two peaks of equal intensity.

(h) Three equal-intensity peaks.

20.78 **(a)** 4 (there are 4 kinds of carbons); **(b)** 5; **(c)** 3.

20.79 In Fig. 20.44, 100 Hz corresponds to a length of 28 mm and J corresponds to a length of $2\frac{1}{4}$ mm, so $J = (2\frac{1}{4}/28)(100\text{ Hz}) = 8$ Hz.

20.80 (a) Unchanged; see Eq. (20.70) and the following paragraph.

(b) Each ν and the difference between the ν's is multiplied by 2; see Eqs. (20.68) and (20.72).

(c) Unchanged, as noted after (20.73).

20.81 From Eq. (20.69), $\sigma_i = 1 - 2\pi\nu_{spec}/|\gamma_i|B_{0,i}$, so $\delta_i = 10^6(\sigma_{ref} - \sigma_i) = 10^6(-2\pi\nu_{spec}/|\gamma_i|)(1/B_{0,ref} - 1/B_{0,i}) = 10^6 B_0(B_{0,ref} - B_{0,i})/B_{0,ref}B_{0,i}$. Since $\sigma_i << 1$, we have $B_{0,i} \approx 2\pi\nu_{spec}/|\gamma_i| = B_0$, and $\delta_i = 10^6(B_{0,ref} - B_{0,i})/B_{0,ref}$.

20.82 **(a)** For a 60-MHz spectrometer, coalescence occurs at 120°C and the formula gives k_c at 120°C as $k_c = \pi|\nu_2 - \nu_1|/2^{1/2}$. From (20.72), $\nu_2 - \nu_1 = 10^{-6}(60 \text{ MHz})(2.94 - 2.79) = 9.0$ Hz and $k_c = \pi(9.0 \text{ s}^{-1})/2^{1/2} = 20 \text{ s}^{-1}$.

(b) Eq. (20.72) and the fact that δ_i is independent of ν_{spec} show that an increase in ν_{spec} increases $\nu_i - \nu_j$. Coalescence occurs when $\nu_{exch} >> |\nu_1 - \nu_2|$, and the increase in $|\nu_1 - \nu_2|$ and the fact that ν_{exch} increases as T increases mean that the coalescence temperature will increase when ν_{spec} is increased.

20.83 $\nu = g\beta_e B_0/h = 2.0026(9.274 \times 10^{-24} \text{ J/T})(2.50 \text{ T})/(6.6261 \times 10^{-34} \text{ J s}) = 7.01 \times 10^{10} \text{ s}^{-1} = 70.1$ GHz.

20.84 There is one set of 4 equivalent protons and a second set of 4 equivalent protons, so there are 5(5) = 25 lines.

20.85 **(a)** 2(7) = 14; **(b)** 3(3) = 9; **(c)** 10; **(d)** 3(2) = 6.

20.86 $[\alpha] = \alpha/[\rho_B/(\text{g/cm}^3)](l/\text{dm}) = 1.90°/(0.0650)(2.00) = 14.6°$.

20.87 The observed α of the mixture is the sum of the α's of the α and β isomers: $\alpha = \alpha_\alpha + \alpha_\beta = [\alpha]_\alpha\rho_\alpha l(\text{cm}^3/\text{dm g}) + [\alpha]_\beta\rho_\beta l(\text{cm}^3/\text{dm g})$ (1). The total solute mass m is $m = m_\alpha + m_\beta$ and division by the solution's volume gives $\rho = \rho_\alpha + \rho_\beta$. Division of Eq. (1) by $\rho l(\text{cm}^3/\text{dm g})$ gives $[\alpha] = [\alpha]_\alpha w_\alpha + [\alpha]_\beta w_\beta$ (2), where $w_\alpha = \rho_\alpha/(\rho_\alpha + \rho_\beta) = m_\alpha/(m_\alpha + m_\beta)$ and w_β are the mass fractions (and also the mole fractions) of the α and β isomers. Equation (2) gives $52.7° = w_\alpha 112.2° + (1 - w_\alpha)17.5°$ and $w_\alpha = 0.372$, or 37.2% α-D-glucose.

20.88 $E = h\nu = hc/\lambda$. The first entry in (20.78) is $E = (6.626 \times 10^{-34} \text{ J s}) \times (2.998 \times 10^8 \text{ m/s})/(200 \times 10^{-9} \text{ m}) = 9.93 \times 10^{-19} \text{ J} = 6.20$ eV; etc. The first entry in the following line is $N_A h\nu = N_A hc/\lambda = (6.022 \times 10^{23}/\text{mol}) \times$ $(6.626 \times 10^{-34}\ \text{J s})(2.998 \times 10^8\ \text{m/s})/(200 \times 10^{-9}\ \text{m}) = 5.98 \times 10^5\ \text{J/mol}$; etc.

20.89 **(a)** I is the energy per unit time that falls on unit cross-sectional area perpendicular to the beam. In time t, the beam travels a distance $c't$ and the photons that pass through cross-sectional area A in time t are contained in a volume $Ac't$. Each photon has energy $h\nu$. If N is the number of photons in the volume $Ac't$, the energy of the photons in this volume is $Nh\nu$. The intensity equals this energy divided by the time t and the cross-sectional area A: so $I = Nh\nu/tA$. So $N = ItA/h\nu$. Then the number of photons per unit volume is $N/V = N/Ac't = (ItA/h\nu)/Ac't = I/h\nu c'$.

(b) $c' = c/1.34 = 2.24 \times 10^8$ m/s. $\nu = c/(488 \times 10^{-9}\ \text{m}) = 6.14 \times 10^{14}\ \text{s}^{-1}$. $N/V = I/h\nu c' = (10^{15}\ \text{W/m}^2)/$ $[(6.63 \times 10^{-34}\ \text{J s})(6.1 \times 10^{14}\ \text{s}^{-1})(2.2 \times 10^8\ \text{m/s})] = 1 \times 10^{25}\ \text{photons/m}^3 =$ $1 \times 10^{19}\ \text{photons/cm}^3$. 18 g of water contains 6×10^{23} molecules in a volume of 18 cm^3, and the number of molecules per cm^3 is 3×10^{22}.

20.90 The energy absorbed is $0.744(0.00155\ \text{J/s})(110\ \text{s}) = 0.126_9$ J. The number of moles of photons absorbed is $(0.1269\ \text{J})/(N_A hc/\lambda) = 4.92 \times 10^{-7}$ mole. So $\Phi =$ $(6.80 \times 10^{-6}\ \text{mole})/(4.92 \times 10^{-7}\ \text{mole}) = 13.8$.

20.91 **(a)** $d[\text{HI}]/dt = -\phi\mathcal{I}_a - k_2[\text{H}][\text{HI}]$. The steady-state approximation for H gives $d[\text{H}]/dt = 0 = \phi\mathcal{I}_a - k_2[\text{H}][\text{HI}]$ and $k_2[\text{H}][\text{HI}] = \phi\mathcal{I}_a$. So $d[\text{HI}]/dt =$ $-2\phi\mathcal{I}_a = -2\mathcal{I}_a$, since $\phi \approx 1$.

(b) The number N of photons absorbed satisfies $4184\ \text{J} = Nh\nu = Nhc/\lambda$ and $N = (4184\ \text{J})(250 \times 10^{-9}\ \text{m})/(6.626 \times 10^{-34}\ \text{J s})(2.998 \times 10^8\ \text{m/s}) =$ 5.27×10^{21}. The number of HI molecules decomposed is $2(5.27 \times 10^{21}) =$ 1.05×10^{22}.

20.92 With the inclusion of reaction (5), Eq. (20.83) becomes $r = k_2[\text{A}^*][\text{A}] - k_4[\text{A}_2] + k_5[\text{A}]^2$. Reaction (5) does not involve A*, so (20.84) still holds and we have $r = k_2[\text{A}]\mathcal{I}_a/(k_2[\text{A}] + k_3) - k_4[\text{A}_2] + k_5[\text{A}]^2$. For the photostationary state, $r = 0$ and we get $[\text{A}_2] = k_5[\text{A}]^2/k_4 + k_2[\text{A}]\mathcal{I}_a/(k_4k_2[\text{A}] + k_3k_4)$. In the absence of radiation, $\mathcal{I}_a = 0$ and the last equation becomes $[\text{A}_2] = k_5[\text{A}]^2/k_4$. The concentration of A_2 is greater in the presence of radiation.

20.93 **(a)** Yes; **(b)** no; **(c)** no. (1 + 1 is not a member of the set.)

20.94 **(a)** Yes; **(b)** no; **(c)** yes; **(d)** no.

20.95 **(a)** We have $AI = IA = A$, since I is the identity element. Since each group element appears exactly once in each row of the multiplication table, we must have $AA = I$. The multiplication table is thus

	I	***A***
I	I	A
A	A	I

(b) Using the property of the identity element, we start with

	I	***A***	***B***
I	I	A	B
A	A	w	x
B	B	y	z

where w, x, y, and z are to be determined. w cannot equal A because this would put two A's in row 2. So w must either equal I or B. If we put $w = I$, then the theorem that each group element appears exactly once in each row means that $x = B$. But with $x = B$, we have B appearing twice in column three, which is not allowed. Hence $w = B$. Filling in the rest of the table, we get

	I	***A***	***B***
I	I	A	B
A	A	B	I
B	B	I	A

This table shows that each element has an inverse, and using the table, one finds that associativity is satisfied, so this is the multiplication table of a group.

20.96 **(a)** F; **(b)** T.

20.97 As in Fig. 20.52, the coordinate axes do not move when a symmetry operation is applied.

(a) They do not commute. **(b)** They do not commute. **(c)** They commute.

20.98 **(a)** $\hat{E}$; **(b)** $\hat{\sigma}$; **(c)** $\hat{i}$; **(d)** $\hat{C}_5^4$; **(e)** $\hat{C}_5^3$; **(f)** $\hat{S}_3^5$. (Since $\hat{S}_3^2$ involves two reflections, which amounts to no reflections, it is not the inverse of $\hat{S}_3$.)

20.99 **(a)** T. **(b)** T. **(c)** T. (If we call the rotation axis the z axis, then the 180° rotation part of $\hat{S}_2$ converts the x and y coordinates to their negatives and the reflection converts the z coordinate to its negative.) **(d)** T.

20.100 **(a)** The structure is trigonal pyramidal. A C_3 axis and three planes of symmetry (each of which contains one of the bonds).

(b) $\hat{C}_3$, $\hat{C}_3^2$, $\hat{E}$, $\hat{\sigma}_a$, $\hat{\sigma}_b$, $\hat{\sigma}_c$.

20.101 Symmetry elements: A C_6 axis that is also an S_6 axis, an S_3 axis and a C_2 axis. A center of symmetry. Seven symmetry planes, six perpendicular to the molecular plane and one coincident with it. Six C_2 axes lying in the molecular plane. Symmetry operations: $\hat{E}$, $\hat{C}_6$, $\hat{C}_6^2$, $\hat{C}_6^3$, $\hat{C}_6^4$, $\hat{C}_6^5$, $\hat{S}_6$, $\hat{S}_6^5$, $\hat{S}_3$, $\hat{S}_3^2$, $\hat{i}$, six $\hat{C}_2$ rotations, and seven $\hat{\sigma}$ reflections, for a total of 24 operations. (The operation $\hat{S}_6^3$ is the same as $\hat{i}$; see the answer to Prob. 20.99c.)

20.102 Four C_3 axes (one along each bond). Three S_4 axes (which are also C_2 axes), one through each pair of opposite faces of the cube in Fig. 20.20. Six symmetry planes, each of which contains two bonds.

20.103 **(a)** T_d; **(b)** $C_{\infty v}$; **(c)** $D_{\infty h}$; **(d)** D_{3h}; **(e)** O_h; **(f)**; C_{4v} **(g)** D_{4h}; **(h)** C_{3v}; **(i)** D_{3h}; **(j)** D_{2h}; **(k)** C_{3v}; **(l)** C_{4v}; **(m)** C_1; **(n)** C_s.

20.104 $\hat{C}_n, \hat{C}_n^2, \hat{C}_n^3, \ldots, \hat{C}_n^{n-1}, \hat{E}$. Order n.

20.105 (a) Yes, since the product $\hat{C}_n \sigma_h$ is a symmetry operation.

(b) When n is even, since $\hat{C}_n^{n/2} = \hat{C}_2$ is then a symmetry operation.

(c) C_{2v}; C_{2h}; C_{2v}.

20.106 D_{3d}; D_{2d}.

20.107 (a) $\begin{pmatrix} 0 & 11 \\ 6 & 6 \end{pmatrix}$ **(b)** $\begin{pmatrix} 3 & 15 \\ 9 & 12 \end{pmatrix}$

20.108 $\mathbf{AB} = \begin{pmatrix} 0.2(4)+4(5) & 0.2(1)+4(8) \\ -1(4)+3(5) & -1(1)+3(8) \end{pmatrix} = \begin{pmatrix} 20.8 & 32.2 \\ 11 & 23 \end{pmatrix}$ $\quad \mathbf{BA} = \begin{pmatrix} -0.2 & 19 \\ -7 & 44 \end{pmatrix}$

20.109 $\begin{pmatrix} 1 & 2 \\ 3 & 4 \end{pmatrix}\begin{pmatrix} 2 \\ 6 \end{pmatrix} = \begin{pmatrix} 1(2)+2(6) \\ 3(2)+4(6) \end{pmatrix} = \begin{pmatrix} 14 \\ 30 \end{pmatrix}$

20.110 Let k be the number of columns in **R** and the number of rows in **S**. (These quantities must be equal or the matrix product is not defined.) The element t_{mn} is calculated using row m of **R** and column n of **S**. So $t_{mn} = \sum_{i=1}^{k} r_{mi} s_{in}$.

20.111 $\mathbf{M}_1\mathbf{N}_1 = \begin{pmatrix} -1 & 5 \\ 3 & 8 \end{pmatrix}\begin{pmatrix} 2 & 9 \\ 4 & 1 \end{pmatrix} = \begin{pmatrix} 18 & -4 \\ 38 & 35 \end{pmatrix}$ $\quad \mathbf{M}_2\mathbf{N}_2 = (7)(6) = (42)$

$$\mathbf{MN} = \begin{pmatrix} -1 & 5 & 0 \\ 3 & 8 & 0 \\ 0 & 0 & 7 \end{pmatrix}\begin{pmatrix} 2 & 9 & 0 \\ 4 & 1 & 0 \\ 0 & 0 & 6 \end{pmatrix} = \begin{pmatrix} 18 & -4 & 0 \\ 38 & 35 & 0 \\ 0 & 0 & 42 \end{pmatrix}$$

20.112 For example, the product of the $\hat{\sigma}(xz)$ and $\hat{\sigma}(yz)$ matrices (in either order) is

$$\begin{pmatrix} 1 & 0 & 0 \\ 0 & -1 & 0 \\ 0 & 0 & 1 \end{pmatrix}\begin{pmatrix} -1 & 0 & 0 \\ 0 & 1 & 0 \\ 0 & 0 & 1 \end{pmatrix} = \begin{pmatrix} -1 & 0 & 0 \\ 0 & -1 & 0 \\ 0 & 0 & 1 \end{pmatrix}$$, which is the matrix that corresponds

to $\hat{C}_2(z)$, in agreement with the product in Table 20.2.

20.113 (a) $\hat{C}_4(z)$ moves the point at (x, y, z) to $(-y, x, z)$ and Eq. (20.92) becomes

$$\begin{pmatrix} x' \\ y' \\ z' \end{pmatrix} = \begin{pmatrix} 0 & -1 & 0 \\ 1 & 0 & 0 \\ 0 & 0 & 1 \end{pmatrix}\begin{pmatrix} x \\ y \\ z \end{pmatrix}$$

(b) $\hat{i}$ moves the point at (x, y, z) to $(-x, -y, -z)$ and the matrix that does this

is $\begin{pmatrix} -1 & 0 & 0 \\ 0 & -1 & 0 \\ 0 & 0 & -1 \end{pmatrix}$.

(c) $\hat{\sigma}(xy)$ moves the point at (x, y, z) to $(x, y, -z)$ and the matrix is

$$\begin{pmatrix} 1 & 0 & 0 \\ 0 & 1 & 0 \\ 0 & 0 & -1 \end{pmatrix}.$$

20.114 (a) Using associativity, we have $(\mathbf{P}^{-1}\mathbf{AP})(\mathbf{P}^{-1}\mathbf{CP}) = (\mathbf{P}^{-1}\mathbf{A})(\mathbf{PP}^{-1})(\mathbf{CP}) = (\mathbf{P}^{-1}\mathbf{A})(\mathbf{I})(\mathbf{CP}) = (\mathbf{P}^{-1}\mathbf{A})(\mathbf{CP}) = \mathbf{P}^{-1}\mathbf{ACP} = \mathbf{P}^{-1}\mathbf{FP}$, since multiplication by a unit matrix has no effect.

(b) The result of part (a) of this problem shows that the transformed matrices $\mathbf{P}^{-1}\mathbf{AP}$, $\mathbf{P}^{-1}\mathbf{BP}$,... multiply the same way as the original matrices **A**, **B**,...; since the original matrices multiply the same way as the group members, so do the transformed matrices.

20.115 As noted a couple of paragraphs before (20.95), each element of a commutative group is in a class by itself, so the number of classes c equals the number of elements h in a commutative group. There are thus h terms on the left side of (20.95). The smallest possible value of each term on the left of (20.95) is 1, and if any representation had a dimension greater than 1, then the value of the left side of (20.95) would exceed h and (20.95) would not hold. So all irreducible representations of a commutative group are one-dimensional.

20.116 For example, the products of the matrices corresponding to $\hat{C}_2(z)$ and $\hat{\sigma}(xz)$ are (1)(–1) = (–1) and (–1)(1) = (–1), and (–1) is the matrix corresponding to $\hat{\sigma}(yz)$, which according to Table 20.2 is the correct product. The other products are verified similarly.

20.117 To find the elements in the same class as E, we form the products $A^{-1}EA$, $B^{-1}EA$,… . But $A^{-1}EA = A^{-1}A = E$ for every element A, so the identity element E is in a class by itself.

20.118 **(a)** $2p_y$; **(b)** $-2p_y$, since the positive and negative lobes are interchanged; **(c)** $2p_y$; **(d)** $-2p_x$; **(e)** $2p_y$; **(f)** $-2p_y$; **(g)** $-2p_y$; **(h)** $-2p_y$.

20.119 (a) The orbitals O2s, O2p_z, and $H_1 1s + H_2 1s$ are each unchanged by each of the four symmetry operations and the C_{2v} character table (Table 20.3) gives the symmetry species as a_1.

(c) The O2p_y orbital is unchanged by $\hat{E}$ and by $\hat{\sigma}(yz)$ and is transformed to –O2p_y by $\hat{C}_2(z)$ and by $\hat{\sigma}(xz)$ and $H_1 1s - H_2 1s$ shows the same behavior. The character table gives the symmetry species as b_2.

20.120 (a) We examine the effects of each symmetry operation on the vectors showing the normal-mode motions of the atoms. The point group is C_{2v} and the z axis coincides with the C_2 axis through the C and O atoms. The $\hat{C}_2(z)$ operation reverses the direction of a vector pointing in the $+x$ or $-x$ direction (and also interchanges the vectors on the H atoms), so for this normal mode the $\hat{C}_2(z)$ character is -1. For $\hat{\sigma}(yz)$, the direction of a vector pointing in the $+x$ or $-x$ direction is reversed and the character is

-1. The $\hat{\sigma}(xz)$ operation interchanges the vectors in the H atoms but leaves the directions of vectors pointing in the $\pm x$ direction unchanged and the character is $+1$. The character table (Table 20.2) gives the symmetry species as b_1.

(b) The point group is C_{3v}. The N-atom vibration vector points along the z axis and is unaffected by each of the six symmetry operations listed in the character table (Table 20.3). The H-atom vibration vectors point along the bonds and although the symmetry operations may interchange or permute these vectors, the direction of each vector on each H is unchanged by each symmetry operation. Thus the characters are all +1 for this mode and the mode has species a_1.

20.121 The representation Γ consists of the matrices in (20.93), and the point group is C_{2v} with order $h = 4$. The irreducible-representation characters χ_i are taken from Table 20.3. The characters χ_Γ of Γ are found by taking the traces (the sums of the diagonal elements) of the matrices in (20.93); these characters are 3, -1, 1, and 1 for $\hat{E}$, $\hat{C}_2(z)$, $\hat{\sigma}(xz)$, and $\hat{\sigma}(yz)$, respectively. Eq. (20.97) gives $a_{A_1(20.93)} = (1/4)[1(3) + 1(-1) + 1(1) + 1(1)] = 1$,

$a_{A_2(20.93)} = (1/4)[1(3) + 1(-1) + (-1)(1) + (-1)(1)] = 0$

$a_{B_1(20.93)} = (1/4)[1(3) + (-1)(-1) + 1(1) + (-1)(1)] = 1$

$a_{B_2(20.93)} = (1/4)[1(3) + (-1)(-1) + (-1)(1) + 1(1)] = 1$ so $\Gamma_{(20.93)} = A_1 \oplus B_1 \oplus B_2$.

20.122 (a) $\hat{E}$ is represented by a unit matrix whose order equals the dimension of the representation. The trace of a unit matrix of order n is n. Hence this representation has dimension 9.

(b) We use Eq. (20.97) with the χ_i values taken from the character table in Table 20.3. We have $a_{A_1\Gamma} = (1/4)[1(9) + 1(-1) + 1(1) + 1(3)] = 3$,

$a_{A_2\Gamma} = (1/4)[1(9) + 1(-1) + (-1)(1) + (-1)(3)] = 1$

$a_{B_1\Gamma} = (1/4)[1(9) + (-1)(-1) + 1(1) + (-1)(3)] = 2$

$a_{B_2\Gamma} = (1/4)[1(9) + (-1)(-1) + (-1)(1) + 1(3)] = 3$

so $\Gamma = 3A_1 \oplus A_2 \oplus 2B_1 \oplus 3B_2$.

20.123 We use Eq. (20.97) with the χ_i values taken from the character table in Table 20.3. The sum in (20.97) is over the h symmetry operations. The characters in

the $C_{3\upsilon}$ character table are listed for each class, and if we take the sum in (20.97) to be over the classes, we must multiply each term in the sum by the number of operations in that class, so as to include each of the h operations in the sum. We have $a_{A_1\Gamma} = (1/6)[1(293) + 2(1)(-118) + 3(1)9] = 14$,
$a_{A_2\Gamma} = (1/6)[1(293) + 2(1)(-118) + 3(-1)9] = 5$
$a_{E\Gamma} = (1/6)[2(293) + 2(-1)(-118) + 3(0)9] = 137$. So $\Gamma = 14A_1 \oplus 5A_2 \oplus 137E$.

20.124 The trace of a matrix such as **M** in (20.90) that is in block-diagonal form is clearly equal to the sum of the traces of each block. (For example, the trace of **M** is –1 + 8 + 7 = 14 and the traces of $\mathbf{M}_1$ and $\mathbf{M}_2$ are 8 – 1 = 7 and 7.) Hence the trace of a matrix of a reducible representation equals the sums of the traces of all the irreducible-representation matrices that it is the direct sum of. So $\chi_\Gamma(\hat{E}) = 4(1) + 1 + 6(2) = 17$, $\chi_\Gamma(\hat{C}_3) = 4(1) + 1 + 6(-1) = -1$,
$\chi_\Gamma(\hat{\sigma}_\upsilon) = 4(1) + (-1) + 6(0) = 3$.

20.125 (a) HBr, H_2S, CH_3Cl, which have nonzero dipole moments.

(b) HBr, CO_2, H_2S, CH_4, CH_3Cl, C_6H_6, which have vibrations that change the dipole moment.

(c) N_2, HBr, CO_2, H_2S, CH_3Cl, and C_6H_6, which are not spherical tops.

20.126 (a) $k_e = E_e''(R_e)$. The function $E_e(R)$ is found by solving the electronic Schrödinger equation, which is the same for H_2 and D_2, so $E_e(R)$ is the same for H_2 and D_2 and this makes k_e, D_e, and R_e the same for H_2 and D_2. (This answer neglects very slight deviations from the Born–Oppenheimer approximation that make E_e, k_e, R_e, and D_e differ extremely slightly for H_2 and D_2.)

(b) $\nu_e = (1/2\pi)(k_e/\mu)^{1/2}$. Since μ is greater for D_2 and k_e is essentially the same, ν_e is smaller for D_2.

(c) $I_e = \mu R_e^2$. Since μ is substantially greater for D_2 and R_e is essentially the same for the two species, I_e is greater for D_2.

(d) $B_e = h/8\pi^2 I_e$. Since I_e is greater for D_2, B_e is greater for H_2.

(e) The same, since $E_e(R)$ is the same for the two.

(f) $D_0 \cong D_e - \frac{1}{2}h\nu_e$. Since ν_e is smaller for D_2 and D_e is essentially the same for the two, D_0 is greater for D_2.

(g) Since ν_e is smaller for D_2 and D_e is the same, D_2 has more bound vibrational levels.

(h) Since ν_e is smaller for D_2, the separation between vibrational levels is smaller for D_2, and more D_2 molecules are in excited vibrational levels at a given T, so the fraction in $\upsilon = 0$ is greater for H_2.

(i) Since I_e is greater for D_2, the separation between D_2 rotational levels is smaller for D_2 and the fraction in $J = 0$ is greater for H_2.

20.127 Since there are $2I + 1$ values of M_I, there are $2I + 1$ two-electron spin functions where electron 1 and electron 2 have the same spin function and so are symmetric functions [similar to $\alpha(1)\alpha(2)$]. The remaining $(2I + 1)^2 - (2I + 1)$ two-electron spin functions are neither symmetric nor antisymmetric, but have forms like $\alpha(1)\beta(2)$ and $\alpha(2)\beta(1)$. These functions must be combined to form symmetric and antisymmetric functions of form similar to $2^{-1/2}[\alpha(1)\beta(2) + \alpha(2)\beta(1)]$ and $2^{-1/2}[\alpha(1)\beta(2) - \alpha(2)\beta(1)]$; the total number of such combined functions will equal the number of uncombined functions used, and so will be $(2I + 1)^2 - (2I + 1)$. Half of these will be symmetric and half antisymmetric. Thus there will be $\frac{1}{2}[(2I + 1)^2 - (2I + 1)] = I(2I + 1)$ combined symmetric functions and $\frac{1}{2}[(2I + 1)^2 - (2I + 1)] = I(2I + 1)$ combined antisymmetric spin functions. Adding in the $2I + 1$ uncombined symmetric spin functions, we get a total of $I(2I + 1) + 2I + 1 = (I + 1)(2I + 1)$ symmetric two-electron spin functions. And we have $I(2I + 1)$ antisymmetric two-electron spin functions.

20.128 (a) joule, newton, watt, pascal, hertz, coulomb, volt, tesla, ohm, kelvin, ampere, siemens;

(b) poise, debye, angstrom, svedberg, dalton, torr, bohr, hartree.

20.129 (a) T.

(b) F. It cannot change its rotational state by absorption or emission of radiation, but can change its rotational state during collisions.

(c) F. Counterexamples are CH_4 and BF_3.

(d) F. The energy might be transferred to another molecule in a collision.

(e) F. A counterexample is H_2O.

(f) T.

(g) F. This formula is only for linear and spherical-top molecules.

(h) T.

(i) T.

(j) F.

(k) This question is too silly to answer.

Chapter 21

21.1 **(a)** T, V, and mole numbers.

(b) T, V, n_B, n_C,...

(c) Over states.

21.2 $Z = \sum_j e^{-E_j/kT}$ is dimensionless.

21.3 The Helmholtz energy A is extensive, so A_2 is 25/10 = 2.5 times A_1. Hence $A_2/A_1 = 2.5 = (-kT \ln Z_2)/(-kT \ln Z_1) = (\ln Z_2)/(\ln Z_1)$.

21.4 $G = A + PV = -kT \ln Z + VkT(\partial \ln Z/\partial V)_{T,N_B} = kTV^2[\partial(V^{-1} \ln Z/\partial V]_{T,N_B}$.

21.5 Let a subscript o denote the partition function and thermodynamic properties before the constant b is added to the levels. Then

$Z = \sum_j e^{-(E_j+b)/kT} = e^{-b/kT} \sum_j e^{-E_j/kT} = e^{-b/kT} Z_o$ and $\ln Z = -b/kT + \ln Z_o$.

(a) $P = kT(\partial \ln Z/\partial V)_{T,N_B} = kT(\partial \ln Z_o/\partial V)_{T,N_B} = P_o$.

(b) $U = kT^2(\partial \ln Z/\partial T)_{T,N_B} = kT^2[b/kT^2 + (\partial \ln Z_o/\partial T)_{V,N_B}] = b + U_o$.

(c) $S = U/T + k \ln Z = b/T + U_o/T - kb/kT + k \ln Z_o = U_o/T + k \ln Z_o = S_o$.

(d) $A = -kT \ln Z = -kT(-b/kT + \ln Z_o) = b - kT \ln Z_o = b + A_o$.

21.6 From $p_j = \exp(-\beta E_j)/Z$ [Eq. (21.15)], we have $\ln p_j = -\beta E_j - \ln Z$. So $-\sum_j p_j \ln p_j = \beta \sum_j p_j E_j + \ln Z \sum_j p_j = (kT)^{-1} U + \ln Z = S/k$, where (21.3), (21.33), and $\sum_j p_j = 1$ were used.

21.7 $z = \sum_r e^{-\varepsilon_r/kT}$ is dimensionless.

21.8 From the equation before (21.53), the number of available translational states is roughly $60V(mkT/h^2)^{3/2} = [60(10\text{ cm}^3)(1\text{ m}^3/10^6\text{ cm}^3)/(6.6 \times 10^{-34}\text{ J s})^3] \times [(0.020\text{ kg/mol})(1.38 \times 10^{-23}\text{ J/K})(300\text{ K})/(6.0 \times 10^{23}/\text{mol})]^{3/2} = 3 \times 10^{27}$.

21.9 **(a)** Each dot with n_x, n_y, and n_z being positive integers corresponds to a stationary state with quantum numbers n_x, n_y, n_z. The states with $\varepsilon_{tr} \le \varepsilon_{max}$ satisfy Eq. (21.52). The distance of a dot from the origin is $r = (n_x^2 + n_y^2 + n_z^2)^{1/2}$, so the the positive square root of (21.52) is $r \le (8mV^{2/3}h^{-2}\varepsilon_{max})^{1/2}$. The region defined by this inequality is a sphere of radius $r_{max} = (8mV^{2/3}h^{-2}\varepsilon_{max})^{1/2}$ and the number of dots in 1/8th of this sphere equals the number of quantum states with energy $\varepsilon_{tr} \le \varepsilon_{max}$; we take 1/8th of the sphere because n_x, n_y, and n_z must each be positive, which is true only in 1/8th of the sphere.

(b) Fig. 23.8 shows the 4 cubes at the same altitude that share a dot, and 4 more cubes above these 4 also share this dot. The number of translational states with $\varepsilon_{tr} \le \varepsilon_{max}$ is then $\frac{1}{8}\,(\frac{4}{3}\pi r_{max}^3) = (\pi/6)[(8mV^{2/3}h^{-2} \cdot 3kT)^{1/2}]^3 = (\pi/6)(24mV^{2/3}h^{-2}kT)^{3/2}$.

21.10 **(a)** T; **(b)** F.

21.11 **(a)** We get 3628810.

(b) $9.332621569 \times 10^{157}$, where we used $100^{100.5} = (10^2)^{100.5} = 10^{201}$.

(c) Using $(10^3)^{1000.5} = 10^{1/2}10^{3001}$ and $e^{-1000} = (e^{-100})^{10} = (3.72007597601 \times 10^{-44})^{10} = 507595.88975 \times 10^{-440}$, we get $4.0238726006 \times 10^{2567}$.

21.12 ln (300!) = ln 1 + ln 2 + ln 3 + · · · + ln 300. A BASIC program is

```
10 S = 0
20 FOR I = 1 TO 300
30 S = S + LOG(I)
40 NEXT I
50 PRINT S
60 END
```

One finds ln (300!) = 1414.905850. Also, $N \ln N - N = 300 \ln 300 - 300 =$ 1411.134742.

21.13 From (21.51) and (21.56), $\ln Z = N_C \ln z_C - \ln N_C! + N_B \ln z_B - \ln N_B! + \cdots = N_C \ln z_C - \ln N_C! + N_B \ln z_B - N_B \ln N_B + N_B + \cdots$ and $(\partial \ln Z/\partial N_B)_{T,V,N_{C\neq B}} = \ln z_B - \ln N_B - 1 + 1 = \ln (z_B/N_B)$. Substitution in Eq. (21.41) gives $\mu_B = -RT \ln (z_B/N_B)$.

21.14 **(a)** F; **(b)** F; **(c)** T; **(d)** T.

21.15 $\varepsilon_x = n_x^2 h^2/8ma^2$ and $\Delta\varepsilon_x = (h^2/8ma^2)[(n_x + 1)^2 - n_x^2] = (h^2/8ma^2)(2n_x + 1)$. From $\varepsilon_x = n_x^2 h^2/8ma^2 = kT$ we get $n_x = a(8mkT)^{1/2}/h =$ $[(0.02 \text{ m})/(6.6 \times 10^{-34} \text{ J s})][8(0.028 \text{ kg}/6.0 \times 10^{23})(1.38 \times 10^{-23} \text{ J/K})(273 \text{ K})]^{1/2}$ and $n_x = 1.1 \times 10^9$. Then $\Delta\varepsilon_x/kT = (h^2/8ma^2)(2n_x + 1)/kT = 1.8 \times 10^{-9}$.

21.16 At room T, $kT = (1.38 \times 10^{-23} \text{ J/K})(298 \text{ K})(1 \text{ eV}/1.6022 \times 10^{-19} \text{ J}) = 0.026 \text{ eV}$; $kT/hc = (1.38 \times 10^{-23} \text{ J/K})(298 \text{ K})/(6.63 \times 10^{-34} \text{ J s})(3 \times 10^{10} \text{ cm/s}) = 207 \text{ cm}^{-1}$; $RT = (1.987 \text{ cal/mol-K})(298 \text{ K}) = 0.59 \text{ kcal/mol} = 2.5 \text{ kJ/mol}$.

21.17 With anharmonicity neglected, $\langle N_1\rangle/\langle N_0\rangle = e^{-(1+1/2)h\nu_0/kT}/e^{-h\nu_0/2kT} = e^{-h\nu_0/kT}$.

(a) $h\nu_0/kT = (6.626 \times 10^{-34} \text{ J s})(6.98 \times 10^{13}/\text{s})/(1.381 \times 10^{-23} \text{ J/K})(298 \text{ K}) =$ 11.24 and $\langle N_1\rangle/\langle N_0\rangle = e^{-11.24} = 0.000013$.

(b) $\langle N_1\rangle/\langle N_0\rangle = e^{-3.121} = 0.044$.

(c) $e^{-1.023} = 0.36$.

21.18 Use Eq. (21.76). Let $x \equiv h\nu/kT = hc\tilde{\nu}/kT = (6.6261 \times 10^{-34} \text{ J s}) \times$ $(2.9979 \times 10^8 \text{ m/s})(2329.8 \text{ cm}^{-1})(100 \text{ cm/m})/(1.38065 \times 10^{-23} \text{ J/K})T =$ $(3352.0 \text{ K})/T$. $z_{\text{vib}} = 1 + e^{-x} + e^{-2x} + \cdots = 1/(1 - e^{-x})$.

(a) $x = (3352 \text{ K})/(298.15 \text{ K}) = 11.243$. $z_{\text{vib}} = 1/(1 - e^{-11.243}) = 1.000013$. $N_0 = (6.0221 \times 10^{23})1/1.000013 = 6.0220 \times 10^{23}$. $N_1 = N_A e^{-11.243}/1.000013 = 7.888 \times 10^{18}$.

(b) $x = (3352 \text{ K})/(1073.15 \text{ K}) = 3.1235$. $z_{\text{vib}} = 1.0460$. $N_0 = N_A(1)/1.046 =$ 5.757×10^{23}. $N_1 = N_A e^{-3.1235}/1.046 = 2.533 \times 10^{22}$.

(c) $x = 1.0241$. $z_{\text{vib}} = 1.5604$. $N_0 = 3.859 \times 10^{23}$. $N_1 = 1.386 \times 10^{23}$.

21.19 With centrifugal distortion neglected, the rotational levels are $B_0hJ(J + 1)$. The degeneracy of each level is $2J + 1$, and $\langle N_1\rangle/\langle N_0\rangle = 3e^{-2B_0h/kT}/e^{-0} =$ $3e^{-2B_0h/kT}$. $2B_0h/k = 2(5.04 \times 10^{10}/\text{s})(6.626 \times 10^{-34} \text{ J s})/(1.381 \times 10^{-23} \text{ J/K}) =$ 4.84 K.

(a) $\langle N_1 \rangle / \langle N_0 \rangle = 3e^{-(4.84\text{ K})/(200\text{ K})} = 2.93$.

(b) $3e^{-(4.84\text{ K})/(600\text{ K})} = 2.98$.

(c) 3.

21.20 The ^{14}N nucleus has nuclear-spin quantum number $I = 1$ As noted near the end of Sec. 20.3, in addition to the $2J + 1$ rotational degeneracy factor, we must include the nuclear-spin degeneracy factor of $(I + 1)(2I + 1) = 6$ for the even-J levels and $I(2I + 1) = 3$ for the odd-J levels. The degeneracy factor is therefore $3(2 \cdot 1 + 1) = 9$ for the $J = 1$ level and is $6 \cdot 1 = 6$ for the $J = 0$ level. With centrifugal distortion neglected, the rotational levels are $B_0hJ(J + 1)$. We have $\langle N_1 \rangle / \langle N_0 \rangle = 9e^{-2B_0h/kT}/6e^{-0} = 1.5e^{-2B_0h/kT}$.

$2B_0h/k = 2(5.96 \times 10^{10}/\text{s})(6.626 \times 10^{-34}\text{ J s})/(1.381 \times 10^{-23}\text{ J/K}) = 5.72$ K. So $\langle N_1 \rangle / \langle N_0 \rangle = 1.5e^{-(5.72\text{ K})/(300\text{ K})} = 1.47$.

21.21 (a) The vibrational levels are nondegenerate and $\langle N_1 \rangle / \langle N_0 \rangle = e^{-\Delta\varepsilon/kT} = e^{-h\nu/kT}$. We have $\langle N_2 \rangle / \langle N_0 \rangle = e^{-2h\nu/kT} = (N_1/N_0)^2$. We find $(0.528)^2 = 0.279$, as is observed. Therefore there is an equilibrium distribution in these levels.

(b) $\ln (N_1/N_0) = -h\nu/kT$ and $T = -h\nu/k \ln (N_1/N_0) = -(6.626 \times 10^{-34}\text{ J s})(6.39 \times 10^{12}\text{ s}^{-1})/(1.381 \times 10^{-23}\text{ J/K})(\ln 0.528) = 480$ K.

(c) $N_3/N_0 = e^{-3h\nu/kT} = (N_1/N_0)^3 = (0.340)^3 = 0.0393$. This is an approximation since anharmonicity has been neglected.

21.22 $\ln Z_{\text{BE}}^{\text{FD}} = -\beta\mu N/N_A + \Sigma_r (\pm 1) \ln [1 \pm e^{\beta(\mu/N_A - \varepsilon_r)}]$, where Eqs. (1.69) and (1.70) were used. Since $1 >> e^{\beta(\mu/N_A - \varepsilon_r)}$, we can use Eq. (8.36) to expand the log and we need include only the first term in the sum; so $\ln Z_{\text{BE}}^{\text{FD}} \approx -\beta\mu N/N_A + \Sigma_r (\pm 1)[\pm e^{\beta(\mu/N_A - \varepsilon_r)}] = -\beta\mu N/N_A + \Sigma_r e^{(\mu/N_A - \varepsilon_r)/kT}$. For $\langle N_r \rangle << 1$, the ± 1 in (21.77) can be neglected to give $\langle N_r \rangle \approx e^{\mu/N_A kT} e^{-\varepsilon_r/kT}$. So $\ln Z_{\text{BE}}^{\text{FD}} \approx -\beta\mu N/N_A + \Sigma_r \langle N_r \rangle = -\mu N/N_A kT + N$. Use of (21.77) with the ± 1 neglected gives $\Sigma_r \langle N_r \rangle = N \approx \Sigma_r e^{\mu/N_A kT} e^{-\varepsilon_r/kT}$, so $e^{\mu/N_A kT} \approx N/(\Sigma_r e^{-\varepsilon_r/kT})$ and $\mu/N_A kT \approx \ln N - \ln(\Sigma_r e^{-\varepsilon_r/kT})$. Hence $\ln Z_{\text{BE}}^{\text{FD}} \approx -N \ln N + N \ln \Sigma_r e^{-\varepsilon_r/kT} + N \approx N \ln z - \ln N!$ and $Z_{\text{BE}}^{\text{FD}} \approx z^N/N!$ for $\langle N_r \rangle << 1$.

21.23 With the ±1 neglected, (21.77) becomes $\langle N_r \rangle_{\text{BE}}^{\text{FD}} = e^{-\varepsilon_r/kT}/e^{-\mu/RT}$. Use of the Prob. 21.13 result gives $-\mu/RT = \ln(z/N)$, so $e^{-\mu/RT} = e^{\ln(z/N)} = z/N$. Hence $\langle N_r \rangle_{\text{BE}}^{\text{FD}} = e^{-\varepsilon_r/kT} N/z$, which is (21.69).

21.24 We have $\exp(\varepsilon_{tr}/kT) = \exp(m\upsilon_x^2/2kT)\exp(m\upsilon_y^2/2kT)\exp(m\upsilon_z^2/2kT)$. However, because of the ±1 term in (21.77), the population $\langle N_r \rangle_{\text{BE}}^{\text{FD}}$ (which is related to the probability of occupation of state r) is not equal to the product of separate factors for υ_x, υ_y, and υ_z; hence, υ_x, υ_y, and υ_z are not statistically independent. When $\langle N_r \rangle << 1$ does hold, the ±1 can be neglected and $\langle N_r \rangle$ becomes the product of factors for the three velocity components. Here, the components are statistically independent.

21.25 **(a)** All; **(b)** z_{tr}.

21.26 **(a)** m; **(b)** I and σ; **(c)** ν.

21.27 $z_{el} < z_{rot} < z_{vib} < z_{tr}$.

21.28 As explained at the end of the problem, we omit the $1/N!$ from Z.

(a) $z = \sum_r e^{-\varepsilon_r/kT} = e^{-0} + e^{-a/kT} = 1 + e^{-a/kT}$. $Z = z^N = (1 + e^{-a/kT})^N$. $\ln Z = N \ln (1 + e^{-a/kT})$. $U = kT^2(\partial \ln Z/\partial T)_{V,N} = kT^2[Ne^{-a/kT}(a/kT^2)/(1 + e^{-a/kT})] = Na/(e^{a/kT} + 1)$. $C_V = (\partial U/\partial T)_V = Na^2e^{a/kT}/kT^2(e^{a/kT} + 1)^2$. $S = U/T + k \ln Z = Na/T(e^{a/kT} + 1) + Nk \ln (1 + e^{-a/kT})$.

(b) $a/kT = 1.81$ and $z = 1 + e^{-1.81} = 1.163$. $U = (6.0 \times 10^{23})(1.0 \times 10^{-20}\text{ J})/(e^{1.81} + 1) = 844\text{ J} = 202\text{ cal}$. $S = U/T + Nk \ln z = (844\text{ J})/(400\text{ K}) + (6.0 \times 10^{23})(1.38 \times 10^{-23}\text{ J/K}) \ln 1.163 = 3.36\text{ J/K} = 0.80\text{ cal/K}$.

(c) As $T \to \infty$, we have $e^{a/kT} \to 1$ and $U \to Na/2$. This is because in the $T = \infty$ limit, the populations of the two levels become equal and $U(T = \infty)$ equals the average energy of the levels multiplied by the number of molecules. As $T \to \infty$, $C_V \to Na^2/4kT^2 \to 0$. At very high T, the populations of the two levels have become essentially equal, so we get no further increase in U as T increases further; hence, $C_V = (\partial U/\partial T)_V$ becomes 0.

(d) As $T \to \infty$, $S \to 0 + Nk \ln 2 = Nk \ln 2$.

(e) Let $t = kT/a$. From the expression in (a), $C_V = Nke^{1/t}/t^2(e^{1/t} + 1)^2 = nRe^{1/t}/t^2(e^{1/t} + 1)^2$ and $C_{V,m}/R = e^{1/t}/t^2(e^{1/t} + 1)$. The plot is

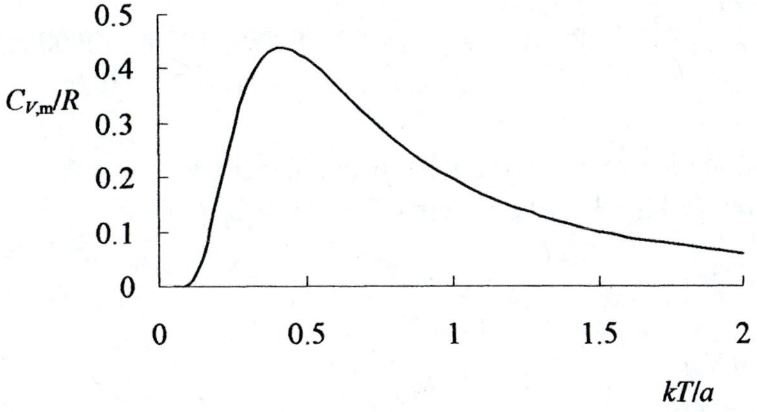

(f) O has only translational and electronic contributions to C_V. O has a very low-lying electronic state whose energy a above the ground electronic state is such that $kT/a > 0.42$ (the maximum on the above graph) for T = 300 K and C_V decreases for temperatures above 300 K.

21.29 Equations (21.76) and (21.90) give $\langle N_\upsilon \rangle/N = e^{-\upsilon\Delta\varepsilon_\upsilon/kT}/z_{\text{vib}} = e^{-0}/z_{\text{vib}} = 1/z_{\text{vib}} = (1 - e^{-h\nu/kT})$.

(a) $1/z_{\text{vib}} = 1 - e^{-10} = 0.999955$. **(b)** $1 - e^{-3} = 0.950$. **(c)** $1 - e^{-2} = 0.865$.

(d) $1 - e^{-1} = 0.632$. **(e)** 0.095.

21.30 (a) $\langle N_\upsilon \rangle/N = e^{-\varepsilon_{\text{vib},\upsilon}/kT}/z_{\text{vib}} = e^{-\upsilon h\nu/kT}/(1 - e^{-\Theta_{\text{vib}}/T})^{-1} = e^{-\upsilon\Theta_{\text{vib}}/T}(1 - e^{-\Theta_{\text{vib}}/T})$, where (21.75), (21.90), and (21.88) were used.

(b) From (21.75) and (21.76) with $\upsilon = 0$: $\langle N_0 \rangle/N = 1/z_{\text{vib}}$ and $z_{\text{vib}} = N/\langle N_0 \rangle$. A similar formula holds for z_{rot}.

(c) $\Theta_{\text{vib}} = h\nu/k = h\tilde{\nu}c/k = (hc/k)(2143\ \text{cm}^{-1}) = 3084$ K, as in the first example in Sec. 21.6. At 25°C: $\Theta_{\text{vib}}/T = 10.344$; $\langle N_0 \rangle = (6.022 \times 10^{23})(1 - e^{-10.344}) = 6.022 \times 10^{23}$.

$\langle N_1 \rangle = N_A e^{-10.344}(1 - e^{-10.344}) = 1.94 \times 10^{19}$.

At 1000°C: $\Theta_{vib}/T = 2.422$. $\langle N_0 \rangle = 5.49 \times 10^{23}$. $\langle N_1 \rangle = 4.87 \times 10^{22}$.

(d) For $\Theta_{vib} = 3352$ K and $\upsilon = 1$, we have $\langle N_1 \rangle / N = e^{-(3352\text{ K})/T}(1 - e^{-(3352\text{ K})/T})$.

We find

T/K	0	2000	4000	6000	8000	10000	12000	15000
$\langle N_1 \rangle /N$	0	0.152	0.245	0.245	0.225	0.204	0.184	0.160

N_2 is not a harmonic oscillator, so anharmonicity and the finite number of vibrational levels make the high-T values inaccurate.

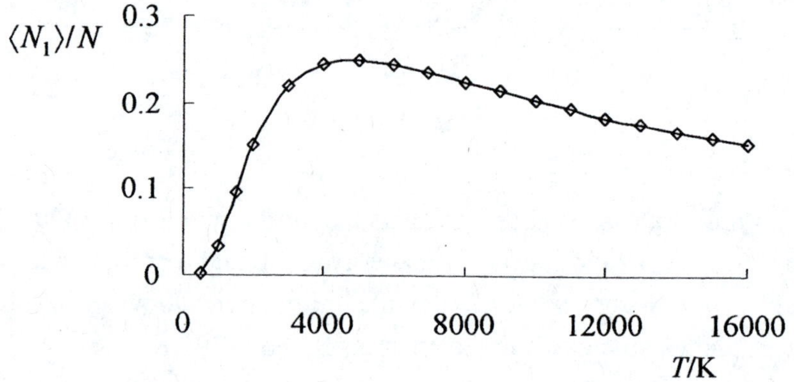

21.31 $d(\langle N_\upsilon \rangle /N)/dT = 0 = (d/dT)(e^{-\upsilon\Theta_{vib}/T} - e^{-(\upsilon+1)\Theta_{vib}/T}) = (\upsilon\Theta_{vib}/T^2)e^{-\upsilon\Theta_{vib}/T} - (\upsilon + 1)(\Theta_{vib}/T^2)e^{-(\upsilon+1)\Theta_{vib}/T} = 0$ and $\upsilon = (\upsilon + 1)e^{-\Theta_{vib}/T}$, so $\Theta_{vib}/T = -\ln[\upsilon/(\upsilon + 1)]$ and $T = \Theta_{vib}/\ln[(\upsilon + 1)/\upsilon] = \Theta_{vib}/\ln(1 + 1/\upsilon)$. For $\Theta_{vib}/T = -\ln[\upsilon/(\upsilon + 1)]$, we get $\langle N_\upsilon \rangle /N = e^{\upsilon \ln[\upsilon/(\upsilon + 1)]} - e^{(\upsilon+1)\ln[\upsilon/(\upsilon+1)]} =$ $\exp\{\ln[\upsilon/(\upsilon + 1)]^\upsilon\} - \exp\{\ln[\upsilon/(\upsilon + 1)]^{\upsilon+1}\} = \upsilon^\upsilon/(\upsilon + 1)^\upsilon - \upsilon^{\upsilon+1}/(\upsilon + 1)^{\upsilon+1} = [(\upsilon + 1)\upsilon^\upsilon - \upsilon^{\upsilon+1}]/(\upsilon + 1)^{\upsilon+1} = \upsilon^\upsilon/(\upsilon + 1)^{\upsilon+1}$.

21.32 Yes. Despite the fact that $\varepsilon_r > \varepsilon_s$ in Eq. (21.72), N_{ε_r} might exceed N_{ε_s} if the degeneracy g_r exceeds g_s. (For example, see Prob. 21.19.)

21.33 **(a)** $z = \sum_{m(\text{levels})} g_m e^{-\varepsilon_m/kT} = 1e^0 + 3e^{-100\text{ K}/200\text{ K}} + 5e^{-300\text{ K}/200\text{ K}} = 3.935$.

(b) Equation (21.71) gives $\langle N(\varepsilon_s) \rangle = Ng_s e^{-\varepsilon_s/kT}/z$.

So $\langle N_1 \rangle = (6.02 \times 10^{23})e^0/3.935 = 1.53 \times 10^{23}$;

$\langle N_2 \rangle = (6.02 \times 10^{23})3e^{-100\text{ K}/200\text{ K}}/3.935 = 2.78 \times 10^{23}$;

$\langle N_3 \rangle = (6.02 \times 10^{23})5e^{-300\text{ K}/200\text{ K}}/3.935 = 1.71 \times 10^{23}$.

(c) As $T \to \infty$, we have $e^{-\varepsilon_m/kT} \to e^{-0} = 1$ and $z \to 1 + 3 + 5 = 9$. So $\langle N_1 \rangle \to (6.02 \times 10^{23})/9 = 0.669 \times 10^{23}$, $\langle N_2 \rangle \to (6.02 \times 10^{23})3/9 = 2.01 \times 10^{23}$, and $\langle N_3 \rangle \to (6.02 \times 10^{23})5/9 = 3.35 \times 10^{23}$.

21.34 (a) $N! = 1 \cdot 2 \cdot 3 \cdots N$ and $\ln N! = \ln 1 + \ln 2 + \ln 3 + \cdots + \ln N = \sum_{x=1}^{N} \ln x$.

(b) For large N, the main contributions to the sum come from the later terms (where x is reasonably close to N); the later terms don't vary greatly as we go from x to $x + 1$. (For example, ln 50 = 3.912 and ln 51 = 3.932.) So Eq. (21.79) can be used.

(c) $\sum_{x=1}^{N} \ln x \approx \int_1^N \ln x\, dx = (x \ln x - x)\big|_1^N = N \ln N - N + 1 \approx N \ln N - N$.

21.35 (a) Let $\theta = \Theta_{\text{rot}}$. Then $\sigma z_{\text{rot}} = \sum_{J=0}^{\infty} (2J + 1)e^{-\theta J(J+1)/T}$. J corresponds to n, $a = 0$, and $f(J) = (2J + 1)e^{-\theta(J^2+J)/T}$. So $f(0) = 1$. Differentiation gives

$f'(J) = [2 - \theta(2J + 1)^2/T]e^{-\theta(J^2+J)/T}$, so $f'(0) = 2 - \theta/T$.

$f''(J) = [-6\theta(2J + 1)/T + \theta^2(2J + 1)^3/T^2]e^{-\theta(J^2+J)/T}$.

$f'''(J) = [-12\theta/T + 12\theta^2(2J + 1)^2/T^2 - \theta^3(2J + 1)^4/T^3]e^{-\theta(J^2+J)/T}$.

$f'''(0) = -12\theta/T + 12\theta^2/T^2 - \theta^3/T^3$.

$f^{iv}(J) = [60(2J + 1)\theta^2/T^2 - 20(2J + 1)^3\theta^3 T^3 + (2J + 1)^5\theta^4/T^4]e^{-\theta(J^2+J)/T}$.

$f^{(v)}(J) =$

$[120\theta^2/T^2 - 180(2J + 1)^2\theta^3/T^3 + 30(2J + 1)^4\theta^4/T^4 - (2J + 1)^6\theta^5/T^5] \times e^{-\theta(J^2+J)/T}$. $f^{(v)}(0) = 120\theta^2/T^2 - 180\theta^3/T^3 + 30\theta^4/T^4 - \theta^5/T^5$.

Noting that $\int_0^{\infty} (2J + 1)e^{-\theta(J^2+J)/T}\, dJ = \int_0^{\infty} e^{-\theta w/T}\, dw = T/\theta$, we have

$\sigma z_{\text{rot}} = T/\theta + \tfrac{1}{2} - (2 - \theta/T)/12 + (-12\theta/T + 12\theta^2/T^2 - \theta^3/T^3)/720 - (120\theta^2/T^2 - 180\theta^3/T^3 + 30\theta^4/T^4 - \theta^5/T^5)/30240 + \cdots$ and

$z_{\text{rot}} = (T/\sigma\Theta_{\text{rot}})(1 + \Theta_{\text{rot}}/3T + \Theta_{\text{rot}}^2/15T^2 + 4\Theta_{\text{rot}}^3/315T^3 + \cdots)$.

(b) $\Theta_{\text{rot}}/T = 85.3/273.15 = 0.3123$. Equation (21.85) gives $z_{\text{rot}} \approx 1/2(0.3123) = 1.601$. Equation (21.86) gives $z_{\text{rot}} =$

$1.601[1 + 0.3123/3 + (0.3123)^2/15 + 4(0.3123)^3/315 + \cdots] = 1.779$ and the error is –10%.

(c) $\Theta_{rot}/T = 2.862/273.15 = 0.01048$. Then $z_{rot} \approx 1/2(0.01048) = 47.72$ and $z_{rot} = 47.72[1 + 0.01048/3 + (0.01048)^2/15 + 4(0.01048)^3/315] = 47.72(1.0035)$ and the error is –0.35%.

21.36 $U_{rot} = nRT^2(d \ln z_{rot}/dT) = nRT^2(d/dT)[\ln T - \ln \sigma\Theta_{rot}] = nRT$. $U_{vib} = nRT^2\, d \ln z_{vib}/dT = -nRT^2[d \ln (1 - e^{-h\nu/kT})/dT] = -nRT^2 e^{-h\nu/kT}(-h\nu/kT^2)/(1 - e^{-h\nu/kT}) = (nRh\nu/k)/(e^{h\nu/kT} - 1)$. $U_{el} = nRT^2(d/dT) \ln g_{el,0} = 0$.

21.37 As $T \to \infty$, the Taylor series for e^x shows that $e^{\Theta_{vib}/T} \to 1 + \Theta_{vib}/T$. Hence, $C_{V,vib} \to nR(\Theta_{vib}/T)^2 e^0/(\Theta_{vib}/T)^2 = nR$.

21.38 $S_{tr} = U_{tr}/T + Nk \ln z_{tr} - Nk(\ln N - 1) = \frac{3}{2}nR + Nk[\frac{3}{2} \ln (2\pi mk/h^2) + \frac{3}{2} \ln T + \ln V] - Nk \ln N + nR = \frac{5}{2}nR + nR \ln [(2\pi mk/h^2)^{3/2} T^{3/2}(NkT/P)(1/N)] = \frac{5}{2}nR + nR \ln [(2\pi m)^{3/2}(kT)^{5/2}/h^3 P]$.
$S_{rot} = U_{rot}/T + Nk \ln z_{rot} = nR + nR \ln [T/\sigma\Theta_{rot}]$. $S_{vib} = U_{vib}/T + Nk \ln z_{vib} = nR\Theta_{vib}/T(e^{\Theta_{vib}/T} - 1) - nR \ln (1 - e^{-\Theta_{vib}/T})$. $S_{el} = U_{el}/T + Nk \ln z_{el} = 0 + nR \ln g_{el,0} = nR \ln g_{el,0}$.

21.39 $S_{tr,m}/R = 2.5 + 2.5 \ln (T/\text{K}) - \ln (P/\text{bar}) + 1.5 \ln M_r + \ln[(2\pi \times 10^{-3}\ \text{kg/mol})^{3/2}(1.38065 \times 10^{-23}\ \text{J/K})^{5/2}\text{K}^{5/2} \times (6.02214 \times 10^{23}/\text{mol})^{-3/2}(6.62607 \times 10^{-34}\ \text{J s})^{-3}(10^5\ \text{N/m}^2)^{-1}] = 2.5 + 2.5 \ln (T/\text{K}) - \ln(P/\text{bar}) + 1.5 \ln M_r - 3.65171 = 2.5 \ln (T/\text{K}) - \ln(P/\text{bar}) + 1.5 \ln M_r - 1.1517$.

21.40 For these gases of closed-shell monatomic molecules, $S = S_{tr}$. Equation (21.108) gives $S^\circ_{m,298}(\text{He}) = (8.3145\ \text{J/mol-K})(1.5 \ln 4.0026 + 2.5 \ln 298.15 - \ln 1 - 1.1517) = 126.15$ J/mol-K. Similarly, $S^\circ_{m,298}(\text{Ne}) = 146.33$ J/mol-K, $S^\circ_{m,298}(\text{Ar}) = 154.85$ J/mol-K, $S^\circ_{m,298}(\text{Kr}) = 164.09$ J/mol-K.

21.41 For H, $g_{el,0} = 2$ and Eqs. (21.107) and (21.108) give $S_m = S_{m,tr} + S_{m,el} = (8.3145\ \text{J/mol-K})(1.5 \ln 1.0079 + 2.5 \ln 298.15 - \ln 1 - 1.1517 + \ln 2) = 114.72$ J/mol-K.

21.42 $C^\circ_{V,\text{m},298} = (3/2)R = 12.47$ J/mol-K. $C^\circ_{P,\text{m},298} = C^\circ_{V,\text{m},298} + R = 20.79$ J/mol-K.

21.43 Equations (21.94) and (21.95) give $U^\circ_{\text{m},298} - U^\circ_{\text{m},0} = U_{\text{tr,m}} = (3/2)RT =$ 1.5(8.3145 J/mol-K)(298.15 K) = 3718 J/mol for each gas.

21.44 $S = (U - U_0)/T + k \ln Z$; $\ln Z = [S - (U - U_0)/T]/k =$ [154.8 J/K − (3718 J)/(298.1 K)](1.3807 × 10^{-23} J/K)$^{-1}$ = 1.03 × 10^{25} = 2.303 log Z. So $\log Z = 4.48 \times 10^{24}$ and $Z = 10^{4.48\times 10^{24}}$ for 1 mole. $Z = z^N/N!$ and $\ln Z = N \ln z - \ln N! = N \ln z - N \ln N + N$. So $\ln z = (\ln Z + N \ln N - N)/N$ = (1.03 × 10^{25})/(6.02 × 10^{23}) + ln (6.02 × 10^{23}) − 1 = 70.9 and $z = 6 \times 10^{30}$.

21.45 **(a)** $\tilde{\nu}_0 = \tilde{\nu}_e - 2\tilde{\nu}_e x_e = 2358.6\ \text{cm}^{-1} - 2(14.3\ \text{cm}^{-1}) = 2330.0\ \text{cm}^{-1}$.

$\tilde{B}_0 = \tilde{B}_e - \frac{1}{2}\tilde{\alpha}_e = 1.998\ \text{cm}^{-1} - \frac{1}{2}(0.017\ \text{cm}^{-1}) = 1.989_5\ \text{cm}^{-1}$.

hc/k = (6.62607 × 10^{-34} J s)(2.99792 × 10^8 m/s)/(1.38065 × 10^{-23} J/K) = 0.0143877 m K = 1.43877 cm K.

$\Theta_{\text{vib}} = \tilde{\nu}_0 hc/k$ = (2330.0 cm^{-1})(1.43877 cm K) = 3352.3 K.

Θ_{rot} = (1.989$_5$ cm^{-1})(1.43877 cm K) = 2.862 K.

(b) $z_{\text{tr}} = (2\pi MkT/N_A)^{3/2}V/h^3$. $V = nRT/P$ = 12310 cm^3. z_{tr} = (0.012310 m^3) × [2π(0.0280 kg/mol)(1.3807 × 10^{-23} J/K)(300 K)/(6.022 × 10^{23}/mol)]$^{3/2}$/ (6.626 × 10^{-34} J s)^3 = 1.78 × 10^{30}.

$z_{\text{rot}} = T/\sigma\Theta_{\text{rot}}$ = (300 K)/2(2.862 K) = 52.4.

$z_{\text{vib}} = [1 - e^{-\Theta_{\text{vib}}/T}]^{-1} = [1 - e^{-3352/300}]^{-1} = 1.000014$. $z_{\text{el}} = g_{\text{el},0} = 1$.

(c) $V = nRT/P$ = 1.026 × 10^5 cm^3 and z_{tr} = 3.57 × 10^{32}; z_{rot} = 437;

z_{vib} = 1.35. z_{el} = 1.

21.46 **(a)** $\Theta_{\text{vib}} = \tilde{\nu}_0 hc/k$ = 5696 K and Θ_{vib}/T = 19.104. $\Theta_{\text{rot}} = \tilde{B}_0 hc/k$ = 29.58 K.

Equations (21.105)–(21.108) give at P = 1 bar:

$S^\circ_{\text{m,tr}}$ = (8.3145 J/mol-K)(1.5 ln 20.0063 + 2.5 ln 298.15 − 1.1517) = 146.22 J/mol-K.

$S^\circ_{\text{m,rot}}$ = (8.3145 J/mol-K){1 + ln[(298.15 K)/1(29.58 K)]} = 27.53 J/mol-K.

$S^\circ_{\text{m,vib}}$ = (8.3145 J/mol-K)[19.104/($e^{19.104}$ − 1) − ln(1 − $e^{-19.04}$)] = 8 × 10^{-7}

J/mol-K. $S^\circ_{\text{m,el}} = 0$ (since all electrons are paired).

So $S^\circ_{\text{m,298}} = S^\circ_{\text{m,tr}} + S^\circ_{\text{m,rot}} + S^\circ_{\text{m,vib}} + S^\circ_{\text{m,el}} = 173.75$ J/mol-K.

(b) Equations (21.100)–(21.102) give: $C^\circ_{V,\text{tr,m}} = 1.5(8.3145$ J/mol-K) = 12.47 J/mol-K. $C^\circ_{V,\text{rot,m}} = 8.3145$ J/mol-K. $C^\circ_{V,\text{vib,m}} =$ (8.3145 J/mol-K)$(19.104)^2e^{19.104}/(e^{19.104} - 1)^2 = 1.5 \times 10^{-5}$ J/mol-K. So $C^\circ_{V,\text{m,298}} = 20.79$ J/mol-K.

(c) $C^\circ_{P,\text{m}} = C^\circ_{V,\text{m}} + R = 29.10$ J/mol-K.

21.47 $\mu = m^2/2m = m/2 = \frac{1}{2}(126.90\text{ g})/(6.022 \times 10^{23}) = 1.0536 \times 10^{-22}$ g.
$I = \mu R^2 = (1.0536 \times 10^{-22}\text{ g})(2.67 \times 10^{-8}\text{ cm})^2 = 7.51 \times 10^{-38}\text{ g cm}^2$.
$\Theta_{\text{rot}} = h^2/8\pi^2 Ik = 0.0536$ K. $\Theta_{\text{vib}} = h\nu_0/k = 306.9$ K and $\Theta_{\text{vib}}/T = 0.6138$.

(a) From Eqs. (21.94)–(21.98) at 500 K: $U^\circ_{\text{m,tr}} = (3/2)RT =$ 1.5(8.3145 J/mol-K)(500 K) = 6236 J/mol. $U^\circ_{\text{m,rot}} = RT = 4157$ J/mol.

$U^\circ_{\text{m,vib}} = (8.3145\text{ J/mol-K})(306.9\text{ K})/(e^{0.6138} - 1) = 3011$ J/mol.

$U^\circ_{\text{m,el}} = 0$. $U^\circ_{\text{m,500}} - U^\circ_{\text{m,0}} = 13404$ J/mol.

(b) $H^\circ_{\text{m,500}} - U^\circ_{\text{m,0}} = U^\circ_{\text{m,500}} + RT - U^\circ_{\text{m,0}} =$
13404 J/mol + (8.3145 J/mol-K)(500 K) = 17561 J/mol.

(c) From (21.105)–(21.108): $S^\circ_{\text{m,tr}} =$
$R(1.5 \ln 253.8 + 2.5 \ln 500 - 1.1517) = 188.65$ J/mol-K.
$S^\circ_{\text{m,rot}} = R\{1 + \ln[500/2(0.0536)]\} = 78.55$ J/mol-K.
$S^\circ_{\text{m,vib}} = R[0.6138/(e^{0.6138} - 1) - \ln(1 - e^{-0.6138})] = 12.50$ J/mol-K.
$S^\circ_{\text{m,el}} = 0$. So $S^\circ_{\text{m,500}} = 279.70$ J/mol-K.

(d) $G^\circ_{\text{m,500}} - U^\circ_{\text{m,0}} = H^\circ_{\text{m,500}} - TS^\circ_{\text{m,500}} - U^\circ_{\text{m,0}} =$
17561 J/mol – (500 K)(279.70 J/mol-K) = –122.29 kJ/mol.

21.48 From Fig. 20.13 the pure-rotational lines are at $2B_0$, $4B_0$, $6B_0$, . . . , so the separation is $2\tilde{B}_0 = 20.9\text{ cm}^{-1}$ and $\tilde{B}_0 = 10.4_5\text{ cm}^{-1}$. Also, $\tilde{\nu}_0 = 2885\text{ cm}^{-1}$. So $\Theta_{\text{rot}} = \tilde{B}_0 hc/k = 15.0_4$ K; $\Theta_{\text{vib}} = \tilde{\nu}_0 hc/k = 4151$ K and $\Theta_{\text{vib}}/T = 13.92$. We shall assume the spectral data are for the predominant species $H^{35}Cl$ and shall calculate $S^\circ_{m,298}$ for $H^{35}Cl$. At 25°C:

$S^\circ_{m,tr} = R(1.5 \ln 36.0 + 2.5 \ln 298.1 - \ln 1 - 1.1517) = 153.5$ J/mol-K.

$S^\circ_{m,rot} = R\{1 + \ln[298.15/1(15.04)]\} = 33.1$ J/mol-K.

$S^\circ_{m,vib} = R[13.92/(e^{13.92} - 1) - \ln(1 - e^{-13.92})] = 0.00011$ J/mol-K. $S^\circ_{m,el} = 0$.

So $S^\circ_{m,298} = 186.6$ J/mol-K.

21.49 **(a)** For this relatively light diatomic molecule, only translation and rotation contribute to C_V and $C^\circ_{V,m} \cong (3/2)R + R = 2.5R = 20.79$ J/mol-K.

$C^\circ_{P,m} = C^\circ_{V,m} + R \cong 3.5R = 29.10$ J/mol-K. The true value is 29.13 J/mol-K.

(b) All gases of relatively light diatomic molecules have $C^\circ_{P,m} \cong 3.5R =$ 29.1 J/mol-K.

21.50 $\varepsilon_{el,1}/k = (0.0149 \text{ eV})(1.6022 \times 10^{-19} \text{ J/eV})/(1.38065 \times 10^{-23} \text{ J/K}) = 172._9$ K. $z_{el} = g_{el,0} + g_{el,1}e^{-\varepsilon_{el,1}/kT} = 2 + 2e^{-(172.9 \text{ K})/T}$. At 30 K, $z_{el} = 2 + 2e^{-172.9/30} = 2.006$. At 150 K, $z_{el} = 2.63$. At 300 K, $z_{el} = 3.12$.

21.51 Using the z_{el} expression found in Prob. 21.50, we have $U_{el} = NkT^2(d \ln z_{el}/dT) = (nRT^2/z_{el})(dz_{el}/dT) = (nRT^2/z_{el})[2(173 \text{ K})T^{-2}e^{-(173 \text{ K})/T}] = nR(346 \text{ K})e^{-(173 \text{ K})/T}/z_{el}$ and $U_{el,m} = (2875 \text{ J/mol})e^{-(173 \text{ K})/T}/[2 + 2e^{-(173 \text{ K})/T}] = (1438 \text{ J/mol})/e^{(173 \text{ K})/T} + 1)$. $C_{V,el,m} = dU_{el,m}/dT = -(1438 \text{ J/mol})[-(173 \text{ K})/T^2]e^{(173 \text{ K})/T}/(e^{(173 \text{ K})/T} + 1)^2 = (2.49 \times 10^5 \text{ J K/mol})T^{-2}e^{(173 \text{ K})/T}/(e^{(173 \text{ K})/T} + 1)^2$. We find

$C_{V,el,m}$/(J/mol-K)	0.86	2.52	3.47	3.42	2.67	2.02	1.30	0.88	0.64
T/K	30	45	60	90	120	150	200	250	300

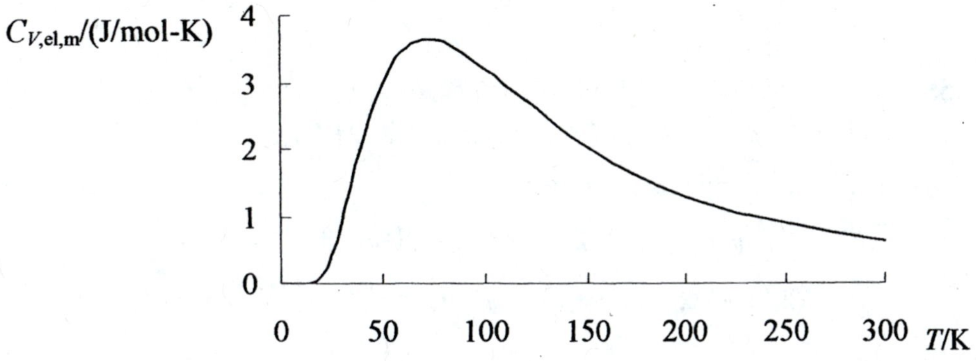

21.52 **(a)** $s - xs = 1 + x + x^2 + x^3 + \cdots - (x + x^2 + x^3 + \cdots) = 1$. So $s = 1/(1 - x)$.

(b) The Taylor series for $1/(1 - x)$ about $x = 0$ is given by Eq. (8.8) as $1/(1 - x) = 1 + x + x^2 + x^3 + \cdots$ for $|x| < 1$.

21.53 **(a)** $z_{\text{vib}} = \sum_{\upsilon} e^{-\varepsilon_{\text{vib},\upsilon}/kT} = \sum_{\upsilon=0}^{\infty} e^{-(\upsilon+1/2)h\nu/kT} = e^{-h\nu/2kT} \sum_{\upsilon=0}^{\infty} e^{-\upsilon h\nu/kT} =$ $e^{-h\nu/2kT}/(1 - e^{-h\nu/kT})$, where Eq. (21.89) was used.

(b) $\ln z_{\text{vib}} = -\frac{1}{2}h\nu/kT - \ln(1 - e^{-h\nu/kT})$.

$U_{\text{vib}} = nRT^2(\partial \ln z_{\text{vib}}/\partial T)_V = nRT^2[\frac{1}{2}h\nu/kT^2 + e^{-h\nu/kT}(h\nu/kT^2)/(1 - e^{-h\nu/kT})]$ $= \frac{1}{2}Nh\nu + nRh\nu/k(e^{h\nu/kT} - 1)$, which differs from Eq. (21.97) by $N(h\nu/2)$; this agrees with the Prob. 21.5b result.

21.54 From Eq. (21.100), $C_{V,\text{tr,m}} = 3R/2$. The translational levels are so closely spaced that this result holds at all temperatures not extremely close to 0 K. From the table after Eq. (21.85), Θ_{rot} is typically of order of magnitude 1 K (except for hydrides). Figure 21.11 shows that $C_{\upsilon,\text{rot,m}} = R$ for T above $1.3\Theta_{\text{rot}}$. From the table after (21.88), Θ_{vib} is typically 10^3 K. Figure 21.10 shows that $C_{V,\text{vib,m}}$ is negligible below $0.1\Theta_{\text{vib}}$ and increases to R as T increases to 1 or 2 times Θ_{vib}. So:

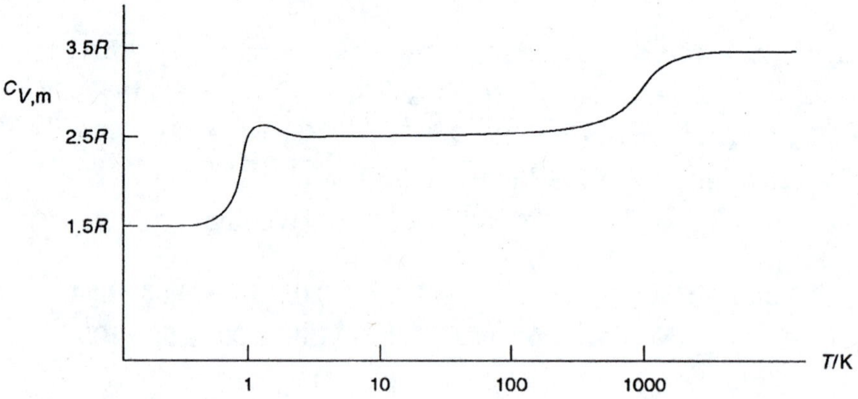

21.55 $z_{\text{rot}} = [(298\text{ K})/1(2.77\text{ K})][1 + (1/3)(2.77/298) + \cdots] = 107._9$.
$\varepsilon_{\text{rot}} = J(J+1)\hbar^2/2I = J(J+1)k\Theta_{\text{rot}}$, where (21.83) was used.

(a) Use of Eq. (21.71) gives

$[\sum_{J=0}^{16} \langle N_J \rangle]/N = z_{\text{rot}}^{-1} \sum_{J=0}^{16} (2J+1)\exp[-J(J+1)\Theta_{\text{rot}}/T] \approx$

$z_{\text{rot}}^{-1} \int_0^{17} (2J+1)\exp[-(J^2+J)\Theta_{\text{rot}}/T]\, dJ = z_{\text{rot}}^{-1} \int_0^{17^2+17} \exp[-(\Theta_{\text{rot}}/T)w]\, dw =$

$z_{rot}^{-1}(-T/\Theta_{rot})[\exp(-306\Theta_{rot}/T) - 1] = (107._9)^{-1}(298/2.77)[1 - e^{-306(2.77/298)}] = 0.93_9$, which is 94%; we took $w \equiv J^2 + J$ and $dw = 2J + 1$. The sum has 17 terms, so 17 was used as the upper limit of the integral.

(b) The required sum is $(107._9)^{-1}\sum_{J=0}^{16}(2J+1)e^{-(J^2+J)(2.77/298)}$. Direct evaluation gives 0.932 = 93.2%. The individual percentages are 0.927, 2.73, 4.38, 5.80, 6.93, 7.71, 8.15, 8.26, 8.07, 7.63, 7.00, 6.25, 5.44, 4.61, 3.82, 3.09, and 2.44 for $J = 0, 1, 2, \ldots$, respectively. A BASIC program is

```
10 S = 0
20 A = 2.77/298.1
30 FOR J = 0 TO 16
40 B = (2*J+1)*EXP(-A*(J*J+J))/107.9
50 P = 100*B
60 PRINT "J=";J;" POP=";P
70 S = S + B
80 NEXT J
90 PRINT "SUM=";S
95 END
```

21.56 $\nu = (k/\mu)^{1/2}/2\pi$. N_2 and F_2 have roughly the same μ values, but N_2 has a triple bond and F_2 has a single bond, so $k_{N_2} >> k_{F_2}$ (cf. Table 20.1). The high k_{N_2} makes ν_{N_2} and Θ_{vib,N_2} high, so the vibrational contribution to S°_{m,N_2} is negligible at room temperature. The light mass of H makes $\mu_{HF} << \mu_{F_2}$, so $\nu_{HF} >> \nu_{F_2}$ and $S^\circ_{vib,m,298,HF} << S^\circ_{vib,m,298,F_2}$.

21.57 **(a)** $z_{vib} \approx \int_0^\infty e^{-\upsilon\Theta_{vib}/T}\,d\upsilon = -(T/\Theta_{vib})e^{-\upsilon\Theta_{vib}/T}\Big|_0^\infty = T/\Theta_{vib}$.

(b) For high T the Taylor series for e^x gives $e^{-h\nu/kT} \approx 1 - h\nu/kT$, so $z_{vib} \to 1/(h\nu/kT) = T/\Theta_{vib}$.

21.58 Equations (21.104) and (21.82) are not applicable at extremely low T since the translational levels are then no longer closely spaced compared with kT and the sum in z_{tr} [Eq. (21.78)] can't be replaced by an integral. Another reason is that at extremely low T, the condition $\langle N_r \rangle << 1$ does not hold, so Bose–Einstein or Fermi–Dirac statistics must be used.

21.59 **(a)** $z_{tr,N_2} \approx z_{tr,CO}$, since $m_{N_2} \approx m_{CO}$. z_{rot,N_2} and $z_{rot,CO}$ differ substantially since $\sigma_{N_2} \approx 2$ but $\sigma_{CO} = 1$. We have $\mu_{N_2} \approx \mu_{CO}$ and $k_{N_2} \approx k_{CO}$ for these isoelectronic triply bonded molecules, so $\nu_{N_2} \approx \nu_{CO}$ (see Table 20.1); therefore $z_{vib,N_2} \approx z_{vib,CO}$. z_{el} for each species is essentially 1.

(b) From (21.64) and (21.104)–(21.107) and the results of part (a), $S^\circ_{m,298,CO(g)} - S^\circ_{m,298,N_2(g)} \approx S^\circ_{298,m,rot,CO(g)} - S^\circ_{m,298,rot,N_2(g)} \approx$ $R \ln(\sigma_{N_2}/\sigma_{CO}) = R \ln 2 = 5.76$ J/mol-K, since we expect Θ_{rot} to be about the same for the two species. The actual difference is 6.06 J/mol-K.

21.60 **(a)** Br_2 is the heaviest molecule and has the largest moment of inertia and the smallest Θ_{rot}. N_2 is the lightest molecule and also has the shortest bond length (since it has a triple bond), so N_2 has the smallest I and the largest Θ_{rot}.

(b) $\nu = (1/2\pi)(k/\mu)^{1/2}$. The large μ of Br_2 makes this singly bonded species have the smallest ν and hence the smallest Θ_{vib}. The high k and small μ of N≡N makes ν_{N_2} and Θ_{vib,N_2} the largest.

(c) The low value of Θ_{rot} for Br_2 makes more of its excited rotational levels populated and makes z_{rot,Br_2} largest.

(d) The small ν_{Br_2} and Θ_{vib,Br_2} make excited vibrational levels populated and z_{vib,Br_2} is largest.

(e) All these species have essentially the same $C_{V,rot,m}$ at room temperature since they all have essentially attained the high-T limit $C_{V,rot,m} = R$ at room temperature.

(f) Br_2, for which excited vibrational levels are accessible at room T.

21.61 z_{tr} is very roughly equal to the number of translational energy levels that have significant populations at temperature T. The particle-in-a-box translational levels have ε_{tr} proportional to $1/m$ (Chap. 18), so an increase in molecular mass m lowers the translational levels and allows more of them to be populated at T, thereby increasing z_{tr} and increasing $S_{tr,m}$.

21.62 **(a)** At 20 K, $z_{rot} \cong T/\sigma\Theta_{rot} = (20\text{ K})/(29.577\text{ K}) = 0.676$. At 30 K, $z_{rot} \cong 1.014$. At 40 K, $z_{rot} \cong 1.352$.

(b) $z_{rot} \cong 0.6762[1 + 1/3(0.6762) + 1/15(0.6762)^2 + 4/315(0.6762)^3] = 1.136$ at 20 K. At 30 K, $z_{rot} \cong 1.426$. At 40 K, $z_{rot} \cong 1.742$.

(c) A BASIC program is

```
10 TH = 29.577
20 S = 0
30 INPUT "T=";T
40 FOR J = 0 TO 10000
50 A = (2*J + 1)*EXP(-J*(J + 1)*TH/T)
60 IF A < 1E-12 THEN 90
70 S = S + A
80 NEXT J
90 PRINT "ZROT=";S
95 GOTO 20
99 END
```

The exact values are $z_{rot} = 1.1565$ at 20 K, $z_{rot} = 1.4312$ at 30 K, $z_{rot} = 1.7439$ at 40 K.

21.63 $S = U/T + k \ln Z = kT(\partial \ln Z/\partial T)_{V,N_B} + k \ln (z^N/N!) =$ $kT(\partial \ln (z^N/N!)/\partial T)_{V,N_B} + Nk \ln z - k \ln N! = NkT(\partial \ln z/\partial T)_{V,N_B} + Nk \ln z -$ $k \ln N! = (NkT/z)(\partial z/\partial T)_{V,N_B} + Nk \ln z \sum_s e^{-\varepsilon_s/kT}/z - k \ln N!$. Use of $\partial z/\partial T =$ $(\partial/\partial T)\sum_s e^{-\varepsilon_s/kT} = (1/kT^2)\sum_s \varepsilon_s e^{-\varepsilon_s/kT}$ gives $S = (N/zT)\sum_s \varepsilon_s e^{-\varepsilon_s/kT} +$ $(Nk/z)\ln z \sum_s e^{-\varepsilon_s/kT} - k \ln N! = Nk \sum_s [(\varepsilon_s/kT + \ln z)(e^{-\varepsilon_s/kT}/z)] - k \ln N!$. From (21.69), $x_s = e^{-\varepsilon_s/kT}/z$ and $\ln x_s = -\varepsilon_s/kT - \ln z$, so $S = -Nk \sum_s x_s \ln x_s - k \ln N!$. The other condition is $\langle N_r \rangle \ll 1$, as in (21.49).

21.64 $\ln z_{rot} = \ln(T/\sigma\Theta_{rot}) + \ln(1 + \Theta_{rot}/3T + \Theta^2_{rot}/15T^2 + \cdots) =$
$\ln T - \ln(\sigma\Theta_{rot}) + (\Theta_{rot}/3T + \Theta^2_{rot}/15T^2 \cdots) - \frac{1}{2}(\Theta_{rot}/3T + \Theta^2_{rot}/15T^2 + \cdots)^2 +$
$\cdots = \ln T - \ln(\sigma\Theta_{rot}) + \Theta_{rot}/3T + (1/15 - 1/18)(\Theta^2_{rot}/T^2) + \cdots$
$= \ln T - \ln(\sigma\Theta_{rot}) + \Theta_{rot}/3T + \Theta^2_{rot}/90T^2 + \cdots$.
$d \ln z_{rot}/dT = 1/T - \Theta_{rot}/3T^2 - \Theta^2_{rot}/45T^3 + \cdots$.
$U_{rot} = nRT^2(d \ln z_{rot}/dT) = nRT - nR\Theta_{rot}/3 - nR\Theta^2_{rot}/45T + \cdots$.
As $T \to \infty$, all terms after the second go to zero and $U_{rot} \to n(RT - R\Theta_{rot}/3)$.

21.65 (a) We have the original orientation plus the following: 120° and 240° rotations about the C_3 axis give 2 indistinguishable orientations; rotation about a C_2 axis followed by rotations about the C_3 axis give 3 more indistinguishable orientations. So $\sigma = 6$.

(b) 2.

(c) 1 (the same as for a heteronuclear diatomic).

(d) 120° and 240° rotations about each of the four C_3 axes produce 2(4) = 8 indistinguishable orientations; 180° rotations about each of the three C_2 axes (these coincide with the S_4 axes—see Fig. 20.22) produce 3 more indistinguishable orientations. Adding in the original orientation, we get $\sigma = 12$.

(e) 2 (the same as for a homonuclear diatomic).

(f) There are three C_2 axes and $\sigma = 4$.

(g) 3.

21.66 $S^\circ_{tr,m} = R(1.5 \ln 34.08 + 2.5 \ln 298.15 - 1.1517) = 152.87$ J/mol-K.

$\tilde{A}_0 = h/8\pi^2 cI_{a,0}$, $\tilde{B}_0 = h/8\pi^2 cI_{b,0}$, $\tilde{C}_0 =$ etc. Equation (21.109) becomes $z_{rot} = (\pi^{1/2}/\sigma)(kT/hc)^{3/2}(1/\tilde{A}_0\tilde{B}_0\tilde{C}_0)^{1/2}$. Using $\sigma = 2$, we get $z_{rot} = 125.9$.

So $S^\circ_{rot,m} = R(1.5 + \ln 125.9) = 52.68$ J/mol-K.

$\Theta_{vib,1}/T = (2615 \text{ cm}^{-1})hc/kT = 12.62$. $\Theta_{vib,2}/T = 5.709$. $\Theta_{vib,3}/T = 12.68$.

$S^\circ_{vib,m} = R[12.62/(e^{12.62} - 1) - \ln(1 - e^{-12.62}) + 5.709/(e^{5.709} - 1) - \ln(1 - e^{-5.709}) + 12.68/(e^{12.68} - 1) - \ln(1 - e^{-12.68})] = 0.186$ J/mol-K. $S^\circ_{el,m} = 0$.

Adding, we get $S^\circ_{298,m} = 205.74$ J/mol-K.

21.67 $S^\circ_{tr,m} = R(1.5 \ln 44.01 + 2.5 \ln 298.15 - \ln 1 - 1.1517) = 156.05$ J/mol-K.

$\Theta_{rot} = \tilde{B}_0 hc/k = 0.561$ K. For this linear molecule, Eq. (21.105) gives $S^\circ_{rot,m} = R\{1 + \ln[(298.15 \text{ K})/2(0.561 \text{ K})]\} = 54.73$ J/mol-K. $\Theta_{vib,1}/T = (1340 \text{ cm}^{-1})hc/kT = 6.466$; $\Theta_{vib,2}/T = (667 \text{ cm}^{-1})hc/kT = 3.219$; $\Theta_{vib,3}/T = 3.219$; $\Theta_{vib,4}/T = 11.34$. Equations (21.118) and (21.106) give $S^\circ_{vib,m} = R[6.466/(e^{6.466} - 1) - \ln(1 - e^{-6.466}) + 2(3.219)/(e^{3.219} - 1) - 2\ln(1 - e^{-3.219}) + 11.34/(e^{11.34} - 1) - \ln(1 - e^{-11.34})] = 0.3616R = 3.01$ J/mol-K. $S^\circ_{el,m} = 0$.

$S^\circ_{m,298} = (156.05 + 54.73 + 3.01 + 0)$ J/mol-K $= 213.79$ J/mol-K.

21.68 $C^\circ_{V,\text{tr,m}} = (3/2)R$. $C^\circ_{V,\text{vib},s,\text{m}} \to R$ at high T for each vibrational mode s.

$C^\circ_{V,\text{rot,m}} \to R$ at high T for linear molecules; $C^\circ_{V,\text{rot,m}} \to (3/2)R$ at high T for nonlinear molecules.

(a) Linear. 3(3) – 5 = 4 normal modes. $C^\circ_{V,\text{m}} \to R(1.5 + 4 + 1) = 6.5R =$ 54.04 J/mol-K. $C^\circ_{P,\text{m}} = C^\circ_{V,\text{m}} + R = 62.36$ J/mol-K.

(b) Nonlinear. 3(3) – 6 = 3 normal modes. $C^\circ_{V,\text{m}} \to R(1.5 + 3 + 1.5) = 6R =$ 49.89 J/mol-K. $C^\circ_{P,\text{m}} \to 58.20$ J/mol-K.

(c) 12 normal modes. $C^\circ_{V,\text{m}} \to R(1.5 + 12 + 1.5) = 15R = 124.72$ J/mol-K. $C^\circ_{P,\text{m}} \to 133.03$ J/mol-K.

21.69 Since z_{tr} is the same as for a gas of diatomic molecules, U_{tr} is given by (21.95). For a gas of linear molecules, z_{rot} is the same as for a gas of diatomics, so (21.96) gives z_{rot}. Use of Eq. (21.109) gives for a gas of nonlinear molecules: $U_{\text{rot}} = nRT^2\, d \ln z_{\text{rot}}/dT = nRT^2\, d(\frac{3}{2}\ln T + \text{const})/dT =$ $\frac{3}{2}nRT$. Equation (21.110) gives $U_{\text{vib}} = -nRT^2(d/dT) \ln \prod_s (1 - e^{-h\nu_s/kT}) =$ $-nRT^2(d/dT) \sum_s \ln (1 - e^{-\Theta_s/T}) = -nRT^2 \sum_s [e^{-\Theta_s/T}(-\Theta_s/T^2)/(1 - e^{-\Theta_s/T})] =$ $nR \sum_s \Theta_s/(e^{\Theta_s/T} - 1)$.

21.70 **(a)** At 0 K, there is no translational, rotational, or vibrational energy (above the zero-point energy) for the H_2 molecules; ΔU°_0 is determined by the change in electronic energy and $\Delta U^\circ_0 = (4.4780 \text{ eV})N_A = (4.4780 \text{ eV}) \times (1.602176 \times 10^{-19} \text{ J/eV})(6.02214 \times 10^{23}/\text{mol}) = 432.06$ kJ/mol.

(b) At 25°C, we need consider only translational and rotational contributions to U_{m} of $H_2(g)$; the translational contribution is $\frac{3}{2}RT$ and the rotational is RT, for a total of $\frac{5}{2}RT$. The translational contribution to U_{m} of 2H(g) is $2(\frac{3}{2}RT) = 3RT$. Hence $\Delta U^\circ_{298} - \Delta U^\circ_0 = 3RT - \frac{5}{2}RT = \frac{1}{2}RT$. Then $\Delta H^\circ_{298} = \Delta U^\circ_{298} + \Delta n_g RT/\text{mol} = \Delta U^\circ_0 + \frac{1}{2}RT + RT = \Delta U^\circ_0 + \frac{3}{2}RT = 432.06$ kJ/mol + 1.5(0.0083145 kJ/mol-K)(298.15 K) = 435.78 kJ/mol. The Appendix gives 2(217.965 kJ/mol) = 435.93 kJ/mol.

21.71 **(a)** $H^\circ_{m,T} - H^\circ_{m,0} = U^\circ_{m,T} + RT - U^\circ_{m,0} = U^\circ_{m,T,tr} - U^\circ_{m,0} + RT = (3/2)RT + RT$
$= 2.5(8.3145 \text{ J/mol-K})(298.15 \text{ K}) = 6.1974$ kJ/mol.

(b) $2.5RT = 20.786$ kJ/mol.

(c) $G^\circ_{m,T} - H^\circ_{m,0} = H^\circ_{m,T} - TS^\circ_{m,T} - H^\circ_{m,0} = H^\circ_{m,T} - H^\circ_{m,0} - TS^\circ_{m,T,tr}$. We take $H^\circ_{m,T} - H^\circ_{m,0}$ from (a) or (b) and use (21.108) for $S^\circ_{m,tr}$. [Alternatively, (21.123) can be used.] $G^\circ_{m,298} - H^\circ_{m,0} = 6.1974$ kJ/mol – (298.15 K)(0.0083145 kJ/mol-K)[1.5 ln 39.948 + 2.5 ln 298.15 – 1.1517] = –39.970 kJ/mol.

(d) $G^\circ_{m,1000} - H^\circ_{m,0} = 20.786$ kJ/mol – 180.00 kJ/mol = –159.22 kJ/mol.

21.72 From (21.123) with $H^\circ_{m,0} = U^\circ_{m,0}$, we have $z = Ne^{-(G_{m,T} - H_{m,0})/RT} =$ $(6.022 \times 10^{23})\exp[(257.7 \text{ J/mol-K})/(8.314 \text{ J/mol-K})] = 1.739 \times 10^{37}$.

21.73 $A - U_0 = -kT \ln Z$ and $\ln Z = -(A - U_0)/kT$. $A - U_0 = U - U_0 - TS \cong H - H_0 - TS$, since there is negligible difference between $H - H_0$ and $U - U_0$ for a solid. So $A - U_0 = [523 \text{ J/mol} - (298.15 \text{ K})(2.377 \text{ J/mol-K})](1 \text{ mol}) = -185.{}_7$ J. $\ln Z = (185.{}_7 \text{ J})/kT = 4.51 \times 10^{22}$.

21.74 Substitution of (21.124) into $\sum_i \nu_i \mu_i = 0$ gives
$0 = \sum_i \nu_i U_{m,0,i} - RT \sum_i \nu_i \ln(z_i / N_i) = \Delta U^\circ_0 - RT \ln[\prod_i (z_i/N_i)^{\nu_i}]$. So
$\Delta U^\circ_0 / RT = \ln\{\prod_i [(z_i/VN_A)/(N_i/VN_A)]^{\nu_i}\} = \ln\{\prod_i [(z_i/VN_A)/c_i]^{\nu_i}\}$. Then
$\exp(\Delta U^\circ_0/RT) = \prod_i [(z_i/VN_A)/c_i]^{\nu_i} = [\prod_i (z_i/VN_A)^{\nu_i}]/[\prod_i (c_i)^{\nu_i}]$. So
$K_c = \prod_i (c_i)^{\nu_i} = \exp(-\Delta U^\circ_0/RT) \prod_i (z_i/VN_A)^{\nu_i}$.

21.75 $S = k \ln W$ and $\Delta S = k \ln (W_{final}/W_{initial})$, so $W_{final}/W_{initial} = e^{\Delta S/k}$.
$\Delta S_{mix} = -R(1.00 \text{ mol} \ln 0.5 + 1.00 \text{ mol} \ln 0.5) = 11.53$ J/K.
So $W_{final}/W_{initial} = \exp[(11.53 \text{ J/K})/(1.381 \times 10^{-23} \text{ J/K})] = \exp(8.35 \times 10^{23})$. Let $10^y = e^{8.35\times10^{23}}$. Taking base-10 logs gives $y = 8.35 \times 10^{23} \log e = 3.63 \times 10^{23}$.
So $W_{final}/W_{initial} = 10^{3.63\times10^{23}}$.

21.76 $S = k \ln W$ and $\ln W = S/k = (191.61 \text{ J/K})/(1.3807 \times 10^{-23} \text{ J/K}) = 1.388 \times 10^{25}$, where the Appendix was used.

21.77 $S_{\text{Tony}} - S_{\text{Rosabella}} = k \ln W_{\text{Tony}} - k \ln W_{\text{Rosabella}} =$
$k \ln [e^{(10^{12})} W_{\text{Rosabella}}] - k \ln W_{\text{Rosabella}} = k \ln e^{(10^{12})} + k \ln W_{\text{Rosabella}} - k \ln W_{\text{Rosabella}}$
$= k(10^{12}) = (1.4 \times 10^{-23} \text{ J/K})10^{12} = 10^{-11}$ J/K, which is utterly negligible.

21.78 **(a)** Yes; **(b)** no; **(c)** no.

21.79 **(a)** Ne.

(b) Each has 2 electrons. The single +2-charged nucleus in He holds the 2 electrons more tightly than the two +1-charged nuclei in H_2, so H_2 is more polarizable.

(c) Cl_2.

(d) Each has 18 electrons. H_2S is more polarizable.

(e) C_2H_6.

21.80 $\upsilon_{d\text{-}d} r^6 = -2[(1.60 \text{ D})(3.336 \times 10^{-30} \text{ C m/D})]^4 /$
$\{3(1.38 \times 10^{-23} \text{ J/K})(298 \text{ K})[4\pi(8.854 \times 10^{-12} \text{ C}^2/\text{N-m}^2)]^2\} = -106 \times 10^{-79} \text{ J m}^6$.
$\upsilon_{d\text{-}id} r^6 = -2[(1.60 \text{ D})(3.336 \times 10^{-30} \text{ C m/D})]^2 (6.48 \times 10^{-30} \text{ m}^3)/$
$[4\pi(8.854 \times 10^{-12} \text{ C}^2/\text{N-m}^2] = -33 \times 10^{-79} \text{ J m}^6$.
$\upsilon_{\text{disp}} r^6 = -[3(11.3 \text{ eV})^2/2(22.6 \text{ eV})](1.602 \times 10^{-19} \text{ J/eV})(6.48 \times 10^{-30} \text{ m}^3)^2 =$
$-570 \times 10^{-79} \text{ J m}^6$.

21.81 $F = -dV/dr = 4\varepsilon(12\sigma^{12}/r^{13} - 6\sigma^6/r^7) = (24\varepsilon/r)[2(\sigma/r)^{12} - (\sigma/r)^6]$.
$\varepsilon = 0.013 \text{ eV} = 2.1 \times 10^{-21}$ J and $\sigma = 3.8$ Å.

(a) $F = [24(2.1 \times 10^{-21} \text{ J})/(8 \times 10^{-10} \text{ m})][2(3.8/8)^{12} - (3.8/8)^6] = -7.1 \times 10^{-13}$ N. **(b)** -1.2×10^{-11} N. **(c)** 5.0×10^{-9} N.

21.82 **(a)** $\upsilon = A/r^{12} - B/r^6$. $0 = A/\sigma^{12} - B/\sigma^6$, so $B\sigma^6 = A$ and $\upsilon = B\sigma^6/r^{12} - B/r^6$.

(b) $d\upsilon/dr = -12B\sigma^6/r^{13} + 6B/r^7 = 0$. So $r_{\text{min}}^6 = 2\sigma^6$ and $r_{\text{min}} = 2^{1/6}\sigma$.

(c) $\varepsilon = \upsilon(\infty) - \upsilon(r_{\text{min}}) = 0 - [B\sigma^6/(2^{1/6}\sigma)^{12} - B/(2^{1/6}\sigma)^6] = B/4\sigma^6$ and $B = 4\sigma^6\varepsilon$.

(d) Substitution of $\sigma = r_{min}2^{-1/6}$ in (21.136) gives
$\upsilon = 4\varepsilon(r_{min}^{12}/4r^{12} - r_{min}^{6}/2r^{6}) = \varepsilon[(r_{min}/r)^{12} - 2(r_{min}/r)^{6}]$.

21.83 **(a)** $r_{min} = 2^{1/6}\sigma = 2^{1/6}(3.5 \text{ Å}) = 3.9 \text{ Å}$.

(b) $r_{min} = 2^{1/6}(8.6 \text{ Å}) = 9.6_5 \text{ Å}$.

21.84 **(a)** As noted in Sec. 21.10. for a nonpolar molecule $\varepsilon \approx 1.3kT_b =$ $1.3(1.38 \times 10^{-23} \text{ J/K})(27.1 \text{ K}) = 4.8_6 \times 10^{-22}$ J.

(b) From Sec. 8.3, $T_c \approx 1.6T_b$, so $\varepsilon \approx 1.3k(T_c/1.6) = 3.4_2 \times 10^{-21}$ J.

21.85 **(a)** We approximate the intermolecular potential by the Lennard-Jones potential. $F(r) = -dV/dr = -4\varepsilon(-12\sigma^{12}/r^{13} + 6\sigma^{6}/r^{7}) =$ $24\varepsilon(2\sigma^{12}/r^{13} - \sigma^{6}/r^{7}) = c/r^{13} - d/r^{7}$. The $F(r)$ curve has the same general appearance as the $\upsilon(r)$ curve, being 0 at $r = \infty$, negative (i.e., attractive) for large r, positive for small r, and ∞ at $r = 0$.

(b) $0 = 24\varepsilon(2\sigma^{12}/b^{13} - \sigma^{6}/b^{7})$, so $2\sigma^{6} = b^{6}$ and $b = 2^{1/6}\sigma$. [*Cf.* Prob. 21.80b; $F(r) = 0$ at the minimum in υ.]

21.86

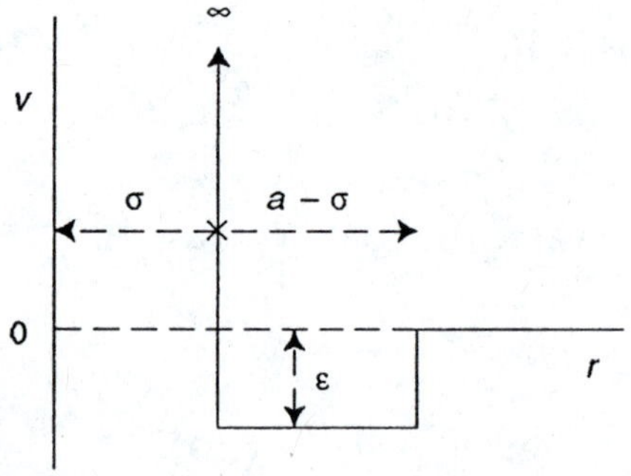

21.87 **(a)** Xe (higher *M*); **(b)** C_2H_5OH (H bonding); **(c)** H_2O.

21.88 **(a)** $\upsilon/\varepsilon = 4[(\sigma/2^{1/6}\sigma)^{12} - (\sigma/2^{1/6}\sigma)^{6}] = 4(\frac{1}{4} - \frac{1}{2}) = -1$ (this is the minimum);

(b) $4[(\sigma/1.5\sigma)^{12} - (\sigma/1.5\sigma)^{6}] = -0.320$;

(c) -0.062; **(d)** -0.016; **(e)** -0.005.

21.89 **(a)** For an ideal gas, $\mathcal{V} = 0$; for linear molecules Eq. (21.145) becomes
$Z_{con} = (4\pi)^{-N} \int_0^\pi \sin\theta_1\, d\theta_1 \cdots \int_0^\pi \sin\theta_N\, d\theta_N \int_0^{2\pi} d\phi_1 \cdots \int_0^{2\pi} d\phi_N \times$
$\int_0^a dx_1 \int_0^b dy_1 \int_0^c dz_1 \cdots \int_0^a dx_N \int_0^b dy_N \int_0^c dz_N = (4\pi)^{-N} 2^N (2\pi)^N (abc)^N =$
V^N. For an ideal gas of nonlinear molecules, $(4\pi)^{-N}$ is replaced by
$(8\pi^2)^{-N} = (2\pi)^{-N}(4\pi)^{-N}$ and we have the additional integral factors
$\int_0^{2\pi} d\chi_1 \cdots \int_0^{2\pi} d\chi_N = (2\pi)^N$, so we still get $Z_{con} = V^N$.

(b) $P = kT(\partial \ln V^N/\partial V)_{T,N} = NkT(\partial \ln V/\partial V) = NkT/V$.

21.90 $B = -2\pi N_A \int_0^\infty (e^{-\mathcal{V}/kT} - 1)r^2\, dr = -2\pi N_A[\int_0^d (e^{-\infty/kT} - 1)r^2\, dr + \int_d^\infty (e^{-0/kT} - 1)r^2\, dr]$
$= -2\pi N_A[-\int_0^d r^2\, dr + 0] = 2\pi N_A d^3/3 = 4N_A[\frac{4}{3}\pi(d/2)^3]$.

21.91 At 100 K, $\log(kT/\varepsilon) = \log[(1.381 \times 10^{-23}$ J/K)(100 K)/$(1.31 \times 10^{-21}$ J)] =
0.0229. For this $\log(kT/\varepsilon)$, Fig. 21.22 gives $B/\sigma^3 N_A \cong -5.4$ and
$B \cong -5.4 \times (3.74 \times 10^{-8}\ \text{cm})^3(6.022 \times 10^{23}/\text{mol}) = -170\ \text{cm}^3/\text{mol}$. At 300 K,
$\log(kT/\varepsilon) = 0.500$; Fig. 21.22 gives $B/\sigma^3 N_A \cong -0.1$ and $B \cong -3\ \text{cm}^3/\text{mol}$.
At 500 K, $\log(kT/\varepsilon) = 0.722$; $B/\sigma^3 N_A \cong 0.5$, and $B \cong 16\ \text{cm}^3/\text{mol}$.

21.92 **(a)** $B = -2\pi N_A \int_0^\sigma (e^{-\infty} - 1)r^2\, dr - 2\pi N_A \int_\sigma^a (e^{\varepsilon/kT} - 1)r^2\, dr - 2\pi N_A \int_a^\infty (e^0 - 1)r^2\, dr$
$= 2\pi N_A \sigma^3/3 - 2\pi N_A(e^{\varepsilon/kT} - 1)(a^3 - \sigma^3)/3 + 0 = 2\pi N_A(\sigma^3 - a^3)e^{\varepsilon/kT}/3 +$
$2\pi N_A a^3/3$.

(b) Substitution of numerical values gives $B(100\ \text{K}) = -163\ \text{cm}^3/\text{mol}$,
$B(300\ \text{K}) = -4\ \text{cm}^3/\text{mol}$, $B(500\ \text{K}) = 17\ \text{cm}^3/\text{mol}$.

(c) From (8.5) and (8.6), $Z \approx 1 + B^\dagger P = 1 + BP/RT \approx$
$1 - (163\ \text{cm}^3/\text{mol})(3.0\ \text{atm})/(82.06\ \text{cm}^3\text{-atm/mol-K})(100\ \text{K}) = 0.94$.

21.93 $U = kT^2(\partial \ln Z/\partial T)_{V,N} = kT^2(\partial \ln Z_{id}/\partial T)_{V,N} + kT^2(\partial \ln Z_{con}/\partial T)_{V,N} =$
$U_{id} + kT^2(-N)[(1/V_m)dB/dT + (1/2V_m^2)dC/dT + \cdots] =$
$U_{id} - nRT^2[(1/V_m)dB/dT + (1/2V_m^2)dC/dT + \cdots]$.
$S = U/T + k \ln Z = U_{id}/T - nRT\,[(1/V_m)dB/dT + (1/2V_m^2)dC/dT + \cdots] + k \ln Z_{id}$
$+ kN \ln V - Nk[(1/V_m)B + (1/2V_m^2)C + \cdots] - Nk \ln V =$
$S_{id} - nR[(1/V_m)(B + T\,dB/dT) + (1/2V_m^2)(C + T\,dC/dT) + \cdots]$, since $S_{id} =$
$U_{id}/T + k \ln Z_{id}$. $G = A + PV = -kT \ln Z + kTV(\partial \ln Z_{con}/\partial V)_{T,N} =$

$-kT \ln Z_{id} - kTN \ln V + kTN(B/V_m + C/2V_m^2 + \cdots) + kTN \ln V +$
$kTVN(1/V + B/VV_m + C/VV_m^2 + \cdots) = G_{id} + nRT(2B/V_m + 3C/2V_m^2 + \cdots)$,
since $G_{id} = A_{id} + (PV)_{id} = -kT \ln Z_{id} + NkT$.

21.94 $A = -kT \ln Z$ and $A_{false} = -kT \ln Z_{false}$, so $A_{false} - A = -kT \ln (Z_{false}/Z) =$
$-(1.38 \times 10^{-23}$ J/K$)(298$ K$) \ln 10^{(\pm 10^{15})} = -(4.11 \times 10^{-21}$ J$)2.30 \log 10^{(\pm 10^{15})} =$
$-(9.5 \times 10^{-21}$ J$)(\pm 10^{15}) = \mp 10^{-5}$ J, which is negligible.

21.95 $\langle N_r \rangle / N = e^{-\varepsilon_r/kT}/z$, so $\langle N_{gs} \rangle / N = e^{-0}/z = 1/z = 1/154.1 = 0.00649$.

21.96 **(a)** Since A is proportional to n and $A = -kT \ln Z$, we see that $\ln Z$ is proportional to n. From (21.81), z_{tr} is proportional to V, which is proportional to n; so z_{tr} is proportional to n. Since z_{tr} is proportional to n, z is proportional to n.

(b) z_{rot}, z_{vib}, and z_{el} are independent of n.

(c) None.

(d) z_{rot}, z_{vib}, and z_{el}.

21.97 Every thermodynamic system in equilibrium.

R21.1 Ensemble and molecular orbitals.

R21.2 **(a)** Same. **(b)** Highest is HF; lowest is H_2. **(c)** Same. **(d)** Highest is CH_4. HF and H_2 are the same. (For H_2, $T/\Theta_{rot} = 3.5$ and Fig. 21.11 shows $C_{V,rot,m}$ differs negligibly from R.) **(e)** Because of its high Θ_{vib}, H_2 is lowest (Fig. 21.10.) Highest is CH_4, since this pentatomic molecule will have some vibrational levels low enough to be significantly populated at room T.)

R21.3 When it has no permanent electric dipole moment. Examples are N_2 and CO_2. (It turns out that O_2 does have a very weak pure rotational absorption spectrum due to its magnetic dipole moment arising from the unpaired electron spins. Amazingly, ground-state CH_4 has a tiny dipole moment of 5×10^{-6} D due to vibration–rotation interaction or centrifugal distortion and this produces an extremely weak pure rotational absorption spectrum.)

R21.4 When the molecule has no normal vibrational modes that produce a change in electric dipole moment. The only examples are homonuclear diatomic molecules.

R21.5 **(a)** T. **(b)** T. **(c)** T. **(d)** F. Equation (21.55) shows that $N!$ and $N^N e^{-N}$ differ greatly from each other. **(e)** T. **(e)** F. For example, see parts (f) and (g). **(f)** T (since $1000 \approx 1001$). **(g)** F. They differ by a factor of 10.

R21.6 **(a)** tr, rot, vib, el.

(b) IR, UV and visible; microwave; radio wave.

R21.7 **(a)** As in Examples 21.2 and 21.3,

$$\tilde{\nu}_0 = \tilde{\nu}_e - 2\tilde{\nu}_e x_e = 2309.01\ \text{cm}^{-1} - 2(39.64\ \text{cm}^{-1}) = 2229.73\ \text{cm}^{-1}$$

$$\tilde{B}_0 = \tilde{B}_e - \tfrac{1}{2}\alpha_e = 6.4264\ \text{cm}^{-1} - \tfrac{1}{2}(0.169\ \text{cm}^{-1}) = 6.3419\ \text{cm}^{-1}$$

$$\Theta_{\text{vib}} = \tilde{\nu}_0 hc/k = \tilde{\nu}_0(1.4388\ \text{cm K}) = 3208.1\ \text{K},$$

$$\Theta_{\text{vib}}/T = (3208.1\ \text{K})/(298.15\ \text{K}) = 10.76$$

$$\Theta_{\text{rot}} = \tilde{B}_0 hc/k = \tilde{B}_0(1.4388\ \text{cm K}) = 9.125\ \text{K}$$

$$S_{\text{tr,m}} = R(1.5\ln 126.91 + 2.5\ln 298.15 - 1.1517) = 169.26\ \text{J/mol-K}$$

$$S_{\text{rot,m}} = R + R\ln(298.15/9.125) = 37.30\ \text{J/mol-K}$$

$$S_{\text{vib,m}} = R(10.76)(e^{10.76} - 1)^{-1} - R\ln(1 - e^{-10.76}) = 0.002\ \text{J/mol-K}$$

$$S_{\text{el,m}} = 0$$

$$S_m = 206.56\ \text{J/mol-K}$$

(b) $C_{V,\text{tr,m}} = \tfrac{3}{2}R = 12.472\ \text{J/mol-K}$ $\quad C_{V,\text{rot,m}} = R = 8.3145\ \text{J/mol-K}$

$$C_{V,\text{vib,m}} = R(10.76)^2 \frac{e^{10.76}}{(e^{10.76} - 1)^2} = 0.020\ \text{J/mol-K}$$

$$C_{V,\text{m}} = 20.806\ \text{J/mol-K}, \qquad C_{P,\text{m}} = C_{V,\text{m}} + R = 29.121\ \text{J/mol-K}$$

R21.8 **(a)** symmetric (one 3-fold axis); **(b)** symmetric (linear); **(c)** symmetric; **(d)** symmetric; **(e)** asymmetric (a twofold axis); **(f)** spherical (more than one fourfold axis); **(g)** symmetric.

R21.9 **(a)** 6 (same as a trigonal planar molecule like BF_3); **(b)** 2; **(c)** 3; **(d)** 3; **(e)** 2; **(f)** This is a tough one. Imagine the octahedral molecule inscribed in a cube, with S at the center of the cube and each F lying at the center of one face of the cube. The FSF line from the center of one face of the cube to the opposite face is a C_4 axis. We can do $\hat{C}_4$, $\hat{C}_4^2$, and $\hat{C}_4^3$ rotations about such an axis to generate equivalent configurations. There are three such C_4 axes between the 6 faces, so the three rotations associated with each such axis give 3(3) = 9 rotations. Let the origin be at the S atom and let the vertical direction be called the z direction. Consider the square formed by the four F atoms that lie in the xy plane. There are two C_2 axes that lie in the xy plane and that bisect opposite sides of this square (these axes do not go through F atoms); we can do a $\hat{C}_2$ symmetry rotation about each such C_2 axis. In addition to the two C axes that lie in the xy plane and go between F atoms, there are two such C_2 axes lying in the xz plane and two more in the yz plane, for a total of six axes and 6 such $\hat{C}_2$ rotations. So far, we have 9 + 6 + 1 indistinguishable configurations brought about by rotations, where the 1 is from the original unrotated molecule. In addition, each diagonal of the cube in which SF_6 is inscribed is a C_3 axis. To see this, note that an endpoint of a diagonal of the cube lies at a corner and three faces of the cube intersect at this corner. Rotation by 120° about a diagonal of the cube moves each of these three faces to the location of one of these same three faces and thus moves the F that is at the center of each face to an F at the center of another face. Thus, for each diagonal of the cube, we can do a $\hat{C}_3$ rotation and a $\hat{C}_3^2$ rotation, and since the cube has 4 diagonals (that each join 2 of the 8 corners), we have 4(2) = 8 rotations associated with the diagonals. Thus the total number of equivalent positions related by rotations is 9 + 6 + 1 + 8 = 24 and the symmetry number is 24. **(g)** 2.

R21.10 See the paragraphs preceding Eq. (21.84).

R21.11 N_2 and O_2 do not absorb IR radiation (see Prob. R21.4) and are not greenhouse gases. CO_2, H_2O, and CH_4 absorb IR radiation and are greenhouse gases.

R21.12 Since K does not change and $\Delta N = +1$ for an absorption we have

$h\nu = E_{\text{upper}} - E_{\text{lower}} = A[(N+1)(N+2) - N(N+1)] + C(K^2 - K^2)$.

$\nu = (A/h)(2N+2)$, $N = 1, 2, 3, \ldots$, so $\nu = 4(A/h), 6(A/h), 8(A/h), \ldots$

The second-lowest and third-lowest absorption frequencies are thus $6/4 = 1.5$ and $8/4 = 2$ times the lowest frequency, so these frequencies are 1.5(456 MHz) = 684 MHz and 912 MHz.

R21.13 **(a)** Two. Three.

(b) A C_3 axis through the CF line. Three symmetry planes, each one containing the nuclei C, F, and one of the H's.

(c) $\hat{E}, \hat{C}_3, \hat{C}_3^2, \hat{\sigma}_a, \hat{\sigma}_b, \hat{\sigma}_c$.

(d) A C_4 axis through the Xe nucleus and perpendicular to the molecular plane. (This axis is also a C_2 axis and an S_4 axis.) A symmetry plane coinciding with the plane of the molecule. Four symmetry planes that are perpendicular to the molecular plane and contain the Xe nucleus; two of these planes contain two diagonally opposite F nuclei; each of the other two bisect two FXeF 90° bond angles. A center of symmetry. Four C_2 axes lying in the molecular plane; two of these go through Xe and two diagonally opposite F nuclei; the other two go through Xe and bisect two FXeF 90° bond angles. The symmetry operations are $\hat{E}, \hat{C}_4, \hat{C}_4^2, \hat{C}_4^3, \hat{S}_4, \hat{S}_4^3, \hat{\sigma}_a, \hat{\sigma}_b, \hat{\sigma}_c, \hat{\sigma}_d, \hat{\sigma}_e, \hat{i}, \hat{C}_{2a}, \hat{C}_{2b}, \hat{C}_{2c}, \hat{C}_{2d}$. ($\hat{S}_4^2$ is the same as $\hat{C}_4^2$.)

(e) C_{3v}, 6.

(f) D_{4h}, 16.

R21.14 See the paragraph after Eq. (21.35).

R21.15 ν = (1540 cm^{-1})(2.998 × 10^{10} cm/s) = 4.617 × 10^{13} s^{-1}.

$\langle N_3 \rangle / \langle N_0 \rangle = 0.0564 = e^{-\Delta\varepsilon/kT} = e^{-(3.5-0.5)(6.626\times10^{-34}\text{ J s})(4.617\times10^{13}\text{ s}^{-1})/(1.3806\times10^{-23}\text{ J/K})T} =$

$e^{-(6648\text{ K})/T}$. ln(0.0564) = –2.875 = –(6648 K)/T and T = 2312 K.

R21.16 T, V, composition.

R21.17 **(a)** Emission of light due to exposure of the system to electromagnetic radiation of frequency corresponding to the energy difference between the initial higher energy level and the lower level.

(b) When a system of identical particles requires a wave function (including spin) that is symmetric with respect to exchange of particles, those particles are called bosons. (By the spin-statistics theorem, bosons have integral spin.)

(c) A nonequilibrium situation in which a higher-energy state has a greater population than a lower-energy state.

(d) Spontaneous emission of light that rapidly follows excitation of a molecule to a higher energy level.

R21.18 **(a)** As noted in Fig. 20.13, the pure-rotational transition frequencies are $2B_0$, $4B_0$, $6B_0$, etc. The spacing between adjacent lines is $2B_0$. The spacings between the lines listed in the problem are 13013.7, 13011.3, 13008.6 MHz. The slight changes are due to centrifugal distortion, which is being neglected. The lines that have the lowest frequencies correspond to the lowest J values and are affected least by centrifugal distortion, so we take 13014 MHz = $2B_0 = 2h/8\pi^2 I_0$ and I_0 = (6.6261 × 10^{-34} J s)/$4\pi^2$(13014 × 10^6 s^{-1}) = 1.2897 × 10^{-45} kg m^2. Then μ = {[(22.990)(34.969)/(22.990 + 34.969)]/(6.02214 × 10^{23})} g = 2.3033 × 10^{-26} kg. $I = \mu d^2$ and $d = (I/\mu)^{1/2}$ = [(1.2897 × 10^{-45} kg m^2)/(2.3033 × 10^{-26} kg)]$^{1/2}$ = 2.366 × 10^{-10} m = 2.366 Å. (see Table 19.3.)

(b) As in Eq. (20.31), $\nu = 2B_0(J+1)$, where J = 0, 1, 2,…. From (a), $2B_0$ = 13014 MHz and for the first line listed, ν= 130224 MHz = (13014 MHz)(J + 1) and J = 9. The first line listed is the J = 9 to 10 transition, and the others are the 10 to 11, the 11 to 12, and the 12 to 13 transitions.

R21.19 (a) At 600 K, occupation of excited electronic states is completely negligible. We know that $\langle N_{\text{vib},\upsilon}\rangle/N = e^{-\varepsilon_{\text{vib},\upsilon}/kT}/z_{\text{vib}}$ [Eq. (21.75)] and $\langle N_{\text{rot},t}\rangle/N = e^{-\varepsilon_{\text{rot},t}/kT}/z_{\text{rot}}$. Here, $\langle N_{\text{vib},\upsilon}\rangle N$ is the probability of being in a particular vibrational state, without regard to the translational, rotational, or electronic state. We are treating the vibrational and rotational motions as independent. The probability of two independent events both occurring is the product of the probabilities of each event, so the probability of simultaneously being in a particular vibrational state and a particular rotational state is the product of the individual probabilities: $\langle N_{\text{vib},\upsilon;\text{rot},t}\rangle/N = [\langle N_{\text{vib},\upsilon}\rangle/N][\langle N_{\text{rot},t}\rangle/N] = (e^{-\varepsilon_{\text{vib},\upsilon}/kT}/z_{\text{vib}})(e^{-\varepsilon_{\text{rot},t}/kT}/z_{\text{rot}})$, where $\langle N_{\text{vib},\upsilon;\text{rot},t}\rangle$ is the number of molecules in a particular vibrational state and in a particular rotational state. The vibrational levels are nondegenerate, but the rotational levels are $(2J + 1)$-fold degenerate. Also, z_{vib} takes the zero level of energy at $\upsilon = 0$. We have $z_{\text{vib}} = (1 - e^{-(3084\text{ K})/(600\text{ K})})^{-1} = 1.006$ and $z_{\text{rot}} = (600\text{ K})/(2.765\text{ K}) = 217$. The fraction of molecules in the $\upsilon = 1$, $J = 2$ vibration–rotation level is $e^{-1(h\nu/kT)}[2(2) + 1]e^{-2(3)\hbar^2/2IkT}/z_{\text{vib}}z_{\text{rot}} = 5\exp(-\Theta_{\text{vib}}/T)\exp(-6\Theta_{\text{rot}}/T)/z_{\text{vib}}z_{\text{rot}}$
$= 5e^{-3084/600}e^{-6(2.765)/600}/(1.006)(217) =$

Chapter 22

22.1 $A = ke^{E_a/RT} = ke^{E_{\text{thr}}/RT}e^{1/2} = N_A \pi d_B^2 e^{1/2}(4RT/\pi M_B)^{1/2}$, where Eqs. (16.69), (22.5), and (22.4) were used.

22.2 **(a)** Equation (22.6) gives $A =$
$(6.0 \times 10^{23}/\text{mol})\pi(3.4 \times 10^{-10}\text{ m})^2[8(8.314\text{ J/mol-K})(500\text{ K})/\pi]^{1/2} \times$
$[\text{mol}/(0.030\text{ kg}) + \text{mol}/(0.048\text{ kg})]^{1/2}(2.72)^{1/2} =$
$3 \times 10^8\text{ m}^3\text{ s}^{-1}\text{ mol}^{-1} = 3 \times 10^{14}\text{ cm}^3\text{ mol}^{-1}\text{ s}^{-1}$.

(b) $8 \times 10^{11}\text{ cm}^3\text{ mol}^{-1}\text{ s}^{-1} = p(3 \times 10^{14}\text{ mol}^{-1}\text{ s}^{-1})$ and $p = 0.003$.

22.3 391 nm.

22.4 $\langle \upsilon_{rc} \rangle = \int_0^\infty \upsilon_{rc} g(\upsilon_{rc})\, d\upsilon_{rc} = 2(m_{rc}/2\pi kT)^{1/2} \int_0^\infty \upsilon_{rc} \exp(-m_{rc}\upsilon_{rc}^2/2kT)\, d\upsilon_{rc} =$
$2(m_{rc}/2\pi kT)^{1/2}(2kT/m_{rc})/2 = (2kT/\pi m_{rc})^{1/2}$.

22.5 The energy per molecule is $(9.6\text{ kcal/mol})/N_A =$
$[(9600\text{ cal/mol})/(6.02 \times 10^{23}/\text{mol})](4.184\text{ J/1 cal}) = 6.7 \times 10^{-20}\text{ J}$.

22.6 **(a)** $\text{D} \overset{d}{-} \text{H} \overset{d}{-} \text{H}$, $d = 0.930$ Å (Example 22.2). Let the origin be at D and let the molecular axis be the x axis. The center of mass is at $x_{\text{cm}} = \sum_i m_i x_i / m_{\text{tot}}$ $= [2.014(0) + 1.008d + 1.008(2d)]/4.030 = 0.698$ Å. The principal axes pass through the center of mass (which is on the molecular axis and 0.698 Å away from D); one principal axis is the molecular axis; the other two are perpendicular to this axis.
$I_{b,\text{DH}_2} = \sum_i m_i r_{b,i}^2 = N_A^{-1}[(2.014)(0.698)^2 + 1.008(0.930 - 0.698)^2 +$
$1.008(1.860 - 0.698)^2](\text{g/mol})\text{ Å}^2 = 2.397 N_A^{-1}\ (\text{g/mol})\text{ Å}^2$.

(b) For H_2, $I_{\text{H}_2} = \mu R^2 = (m_\text{H}^2/2m_\text{H})R^2 = \frac{1}{2}m_\text{H}R^2 =$
$N_A^{-1}\frac{1}{2}(1.008)(0.741)^2(\text{g/mol})\text{ Å}^2 = 0.2767 N_A^{-1}(\text{g/mol})\text{ Å}^2$. Then
$I_{b,\text{DH}_2}/I_{\text{H}_2} = 2.397/0.2767 = 8.66$. Since $\sigma_{\text{H}_2} = 2$ and $\sigma_{\text{DH}_2} = 1$, the rotational-partition-function ratio is $2(8.66) = 17.3$.

22.7 **(a)** $z_{\text{vib},H_2} = [1 - \exp(-h\tilde{\nu}c/kT)]^{-1}$; $h\tilde{\nu}c/kT = (6.626 \times 10^{-34}$ J s) $\times$ (4400 cm^{-1})(2.998 $\times$ 10^{10} cm/s)/(1.3807 $\times$ 10^{-23} J/K)(450 K) = 14.07; $z_{\text{vib},H_2} = (1 - e^{-14.07})^{-1} = 1.000$. The activated complex has 3 ordinary vibrations, with wave numbers 1766, 840, and 840 cm^{-1} and $z^{\ddagger\prime}_{\text{vib}}$ is the product of factors for each of these vibrations. We have $h\tilde{\nu}_{\text{str}}c/kT = h(1766 \text{ cm}^{-1})c/kT = 5.646$, $h\tilde{\nu}_{\text{bend}}c/kT = h(840 \text{ cm}^{-1})c/kT = 2.686$. Then $z^{\ddagger\prime}_{\text{vib}} = (1 - e^{-5.646})^{-1}(1 - e^{-2.686})^{-2} = 1.156$. From (21.81) we find: $(z^{\ddagger}_{\text{tr}}/V)/[(z_{\text{tr,D}}/V)(z_{\text{tr},H_2}/V)] = (m_{DH_2}/m_D m_{H_2})^{3/2}(h^2/2\pi kT)^{3/2} =$ [4.030/(2.014)2.016]$^{3/2}$(mol/10^{-3} kg)$^{3/2}$(6.022 $\times$ 10^{23}/mol)$^{3/2}$ $\times$ (6.626 $\times$ 10^{-34} J s)^3[2π(1.3807 $\times$ 10^{-23} J/K)(450 K)]$^{-3/2}$ = 5.51 $\times$ 10^{-31} m^3 = 5.51 $\times$ 10^{-25} cm^3.

(b) $N_A kTh^{-1}\exp(-\Delta\varepsilon^{\ddagger}_0/kT) =$ (6.022 $\times$ 10^{23}/mol)(1.381 $\times$ 10^{-23} J/K)(450 K)(6.626 $\times$ 10^{-34} J s)$^{-1}$ $\times$ exp{(−5.73 $\times$ 10^{-20} J)/[(450 K)(1.3807 $\times$ 10^{-23} J/K)]} = 5.5$_8$ $\times$ 10^{32}/mol-s. Then k_r = (5.5$_8$ $\times$ 10^{32}/mol-s)(5.51 $\times$ 10^{-25} cm^3)(17.3)(1.156)(1) = 6.1$_5$ $\times$ 10^9 cm^3 mol^{-1} s^{-1}.

22.8 Replacing 450 K by 600 K in Prob. 22.7b, we find $N_A kTh^{-1}\exp(-\Delta\varepsilon^{\ddagger}_0/kT) =$ 7.4$_6$ $\times$ 10^{33}/mol-s. Replacing 450 K by 600 K in Prob. 22.7b, we find $(z^{\ddagger}_{\text{tr}}/V)/[(z_{\text{tr,D}}/V)(z_{\text{tr},H_2}/V)] = 3.58 \times 10^{-31}$ m^3, $z_{\text{vib},H_2} = 1.000$, $z^{\ddagger\prime}_{\text{vib}} = 1.351$. So k_r = (7.4$_6$ $\times$ 10^{33}/mol-s)(3.58 $\times$ 10^{-25} cm^3)(17.3)(1.351)(1) = 6.2 $\times$ 10^{10} cm^3/mol-s. The tunneling correction is 7.5/6.2 = 1.2, which is less than that at 450 K. Tunneling becomes less important as T increases.

22.9 $\tilde{\nu} = \nu/c = (1/2\pi c)(k/\mu)^{1/2}$. k is the same for H_2 and D_2 so $\tilde{\nu}_{D_2}/\tilde{\nu}_{H_2} = (\mu_{H_2}/\mu_{D_2})^{1/2} = (\frac{1}{2}m_H/\frac{1}{2}m_D)^{1/2} = (1.0078/2.0141)^{1/2} = 0.7074$ and $\tilde{\nu}_{D_2} = 3112$ cm^{-1}. The ZPE of the activated complex is $\frac{1}{2}hc(1765 + 671 + 671)$ cm^{-1} = 3.09 $\times$ 10^{-20} J. The ZPE of H is 0 and that of D_2 is $\frac{1}{2}hc\tilde{\nu}_{D_2} = 3.09 \times 10^{-20}$ J. Δ(ZPE) = 0. The classical barrier height is the same as that for D + H_2, namely, 6.68 $\times$ 10^{-20} J. So $\Delta\varepsilon^{\ddagger}_0 = 6.68 \times 10^{-20}$ J. Also, $z_{\text{el,H}} = 2$, $z_{\text{el},D_2} = 1$, $z^{\ddagger}_{\text{el}} = 2$. $z^{\ddagger}_{\text{rot}}/z_{\text{rot},D_2} = I^{\ddagger}\sigma_{D_2}/I_{D_2}\sigma^{\ddagger} = 2I^{\ddagger}/I_{D_2}$. $I_{D_2} = \mu R^2 = m_D R^2/2 = N_A^{-1}\frac{1}{2}(2.014)(0.741)^2$(g/mol) Å^2 = $0.553N_A^{-1}$ (g/mol) Å^2.

$\mathrm{H} \overset{d}{-} \mathrm{D} \overset{d}{-} \mathrm{D}$, $d = 0.930$ Å (Example 22.2). The procedure of Prob. 22.6a gives (with the origin at H) $x_{cm} = 1.116$ Å, $I^{\ddagger} = 2.440$ (g/mol) Å^2, and $z^{\ddagger}_{rot}/z_{rot,D_2} = 8.82$. Proceeding as in Prob. 22.7a, we find $z_{vib,D_2} = 1.001$, $z^{\ddagger\prime}_{vib} = 1.586$, $(z^{\ddagger}_{tr}/V)/[(z_{tr,H}/V)(z_{tr,D_2})] = 5.00 \times 10^{-25}$ cm^3.
$N_A kTh^{-1}\exp(-\Delta\varepsilon^{\ddagger}_0/kT) = 2.3_7 \times 10^{33}$/mol-s. So $k_r = (2.3_7 \times 10^{33}$/mol-s$) \times (5.00 \times 10^{-25}$ cm$^3)(8.82)(1.586)(1) = 1.6_6 \times 10^{10}$ cm^3 mol^{-1} s^{-1}.

22.10 The Taylor series for e^x gives at high T, $e^{-h\nu_s/kT} \approx 1 - h\nu_s/kT$ and $z_{vib,s} \approx 1/(h\nu_s/kT) = kT/h\nu_s$. There are f_{vib} factors in z_{vib} in Eq. (21.110), so z_{vib} is proportional to $T^{f_{vib}}$ at high T.

22.11 Let lin, nonlin, lin‡, and nonlin‡ denote linear and nonlinear reactant molecules and linear and nonlinear transition states.

(a) First consider the reaction between an atom and a molecule. Let the reaction be atom$_A$ + lin$_B$ → lin‡. The ratio $z^{\ddagger}_{tr}/z_{tr,A}z_{tr,B}$ is proportional to $T^{3/2}/T^{3/2}T^{3/2} = T^{-3/2}$ for any bimolecular reaction. Also, $z^{\ddagger}_{rot}/z_{B,rot} \propto T/T = T^0$. Let lin$_B$ have w atoms. Then lin$_B$ has $3w - 5$ normal modes. Further, lin‡ has $w + 1$ atoms and $3(w + 1) - 5 - 1 = 3w - 3$ "ordinary" normal modes and one anomalous mode (the "vibration" along the reaction coordinate). So lin‡ has 2 more ordinary normal modes than does lin$_B$. Also, the vibrational frequencies of lin‡ and lin$_B$ should be of the same order of magnitude in most cases, so we expect that $z^{\ddagger\prime}_{vib}/z_{vib,B}$ is proportional to T^j, where $0 \le j \le \frac{1}{2}(f^{\ddagger\prime}_{vib} - f_{vib,B})$. Hence, $z^{\ddagger\prime}_{vib}/z_{vib,B} \propto T^c$, where $0 \le c \le 1$. Equation (22.23) then gives $A \propto T^1 T^{-3/2} T^0 T^c = T^{c-1/2} = T^m$, where $-\frac{1}{2} \le m \le \frac{1}{2}$. Now consider atom$_A$ + lin$_B$ → nonlin‡. We have $z^{\ddagger}_{rot}/z_{rot,B} \propto T^{3/2}/T = T^{1/2}$. Reasoning similar to the above shows that the species nonlin‡ has 1 more ordinary normal mode than lin$_B$, so $z^{\ddagger\prime}_{vib}/z_{vib,B} \propto T^b$, where $0 \le b \le \frac{1}{2}$. Hence, $A \propto T^1 T^{-3/2} T^{1/2} T^b = T^b = T^m$, where $0 \le m \le \frac{1}{2}$. For atom$_A$ + nonlin$_B$ → nonlin‡, we find similarly that $A \propto T^1 T^{-3/2} T^0 T^a = T^{a-1/2}$, where $0 \le a \le 1$. So $-\frac{1}{2} \le m \le \frac{1}{2}$.

(b) Now let the reaction be between two molecules. First consider nonlin$_A$ + nonlin$_B$ → nonlin‡. Let A and B have w_A and w_B atoms. Then

A and B have a total of $3w_A + 3w_B - 12$ normal modes and nonlin‡ has $3(w_A + w_B) - 6 - 1 = 3w_A + 3w_B - 7$ ordinary normal modes, which is 5 more than the total of A and B. So $z^{\ddagger\prime}_{\text{vib}}/z_{\text{vib,A}}z_{\text{vib,B}} \propto T^d$, where $0 \le d \le 2\frac{1}{2}$. Hence, $A \propto T^1 T^{-3/2} T^{-3/2} T^d = T^{d-2} = T^m$, where $-2 \le m \le \frac{1}{2}$.

For lin$_A$ + nonlin$_B \to$ nonlin‡, we find similarly that nonlin‡ has 4 more ordinary normal modes than A + B, so $A \propto T^1 T^{-3/2} T^{-1} T^e = T^{e-3/2} = T^m$, where $0 \le e \le 2$ and $-\frac{3}{2} \le m \le \frac{1}{2}$. For lin$_A$ + lin$_B \to$ nonlin‡, nonlin‡ has 3 more ordinary normal modes than A and B together, so $A \propto T^1 T^{-3/2} T^{-1/2} T^g = T^{g-1} = T^m$, where $0 \le g \le \frac{3}{2}$ and $-1 \le m \le \frac{1}{2}$.

22.12 From the Example 22.2 in Sec. 22.4, $\Delta\varepsilon_0^\ddagger = 5.73 \times 10^{-20}$ J. So $\Delta E_0^\ddagger =$ $(5.73 \times 10^{-20}\text{ J})(6.02 \times 10^{23}/\text{mol}) = (3.45 \times 10^4\text{ J/mol})(1\text{ cal}/4.184\text{ J}) =$ 8.25 kcal/mol. m is the exponent of T in k_r. The example in Sec. 22.4 shows the z_{tr} ratio and the z_{rot} ratio to be proportional to $T^{-3/2}$ and T^0, respectively. At 300 K, the z_{vib} ratio is nearly equal to 1 and its temperature dependence is essentially negligible. z_{el} is independent of T at 300 K. So Eq. (22.19) gives $k_r \propto T^1 T^{-3/2} T^0 = T^{-1/2}$ and $m = -\frac{1}{2}$. Then $E_a = \Delta E_0^\ddagger + mRT =$

8250 cal/mol $- \frac{1}{2}R(300\text{ K}) = 7.95$ kcal/mol $= 33.3$ kJ/mol..

22.13 The activated complex $A^\ddagger$ is an A molecule in the hole or within a distance δ beyond the hole. We have $\Delta\varepsilon_0^\ddagger = 0$. Also, z_{vib}, z_{rot}, and z_{el} are the same for A and $A^\ddagger$. The volume factor in $z_{\text{tr}}^\ddagger$ is $\mathscr{A}\delta$, where $\mathscr{A}$ is the hole area; also, $z_{\text{tr}}^\ddagger$ contains an extra factor of ½ since $A^\ddagger$ is moving only forward along the reaction coordinate (which is the direction perpendicular to the hole). Equations (22.16), (22.18), and (21.81) give $r = \delta^{-1}(2kT/\pi m)^{1/2}(z_{\text{tr}}^\ddagger/z_{\text{tr,A}})[\text{A}] =$ $\delta^{-1}(2RT/\pi M)^{1/2}\frac{1}{2}(2\pi mkT/h^2)^{3/2}\mathscr{A}\delta[(2\pi mkT/h^2)^{3/2}V]^{-1}[\text{A}] = (RT/2\pi M)^{1/2}P\mathscr{A}/VRT$, since $[\text{A}] = P/RT$. So $J = rV = -dn_A/dt = P\mathscr{A}/(2\pi MRT)^{1/2}$.

22.14 From Sec. 20.9, $\tilde{\nu}_{\text{CH}} \approx 3000\text{ cm}^{-1}$ and (see the Isotope Effects subsection) $\exp[-\Delta(\Delta\varepsilon_0^\ddagger)/kT] =$

$\exp[-0.146(6.63 \times 10^{-34}\text{ J s})(300000\text{ m}^{-1})(3.0 \times 10^8\text{ m/s})/(1.38 \times 10^{-23}\text{ J/K}) \times (300\text{ K})] = 0.122 = 1/8.2$. As T increases, the exponential comes closer to 1 and the isotope effect decreases.

22.15 $\nu = (1/2\pi)(k/\mu)^{1/2}$. $\nu_{CT}/\nu_{CH} = (\mu_{CH}/\mu_{CT})^{1/2} \approx (m_H/m_T)^{1/2} = (1/3)^{1/2}$ and $\nu_{CT} = \nu_{CH}/\sqrt{3}$. The ZPE of the reactant is lowered by $\frac{1}{2}h(\nu_{CH} - \nu_{CT}) = \frac{1}{2}h\nu_{CH}(1 - 1/\sqrt{3}) = 0.211h\nu_{CH}$ and the ZPE of the activated complex is unaffected. Replacement of 0.146 with 0.211 in Prob. 22.14 shows k_r is multiplied by 0.048 = 1/21.

22.16 The rate-determining step doesn't involve breaking the Ar—H bond, so (a) and (c) are ruled out.

22.17 **(a)** $\Delta H_c^{\circ\ddagger} = \Delta H^{\circ\ddagger} = E_a - nRT$ = 2500 cal/mol – 2(1.987 cal/mol-K)(270 K) = 1.4_3 kcal/mol. From Eq. (22.35), $\Delta S_c^{\circ\ddagger} = R\{\ln[Ah(c^\circ)^{n-1}/kT] - n\}$ = $R\{\ln[(6 \times 10^8\ \text{dm}^3/\text{mol-s})(6.6 \times 10^{-34}\ \text{J s})(\text{mol/dm}^3)/(1.38 \times 10^{-23}\ \text{J/K})(270\ \text{K})] - 2\}$ = –22.1 cal/mol-K.
$\Delta G_c^{\circ\ddagger}$ = (1.4_3 kcal/mol) – (270 K)(–0.0221 kcal/mol-K) = 7.4 kcal/mol.

(b) The equations used in (a) give $\Delta H_c^{\circ\ddagger}$ = 40.3 kcal/mol, $\Delta S_c^{\circ\ddagger}$ = –23.2 cal/mol-K, and $\Delta G_c^{\circ\ddagger}$ = 103.0 kcal/mol.

22.18 **(a)** Equation (22.44) gives $\log (k_r/k_r^\infty) = 1.02(-6)[I^{1/2}/(1 + I^{1/2}) - 0.30I]$.
For $I = 10^{-3}$, $\log (k_r/k_r^\infty) = -0.1858$ and $k_r/k_r^\infty = 0.65$.
For $I = 10^{-2}$, $k_r/k_r^\infty = 0.29$. For $I = 10^{-1}$, $k_r/k_r^\infty = 0.052$.

(b) Replacement of the factor –6 by +6 gives k_r/k_r^∞ = 1.53, 3.4_5, and $19._4$ for $I = 10^{-3}, 10^{-2}, 10^{-1}$.

22.19 **(a)** Equation (22.44) shows that for an elementary reaction in very dilute solutions, k_r increases as I increases if $z_B z_C > 0$ and decreases as I increases if $z_B z_C < 0$. (i) Little change. (ii) Increases. (iii) Decreases.

(b) As I increases from zero, the function $I^{1/2}/(1 + I^{1/2}) - 0.30I$ increases to a maximum and then decreases. Hence (22.44) predicts that k_r will go through a maximum if $z_B z_C > 0$ and will go through a minimum if $z_B z_C < 0$ [provided (22.44) holds at least qualitatively at moderate ionic strengths].

22.20 The slope of a plot of log (k_r/k°) vs. $I^{1/2}/(1 + I^{1/2}) - 0.30I$ is $1.02z_Cz_B$. The data are

$\log(k_r/k^\circ)$	–0.979	–0.951	–0.928	–0.900	–0.854
$I^{1/2}/(1 + I^{1/2}) - 0.30I$	0.0464	0.0559	0.0724	0.0817	0.0966

The straight-line fit is only fair; the slope is $2.3_7 = 1.02z_Cz_B$ and $z_Cz_B = 2.3 \approx 2$.

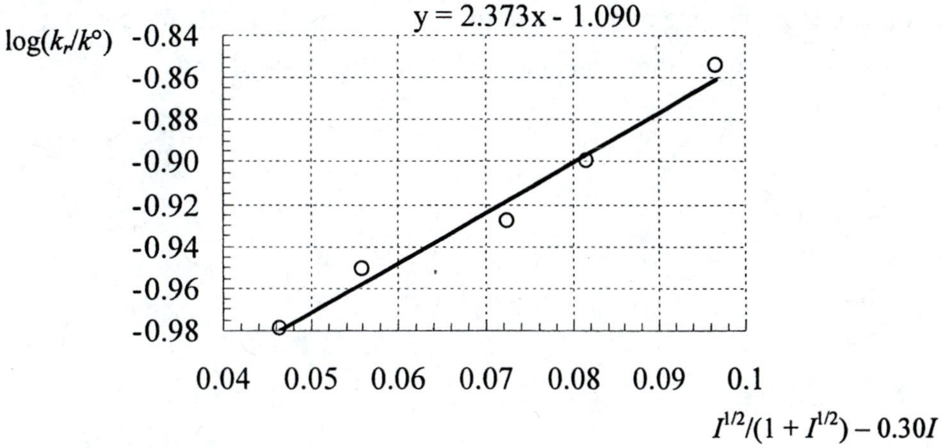

22.21 Equation (22.49) gives $\Delta S^{\circ\ddagger}/R = \ln (Ahc^\circ/kTe) = \ln 0.00011_8 = -9.0_5$. So $\Delta S^{\circ\ddagger} = R(-9.0_5) = -18._0$ cal/mol-K. From (22.48), $\Delta H^{\circ\ddagger} = E_a - RT =$ 15700 cal/mol – R(300 K) = 15.1 kcal/mol. $\Delta G^{\circ\ddagger} = \Delta H^{\circ\ddagger} - T\,\Delta S^{\circ\ddagger} =$ (15.1 kcal/mol) – (300 K)(–0.0180 kcal/mol-K) = $20._5$ kcal/mol.

22.22 **(a)** We get $[C]_{R=r_B+r_C} = [C][1 - k_{chem}/(k_{diff} + k_{chem})]$. For diffusion-controlled reactions, $k_{chem} >> k_{diff}$ and $[C]_{R=r_B+r_C}/[C] = 1 - 1/(1 + k_{diff}/k_{chem}) \approx 1 - (1 - k_{diff}/k_{chem}) = k_{diff}/k_{chem}$, where Eq. (8.8) was used. For chemically controlled reactions, $k_{chem} << k_{diff}$ and $[C]_{R=r_B+r_C} \approx [C]$.

(b) This ratio is given by (a) as $1 - k_{chem}/(k_{diff} + k_{chem}) = 1 - 1/(k_{diff}/k_{chem} + 1)$. We find 0.091, 0.5, and 0.91 for $k_{chem}/k_{diff} = 10$, 1, and 0.1, respectively.

22.23 1. A unimolecular reaction in the falloff region. 2. A diffusion controlled reaction, where the depletion of C molecules around a given B molecule is a departure from the Boltzmann distribution. 3. A photochemical reaction. 4. A laser-illuminated reaction. 5. A reaction with a very low E_a (see Sec. 22.1).

22.24 In the unimolecular decomposition of C_2H_5Cl, the transition state of Fig. 22.20b has two reactant bonds partially weakened and an equal number of new bonds partially formed; moreover, the transition state and the reactant have similar sizes. Thus the degree of disorder in the activated complex is similar to that in the reactant molecule, and we expect $\Delta S_c^{\circ\ddagger}$ to be close to zero for this and similar decompositions. In the C_2H_6 decomposition (Sec. 22.6) and similar decompositions, the bond elongation in the transition state produces a "disordered" transition state, and we expect $\Delta S_c^{\circ\ddagger} >> 0$ here.

Chapter 23

23.1 **(a)** Metallic; **(b)** molecular; **(c)** covalent; **(d)** ionic; **(e)** molecular; **(f)** ionic; **(g)** metallic.

23.2 **(a)** For C(graphite) $\rightarrow$ C(g), $E_{c,298} = \Delta H^\circ_{298} = (716.7 - 0)$ kJ/mol = 716.7 kJ/mol.

(b) For SiC(c) $\rightarrow$ Si(g) + C(g), $E_{c,298} = (455.6 + 716.7 + 65.3)$ kJ/mol = 1237.6 kJ/mol.

(c) For SiO_2(c) $\rightarrow$ Si(g) + 2 O(g), $E_{c,298} =$ [455.6 + 2(249.2) – (–910.9)] kJ/mol = 1864.9 kJ/mol.

23.3 $E_c = (411.15 + 107.32 + 121.68)$ kJ/mol + $(6.022 \times 10^{23}$/mol) × (5.139 – 3.614)eV$(1.6022 \times 10^{-19}$ J/1 eV)(1 kJ/1000 J) = 787.3 kJ/mol.

23.4 **(a)** I_2 (greater M); **(b)** NH_3 (H bonding); **(c)** SiO_2 (covalent solid); **(d)** MgO (greater ionic charges).

23.5 For $H_2O(\ell) \rightarrow H_2O(g)$, $E_{c,298} = \Delta H^\circ_{298} = 44.0$ kJ/mol.

23.6 $E_{\text{Coul}} = -(1.6022 \times 10^{-19}\text{ C})^2(1.74756)(6.0221 \times 10^{23}\text{/mol})/$ $4\pi(8.8542 \times 10^{-12}\text{ C}^2\text{/N-m}^2)(2.798 \times 10^{-10}\text{ m}) = -867.8$ kJ/mol.

23.7 $n = 1 + 72\pi(8.854 \times 10^{-12}\text{ C}^2\text{/N-m}^2)(2.798 \times 10^{-10}\text{ m})^4/$ $[3.7 \times 10^{-11}\text{ (m}^2\text{/N)}](1.602 \times 10^{-19}\text{ C})^2 1.74756 = 8.4$.

23.8 **(a)** From (23.11) with R_0 at 0 K approximated by R_0 at 25°C, we have $n \cong 1 + 72\pi(8.854 \times 10^{-12}\text{ C}^2\text{/N-m}^2)(3.299 \times 10^{-10}\text{ m})^4(10^5\text{ Pa})/$ $(5.5 \times 10^{-6})(1.602 \times 10^{-19}\text{ C})^2 1.74756 \cong 10.6$. From (23.9), $E_c \cong (1.602 \times 10^{-19}\text{ C})^2 1.74756(6.022 \times 10^{23}\text{/mol})(1 - 1/10.6)/$ $4\pi(8.854 \times 10^{-12}\text{ C}^2\text{/N-m}^2)(3.299 \times 10^{-10}\text{ m}) = 666$ kJ/mol.

(b) From Fig. 23.17, the nearest-neighbor distance R_0 is one-half the length d of the diagonal of the unit cell. As in Fig. 14.1 and Eq. (14.1), $d^2 = a^2 + a^2 + a^2$ and $d = 3^{1/2}a$, where a is the edge length. So $R_0 = \frac{1}{2}(3^{1/2})a = \frac{1}{2}(3^{1/2})(4.123 \times 10^{-10}\ \text{m}) = 3.571 \times 10^{-10}\ \text{m}$. Fig 23.17 shows that there is one CsCl ion pair per unit cell, so the molar volume is $V_{m,0} = N_A a^3 = N_A(2R_0/3^{1/2})^3 = 8R_0^3 N_A/3^{3/2}$ and Eq. (23.10) becomes $n = 1 + 32\pi\varepsilon_0(3^{1/2})R_0^4/\kappa e^2\mathfrak{M} \cong 1 + 32\pi(8.854 \times 10^{-12}\ \text{C}^2/\text{N-m}^2)(3^{1/2}) \times (3.571 \times 10^{-10}\ \text{m})^4(10^5\ \text{Pa})/(6 \times 10^{-6})(1.602 \times 10^{-19}\ \text{C})^2 1.762675 \cong 10._2$. Eq. (23.9) gives $E_c \cong (1.602 \times 10^{-19}\ \text{C})^2 1.762675(6.022 \times 10^{23}/\text{mol}) \times (1 - 1/10.2)/4\pi(8.854 \times 10^{-12}\ \text{C}^2/\text{N-m}^2)(3.571 \times 10^{-10}\ \text{m}) = 620\ \text{kJ/mol}$. In both (a) and (b), the theoretical value is less than the experimental value due to neglect of the dispersion energy.

23.9 $V_m = 2N_A R^3$, so $(\partial V_m/\partial R)_{T,P} = 6N_A R^2$. Taking P as 1 atm = 101325 N/m² and putting R equal to the NaCl equilibrium value 2.80 Å, we have $P(\partial V_m/\partial R)_{T,P} = (101325\ \text{N/m}^2)6(6.02 \times 10^{23}/\text{mol})(2.80 \times 10^{-10}\ \text{m})^2 = 2.9 \times 10^{10}\ \text{N/mol}$. Taking $U_m \cong E_p$, we have $(\partial U_m/\partial R)_{T,P} \cong (\partial E_p/\partial R)_{T,P} = (e^2/4\pi\varepsilon_0)\mathfrak{M}N_A R_0^{-1}(-R_0/R^2 + R_0^n/R^{n+1}) = (e^2/4\pi\varepsilon_0)\mathfrak{M}N_A(R_0^{n-1}/R^{n+1} - 1/R^2)$. At the equilibrium value $R = R_0$, $\partial U/\partial R$ is zero, but for a value $R = 2.81$ Å that deviates by 0.01 Å from R_0, we have, using $n = 8.4$ and $e^2/4\pi\varepsilon_0 = (1.6 \times 10^{-19}\ \text{C})^2/4\pi(8.85 \times 10^{-12}\ \text{C}^2\ \text{N}^{-1}\ \text{m}^{-2}) = 2.3 \times 10^{-28}\ \text{N m}^2$, $\partial U_m/\partial R = (2.3 \times 10^{-28}\ \text{N m}^2)(1.75)(6.02 \times 10^{23}/\text{mol}) \times [(2.80 \times 10^{-10}\ \text{m})^{7.4}/(2.81 \times 10^{-10}\ \text{m})^{9.4} - 1/(2.81 \times 10^{-10}\ \text{m})^2] = -8.0 \times 10^{13}\ \text{N/mol}$, and $P(\partial V_m/\partial R)$ is negligible compared with $\partial U_m/\partial R$.

23.10 (a) $(\partial^2 U_m/\partial V_m^2)_T = T(\partial^2 P/\partial V_m \partial T) - (\partial P/\partial V_m)_T$. As $T \to 0$, the first term on the right side of this equation goes to zero and $(\partial^2 U_m/\partial V_m^2)_T \to -(\partial P/\partial V_m)_T = 1/\kappa V_m$, where $\kappa \equiv -V_m^{-1}(\partial V_m/\partial P)_T$ was used.

(b) Use of $V_m^{1/3} = c^{1/3}R$ and $V_{m,0}^{1/3} = c^{1/3}R_0$ in (23.8) gives the desired equation for $-E_p$; differentiation of this equation gives $-\partial^2 E_p/\partial V_m^2 = (e^2/4\pi\varepsilon_0)\mathfrak{M}N_A R_0^{-1}[(-\frac{1}{3})(-\frac{4}{3})V_{m,0}^{1/3}V_m^{-7/3} - n^{-1}(-\frac{n}{3})(-\frac{n}{3} - 1)V_{m,0}^{n/3}V_m^{-n/3-2}]$. Setting $V_m = V_{m,0}$, we get $-\partial^2 E_p/\partial V_m^2\big|_{R=R_0} = (e^2/4\pi\varepsilon_0)\mathfrak{M}N_A R_0^{-1}V_{m,0}^{-2} \times (1 - n)/9$. Substitution of this result into the 0-K equations

$\partial^2 E_p/\partial V_m^2\big|_{R=R_0} = \partial^2 U_m/\partial V_m^2\big|_{R=R_0} = 1/\kappa V_{m,0}$ gives $(e^2/4\pi\varepsilon_0)\mathfrak{M}N_A R_0^{-1} \times V_{m,0}^{-2}(n-1)/9 = 1/\kappa V_{m,0}$; solving for n, we get (23.10).

23.11 A given positive ion has two negative ions at a distance R, two positive ions at $2R$, two negative ions at $3R$, two positive at $4R$, etc. The potential energy of interaction between one positive ion and all the other ions is $(e^2/4\pi\varepsilon_0)(-2/R + 2/2R - 2/3R + 2/4R - \cdots) = -(e^2/4\pi\varepsilon_0)\mathfrak{M}/R$, where $\mathfrak{M} = 2(1 - 1/2 + 1/3 - 1/4 + 1/5 - \cdots)$. By symmetry, the potential energy of interaction between a given negative ion and all the other ions is $-(e^2/4\pi\varepsilon_0)\mathfrak{M}/R$. Multiplication of $-2(e^2/4\pi\varepsilon_0)\mathfrak{M}/R$ by N_A and division by 2 (to avoid counting each interionic interaction twice) gives $E_{Coul} = -(e^2/4\pi\varepsilon_0)\mathfrak{M}N_A/R$, as in Eq. (23.4). Equation (8.36) with $x = 2$ gives $\ln 2 = 1 - 1/2 + 1/3 - 1/4 + \cdots$, so $\mathfrak{M} = 2 \ln 2 = 1.38629$.

23.12 **(a)** $-E_c = (6.022 \times 10^{23}/\text{mol})(118\ \text{K})(1.381 \times 10^{-23}\ \text{J/mol-K}) \times [24.264(3.50\ \text{Å}/3.75\ \text{Å})^{12} - 28.908(3.50\ \text{Å}/3.75\ \text{Å})^6]$ and $E_c = 8.3$ kJ/mol.

(b) At equilibrium, $\partial E_p/\partial R = -\partial E_c/\partial R = 0 = N_A\varepsilon[-12(24.264)\sigma^{12}/R^{13} + 6(28.908)\sigma^6/R^7]$;
so $R_0^6 = [12(24.264)/6(28.908)]\sigma^6$ and $R_0/\sigma = 1.09$. The experimental value is $R_0/\sigma = (3.75\ \text{Å})/(3.50\ \text{Å}) = 1.07$.

23.13 $E_c \approx -24(6.022 \times 10^{23}/\text{mol})(0.0101\ \text{eV})(1.602 \times 10^{-19}\ \text{J/eV}) \times [(3.50/3.75)^{12} - (3.50/3.75)^6] = 5240\ \text{J/mol} = 1.25\ \text{kcal/mol}$.

23.14 **(a)** Each of the 8 points at the corners is shared with a total of 8 unit cells. The point within the unit cell is not shared. So each unit cell has $8/8 + 1 = 2$ lattice points.

(b) $8/8 + 2/2 = 2$

23.15 **(a)** The unit cell has $8/8 + 6/2 = 4$ lattice points. There is one basis group at each lattice point, so each unit cell has 4 basis groups.

(b) Each unit cell has $8/8 = 1$ lattice point and therefore has 1 basis group.

23.16 An orthorhombic lattice has 90° angles and Eq. (23.12) gives $Z = \rho abcN_A/M = (2.93\ \text{g/cm}^3)(4.94 \times 10^{-8}\ \text{cm})(7.94 \times 10^{-8}\ \text{cm}) \times (5.72 \times 10^{-8}\ \text{cm})(6.022 \times 10^{23}/\text{mol})/(100.09\ \text{g/mol}) = 3.96 \approx 4$. There are 4 formula units and hence 4 Ca^{2+} ions per unit cell.

23.17 A tetragonal lattice has 90° angles and has $a = b$. Equation (23.12) gives $\rho = MZ/abcN_A = (79.899\ \text{g/mol})2/(4.594 \times 10^{-8}\ \text{cm})^2(2.959 \times 10^{-8}\ \text{cm}) \times (6.022 \times 10^{23}/\text{mol}) = 4.249\ \text{g/cm}^3$.

23.18 The c intercepts are all at ∞, so the Miller l index is 0 in each case. With origin at the leftmost dot in the third row from the bottom, the leftmost p_2 plane intercepts the a axis at 1 and the b axis at $-\frac{1}{2}$. The reciprocals of these intercepts give the Miller indices as $(1\bar{2}0)$. With origin at the sixth dot in the bottom row, the s_3 surface intersects the a axis at 1 and is parallel to the b axis (intercept at ∞). The reciprocals give (100) as the Miller indices. [The p_2 planes can also be called $(\bar{1}20)$.]

23.19

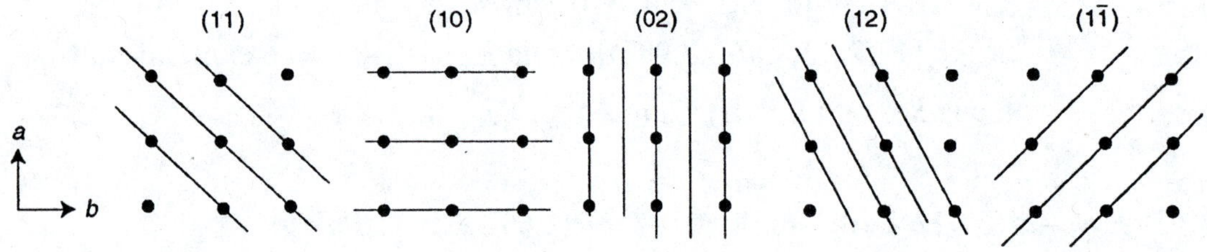

23.20 A body-centered lattice has $8/8 + 1 = 2$ lattice points per unit cell. There is one basis group per lattice point, so there are 2 basis groups per unit cell. The basis therefore has $16/2 = 8$ molecules of $COCl_2$.

23.21 For a spherical atom of radius r inscribed in a cubic unit cell of edge length $2r$, the atom's volume is $\frac{4}{3}\pi r^3$ and the unit cell's volume is $(2r)^3 = 8r^3$. The percentage of occupied space is $(\frac{4}{3}\pi r^3/8r^3)100\% = (\pi/6)100\% = 52.4\%$ and there is 47.6% empty space.

23.22

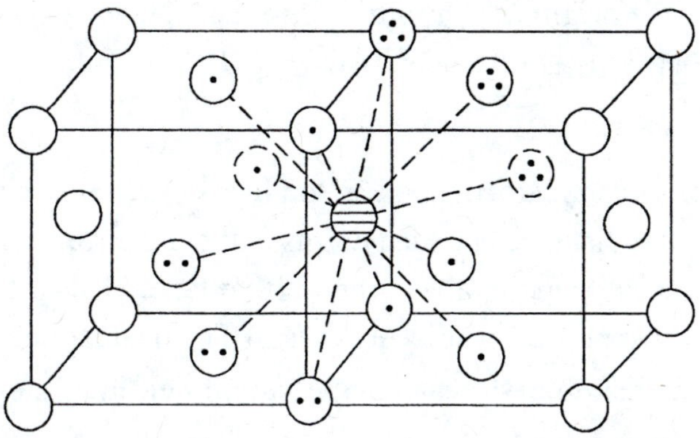

The shaded atom touches the six atoms that have a single dot; all seven of these atoms lie in the (111) plane. The shaded atom also touches the three atoms with two dots [which lie below the (111) plane of the shaded atom] and touches the three atoms with three dots [which lie above the (111) plane of the shaded atom]. (The two atoms drawn with broken circles lie on the back faces of the unit cells.)

23.23 A cubic unit cell has right angles and has $a = b = c$, so Eq. (23.12) gives $\rho = MZ/abcN_A = (58.10\text{ g/mol})\,4/a^3(6.022 \times 10^{23}\text{/mol}) = 2.48\text{ g/cm}^3$. We get $a = 5.38 \times 10^{-8}$ cm = 5.38 Å. As is clear from Fig. 23.16b, the nearest-neighbor distance is $\tfrac{1}{2}a = 2.69$ Å.

23.24 The CsCl space lattice is simple cubic with $Z = 1$; Eq. (23.12) gives $a^3 = MZ/\rho N_A = (212.8\text{ g/mol})1/(4.44\text{ g/cm}^3)(6.022 \times 10^{23}\text{/mol}) = 7.96 \times 10^{-23}\text{ cm}^3$, so $a = 4.30$ Å. From Fig. 23.17a, the nearest-neighbor distance is half the length of the diagonal of the cubic unit cell, which is $\frac{1}{2}\sqrt{3}\,a = 3.72$ Å.

23.25 From Fig. 23.18 and the associated discussion, the lattice is face-centered cubic with $a = b = c$; there are 8 F^- ions and $8/8 + 6/2 = 4$ Ca^{2+} ions per unit cell, so $Z = 4$. Equation (23.12) gives $a^3 = MZ/\rho N_A = (78.08\text{ g/mol})4/(3.18\text{ g/cm}^3)(6.022 \times 10^{23}\text{/mol}) = 1.63 \times 10^{-22}\text{ cm}^3$ and $a = 5.46$ Å.

23.26 The lattice is face-centered cubic with 90° angles and $a = b = c$. Equation (23.12) gives $a = (MZ/\rho N_A)^{1/3} =$
$[(12.011\text{ g/mol})8/(3.51\text{ g/cm}^3)(6.022 \times 10^{23}\text{/mol})]^{1/3} = 3.56_9$ Å. Nearest-neighbor atoms are at points 0 0 0 and $\frac{1}{4}a\ \frac{1}{4}a\ \frac{1}{4}a$. The distance between the

point (x, y, z) and the origin $(x^2 + y^2 + z^2)^{1/2}$, so the distance between nearest-neighbor atoms is $(a^2/16 + a^2/16 + a^2/16)^{1/2} = \frac{1}{4}\sqrt{3}\, a = 1.54_5$ Å.

23.27 There is one atom at each lattice point. In the face-centered cubic unit cell in Fig. 23.7, the closest distance between points is the distance between the point at a center of a face and a point at a corner of that face. (This also equals the distance between two points on adjacent faces.) The nearest-neighbor distance is thus one-half the length of the diagonal of a unit-cell face, namely, $\frac{1}{2}\sqrt{2}\, a = \frac{1}{2}\sqrt{2}\,(5.311 \text{ Å}) = 3.755$ Å.

23.28 **(a)** From Fig. 23.10b, the (100) planes are spaced by $a = 4.70$ Å; Eq. (23.13) gives $\sin\theta = n\lambda/2d_{hkl} = n(1.54 \text{ Å})/2(4.70 \text{ Å}) = 0.1638n = 0.1638, 0.3276, 0.4914, \ldots$. We get $\theta = 9.4°, 19.1°, 29.4°, 40.9°, 55.0°$, and $79.4°$.

(b) Planes s and u in Fig. 23.10a are (110) planes. We see that the distance between these planes is half the length of the diagonal of the bottom face of the cubic unit cell, namely, $d_{110} = \frac{1}{2}\sqrt{2}\, a = a/\sqrt{2}$. [This also follows from the formula $d_{hkl} = a/(h^2 + k^2 + l^2)^{1/2}$ in Sec. 23.9.]
So $\sin\theta = n(1.54 \text{ Å})/2(3.32_3 \text{ Å}) = 0.2317n$ and $\theta = 13.4°, 27.6°, 44.0°$, and $67.9°$.

23.29 **(a)** The $\sin^2\theta$ values are 0.1069, 0.1424, 0.2849, 0.3916, 0.4273, 0.5696, and 0.6767. The ratios of these $\sin^2\theta$ values are
1 : 1.33 : 2.67 : 3.66 : 4.00 : 5.33 : 6.33. So the lattice is face-centered (F).

(b) From Sec. 23.9, these are the 111, 200, 220, 311, 222, 400, and 331 reflections.

(c) $a = \lambda(h^2 + k^2 + l^2)^{1/2}/2\sin\theta = (1.542 \text{ Å})(1^2 + 1^2 + 1^2)^{1/2}/2\sin 19.08° = 4.085$ Å. Similarly, the other angles give 4.086, 4.086, 4.086, 4.086, 4.086, and 4.085 Å.

23.30 The band extends from 392 Å to 422 Å, which (using $\nu = c/\lambda$) is from 7.65×10^{15} Hz to 7.10×10^{15} Hz. So the band width is $\Delta E = h\,\Delta\nu = (6.626 \times 10^{-34} \text{ J s})(0.55 \times 10^{15} \text{ s}^{-1})(1 \text{ eV}/1.60 \times 10^{-19} \text{ J}) = 2.3$ eV.

23.31 **(a)** Equations (23.21) and (21.38) give $U = kT^2(\partial \ln Z/\partial T)_{V,N} =$
$kT^2[U_0/kT^2 - 3N(-\Theta_E/T^2)e^{-\Theta_E/T}/(1 - e^{-\Theta_E/T})] =$
$U_0 + 3Nk\Theta_E/(e^{\Theta_E/T} - 1)$.

(b) $C_V = (\partial U/\partial T)_V = 3Nk(\Theta_E/T)^2 e^{\Theta_E/T}/(e^{\Theta_E/T} - 1)^2$.

23.32 $A = -kT \ln Z = U_0 + 3NkT \ln(1 - e^{-\Theta_E/T})$.

23.33 $U = U_0 + 3Nk\Theta_E/[\exp(\Theta_E/T) - 1]$ [Eq. (23.22)].

(a) In the high-T limit, Θ_E/T goes to 0 and we can use the e^x Taylor series to write $\exp(\Theta_E/T) - 1 \approx (1 + \Theta_E/T + \cdots) - 1 = \Theta_E/T$.
Then $U \to U_0 + 3NkT$.

(b) In the low-T limit, $\exp(\Theta_E/T)$ is very large and the -1 in the denominator can be neglected to give $U \to U_0 + 3Nk\Theta_E/e^{\Theta_E/T}$, which becomes U_0 at $T = 0$.

23.34 **(a)** $S = U/T + k \ln Z = U_0/T + 3Nk\Theta_E/T(e^{\Theta_E/T} - 1) - U_0/T -$
$3Nk \ln(1 - e^{-\Theta_E/T}) = 3Nk(\Theta_E/T)/(e^{\Theta_E/T} - 1) - 3Nk \ln(1 - e^{-\Theta_E/T})$.

(b) For Al, $N/n = N_A$, $Nk/n = R$, $S_m = 3(1.987 \text{ cal/mol-K})(240 \text{ K}/298 \text{ K})/(e^{240/298} - 1) - 3(1.987 \text{ cal/mol-K}) \ln(1 - e^{-240/298}) = 7.41$ cal/mol-K. For diamond, replacement of 240 by 1220 gives $S_m = 0.514$ cal/mol-K. Agreement with experiment is fair.

23.35 **(a)** $\Theta_E/T = (240 \text{ K})/(50 \text{ K}) = 4.8$. Dividing Eq. (23.23) by n and using $Nk/n = N_A k = R$, we have $C_{V,m} = 3(1.987 \text{ cal/mol-K})(4.8)^2 e^{4.8}/(e^{4.8} - 1)^2 =$ 1.15 cal/mol-K = 4.81 J/mol-K.

(b) $\Theta_E/T = (240 \text{ K})/(100 \text{ K}) = 2.40$, and we get $C_{V,m} = 3.77$ cal/mol-K.

(c) $\Theta_E/T = 1.00$ and $C_{V,m} = 5.49$ cal/mol-K.

(d) $\Theta_E/T = 0.600$ and $C_{V,m} = 5.79$ cal/mol-K.

23.36 The Solver is set up to minimize the sum of the squares of the deviations of Einstein values from experimental values. The Solver converges for any reasonable initial guess for Θ_E to give $\Theta_E = 229.1$ K and $b = 0.000729$ J/mol-K^2. The Einstein curve is pretty accurate except that it shows a significant deviation from the 50 K value.

23.37 (a) Using the equation in Problem 14.19, we have $U = kT^2(\partial \ln Z/\partial T)_{V,N} =$ $kT^2\{U_0/kT^2 - (9N/\nu_m^3) \int_0^{\nu_m} [e^{-h\nu/kT}(-h\nu/kT^2)/(1 - e^{-h\nu/kT})]\nu^2\, d\nu\} =$ $U_0 + (9Nh/\nu_m^3) \int_0^{\nu_m} [\nu^3/(e^{h\nu/kT} - 1)]\, d\nu$.

(b) $C_V = (\partial U/\partial T)_{V,N} = (9Nh/\nu_m^3) \int_0^{\nu_m} [(h\nu^4/kT^2)e^{h\nu/kT}/(e^{h\nu/kT} - 1)^2]\, d\nu$. Let $x = h\nu/kT$. Then $d\nu = (kT/h)\, dx$ and $C_V = (9Nh^2/kT^2\nu_m^3) \int_0^{h\nu_m/kT} [(kTx/h)^4 e^x/(e^x - 1)^2](kT/h)\, dx =$ $9Nk(T/\Theta_D)^3 \int_0^{\Theta_D/T} [x^4 e^x/(e^x - 1)^2]\, dx$, where $\Theta_D = h\nu_m/k$.

23.38 One mole of a metallic element has N_A molecules, and the Einstein and Debye theories show that the limiting high-temperature $C_{V,m}$ for a solid metallic element is $3N_Ak = 3R$. The Debye temperature for most metals is not high, so $C_{V,m}$ of most metals is reasonably close to $3R$ for temperatures near room temperature. Thus $C_{V,m} \cong C_{P,m} \cong 3R = 6$ cal/mol-K. For a metallic element, $C_{P,m} = c_P(A_r$ g/mol), where c_P is the specific heat (capacity) and A_r is the (dimensionless) atomic weight. Hence, $c_PA_r = C_{P,m}$ mol/g ≈ 6 cal/(g K).

23.39 (a) T is quite low; Eq. (23.31) gives $\Theta_D^3 = T^3(12\pi^4 N_Ak/5C_{V,m}) =$ $(10\text{ K})^3 12\pi^4(6.022 \times 10^{23}/\text{mol})(1.381 \times 10^{-23}\text{ J/K})/$ $5(0.96 \times 4.184\text{ J/mol-K}) = 4.84 \times 10^5\text{ K}^3$ and $\Theta_D = 78._5$ K.

(b) Equation (23.31) shows that C_V is proportional to T^3 at low temperatures, so at 12 K we have $C_{V,m} = (12/10)^3(0.96\text{ cal/mol-K}) = 1.66$ cal/mol-K.

23.40 (a) $T/\Theta_D = 298\text{ K}/320\text{ K} = 0.931$. Figure 23.28 gives $C_V/3Nk = 0.94$ at $T/\Theta_D = 0.93$. NaCl has $2N_A$ particles per mole and $Nk/n = 2N_Ak = 2R$, so $C_{V,\text{m,Debye}} = 3(2R)(0.94) = 47$ J/mol-K. $C_{V,m}$ and $C_{P,m}$ don't differ greatly for solids at room temperature (Sec. 4.5) and the Appendix $C_{P,m}$ value is 50.5 J/mol-K.

(b) $T/\Theta_D = 298/2230 = 0.134$. Figure 23.28 gives $C_V/3Nk = 0.17$ at $T/\Theta_D = 0.134$, so $C_{V,m} = 3R(0.17) = 4.2$ J/mol-K compared with the experimental $C_{P,m} = 6.1$ J/mol-K.

23.41 **(a)** Adding Eq. (23.32) to (23.31), we have at low T for a metal: $C_{V,m} = (12\pi^4 N_A k/5)(T/\Theta_D)^3 + bT$ and $C_{V,m}/T = b + 12\pi^4 R/5\Theta_D^3)T^2$. A plot of $C_{V,m}/T$ vs. T^2 should be linear with intercept b and slope $12\pi^4 R/5\Theta_D^3$.

(b) The data are:

$10^3(C_{V,m}/T)/(\text{cal/mol-K}^2)$	0.188	0.313	0.523	0.757
T^2/K^2	1.82	4.00	9.00	16.00

The intercept is 1.4×10^{-4} cal mol^{-1} K^{-2} = b; the slope is 3.95×10^{-5} cal/mol-K^4. So $\Theta_D^3 = 12\pi^4(1.987 \text{ cal/mol-K})/5(3.95 \times 10^{-5} \text{ cal/mol-K}^4) = 1.176 \times 10^7 \text{ K}^3$ and $\Theta_D = 227$ K.

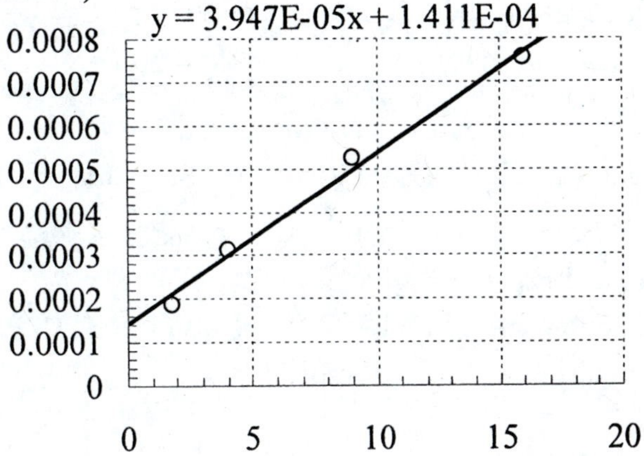

23.42 **(a)** $S_{\text{solid}} - S_{\text{solid},0} = S_{\text{solid}} = \int_0^T (C_P/T')\,dT' \approx \int_0^T (C_V/T')\,dT' = (12\pi^4 Nk/5\Theta_D^3)\int_0^T T'^2\,dT' = 4\pi^4 NkT^3/5\Theta_D^3$.

(b) $S_{\text{gas}} - S_{\text{solid}} = S_{\text{tr}} + S_{\text{el}} - S_{\text{solid}} = nR \ln g_{\text{el},0} + 2.5nR + nR\ln[(2\pi m)^{3/2}(kT)^{5/2}/h^3P] - 4\pi^4 NkT^3/5\Theta_D^3 = \Delta_{\text{sub}}H/T$. Solving this equation for P, we get $P = g_{\text{el},0}(2\pi m)^{3/2}(kTe)^{5/2}h^{-3} \times \exp[-4\pi^4 NkT^3/5nR\Theta_D^3]\exp(-\Delta_{\text{sub}}H/nRT)$. At very low T ($T \ll \Theta_D$), the first exponential is accurately approximated as 1, and $P = g_{\text{el},0}(2\pi m)^{3/2} \times (kTe)^{5/2}h^{-3}e^{-\Delta_{\text{sub}}H/nRT}$, which has the form $aT^{5/2}e^{-c/T}$, since $\Delta_{\text{sub}}H$ varies only slowly with T. As $T \to 0$, both $T^{5/2}$ and $e^{-c/T}$ go to 0, so $P \to 0$.

23.43 **(a)** For a gas of hard spheres of diameter d, there is zero probability for a pair of molecules to be closer than d, and (since there are no intermolecular forces for distances greater than *d*) there is equal probability for all distances greater than *d* to occur.

(b) For a solid, the atoms, molecules, or ions vibrate about fixed equilibrium locations and $g(r)$ shows narrow peaks at the various equilibrium separations.

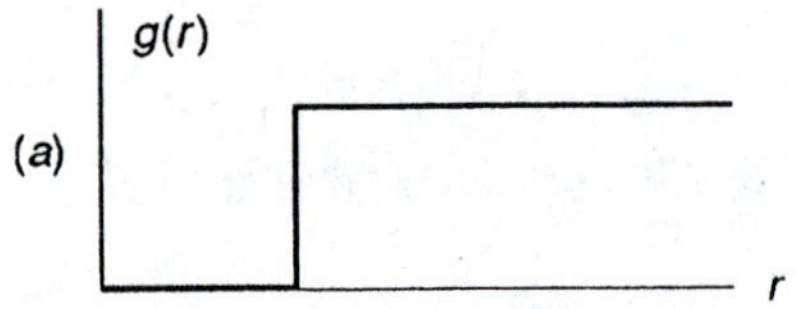

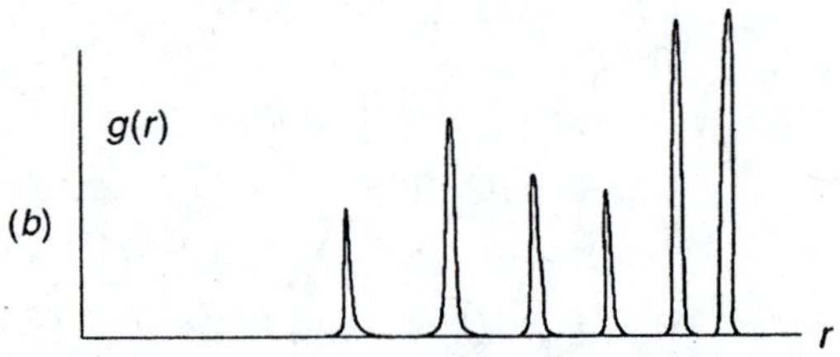

23.44 U of an ideal gas is independent of pressure, so $U_{m,liq} - U_{m,gas} = -\Delta_{vap}U^\circ_{298} = -(\Delta H^\circ_{298} - P^\circ\Delta V^\circ) \approx -\Delta H^\circ_{298} + P^\circ V^\circ_{m,gas} = -\Delta H^\circ_{298} + RT = -44012$ J/mol + RT $= -41.53$ kJ/mol. S of an ideal gas depends on P. The 25°C molar volume of liquid water (density 0.99704 g/cm^3) is 18.07 cm^3. We use this 25°C path for one mole: liq(P°, 18.07 cm^3) $\overset{1}{\rightarrow}$ ideal gas(P°) $\overset{2}{\rightarrow}$ ideal gas(18.07 cm^3). The Appendix and Eq. (3.30) give $S_{liq} - S_{gas} = -(\Delta S_1 + \Delta S_2) =$ -118.92 J/mol-K $- R \ln [(18.07 \text{ cm}^3)P^\circ/RT] = -(118.92 - 60.06)$ J/mol-K $=$ -58.86 J/mol-K $= -14.07$ cal/mol-K.

23.45 **(a)** Each molecule has 6 bonds and V_{str} has 6(300) = 1800 terms.

(b) Each methyl hydrogen atom has two 1,4 van der Waals interactions with other atoms in the same molecule and so there are 3(2) = 6 van der Waals interactions in each molecule. Multiplication by the number of molecules gives 6(300) = 1800 terms for intramolecular van der Waals interactions. Each atom in a given molecule has an intermolecular van der Waals interaction with 7(299) = 2093 atoms in other molecules. The number of intermolecular van der Waals interactions involving the atoms of a particular molecule is then 7(2093) = 14651. Multiplication by the total number of molecules and division by 2 to avoid counting each interaction twice gives 14651(300)/2 = 2197650 intermolecular van der Waals interactions. Adding in the intramolecular interactions, we have 2197650 + 1800 = 2199450 van der Waals terms.

23.46 The initial configuration of the molecules may be far from equilibrium, so one allows the system to reach a near-equilibrium configuration before one uses the MD data.

23.47 **(a)** Special theory of relativity, Brownian motion theory, distance traveled by diffusing molecules, photon explanation of photoelectric effect, quantum theory of C_V of solids, Bose–Einstein statistics.

(b) Probability interpretation of wave function, work on matrix mechanics form of quantum mechanics, calculation of E_c of ionic solids, Born–Oppenheimer approximation.

(c) Debye–Hückel theory, Debye theory of C_V of solids, Debye–Langevin equation, Debye equation for ionic diffusion-controlled reactions, work on electrical conductivity of solutions.

23.48 **(a)** One gram of solid Ar has a volume of $(1.00 \text{ g})/(1.59 \text{ g/cm}^3) = 0.629 \text{ cm}^3$. The hard-sphere atoms occupy $(100 - 26)\% = 74\%$ of this volume, which is 0.46_5 cm^3. One gram of liquid occupies $(1.00 \text{ g})/(1.42 \text{ g/cm}^3) = 0.704 \text{ cm}^3$. The empty-space volume in the liquid is $(0.704 - 0.46_5) \text{ cm}^3 = 0.23_9 \text{ cm}^3$, so the percent empty space is $(0.23_9/0.704)100\% = 34\%$.

(b) The volume of 1 g of the gas is $V \cong (1\text{g}/39.9 \text{ g mol}^{-1})RT/P = 179 \text{ cm}^3$. The percent of filled space in the gas is $(0.46_5/179)100\% = 0.26\%$ and the empty space is 99.74%.

23.49 **(a)** T. **(b)** T. **(c)** T. **(d)** T.

Notes